AF819517

Advances in Experimental Medicine and Biology

Volume 1509

Series Editors

Nima Rezaei, Research Center for Immunodeficiencies, Children's Medical Center, Tehran University of Medical Sciences, Tehran, Iran

Ortrud Steinlein, Institute of Human Genetics, LMU University Hospital, Munich, Germany

Avia Rosenhouse-Dantsker, Department of Chemistry, University of Illinois at Chicago, Chicago, IL, USA

Robert Gerlai, Department of Psychology, University of Toronto, Mississauga, ON, Canada

Advances in Experimental Medicine and Biology provides a platform for scientific contributions in the main disciplines of the biomedicine and the life sciences. This series publishes thematic volumes on contemporary research in the areas of microbiology, immunology, neurosciences, biochemistry, biomedical engineering, genetics, physiology, and cancer research. Covering emerging topics and techniques in basic and clinical science, it brings together clinicians and researchers from various fields.

Advances in Experimental Medicine and Biology has been publishing exceptional works in the field for over 40 years, and is indexed in SCOPUS, Medline (PubMed), EMBASE, Reaxys, EMBiology, the Chemical Abstracts Service (CAS), and Pathway Studio.

2023 CiteScore: 5.9

Julian Weiler • Mareike Sieler • Thomas Dittmar
Editors

Cell Fusion

New Insights into the Mechanism of Cell Fusion and Its Biological Consequences

Springer

Editors
Julian Weiler
Immunology and Tumour Biology
Center for Biomedical Education
and Research (ZBAF)
Witten/Herdecke University
Witten, Germany

Mareike Sieler
Immunology and Tumour Biology
Center for Biomedical Education
and Research (ZBAF)
Witten/Herdecke University
Witten, Germany

Thomas Dittmar
Immunology and Tumour Biology
Center for Biomedical Education
and Research (ZBAF)
Witten/Herdecke University
Witten, Germany

ISSN 0065-2598 ISSN 2214-8019 (electronic)
Advances in Experimental Medicine and Biology
ISBN 978-3-032-22636-5 ISBN 978-3-032-22637-2 (eBook)
https://doi.org/10.1007/978-3-032-22637-2

© The Editor(s) (if applicable) and The Author(s), under exclusive license to Springer Nature Switzerland AG 2026

This work is subject to copyright. All rights are solely and exclusively licensed by the Publisher, whether the whole or part of the material is concerned, specifically the rights of translation, reprinting, reuse of illustrations, recitation, broadcasting, reproduction on microfilms or in any other physical way, and transmission or information storage and retrieval, electronic adaptation, computer software, or by similar or dissimilar methodology now known or hereafter developed.
The use of general descriptive names, registered names, trademarks, service marks, etc. in this publication does not imply, even in the absence of a specific statement, that such names are exempt from the relevant protective laws and regulations and therefore free for general use.
The publisher, the authors and the editors are safe to assume that the advice and information in this book are believed to be true and accurate at the date of publication. Neither the publisher nor the authors or the editors give a warranty, expressed or implied, with respect to the material contained herein or for any errors or omissions that may have been made. The publisher remains neutral with regard to jurisdictional claims in published maps and institutional affiliations.

This Springer imprint is published by the registered company Springer Nature Switzerland AG
The registered company address is: Gewerbestrasse 11, 6330 Cham, Switzerland

If disposing of this product, please recycle the paper.

Contents

Editors and Contributors

About the Editors

Julian Weiler studied biology at Heinrich Heine University in Düsseldorf, where he graduated with a focus on immunology and wound healing. For his PhD thesis, he joined the Institute of Immunology at Witten/Herdecke University, where his research focused on the investigation of fusion-relevant factors and signaling pathways that are upregulated under inflammatory conditions, such as those found in the tumor microenvironment. Since 2020, he has been working as a postdoctoral researcher, deciphering the relationship between aberrant cell fusion and tumor development.

Mareike Sieler studied Chemical Biology at the Technical University of Dortmund and went to the institute of Immunology at Witten/Herdecke University for her Dr. rer. nat thesis in 2019. Her research focused primarily on characterizing cell fusion in breast cancer cells and the role of fusogenic proteins. After completing her doctorate in 2023, she spent another two years as a postdoctoral researcher investigating the mechanisms of cell fusion, but with a focus on the role of the cell membrane and lipid-distributing proteins.

Thomas Dittmar studied chemistry at the Ruhr-University at Bochum, and in 1995, he went to Institute of Immunology at Witten/Herdecke University for his Dr. rer. nat. thesis, which he received in 1999. In 2003, he became a Junior Professor of Tumor Biology, and since 2010, he is a full professor at Witten/Herdecke University. His main focus of research is breast cancer with a special emphasis on the role of cell-cell fusion events in (breast) cancer progression.

Contributors

M. Amin Abdolkhani Department of Biological and Biomedical Engineering, McGill University, Montréal, QC, Canada

Ashley N. Anderson Department of Cell, Developmental and Cancer Biology, OHSU, Portland, OR, USA

Marta Cadevall Angles Center for Genomic Regulation (CRG), The Barcelona Institute of Science and Technology, Barcelona, Spain

Universitat Pompeu Fabra (UPF), Barcelona, Spain

Soham Bhattacharyya Department of Biomedical Science and Technology, School of Biological Sciences, Ramakrishna Mission Vivekananda Educational Research Institute (RKMVERI), Kolkata, India

Bo Li State Key Laboratory of Oral Diseases, National Center for Stomatology, National Clinical Research Center for Oral Diseases, West China Hospital of Stomatology, Sichuan University, Chengdu, China

Department of Orthodontics, West China Hospital of Stomatology, Sichuan University, Chengdu, China

Gloria C. Bonel-Pérez The Innate Immune Response Group, IdiPAZ, La Paz University Hospital, Madrid, Spain

Tumour Immunology Laboratory, IdiPAZ, La Paz University Hospital, Madrid, Spain

Eimear M. Byrne Center for Genomic Regulation (CRG), The Barcelona Institute of Science and Technology, Barcelona, Spain

Hirak Chakraborty School of Chemistry, Sambalpur University, Burla, Odisha, India

Chung Yu Chan National Institute of Neurological Disorders and Stroke, Bethesda, MD, USA

Ariane M. Chitoh Department of Biology, Jackson State University, Jackson, MS, USA

Environmental Science PhD Program, Jackson State University, Jackson, MS, USA

Debopriya Choudhury Department of Biomedical Science and Technology, School of Biological Sciences, Ramakrishna Mission Vivekananda Educational Research Institute (RKMVERI), Kolkata, India

Maria Pia Cosma Center for Genomic Regulation (CRG), The Barcelona Institute of Science and Technology, Barcelona, Spain

Universitat Pompeu Fabra (UPF), Barcelona, Spain

ICREA, Barcelona, Spain

Medical Research Institute, Guangdong Provincial People's Hospital (Guangdong Academy of Medical Sciences), Southern Medical University, Guangzhou, China

Yu-Bin Ding Department of Toxicology, Joint International Research Laboratory of Reproduction and Development of the Ministry of Education of China, School of Public Health, Chongqing Medical University, Chongqing, China

Thomas Dittmar Immunology and Tumour Biology, Center for Biomedical Education and Research (ZBAF), Witten/Herdecke University, Witten, Germany

Hana M. Dobrovolny Department of Physics & Astronomy, Texas Christian University, Fort Worth, TX, USA

Taelor J. Ekstrom Department of Cell, Developmental and Cancer Biology, OHSU, Portland, OR, USA

Alejandro Forigua Department of Chemical Engineering, McGill University, Montréal, QC, Canada

Kathryn L. Fowler Department of Surgery, Oregon Health & Science University (OHSU), Portland, OR, USA

Taylor Gierke Department of Molecular Biophysics and Biochemistry, Yale University, New Haven, CT, USA

Nicole R. Giske Department of Cell, Developmental and Cancer Biology, OHSU, Portland, OR, USA

Xiaoli Guo School of Life Sciences, Yunnan Normal University, Kunming, Yunnan, China

National Institute of Neurological Disorders and Stroke, Bethesda, MD, USA

Laura Hurtado-Navarro The Innate Immune Response Group, IdiPAZ, La Paz University Hospital, Madrid, Spain

Tumour Immunology Laboratory, IdiPAZ, La Paz University Hospital, Madrid, Spain

UNIE University, Madrid, Spain

Malgorzata Kloc The Houston Methodist Research Institute, Transplant Immunology, Houston, TX, USA

Department of Surgery, The Houston Methodist Hospital, Houston, TX, USA

Kayode Komolafe Department of Biology, Jackson State University, Jackson, MS, USA

Environmental Science PhD Program, Jackson State University, Jackson, MS, USA

Pulan Liu Department of Molecular Biophysics and Biochemistry, Yale University, New Haven, CT, USA

Eduardo López-Collazo The Innate Immune Response Group, IdiPAZ, La Paz University Hospital, Madrid, Spain

Tumour Immunology Laboratory, IdiPAZ, La Paz University Hospital, Madrid, Spain

CIBER of Respiratory Diseases (CIBERES), Madrid, Spain

UNIE University, Madrid, Spain

Yonggang Lu Premium Research Institute for Human Metaverse Medicine (WPI-PRIMe), The University of Osaka, Osaka, Japan

Pushkar Malakar Department of Biomedical Science and Technology, School of Biological Sciences, Ramakrishna Mission Vivekananda Educational Research Institute (RKMVERI), Kolkata, India

Tomonori Matsumoto Laboratory of Ploidy Pathology, Graduate School of Frontier Biosciences, The University of Osaka, Suita, Osaka, Japan

Smruti Mishra School of Chemistry, Sambalpur University, Burla, Odisha, India

Joanna N. Modi Department of Pathology and Cell Biology, Herbert Irving Comprehensive Cancer Center, Columbia University, New York, NY, USA

Abigail Moore Department of Cell, Developmental and Cancer Biology, OHSU, Portland, OR, USA

Christopher Moraes Department of Biological and Biomedical Engineering, McGill University, Montréal, QC, Canada

Department of Chemical Engineering, McGill University, Montréal, QC, Canada

Rosalind and Morris Goodman Cancer Institute, McGill University, Montréal, QC, Canada

Division of Experimental Medicine, McGill University, Montréal, QC, Canada

School of Biomedical Engineering, The University of British Columbia, Vancouver, BC, Canada

International Collaboration on Repair Discoveries (ICORD), Vancouver Coastal Health Research Institute, Vancouver, BC, Canada

Felicite K. Noubissi Department of Biology, Jackson State University, Jackson, MS, USA

Research Centers in Minority Institutions (RCMI), Center for Health Disparity Research (RCMI-CHDR), Jackson State University, Jackson, MS, USA

Oluwatoyin V. Odubanjo Department of Biology, Jackson State University, Jackson, MS, USA

Research Centers in Minority Institutions (RCMI), Center for Health Disparity Research (RCMI-CHDR), Jackson State University, Jackson, MS, USA

Manbit Subhadarsi Panda School of Chemistry, Sambalpur University, Burla, Odisha, India

Ranish K. Patel Department of Surgery, Oregon Health & Science University (OHSU), Portland, OR, USA

Xianghe Qiao State Key Laboratory of Oral Diseases, National Center for Stomatology, National Clinical Research Center for Oral Diseases, West China Hospital of Stomatology, Sichuan University, Chengdu, China

Department of Oral and Maxillofacial Surgery, The First Affiliated Hospital of Zhengzhou University, Zhengzhou, China

Mareike Sieler Immunology and Tumour Biology, Center for Biomedical Education and Research (ZBAF), Witten/Herdecke University, Witten, Germany

Shaogeng Tang Department of Molecular Biophysics and Biochemistry, Yale University, New Haven, CT, USA

Alison M. Taylor Department of Pathology and Cell Biology, Herbert Irving Comprehensive Cancer Center, Columbia University, New York, NY, USA

Brahmachari Vedeshachaitanya Department of Biomedical Science and Technology, School of Biological Sciences, Ramakrishna Mission Vivekananda Educational Research Institute (RKMVERI), Kolkata, India

Julian Weiler Immunology and Tumour Biology, Center for Biomedical Education and Research (ZBAF), Witten/Herdecke University, Witten, Germany

Melissa H. Wong Department of Cell, Developmental and Cancer Biology, OHSU, Portland, OR, USA

Knight Cancer Institute, OHSU, Portland, OR, USA

Ling-Gang Wu National Institute of Neurological Disorders and Stroke, Bethesda, MD, USA

Yan Xiong Department of Toxicology, Joint International Research Laboratory of Reproduction and Development of the Ministry of Education of China, School of Public Health, Chongqing Medical University, Chongqing, China

Amanda Zucker Department of Cell, Developmental and Cancer Biology, OHSU, Portland, OR, USA

Chapter 1
A Brief Introduction into the Biological Phenomenon of Cell–Cell Fusion

Mareike Sieler, Julian Weiler, and Thomas Dittmar

Abstract Even though the merger of two (or more) cells appears simple, the whole process is rather complex and not yet fully understood. While it is clear that cell–cell fusion and virus–cell fusion depends on the expression and presence of specific fusion-relevant proteins and molecules, such as fusogens in general and their receptors, cell–cell adhesion molecules, proteases, and phospholipids like phosphatidylserine (PS), the underlying mechanisms that regulate and control cell–cell fusion still remains to be elucidated. In this regard, it is necessary to discriminate between physiological and pathophysiological cell–cell fusion processes. In physiological cell–cell merger, the fusion machinery, which is composed of all fusion-relevant proteins and molecules, is initiated in individual cells in an inherent and even evolutionarily conserved time- and development-dependent manner. The sperm and the oocyte have to fuse during fertilization, the merger of trophoblasts into multinucleated syncytiotrophoblasts is important for proper placenta formation and function, and multinucleated myofibers cannot originate without the fusion of single-nucleated myoblasts. In contrast, cell–cell fusion events that occur in adult organisms in a rather non-evolutionarily conserved manner, such as the merger of cells in tissue regeneration processes or in tumor tissues, remain ambiguous. While it is clear that the fusion machinery must be active, it is not completely understood how the merger machinery is ultimately initiated and controlled in these cells, which particularly applies to tumor cells. In fact, it is well-known that tumor cells can fuse with other normal cells, such as macrophages, fibroblasts, and stem cells, but it is completely unclear why tumor cells and normal cells do this. Is this the reactivation of probably evolutionarily conserved mechanisms that have occurred in cancer cells and normal cells, or is this purely random? And if so, does the merging of cancer cells and normal cells have any benefits for the organism, for example, by enabling the immune system to better recognize and eliminate the fused tumor cells? Or, on

M. Sieler · J. Weiler · T. Dittmar (✉)
Immunology and Tumour Biology, Center for Biomedical Education and Research (ZBAF), Witten/Herdecke University, Witten, Germany
e-mail: Thomas.Dittmar@uni-wh.de

© The Author(s), under exclusive license to Springer Nature Switzerland AG 2026

J. Weiler et al. (eds.), *Cell Fusion*, Advances in Experimental Medicine and Biology 1509, https://doi.org/10.1007/978-3-032-22637-2_1

the other hand, does illicit tumor cell fusion compound tumor growth and evolution via genomic rewiring/reprogramming and the inheritance of genomic instability to the next generation?

Thus, the aim of this book is to bring more light into the complex, tightly regulated, and fascinating process of cell–cell fusion.

Keywords Cell–cell fusion · Heterokaryon · Synkaryon · Syncytialization · Fusogens · Phosphatidylserine

Background The biological phenomenon of cell fusion appears to be a simple process, like the merger of two soap bubbles. However, what apparently appears simple is in reality a highly complex and tightly regulated process that is still not fully understood. This applies not only to the fusion of two (or more) cells but also to the merger of enveloped viruses with host cells, the fusion of mitochondria, and the fusion of vesicles.

The process of membrane merger can simply be subdivided into three discrete steps. The first step is characterized by the recognition of the membranes that are destined to fuse and their tight adhesion (for example, two cells or two vesicles) (Hernandez and Podbilewicz 2017; Brukman et al. 2019; Whitlock and Chernomordik 2024; Dittmar and Hass 2022, 2023). At this stage, the membranes are positioned at a distance not closer than approximately 10 nm (Hernandez and Podbilewicz 2017; Brukman et al. 2019; Whitlock and Chernomordik 2024; Dittmar and Hass 2022, 2023). Moreover, this stage is further accompanied by the clearing of plasma membrane-bound proteins, glycoproteins, and glycolipids from the site of membrane fusion (Lee et al. 2025). Indeed, an increasing cell surface crowding with proteins, glycoproteins, and glycolipids reduces the fusion efficiency of polyethylen glycol (PEG)-mediated and fusogen-mediated cell–cell fusion, as well as synthetic membranes under force (Lee et al. 2025). The second step is the fusion of the membranes themselves, and the third step is characterized by the fact that the fusion process is thereby completed (Hernandez and Podbilewicz 2017; Brukman et al. 2019; Whitlock and Chernomordik 2024; Dittmar and Hass 2022, 2023). In particular, a pre-fusion (step 1) and a post-fusion (step 3) state have been postulated for cells. The pre-fusion state means that cells that want to fuse must first assume a state that enables them to merge with other cells (Zhou and Platt 2011). The post-fusion state, on the other hand, means that after successful fusion, the cellular mechanisms that enable cells to fuse are stopped in order to prevent further unwanted fusion processes (Zhou and Platt 2011). However, neither the pre-fusion state nor the post-fusion state has been precisely defined to date. It is therefore unclear, for example, which molecules characterize the pre-fusion or post-fusion state or which molecules cause cells to transition to the pre-fusion or post-fusion state.

Although the merger of vesicles appears to be different from cell–cell merger or enveloped virus × host cell merger, it is assumed that the thermodynamic and biophysical requirements that are encountered during the membrane fusion (step 2) are expected to be the same (Hernandez and Podbilewicz 2017). In this regard,

Hernandez and Podbilewicz postulated the hallmarks of cell–cell fusion (or, in general, membrane merger). These are (i) the dehydration of contacting plasma membranes and bringing the phospholipid heads to distances of close to 0 nm, (ii) the merger of the outer monolayers, or hemifusion, via a stalk and/or diaphragm intermediates, and (iii) opening and expansion of fusion pore(s) from nanometer diameter to multiple microns (Hernandez and Podbilewicz 2017). Furthermore, it must be borne in mind that various energy barriers must be overcome during the fusion of membranes (Hernandez and Podbilewicz 2017). One example of this is the mutual repulsion of the membranes caused by the negative charge of the phospholipid heads in both membranes (Hernandez and Podbilewicz 2017; Brukman et al. 2019; Whitlock and Chernomordik 2024; Dittmar and Hass 2022, 2023).

Fusogens and fusogen complexes are well-known proteins/protein complexes that are essential for membrane fusion in order to overcome the energy barriers associated with cell/membrane fusion (Hernandez and Podbilewicz 2017; Brukman et al. 2019; Whitlock and Chernomordik 2024; Dittmar and Hass 2022, 2023). Over the course of evolution, different types of fusogens have developed independently of one another (Hernandez and Podbilewicz 2017; Brukman et al. 2019; Whitlock and Chernomordik 2024). Some share structural similarities (although of different origin), such as EFF-1 from Caenorhabditis elegans *(C. elegans)*, the sexual gamete fusogen HAP2 from *Clamydomonas reinhardtii* (*C. reinhardtii*), and the tick-borne encephalitis virus E glycoprotein (TBEV E) (Hernandez and Podbilewicz 2017; Brukman et al. 2019; Whitlock and Chernomordik 2024), while other fusogens are structurally completely different. For example, syncytin-1, which causes the fusion of human trophoblasts into syncytiotrophoblasts during placentation, is of retroviral origin and belongs to the family of class I viral fusogens, whereas myomaker and myomerger, which induce the fusion of human myoblasts during myogenesis, are multi-pass and single-pass transmembrane proteins, respectively (Hernandez and Podbilewicz 2017; Brukman et al. 2019; Whitlock and Chernomordik 2024; Chen et al. 2020; Leikina et al. 2018; Durnaoglu et al. 2021; Mi et al. 2000). Soluble N-ethylmaleimide-sensitive factor attachment receptor (SNARE) proteins are a highly conserved protein superfamily across the eukaryotic kingdom (both plants and animals), whose members are responsible for almost all fusion events of both the exocytotic and endocytic branches of the secretory pathway (Jahn et al. 2024). The SNARE core complex consists of four different SNARE proteins that are directly or indirectly anchored to membranes, and the zippering of these proteins promotes membrane fusion (Jahn et al. 2024). Interestingly, after mediating vesicle fusion, SNARE proteins disassemble and, thus, can be used for further vesicle merger (Jahn et al. 2024). This property of SNARE proteins differs from classic fusion proteins, which mediate the fusion of plasma membranes only once.

In addition to fusogen/fusogen complexes, the phospholipid phosphatidylserine (PS) has been identified as another molecule that appears to be indispensable for plasma membrane merger (Whitlock and Chernomordik 2021). The translocation of PS from the inner leaflet to the outer leaflet of the plasma membrane is a well-known process in early apoptosis, and because of that, PS is rather known as a marker for apoptotic cells (Nagata et al. 2020; Martin et al. 1995). Interestingly, the inhibition of

PS translocation to the outer leaflet of the plasma membrane by blocking or knocking out PS-specific scramblases resulted in markedly impaired fusion frequencies of different cell types (Rival et al. 2019; Shabo et al. 2025; Tsuchiya et al. 2018; Verma et al. 2018; Zhang et al. 2020; Uygur et al. 2019). Interestingly, even the infection of enveloped viruses with host cells apparently depends on the cell surface exposure of PS (Zaitseva et al. 2017; Braga et al. 2021). However, the exact role of PS in the fusion of plasma membranes still needs to be clarified in ongoing studies.

Another complexity in membrane merger is the regulation of this process, which particularly applies to cell–cell fusion. Fertilization, placentation, myogenesis, and osteoclastogenesis are examples of well-controlled and tightly regulated physiological cell–cell fusion processes (Hernandez and Podbilewicz 2017; Brukman et al. 2019; Whitlock and Chernomordik 2024; Dittmar and Hass 2022, 2023). In this regard, placentation and myogenesis only occur at certain time points during development. These cell–cell fusion events are attributed to evolutionarily conserved intrinsic epigenetic/genetic processes, which lead to the induction of fusogen expression at specific times/periods. For example, myomaker and myomerger, which facilitate myoblast fusion in myogenesis, are only expressed at discrete time points during embryogenesis (Chen et al. 2020). Osteoclastogenesis represents another well-controlled and tightly regulated physiological cell–cell fusion process, which, however, occurs in a lifelong manner in bone tissues (Helming and Gordon 2009). Bone-resorbing osteoclasts are the counterparts to bone-forming osteoblasts, and the balance between these two cell populations is essential for bone homeostasis (Kim et al. 2020). A disturbance of this balance can lead to either increased bone formation (too many osteoblasts and too few osteoclasts) or increased bone resorption (too many osteoclasts and too few osteoblasts) (Kim et al. 2020). This shows that the genesis of cell–cell fusion-derived osteoclasts must be strictly controlled.

In contrast to these well-controlled and strictly regulated cell–cell fusion processes, two other cell–cell fusion events have been identified that are still poorly understood. This is the cell–cell fusion-mediated regeneration of damaged cells/tissues and the homotypic and heterotypic merger of cancer cells. Numerous studies have demonstrated that bone marrow-derived cells (BMDCs) can fuse with various somatic cell types, such as skeletal muscle, liver, lung, intestinal, skin, heart, brain, and retinal cells, thereby restoring tissue function (for review, see Alvarez-Dolado and Martinez-Losa (2011), Dornen and Dittmar (2021), and Pesaresi et al. (2019)). However, it remains unclear how the process of cell fusion-mediated tissue regeneration is regulated. Do damaged cells send out specific signals or express specific markers that identify them as damaged and in need of regeneration? And if so, what signals/markers could these be? Similarly, the question arises as to which processes are necessary for BMDCs to fuse with the damaged cells in the first place. As mentioned above, a pre-fusion state has been postulated for fusogenic cells, which they must first assume in order to be able to fuse with other cells. Inflammation appears to be a positive trigger for cell fusion-mediated tissue regeneration (Davies et al. 2009; Johansson et al. 2008; McNally and Anderson 2011). However, it is unclear what effect pro-inflammatory conditions/cytokines have on damaged tissue cells and BMDCs and whether or how they are involved in the transition to a profusogenic state of the cells.

The same applies to the fusion of cancer cells. Here, too, it is completely unclear why cancer cells fuse with each other or with other cells, such as macrophages or stem cells (Clawson 2013; Hass et al. 2021; Lopez-Collazo and Hurtado-Navarro 2025; Raj et al. 2019; Zhang et al. 2021; Dittmar et al. 2023; Bastida-Ruiz et al. 2016). Since the tumor microenvironment resembles chronically inflamed tissue (Dvorak 1986; Balkwill and Mantovani 2001), it could be assumed that similar processes to cell fusion-mediated tissue regeneration by BMDCs take place here. In the case of tumor cells, it is also known that they may have an inherent fusogenicity. This is suggested by various in vitro studies in which the spontaneous fusion of cancer cells was observed (Dittmar et al. 2011; Dornen et al. 2020; Li et al. 2014; Lindstrom et al. 2017; Melzer et al. 2021; Merle et al. 2021; Zhang et al. 2018; Wakeling et al. 1994). The phenomenon of spontaneous fusion of cancer cells is also not understood. Studies suggest that cancer cells can express the fusogen syncytin-1 (Bjerregaard et al. 2006; Fu et al. 2021; Benesova et al. 2017; Yu et al. 2014). However, why cancer cells express syncytin-1 is just as unclear as the fact that only very few (approx. 0.1–1%) cancer cells actually possess fusogenic properties. In any case, several studies in recent years have clearly demonstrated the presence of tumor hybrids in both the primary tumor and peripheral blood of various types of cancer (Dietz et al. 2021; Henn et al. 2021; Parappilly et al. 2022; Patel et al. 2024; LaBerge et al. 2021; LaBerge et al. 2017; Lazova et al. 2013; Gast et al. 2018). In particular, the number of circulating tumor hybrids correlated with a significantly worse prognosis for cancer patients (Dietz et al. 2021; Henn et al. 2021; Parappilly et al. 2022; Patel et al. 2024). Therefore, further analysis of the process of cell–cell fusion, particularly in cancer cells, is necessary to further elucidate this mechanism, which could lead to the development of novel strategies that could prevent the formation of tumor hybrids and, thus, also cancer progression.

This (short) background indicates that cell–cell fusion (or membrane merger in general) is a common biological phenomenon that plays a pivotal role in both health and disease.

Fifteen years ago, Prof. Kurt Zänker and I, Thomas Dittmar, published the two-volume book *Cell Fusion in Health and Disease* with Springer Verlag, which was very well received by the scientific community. It is remarkable to see what has happened in the field of cell–cell fusion over the past 15 years and what new insights have been gained. For example, the phospholipid PS has been identified as an essential molecule for cell–cell fusion. Similarly, tumor hybrids have been detected in primary tumors and in the peripheral blood of cancer patients. However, much remains unknown, such as how cell–cell fusion is specifically induced in the case of tissue regeneration or the fusion of tumor cells. Is there a pre-fusion state of cells, and if so, how is it induced, and what characterizes it? Why does fusion in the context of tissue regeneration and tumor cells tend to result in the formation of bikaryotic heterokaryons rather than syncytia? These and other questions need to be investigated.

We are grateful that so many internationally recognized experts accepted our invitation to participate in this exciting book project. We would like to thank them all for their interest in this important topic and for finding the time, despite their other commitments and responsibilities, to present excellent and comprehensive overviews of the most important current findings in their scientific fields on this topic. We would also like to take this opportunity to thank Dr. Ina Stoeck and Bibhuti Bhusan Sharma from Springer Publishers (Heidelberg, Germany) for their kind support and excellent cooperation on this project, as well as for making this book project possible.

We hope that the book *Cell Fusion 2.0* will be as successful as the two-volume work *Cell Fusion in Health and Disease*, which we published 15 years ago. Unfortunately, Prof. Zänker will not be able to see this happen, as he passed away in 2021.

Furthermore, we hope that this book will promote new scientific approaches in the field of cell fusion in health and disease, as well as closer interdisciplinary collaboration on this fascinating and important topic in the future.

References

Alvarez-Dolado M, Martinez-Losa M (2011) Cell fusion and tissue regeneration. Adv Exp Med Biol 713:161–175

Balkwill F, Mantovani A (2001) Inflammation and cancer: back to Virchow? Lancet 357(9255): 539–545

Bastida-Ruiz D, Van Hoesen K, Cohen M (2016) The dark side of cell fusion. Int J Mol Sci 17(5)

Benesova M, Trejbalova K, Kovarova D, Vernerova Z, Hron T, Kucerova D et al (2017) DNA hypomethylation and aberrant expression of the human endogenous retrovirus ERVWE1/ syncytin-1 in seminomas. Retrovirology 14(1):20

Bjerregaard B, Holck S, Christensen IJ, Larsson LI (2006) Syncytin is involved in breast cancer-endothelial cell fusions. Cell Mol Life Sci 63(16):1906–1911

Braga L, Ali H, Secco I, Chiavacci E, Neves G, Goldhill D et al (2021) Drugs that inhibit TMEM16 proteins block SARS-CoV-2 spike-induced syncytia. Nature 594(7861):88–93

Brukman NG, Uygur B, Podbilewicz B, Chernomordik LV (2019) How cells fuse. J Cell Biol 218(5):1436–1451

Chen B, You W, Wang Y, Shan T (2020) The regulatory role of Myomaker and Myomixer-Myomerger-minion in muscle development and regeneration. Cell Mol Life Sci 77(8): 1551–1569

Clawson GA (2013) Cancer. Fusion for moving. Science 342(6159):699–700

Davies PS, Powell AE, Swain JR, Wong MH (2009) Inflammation and proliferation act together to mediate intestinal cell fusion. PLoS One 4(8):e6530

Dietz MS, Sutton TL, Walker BS, Gast CE, Zarour L, Sengupta SK et al (2021) Relevance of circulating hybrid cells as a non-invasive biomarker for myriad solid tumors. Sci Rep 11(1): 13630

Dittmar T, Hass R (2022) Extracellular events involved in cancer cell-cell fusion. Int J Mol Sci 23(24):16071

Dittmar T, Hass R (2023) Intrinsic signalling factors associated with cancer cell-cell fusion. Cell Commun Signal 21(1):68

Dittmar T, Schwitalla S, Seidel J, Haverkampf S, Reith G, Meyer-Staeckling S et al (2011) Characterization of hybrid cells derived from spontaneous fusion events between breast epithelial cells exhibiting stem-like characteristics and breast cancer cells. Clin Exp Metastasis 28(1): 75–90

Dittmar T, Sieler M, Hass R (2023) Why do certain cancer cells alter functionality and fuse? Biol Chem 404(10):951–960
Dornen J, Dittmar T (2021) The role of MSCs and cell fusion in tissue regeneration. Int J Mol Sci 22(20):10980
Dornen J, Myklebost O, Dittmar T (2020) Cell fusion of mesenchymal stem/stromal cells and breast cancer cells leads to the formation of hybrid cells exhibiting diverse and individual (stem cell) characteristics. Int J Mol Sci 21(24):9636
Durnaoglu S, Lee SK, Ahnn J (2021) Syncytin, envelope protein of human endogenous retrovirus (HERV): no longer 'fossil' in human genome. Anim Cells Syst (Seoul) 25(6):358–368
Dvorak HF (1986) Tumors: wounds that do not heal. Similarities between tumor stroma generation and wound healing. N Engl J Med 315(26):1650–1659
Fu Y, Zhuang X, Xia X, Li X, Xiao K, Liu X (2021) Correlation between promoter hypomethylation and increased expression of Syncytin-1 in non-small cell lung cancer. Int J Gen Med 14:957–965
Gast CE, Silk AD, Zarour L, Riegler L, Burkhart JG, Gustafson KT et al (2018) Cell fusion potentiates tumor heterogeneity and reveals circulating hybrid cells that correlate with stage and survival. Sci Adv 4(9):eaat7828
Hass R, von der Ohe J, Dittmar T (2021) Hybrid formation and fusion of cancer cells in vitro and in vivo. Cancers (Basel) 13(17):4496
Helming L, Gordon S (2009) Molecular mediators of macrophage fusion. Trends Cell Biol 19:514–522
Henn TE, Anderson AN, Hollett YR, Sutton TL, Walker BS, Swain JR et al (2021) Circulating hybrid cells predict presence of occult nodal metastases in oral cavity carcinoma. Head Neck 43(7):2193–2201
Hernandez JM, Podbilewicz B (2017) The hallmarks of cell-cell fusion. Development 144(24): 4481–4495
Jahn R, Cafiso DC, Tamm LK (2024) Mechanisms of SNARE proteins in membrane fusion. Nat Rev Mol Cell Biol 25(2):101–118
Johansson CB, Youssef S, Koleckar K, Holbrook C, Doyonnas R, Corbel SY et al (2008) Extensive fusion of haematopoietic cells with Purkinje neurons in response to chronic inflammation. Nat Cell Biol 10(5):575–583
Kim JM, Lin C, Stavre Z, Greenblatt MB, Shim JH (2020) Osteoblast-osteoclast communication and bone homeostasis. Cells 9:9
LaBerge GS, Duvall E, Grasmick Z, Haedicke K, Pawelek J (2017) A melanoma lymph node metastasis with a donor-patient hybrid genome following bone marrow transplantation: a second case of leucocyte-tumor cell hybridization in cancer metastasis. PLoS One 12(2):e0168581
LaBerge G, Duvall E, Grasmick Z, Haedicke K, Galan A, Pawelek J (2021) A melanoma patient with macrophage-cancer cell hybrids in the primary tumor, a lymph node metastasis and a brain metastasis. Cancer Genet 256–257:162–164
Lazova R, Laberge GS, Duvall E, Spoelstra N, Klump V, Sznol M et al (2013) A melanoma brain metastasis with a donor-patient hybrid genome following bone marrow transplantation: first evidence for fusion in human cancer. PLoS One 8(6):e66731
Lee DSW, Oster LF, Son S, Fletcher DA (2025) Cell surface crowding is a tunable energetic barrier to cell-cell fusion. Nat Commun 16(1):7158
Leikina E, Gamage DG, Prasad V, Goykhberg J, Crowe M, Diao J et al (2018) Myomaker and Myomerger work independently to control distinct steps of membrane remodeling during myoblast fusion. Dev Cell 46(6):767–780
Li H, Feng Z, Tsang TC, Tang T, Jia X, He X et al (2014) Fusion of HepG2 cells with mesenchymal stem cells increases cancerassociated and malignant properties: an in vivo metastasis model. Oncol Rep 32(2):539–547
Lindstrom A, Midtbo K, Arnesson LG, Garvin S, Shabo I (2017) Fusion between M2-macrophages and cancer cells results in a subpopulation of radioresistant cells with enhanced DNA-repair capacity. Oncotarget 8(31):51370–51386

Lopez-Collazo E, Hurtado-Navarro L (2025) Cell fusion as a driver of metastasis: re-evaluating an old hypothesis in the age of cancer heterogeneity. Front Immunol 16:1524781

Martin SJ, Reutelingsperger CP, McGahon AJ, Rader JA, van Schie RC, LaFace DM et al (1995) Early redistribution of plasma membrane phosphatidylserine is a general feature of apoptosis regardless of the initiating stimulus: inhibition by overexpression of Bcl-2 and Abl. J Exp Med 182(5):1545–1556

McNally AK, Anderson JM (2011) Macrophage fusion and multinucleated giant cells of inflammation. Adv Exp Med Biol 713:97–111

Melzer C, Ohe JV, Luo T, Hass R (2021) Spontaneous fusion of MSC with breast cancer cells can generate tumor dormancy. Int J Mol Sci 22(11):5930

Merle C, Lagarde P, Lartigue L, Chibon F (2021) Acquisition of cancer stem cell capacities after spontaneous cell fusion. BMC Cancer 21(1):241

Mi S, Lee X, Li X, Veldman GM, Finnerty H, Racie L et al (2000) Syncytin is a captive retroviral envelope protein involved in human placental morphogenesis. Nature 403(6771):785–789

Nagata S, Sakuragi T, Segawa K (2020) Flippase and scramblase for phosphatidylserine exposure. Curr Opin Immunol 62:31–38

Parappilly MS, Chin Y, Whalen RM, Anderson AN, Robinson TS, Strgar L et al (2022) Circulating neoplastic-immune hybrid cells predict metastatic progression in uveal melanoma. Cancers (Basel) 14(19)

Patel RK, Parappilly M, Farley HC, Latour EJ, Wang LG, Nair AM et al (2024) Circulating neoplastic-immune hybrid cells are biomarkers of occult metastasis and treatment response in pancreatic cancer. Cancers (Basel) 16(21)

Pesaresi M, Sebastian-Perez R, Cosma MP (2019) Dedifferentiation, transdifferentiation and cell fusion: in vivo reprogramming strategies for regenerative medicine. FEBS J 286(6):1074–1093

Raj AT, Kheur S, Patil VR, Gupta AA (2019) Assessing the role of cell fusion in cancer metastasis. Oral Oncol 90:124

Rival CM, Xu W, Shankman LS, Morioka S, Arandjelovic S, Lee CS et al (2019) Phosphatidylserine on viable sperm and phagocytic machinery in oocytes regulate mammalian fertilization. Nat Commun 10(1):4456

Shabo I, Midtbo K, Branstrom R, Lindstrom A (2025) Monocyte-cancer cell fusion is mediated by phosphatidylserine-CD36 receptor interaction and induced by ionizing radiation. PLoS One 20(1):e0311027

Tsuchiya M, Hara Y, Okuda M, Itoh K, Nishioka R, Shiomi A et al (2018) Cell surface flip-flop of phosphatidylserine is critical for PIEZO1-mediated myotube formation. Nat Commun 9(1):2049

Uygur B, Leikina E, Melikov K, Villasmil R, Verma SK, Vary CPH et al (2019) Interactions with muscle cells boost fusion, stemness, and drug resistance of prostate cancer cells. Mol Cancer Res 17(3):806–820

Verma SK, Leikina E, Melikov K, Gebert C, Kram V, Young MF et al (2018) Cell-surface phosphatidylserine regulates osteoclast precursor fusion. J Biol Chem 293(1):254–270

Wakeling WF, Greetham J, Bennett DC (1994) Efficient spontaneous fusion between some co-cultured cells, especially murine melanoma cells. Cell Biol Int 18(3):207–210

Whitlock JM, Chernomordik LV (2021) Flagging fusion: phosphatidylserine signaling in cell-cell fusion. J Biol Chem 296:100411

Whitlock JM, Chernomordik LV (2024) Cell-cell fusion: to lose one life and begin another. BioEssays.:e2400206

Yu H, Liu T, Zhao Z, Chen Y, Zeng J, Liu S et al (2014) Mutations in 3′-long terminal repeat of HERV-W family in chromosome 7 upregulate syncytin-1 expression in urothelial cell carcinoma of the bladder through interacting with c-Myb. Oncogene 33(30):3947–3958

Zaitseva E, Zaitsev E, Melikov K, Arakelyan A, Marin M, Villasmil R et al (2017) Fusion stage of HIV-1 entry depends on virus-induced cell surface exposure of phosphatidylserine. Cell Host Microbe 22(1):99–110

Zhang LN, Kong CF, Zhao D, Cong XL, Wang SS, Ma L et al (2018) Fusion with mesenchymal stem cells differentially affects tumorigenic and metastatic abilities of lung cancer cells. J Cell Physiol:3570–3582

Zhang Y, Le T, Grabau R, Mohseni Z, Kim H, Natale DR et al (2020) TMEM16F phospholipid scramblase mediates trophoblast fusion and placental development. Sci Adv 6(19):eaba0310

Zhang H, Ma H, Yang X, Fan L, Tian S, Niu R et al (2021) Cell fusion-related proteins and signaling pathways, and their roles in the development and progression of cancer. Front Cell Dev Biol 9:809668

Zhou X, Platt JL (2011) Molecular and cellular mechanisms of mammalian cell fusion. Adv Exp Med Biol 713:33–64

Chapter 2
The Interplay of Lipids and Peptides in Membrane Merger

Manbit Subhadarsi Panda, Smruti Mishra, and Hirak Chakraborty

Abstract Membrane fusion is an important process for the survival of eukaryotes. Moreover, enveloped viruses enter the host cell utilizing membrane fusion. Therefore, it is essential to comprehend the mechanistic details of this intrinsic and crucial process. In this chapter, we will discuss several aspects of membrane fusion, including intermediate states and their energetics. We will further discuss the role of lipids in the stabilization of intermediate structures that lead to fusion pore opening. We will elaborate on the different existing models of membrane merger or fusion, with a major focus on the importance of lipid–peptide interplay for the completion of the process.

Keywords Membrane merger · Lipid-stalk model · Membrane curvature · Hemifusion · Lipid–peptide interaction · Energetics of fusion · Fusion models

Abbreviations

DAG	Diacylglycerol
DOPC	Dioleoylphosphatidylcholine
DOPE	Dioleoylphosphatidylethanolamine
DOPG	Dioleoylphosphatidylglycerol
DOPS	Dioleoylphosphatidylserine
HA	Hemagglutinin
HIV	Human immunodeficiency virus
LPA	Lysophosphatidic acid
LPC	Lysophosphatidylcholine
OA	Oleic acid

Authors 'Manbit Subhadarsi Panda' and 'Smruti Mishra' have been equally contributed for this chapter.

M. S. Panda · S. Mishra · H. Chakraborty (✉)
School of Chemistry, Sambalpur University, Burla, Odisha, India
e-mail: hirak@suniv.ac.in

© The Author(s), under exclusive license to Springer Nature Switzerland AG 2026

J. Weiler et al. (eds.), *Cell Fusion*, Advances in Experimental Medicine and Biology 1509, https://doi.org/10.1007/978-3-032-22637-2_2

PEG	Polyethylene glycol
PI	Phosphatidylinositol
POPC	Palmitoylphosphatidylcholine
POPE	Palmitoylphosphatidylethanolamine
POPG	Palmitoylphosphatidylglycerol
SM	Sphingomyelin
SNARE	Soluble N-ethylmaleimide-sensitive-factor attachment protein receptor

2.1 Introduction to Membrane Merger or Fusion

The membrane plays a crucial role in multiple biological functions and serves as a meeting point for both lipids and proteins. The environmental conditions modulate the lipid–protein interaction by regulating protein expression and modulating membrane lipid composition (Escribá et al. 2008). Biological membranes exhibit intrinsic lateral and transbilayer heterogeneity, originating from the lipid composition, physical properties of lipids such as polarity and unsaturation in the acyl chain, and their packing ability (Haldar et al. 2012; Kinnun et al. 2020). Thus, the composition, shape, and organization of lipids in the membrane modulate the membrane properties and the interactions of the membrane with proteins or peptides. Some proteins prefer a certain lipid in their vicinity for their appropriate structure and function (Kumar et al. 2021; Majumder et al. 2008). Membrane merger involves the fusion of two lipid bilayers, leading to intermixing of their hydrophobic cores. The process requires overcoming a significant activation energy barrier for the mixing of lipids and contents of two fusing bilayers. Membrane merger, a fundamental biological process, is essential for several cellular events or functions such as vesicle trafficking, exocytosis, endocytosis, tissue development, neurotransmission, signal transduction, fertilization, import of nutrients, export of wastes, formation of organelles, and entry of enveloped virus into host cell, which can vary significantly in space and time and with specific fusogens (Chernomordik and Kozlov 2008). Membrane fusion can occur in specific regions of a cell, such as at the plasma membrane, with endosomes, or at the nuclear membrane. The interplay between lipids and peptides plays a critical role in facilitating membrane merger, with lipids providing the structural framework and peptides acting as molecular catalysts or modulators of membrane dynamics (Jahn and Scheller 2006; Lentz et al. 2000). Lipids are the primary constituents of biological membranes, forming a dynamic bilayer that serves as a barrier to ions and molecules. The fusion intersection of two membranes may possess distinct lipid compositions and protein machinery, which influence membrane properties such as curvature, fluidity, and propensity for membrane fusion (van Meer et al. 2008). For instance, lipids like phosphatidylethanolamine (PE) and negatively charged phosphatidylserine (PS) promote membrane curvature and destabilization, which are critical for initiating fusion (Chernomordik and Kozlov 2003). Additionally, lipid microdomains, enriched in sphingolipids and cholesterol, can organize membrane components to facilitate localized fusion events (Simons and Sampaio 2011; Yang et al. 2016a). The fusion process demonstrates spatial and temporal differences depending on the properties of the membrane. The contact area for the fusion of larger organelles, like yeast vacuoles, is approximately 10,000 times greater than that of synaptic vesicles. Moreover, the fusion of yeast vacuoles takes several minutes to complete, whereas synaptic vesicles fuse in a millisecond (Jahn et al. 2003). Likewise, extracellular fusion events vary widely from the formation of large syncytia in muscle tissue to the entry of small viruses into host cells. Despite this diversity, all fusion reactions proceed through elementary processes, including membrane contact (docking), membrane merger (stalk and transmembrane contact), and the opening of an aqueous fusion pore (pore formation) (Chernomordik and Kozlov 2008; Jahn et al. 2003). Fusogenic substances, including proteins and peptides, regulate these activities by their structural reorganization that supplies the necessary energy for overcoming the activation barrier of intermediate steps. Fusion peptides, often derived from the N-terminus of viral or cellular fusion proteins, are inserted into target membranes, destabilizing the lipid bilayer and reducing the energy required for fusion (Harrison 2015). For example, the influenza virus HA fusion peptide interacts with lipid headgroups and acyl chains, inducing non-lamellar structures that promote membrane merging (Chakraborty et al. 2013; Tamm et al. 2002; Tenchov et al. 2013). The interaction between lipids and peptides is highly dynamic and context-dependent. Peptides often exhibit selectivity for specific lipid compositions, enabling them to target particular membranes or regions within a membrane (Epand and Epand 2009). Conversely, lipids can modulate the structure and activity of peptides, influencing their insertion depth, orientation, and ability to induce membrane curvature (Gong et al. 2023). For instance, cholesterol can enhance the binding and activity of certain fusion peptides, while anionic lipids can attract cationic peptides, facilitating their interaction with the membrane (Giraldo-Lorza et al. 2024). Understanding the relationship between lipids and peptides in membrane merger has significant implications for biomedical research. Insights into these mechanisms can inform the development of antiviral therapies, drug delivery systems, and synthetic membranes for biotechnology applications (White and Whittaker 2016). Moreover, dysregulation of membrane fusion processes is implicated in diseases such as cancer and neurodegenerative disorders, highlighting the importance of elucidating the molecular details of lipid–peptide interactions (Südhof and Rothman 2009). Taken together, the interplay between lipids and peptides is important for membrane fusion, with lipids providing the structural context and driving the dynamic changes necessary for membrane merger. This intricate collaboration underscores the complexity of biological membranes and their role in cellular function and diseases. Fusion induced by small molecules or ions is expected to be different from that caused by proteins or peptides, as they do not share these benefits (Mondal Roy and Sarkar 2011).

2.2 Membrane Fusion

Membrane fusion is a crucial biological process in eukaryotic cells that allows the merging of two separate lipid bilayers into a single continuous bilayer, enabling the mixing of the internal contents of the fused membranes. Fusion has been distinguished into three types throughout evolution: (i) Extracellular and intracellular fusion of enveloped viruses with host cells, in which the entire process is carried out by a single protein, causing infection: This process is initiated when viral envelope proteins interact with specific receptors on the surface of the host cell. This interaction triggers conformational changes in the viral fusion protein, leading to the merging of the viral envelope with the host cell membrane (Jahn et al. 2003). (ii) Extracellular fusion of eukaryotic cells: This includes the fusion of sperm with oocytes during fertilization and formation of syncytia in muscle cells (myoblast fusion) (Jahn et al. 2003; Wassarman and Litscher 2008). (iii) Intracellular fusion of organelles: This involves the merging of membrane-bound organelles within cells. This process is vital for the exchange of materials and functions between organelles, maintaining cellular homeostasis and function. The fusion of organelles is generally mediated by dynamic supramolecular assemblies involving various conserved protein families, such as SNARE proteins, tethering factors, and lipids. Some SNARE proteins (syntaxins, SNAP-25, and VAMPs) play a crucial role in docking and catalyzing membrane fusion events, allowing for precise targeting and specificity (Söllner and Rothman 1994). Examples of organelle fusion include phagocytosis and mitochondrial fusion (Jahn et al. 2003). In endocrine cells, the fusion of secretory vesicles with the plasma membrane releases hormones into the bloodstream. The expression patterns of calpain and synaptotagmin isoforms can indeed reflect diverse mechanisms of stimulus–secretion coupling in excitable endocrine cells. Understanding these proteins and their roles can provide insights into the complex processes that underlie hormone and neurotransmitter release (Aganna et al. 2006).

2.3 Models of Membrane Fusion

To understand the fusion mechanism and correlate experimental data, different models of membrane fusion have been proposed. Nevertheless, no consistent model exists that considers the molecular characteristics of proteins and lipids, as well as the macroscopic membrane features.

2.3.1 Helfrich Model

The Helfrich model views lipid bilayers as infinitely thin, structureless elastic films that are mainly controlled by their bending deformations. The bending resistance of the membrane is described by the Helfrich elastic energy function, which measures the energetic cost of curving the membrane away from its preferred (spontaneous) curvature. The classical Helfrich model was, in fact, not a very popular model in membrane fusion because it treated membranes as homogeneous entities and was only able to focus on minor deformations, but the internal structure and asymmetry between monolayers remained unaddressed. This presented challenges, particularly because fusion intermediates (stalk and hemifusion diaphragm (HD)) and fusion pore involve regions of high curvature and unique lipid environments in the distal and apposed monolayers (Akimov et al. 2020; Campelo et al. 2014; Markin and Albanesi 2002). To address these limitations, the model was generalized, often called the Kozlov–Markin modification, which considers the separate treatment of each monolayer's elastic properties during fusion. This approach enables the quantitative calculation of bending energies for fusion intermediate structures, elucidating how membrane shape and local composition influence fusion pathways.

2.3.2 Lipid-Stalk Model

In this model, a set of intermediate structures (stalk and hemifusion diaphragm) and a fusion pore were postulated. The model also includes equations for Gaussian curvature to account for topological changes, which are especially important for membrane fusion (Akimov et al. 2020; Bottacchiari et al. 2024; Campelo et al. 2014). The development of the classical fusion model began with the groundbreaking research of Kozlov and Markin. According to this hypothesis, two bilayer membranes combine by the sequential fusing of monolayers, first the contact ones (outer leaflet) and subsequently the distal ones (inner leaflet). Monolayers placed into close contact with one another may rupture locally, exposing the membrane core to water and forming opposing hydrophobic patches. An hourglass-shaped local connection between the contact monolayers, known as the stalk (Kozlov and Markin 1983; Kozlov and Markin 1984), is formed when the patches eventually merge after being drawn to one another. This membrane remodeling leaves the distal (inner) monolayers flat. Therefore, the first intermediate structure in the fusion process is the stalk. The radial expansion of the stalk enables the formation of the joint bilayer by bringing the lipids of inner monolayers into tail-to-tail contact (Kozlovsky and Kozlov 2002; Kuzmin et al. 2001). This structure is known as the trilaminar structure or hemifusion diaphragm (HD). The longitudinal expansion of the diaphragm leads to the formation of a fusion pore that allows two water volumes surrounded by the fusing membranes to merge (Kozlovsky et al. 2002), as shown in Fig. 2.1.

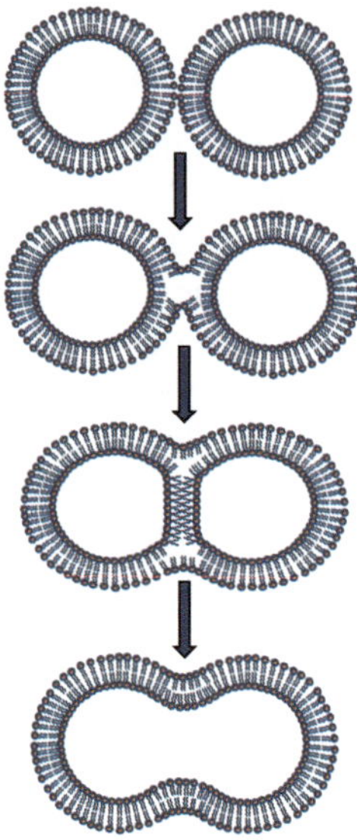

Fig. 2.1 Schematic representation of intermediate states during membrane fusion according to the lipid-stalk model

Based on the evidence from the literature, Markin and Kozlov have proposed two main fusion mechanisms, i.e., adhesion and stalk mechanisms. According to the adhesion mechanism, the tight connection of membranes involves bilayer rearrangement, whereas in the stalk mechanism, a bridge is developed between approaching membranes, and this intermediate is referred to as the "stalk" (Markin et al. 1984). The adhesion mechanism shows an inverted micellar structure at the bilayer's merging point and entails the conversion of lipid molecules at the contact region from liquid to crystalline phase. The two approaching membranes' outer layers rupture as a result of the stress created by the crystallization of lipid molecules, which also decreases the area per molecule (Kozlov and Markin 1984). Nevertheless, the stalk-like intermediate is more energetically feasible than the inverted micelle-type intermediate, as the latter is only generated under certain environments close to the bilayer phase transition into the inverse hexagonal (H_{II}) phase. As a result, it involves more lipid molecules compared to the stalk (Siegel 1993).

2.3.3 Mattress Model

This model is a thermodynamic model that describes the phase diagrams of mixtures of lipid bilayers and amphiphilic proteins or polypeptides in aqueous solution. The thickness of the hydrophobic region of the lipid bilayer and the length of the hydrophobic region of the proteins are the basic geometrical variables of this model. The model assumes the elastic properties of proteins and the lipid bilayer, in addition to direct and indirect lipid–protein interactions. The key concept of this model is the mismatch of the hydrophobic regions of the lipids and proteins. The phase behavior is important to explain the experiments on planar aggregation of proteins in lipid bilayers (Mouritsen and Bloom 1984).

2.3.4 *Protein Coat Model*

The protein coat model assumes that before the fusion reaction, the membrane containing the anchored fusion proteins is attached; however, there is no interaction between fusion proteins and the target membrane at the top, as shown in Fig. 2.2a. These fusion proteins interact with each other and form a dense protein coat, which triggers a change in the effective shape of the protein molecules. As a result, the protein coat bends toward the interior of the vesicle, forming a negative spontaneous curvature, as shown in Fig. 2.2b. This bending of the coat results in a funnel-like shape with the protein-depleted lipid bilayer at the top of the funnel, protruding toward the target membrane and primed for hemifusion by bending stresses (Fig. 2.2c). A hemifusion diaphragm is created when the bulged bilayer hemifuses with the target membrane, as shown in Fig. 2.2d. The protein funnel becomes an expanding semi-toroid when the coat is bent further, whereas the HD enlarges, concentrating the intense deformation along its edge and creating lateral stress (Fig. 2.2e). A fusion pore forms and expands as a result of the lateral tension and strains around the HD's perimeter (Fig. 2.2f) (Kozlov and Chernomordik 2002).

2.3.5 *Kiss-and-Run Fusion Mechanism*

The "kiss-and-run" fusion mechanism is an alternative pathway to the classic "full fusion" model traditionally associated with exocytosis in various types of cells, including endocrine cells, chromaffin cells, and neurons. In this fusion mechanism, vesicles transiently dock with the plasma membrane, forming a small pore through which some contents (like neurotransmitters or hormones) can be released into the

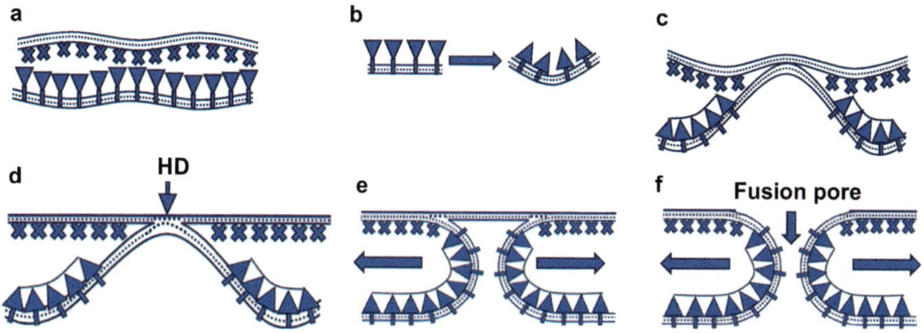

Fig. 2.2 Schematic representation of the mechanism of the protein coat model

extracellular space. After the release, the vesicle can then reorganize and return to its original state, retaining its shape and membrane integrity (Alabi and Tsien 2013; Harata et al. 2006; Wen et al. 2017). Calcium ions (Ca^{2+}) play a pivotal role in regulating vesicle fusion, affecting both the classical full fusion mechanism and the kiss-and-run fusion mechanism (Alés et al. 1999).

2.3.6 Meshless Membrane Model

A particle-based, meshless membrane model is suggested, which is ideal for modeling membrane dynamics when topological changes occur. The paradigm is associated with the Brownian motion of bilayer fusion, which provides the mobility and thermal energy that drive interactions across lipid membranes. The randomness of this motion aids in the bilayer approach, helps overcome energetic barriers to fusion, and improves membrane fluidity—all of which are necessary for the merging process (Noguchi and Takasu 2001).

2.3.7 Müller and Schick Model of Fusion

In this model, the fusion of the tensed apposed bilayer of amphiphilic molecules has been studied by using Monte Carlo simulations, which were based on a coarse-grained lattice model. The typical stalk mechanism is different from the fusion process. A stalk intermediate is formed between the apposing bilayers, which, instead of spreading out radially to produce an axially symmetric hemifusion diaphragm of the inner leaflets of both bilayers, promotes the nucleation of tiny holes in the vicinity of the bilayers. In one pathway, the stalk encloses a hole in one bilayer, forming a diaphragm made of both leaflets of the unbroken bilayer, which bursts to form the fusion pore. In another pathway, a second hole appears in the other bilayer before the stalk can wrap a hole in the first one. The stalk aligns and encircles both holes to complete the fusion pore. The outer and inner leaflets of the bilayer merge as a result of both pathways, which also permit temporary leakage (Müller et al. 2003).

2.3.8 Flexible Surface Model

This model is a new biomembrane model that describes the balance between the spontaneous curvature and hydrophobic forces in lipid–peptide interactions. The conformational energetics of membrane proteins are modulated by the chemically nonspecific properties of the lipid bilayer. This model relates lipid polymorphism and membrane remodeling to biological function by frustration of the monolayer curvature-free energy. The curvature-free energy of the membrane film balances the solvation energy by the acyl chains, which is impacted by changes to the intramembranous hydrophobic surface of the protein. Using rhodopsin as a prototype, which becomes a sensor of the spontaneous (intrinsic) curvature of the monolayer when exposed to light, explains how bilayer thickness, cholesterol, non-lamellar-forming lipids, and osmotic pressure affect its activation. Rhodopsin becomes a sensor of the negative spontaneous curvature of the membranes when light is absorbed (Wiedmann et al. 1988). This concept of rhodopsin aids in connecting theory and experiment. When a photosensitive protein, rhodopsin, absorbs light, it results in the formation of two metastable equilibrium intermediates, Metarhodopsin I (Meta I) and Metarhodopsin II (Meta II) (Soubias et al. 2010). The equilibrium is shifted toward the inactive Meta I state of rhodopsin by lipids with roughly zero spontaneous curvature, while it is shifted toward the active Meta II state by lipids with negative spontaneous curvature. Bending energy is frustrated when a fluid membrane deforms away from the spontaneous curvature of the monolayer, resulting in changes in protein conformation. The new biomembrane model shows how the material properties of the lipid bilayer affect the energetics of membrane proteins. The fluid mosaic model, which is the conventional model found in biochemistry textbooks, is challenged by this emerging biomembrane model (Brown 2012).

2.4 Determination of Fusion Parameters

2.4.1 Weinreb and Lentz Analysis

Weinreb and Lentz proposed a model that utilizes the fusion observables (lipid mixing (LM), content mixing (CM), and content leakage kinetics) to account for the time courses of PEG-induced vesicle fusion. The suggested model is based on the lipid-stalk hypothesis and assumes how the fusion pore evolved from the aggregated vesicles via one or two intermediate states. Generally, the individual fusion events are considered to proceed through two semi-stable non-lamellar structural intermediates (I_1 and I_2) to form a final fusion pore (Scheme 2.1) (Chakraborty et al. 2012;

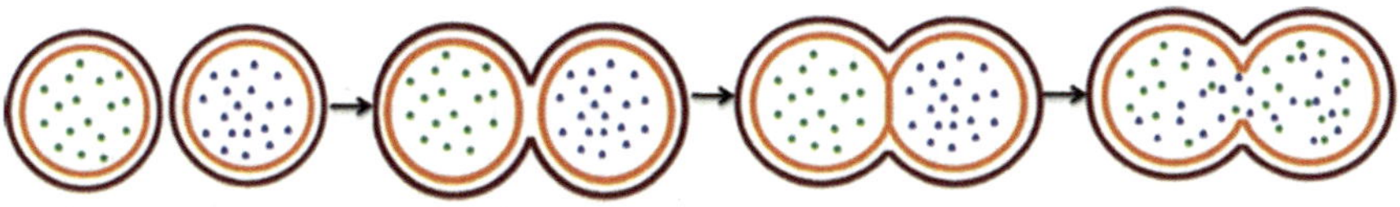

Scheme 2.1 Schematic representation of the membrane fusion process based on the lipid-stalk model

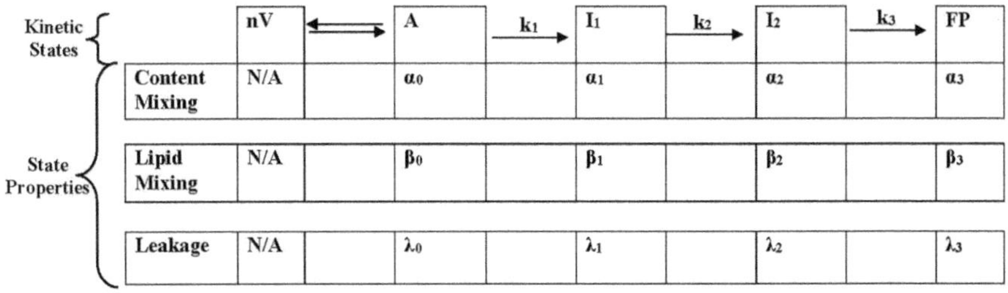

Scheme 2.2 Schematic representation of the fusion reaction, including the rate constants (k_i) of different steps, probabilities of content mixing (α_i) and lipid mixing (β_i), and the rate constant of leakage (λ_i) at different states

Weinreb and Lentz 2007). Lipid mixing primarily governs the formation of the stalk and extended transmembrane contact, providing information about the intermixing of lipids between two different vesicles. Content mixing provides information about the merging of the inner contents of two fusing membranes, which leads to pore formation. However, because the tiny fluorophores utilized in the experiment of content mixing can pass through the lipid bilayer of two apposing membranes through thermal fluctuation, there is a possibility of observing content mixing in the early stage of the fusion process. Content leakage aids in confirming the integrity of the membrane and is a crucial parameter to balance the content mixing signal since it decreases as a result of spontaneous leakage (Joardar et al. 2022b; Weinreb and Lentz 2007). The states in the analytical model of the membrane fusion process are the thermodynamic states made up of ensembles of microstructures that closely mimic the structures found in the lipid-stalk model.

Upon the addition of PEG, the aggregation of small unilamellar vesicles (SUVs) (A) takes place immediately, and the rate of PEG-mediated aggregation is around ten times faster than the rate of the fusion event, as suggested by the structural model. In the presence of PEG, the aggregation process remains irreversible, preventing the vesicles from diffusing. This aggregated form then gets converted to a stalk state (I_1) and expanded transmembrane contact or hemifusion diaphragm (I_2) with rate constants k_1 and k_2 and is then transformed into the fusion pore state (FP State) with rate constant k_3. The probabilities of lipid mixing (β_i), content mixing (α_i), and average rate constant of leakage (λ_i) at different states of membrane fusion reflect the possibilities of these events occurring throughout the fusion process (Scheme 2.2).

At the aggregated state (A), the probability of CM and LM is assumed to be zero ($\alpha_0 = \beta_0 = 0$), and at the fusion pore state (FP State), the CM predominates over the negligible LM ($\beta_3 = 0$). The three intrinsic rate constants (k_1, k_2, and k_3) and eight comprehensive parameters (α_1, α_2, β_1, β_2, λ_0, λ_1, λ_2, and λ_3) are used to explain the three independent time courses represented by the double exponential. The loss of fluorescence signal due to leakage was counter-balanced in the content mixing

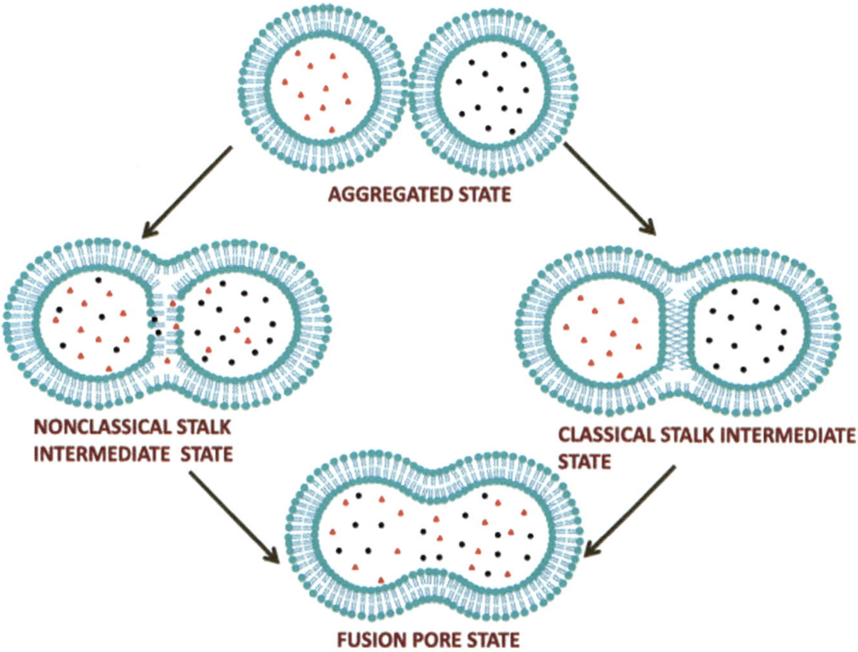

Scheme 2.3 Schematic representation of the classical and nonclassical stalk models of membrane fusion. (The scheme has been adapted from Joardar et al. (2022a) with permission)

signal. Therefore, it would be convenient to utilize the probabilities of lipid and content mixing as a marker for the classical stalk model. In this model, the value of the probability of lipid mixing must be higher than the probability of content mixing at the stalk state. Alternatively, the probability of content mixing is likely to be higher than the probability of lipid mixing at the fusion pore state. A deviation from the classical stalk model is referred to as a nonclassical stalk model (Scheme 2.3) (Joardar et al. 2022a).

2.4.2 *Other Treatments*

To understand the membrane fusion on a molecular level, different fusion models were proposed, based on either a continuum model (Kasson and Pande 2007; Malinin and Lentz 2004) of membrane deformation or discrete shape transition of supramolecular assemblies (Müller et al. 2003). A quantitative model has been proposed based on low-energy intermediates, which suggests the direct opening of the fusion pore from the very beginning of the connecting bilayers between the two membranes (Kuzmin et al. 2001; Siegel 1993). Despite the existence of various models, there are some gaps in understanding the intermediate steps, and the transient and unfavorable fluctuation of lipid and water rearrangement in the steps brings difficulties in experimental observations (Chakraborty et al. 2012). This lipid and water rearrangement is supposed to contribute to an activation barrier, and fusion proteins or peptides function as machines that store conformational energy to promote the formation of fusion pores between apposing bilayers through an array of non-lamellar intermediates (Lentz et al. 2000). Molecular dynamics simulation study states that vesicles containing high phosphatidylcholine (PC) content enhanced the activation energies for the formation of initial stalk-like intermediate and hemifusion intermediates for fusion. Different lipid compositions show a significant difference in fusion kinetics and fusion mechanism (Kasson and Pande 2007). The lipid-stalk hypothesis is an important concept that explains how certain pathogens, regardless of their divergent morphologies, genome sequences, family, class, and structures, utilize a common mechanism for membrane fusion (Pattnaik and Chakraborty 2020). This mechanism is particularly relevant in the context of viral entry into host cells and the fusion of membranous compartments during infection and is believed to be the most accepted hypothesis.

2.5 Steps of Membrane Merger or Fusion

Membrane merger typically describes the merging of two membranes to form a single continuous membrane, and the inner contents mix through the formation of a fusion pore. The architecture and energy of the intermediate structures that are formed during the membrane fusion process have a major role in determining the feasibility and kinetics of membrane merger (Chernomordik and Kozlov 2003). The fusion stalk is an intermediate structure that is frequently shared by biological and model membrane fusion. The first hemifusion intermediate is the stalk, which develops into a fusion pore to complete the fusion process (Chernomordik et al. 1998; Melikyan et al. 1995; Zimmerberg et al. 1994). Each stage of membrane merger has been discussed below:

2.5.1 Proximity of Membranes or Docking

The first requirement of membrane merger is to establish contact between the apposed bilayers when present close to each other, as shown in Fig. 2.3a. In the absence of any electric charge on these bilayers, they can approach each other spontaneously up to an equilibrium distance of 2–3 nm (Rand and Parsegian 1989), but the actual distance between the membranes is much larger than this due to electrostatic repulsion and steric interaction between the bilayers due to membrane proteins. Thus, it is important to bring the membrane bilayers within such a range that membranes can merge easily. Mostly, the integral proteins perform the function of either pushing or pulling the membranes toward each other, thereby inducing bending to minimize the areas of strong intermembrane repulsion (Leikin et al.

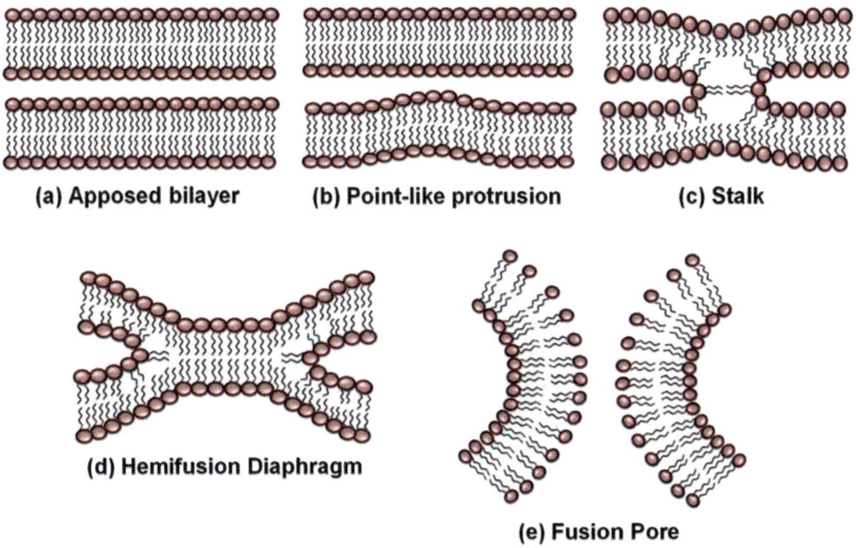

Fig. 2.3 Schematic representation of the different stages of membrane fusion of lipid bilayers. (**a**) Apposed bilayer (**b**) Point-like protrusion (**c**) Stalk (**d**) Hemifusion diaphragm (**e**) Fusion pore

1987). The protein–protein interaction is the crucial parameter that brings two membranes close to each other. Enveloped viruses, surrounded by a lipid membrane, utilize the fusion protein–host cell receptor interaction to bring two membranes into proximity, followed by the proteolytic cleavage of the fusion protein. The fusion domain undergoes conformational changes that expose a hydrophobic fusion peptide, which helps pull the viral and host membranes into proximity, leading to fusion and allowing the viral contents to enter the host cell (White and Whittaker 2016).

2.5.2 *Pre-fusion Stage*

In this stage, a point-like protrusion is generated in one of the bilayers, which must exhibit the least repulsion from the apposing membrane, delivering a minimal contribution to the energy, as shown in Fig. 2.3b. This point-like contact between the membranes arises due to dehydration, which reduces the hydrophobic energy of the monolayer rupture. Because of the apparent high energy of sharp bending at the top of the protrusion, point-like protrusions have not previously been taken into consideration. Nonetheless, recent advancements in the elastic model of membrane rearrangements, involving tilt deformation (Kozlovsky and Kozlov 2002), have shown that such structures are feasible. Proteins play a vital role in this stage by dehydrating the local membrane contact and elastic stresses, endorsing membrane fluctuations.

2.5.3 Fusion Stalk Intermediate

The stalk is the first intermembrane lipid connection, formed by the junction of defects in the contacting monolayer of two membranes (Gingell and Ginsberg 1978). When two lipid bilayers come into close contact, hydration repulsion increases exponentially due to the larger surface area. To reduce this hydration repulsion force between the polar headgroups of lipid membranes linked by surface water, a point-like protrusion occurs in one membrane, connecting the approaching outer leaflet of the other membrane to form a continuous structure that resembles a semi-stable hourglass-like intermediate or stalk state, as shown in Fig. 2.3c. The lipid composition also influences this hydration force. Lipids with intrinsic negative curvature facilitate the formation of protrusions or bulges, which are necessary for producing the stalk intermediate. This negative spontaneous curvature of the contacting monolayer lowers the energy required for stalk formation. Fusion proteins are also crucial in forming the stalk, as they provide the energy needed for this intermediate formation.

2.5.4 Hemifusion Diaphragm

It is well-known that the lipid membrane shifts from a lamellar to a hexagonal phase during stalk formation. As a result, the stalk intermediate expands radially, and the inner monolayers of two approaching membranes converge to form a bilayer by connecting the hydrophobic tails of the two approaching membranes. A short-lived, metastable, trilaminar structure called the hemifusion diaphragm is created by this lipid rearrangement in the stalk, which then expands longitudinally to open the pore, as shown in Fig. 2.3d. The expansion of the stalk is caused by tension in the outer leaflets and excess lipids in the inner leaflet. A circular HD has always been shown to be more energetically favorable than an extended stalk. An important factor that determines the expansion of the stalk into HD is the spontaneous curvature of the merging membranes (Kozlovsky et al. 2002).

2.5.5 Fusion Pore

Extremely high mechanical tension in the hemifusion diaphragm causes the core bilayer to rupture, leading to curve the edge of the pores longitudinally, covered by polar headgroups, preventing contact of the hydrophobic interior of the pores, which leads to the opening of the pore (Risselada et al. 2014), as shown in Fig. 2.3e. The rim size of the hemifusion diaphragm or the perimeter of the core bilayer generates this tension, which determines the pore size; low tension creates a small or flickering fusion pore, while high tension causes a massive pore opening (Katsov et al. 2004).

Higher activation energy is needed for fusion pore opening than for stalk formation, which is regulated by the curvature produced by various lipidic shapes or the lateral tension produced by proteins (Chernomordik et al. 1995).

2.6 Energetics of Fusion Intermediates

The amount of free energy needed to generate a stalk was determined by the energy essential for bending the monolayer surface (Siegel 1993). Chernomordik and Kozlov proposed that the fusion stalk is an intermediate structure that limits the fusion rate by establishing an energy barrier for stalk formation (Chernomordik and Kozlov 2003). The saddle splay (Gaussian) modulus of the lipid monolayers, as well as the hydration repulsion between the apposing membranes, can have a significant impact on the stalk energy, in addition to the spontaneous curvature of the lipid. The key factors that contribute to the energy of stalk formation are the deformation of membrane monolayers from their original flat shape into the stalk structure, resulting in a change in their elastic energy, and the shift in energy due to intermembrane interactions. Due to the stalk formation, the area of the water-exposed membrane decreases, resulting in a reduction of the hydration energy of the system (Helfrich 1973; Kozlovsky et al. 2004). The factors affecting the energy of fusion intermediates are discussed below:

2.6.1 Elastic Energy of Bending of Membrane Monolayers

Membrane bending refers to the deformation of a biological or model membrane into a curved shape. In general, several factors like pH, temperature, and salt concentration affect membrane bending (Lähdesmäki et al. 2010). Membrane bending results in the formation of structures called nipples, which are formed from the point of protrusion of both monolayers when two closely apposed monolayers meet each other. A balance between the physical forces, repulsion due to hydration, and the attraction created by fusion proteins corroborates the equilibrium distance between the tips of the two nipples. This nipple formation triggers the formation of hydrophobic patches at the tip of a nipple due to temporary displacements of polar headgroups from each other. Outer leaflets of the bilayer merge to form the stalk without undergoing substantial additional bending as hydrophobic surfaces attract each other. A void is generated in the stalk structure that gives rise to hydrocarbon–vacuum interfaces (Kuzmin et al. 2001; Leikin et al. 1987; Siegel 1993). To reduce this hydrophobic void energy, the inner and outer monolayers should make more contact, and the inner leaflets deform by tilting their lipids relative to the monolayer surface, in addition to bending, which reduces the free energy of stalk expansion. When contact occurs, the nearby lipids easily spread out to the unchanged area of the nipple, while the core molecules are aligned parallel to the axis of symmetry of the stalk, and this structure is known as "modified stalk" (Hui et al. 1981; Kuzmin et al. 2001). This alteration in the stalk structure is due to the combined effect of membrane surface bending, splay, and tilting of the hydrocarbon chain (Kozlovsky and Kozlov 2002). The lipids of the modified stalk can reorganize into a fusion pore at an energy of about 40 k_BT higher than that of the original stalk. The inner layers of the membranes come together and merge, creating the smallest water-filled channel between the two compartments. This channel remains stable because forces that usually push the membranes apart, known as hydration forces, are balanced, allowing the channel to stay open without collapsing. The pre-pore is a temporary structure formed when the acyl chains in both monolayers of the membrane are tightly packed, especially in the middle part of the membrane, as it has a smaller radius. As the pre-pore expands to a fusion pore, the chains spread out and make the pore larger, resulting in lowering the energy of the system (Kuzmin et al. 2001). The type of lipid headgroup, chain length, and chain unsaturation affect the bending modulus (Rawicz et al. 2000). Lipid bilayers and surrounding membrane proteins interact hydrophobically, which couples protein conformation to the mechanical properties of the bilayer. The interplay between lipids and membrane proteins assists in forming non-bilayer phases that regulate protein function and generate the curvature stress. Therefore, the curvature stress also influences the protein structure, in addition to the electrostatic energy of the bilayer. Changes in the electrostatic energy of DOPS bilayers cause changes in curvature stress, which affect the structurally distinct gramicidin A monomer $\leftrightarrow$ dimer equilibrium, leading to a reduction in the electrostatic energy of the unaltered bilayer and encouraging channel dissociation (Lundbaek et al. 1997).

2.6.2 Splay and Tilt of the Hydrocarbon Chain

According to continuum elasticity theory, splay (a generalization of bending), tilt, and area compression are the three basic lipid monolayer deformations that occur via any arbitrary small perturbation of the monolayer. However, the energy required to deform a monolayer by splay and tilt is far lesser than that required to deform via area compression (Hamm and Kozlov 2000; Helfrich 1973; Kuzmin et al. 2005). In membrane biophysics, the splay and tilt of a membrane monolayer are important concepts that explain how the lipid molecules within a monolayer are organized and oriented. They are mostly essential while studying the mechanical properties of lipid membranes, curvature generation, and interactions with proteins or peptides. Splay refers to the spatial variation or divergence of the lipidic tail orientations within the monolayer and measures how the lipid molecules deviate from being parallel to the other leaflet. In vesicles, the lipid tails must accommodate curvature, leading to lipid splay and the modulation of angle between neighboring lipid tails. Splay contributes to membrane elasticity, bending energy, and developing curvature energies using the Helfrich Hamiltonian. Tilt illustrates the angle between the lipidic tail (molecular axes) and the local normal vector (N) to the membrane surface when the lipidic tails lean away from the normal, either due to crowding, curvature stress, or protein

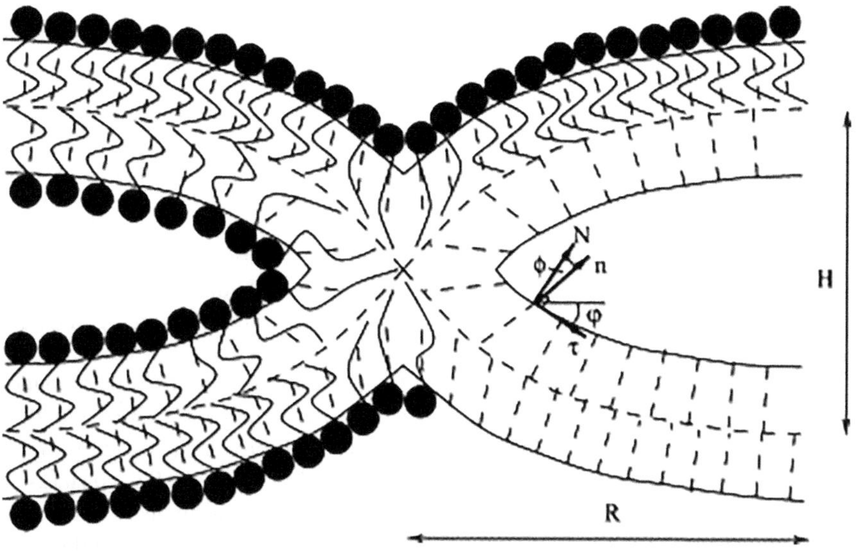

Fig. 2.4 Schematic representation of the tilt model for stalk and transmembrane contact. The shape is round and symmetrical. The black circles are the heads of the lipid molecules, and the wavy lines show the tilted hydrocarbon chains inside the membrane. The stalk area in the middle has a sharp corner where the membrane layers meet. The labels show different angles and directions: N is the unit vector of the normal to the direction of the monolayer surface, n is the direction of a lipidic chain, τ is a tangential unit vector, and φ and ϕ are angles describing how much the chains tilt. H and R represent the distance between membranes and the width of the stalk, respectively. (The figure has been adapted from Kozlovsky and Kozlov (2002) with permission)

interactions (Hamm and Kozlov 2000; Helfrich 1973). The hydrocarbon chains are arranged at an angle with respect to the surface of the membrane, which fills the hydrophobic void space, resulting in a sharp corner in the monolayer. The corner is located between two neighboring molecules in the cross section of the stalk so that lipid molecule does not need to bend much, as shown in Fig. 2.4. Near the sharp corner, the membrane is curved along with the tilt of the hydrocarbon chains while moving away from the corner, the chains gradually straighten out, and the membrane surface becomes flat again with the disappearance of the tilt. Figure 2.4 suggests the initial merging of the outer monolayers of the fusing membranes, whereas the inner layers are already bent and deformed enough that they meet each other at certain points (Kozlovsky and Kozlov 2002).

In the absence of tilt deformation of the lipidic chains, a specific kind of bending called saddle splay deformation arises, which takes the shape of the Gaussian curvature of the surface. However, the saddle splay energy is altered by a topological rearrangement of the membrane monolayers brought on by stalk formation and pore formation (Kozlovsky et al. 2004). Saddle splay in a membrane monolayer refers to negative Gaussian curvature, describing regions where the monolayer bends in opposite directions along two axes, like a saddle. The lipids bend in a way that helps bring both the monolayers together. This idea correlates to the earlier scientific calculations and studies of how forces are distributed within the monolayers, confirming that lipids with bending property are more likely to form the stalk intermediate (Szleifer et al. 1990; Templer et al. 1998).

2.6.3 *Hydration Repulsion*

The hydrated surfaces of two lipid bilayers experience a strong repulsive force when they are within around 1–3 nm, which is referred to as hydration repulsion. The source of this repulsive force is the energy required to dehydrate polar lipid headgroups, such as those of PC and PE. At very short distances (less than 2–3 nm), water molecules close to lipid headgroups form ordered, hydrogen-bonded networks that dominate over the van der Waals force of attraction. The hydration repulsion affects bending rigidity and interactions with proteins/ions and prevents membrane fusion from occurring. Viral fusion proteins, such as HIV gp41 fusion peptide and peptides like SNAREs and synaptotagmin, contribute to promoting membrane fusion by overcoming this barrier (Parnas and Parnas 1994; Südhof and Rothman 2009; Swarnakar et al. 2025; Xiao et al. 2021). Many reports discuss the role of membrane-merging-bridging ions like Ca^{2+} or some dehydrating agents like PEG, which cause excellent membrane fusion (Hong et al. 2018; Lentz 2007), leading to the fact that the most energy-intensive steps of the fusion process are bringing two membranes close together and overcoming the hydration repulsion. The close apposition of two bilayers in a low-hydrated environment allows lipid fluctuation between the outer leaflets of two bilayers. The fluctuations result in the formation of close (< 0.5 nm) contact between the membranes within a small area (~10 nm^2). The rupture of the interacting monolayers in this area is caused by an increasing hydration repulsion between apposing polar heads of the lipid molecules. Membranes fuse due to this rupture, creating a bridge that connects the monolayers and is commonly referred to as the monolayer stalk (Leikin et al. 1987).

Klaus Arnold and others demonstrated that PEG works by volume exclusion, which generates an osmotic pressure to push membranes into close contact in a dehydrated region. Using PEG to bring membranes into contact allows for a detailed analysis of the function of various lipids and fusion proteins in membrane fusion. The three-step sequential kinetic model, as proposed by Weinreb and Lentz, describes the PEG-triggered fusion of many model membranes and the role of fusion peptides on membrane fusion in quantitative detail (Lentz 2007; Weinreb and Lentz 2007). One can investigate the potential method of PEG-mediated fusion by monitoring changes in rates and extents of lipid mixing and content mixing associated with the presence of whole fusion proteins or peptides derived from them. Additionally, this approach provides a means to identify alterations in the fusion mechanism following analysis using the two-intermediate model by Weinreb and Lentz (Chakraborty et al. 2012; Chakraborty et al. 2013; Weinreb and Lentz 2007). Using this method, it is revealed that fusion is catalyzed by both the fusion peptide and the transmembrane domains (TMDs) of fusion proteins (Chakraborty et al. 2013;

Sengupta et al. 2014). At low concentrations, monovalent cations had little effect on the vesicle morphology due to the weak binding between the ions and the phospholipid headgroups. In contrast to monovalent ions, only a small concentration of added divalent cations was required to promote vesicle fusion. The ability to destroy the vesicle morphology followed the order $Cu^{2+} > Mg^{2+} > Ca^{2+} > Na^{+} > K^{+}$ (Hong et al. 2018). Polarized infrared spectra demonstrate that complex formation of the phosphate groups with divalent ions results in headgroup immobilization and conformational changes. Lipids with non-lamellar phases reduce hydration repulsion by promoting curvature stress, which initiates the merging of membranes (Binder and Zschörnig 2002). In PC vesicles, since the phospholipid membrane in pure water has no net charge, there are no electrostatic interactions. Therefore, PC vesicles exist in pure water for a long time in a stable form without aggregating, even in the absence of electrostatic repulsion. A short-range hydration repulsive force between the adjacent vesicles, originating from strong hydration by water molecules around the PC headgroups, is responsible for the stability of the lipid vesicles (Gamon et al. 1989; Inoue et al. 1992; Israelachvili and McGuiggan 1988). The hydration repulsion is somewhat eliminated in the presence of positively charged ions on the membrane surface. Mainly, the intervesicular interaction potential between PC vesicles is made up of three forces: the short-range strongly repulsive steric hydration force that is reduced by the metal ions adsorbed on the vesicular surface, the long-range repulsive electrostatic force caused by the binding of the metal ions, and the van der Waals attraction (Marra and Israelachvili 1985). The effects of cations on the morphology of lipid vesicles may depend on the competition between the destruction of the hydration force, which destabilizes the vesicles, and the formation of an electric charge on the vesicular surface, which stabilizes the vesicles. A certain amount of lipid vesicle aggregation will occur when the electrostatic repulsion is overcome by the reduction of hydration force (Hong et al. 2018).

2.7 Membrane Curvature: Implications of Lipidic Shape in Membrane Merger

Membranes must deform to generate local shapes. The structure and elasticity of the membrane dictate how much work the cell must perform to deform it. Membrane curvature refers to the bending or deformation of a biological membrane from a flat surface into a curved shape, and it plays a crucial role in several cellular processes, such as membrane trafficking, viral entry, and defining the morphology of the cell and organelles (McMahon and Boucrot 2015). Mathematically, curvature is a quantitative measure of the deviation of an edge from a straight line. For a circle, the measure of curvature is the inverse of its radius (R), i.e., 1/R, and, thus, has units of m^{-1}. The curvatures of the plane between the membrane monolayers describe how a bilayer bends. Even though a surface bending is defined by two major curvatures (C_1 and C_2), where $C_1 = 1/R_1$ and $C_2 = 1/R_2$, and their combination is

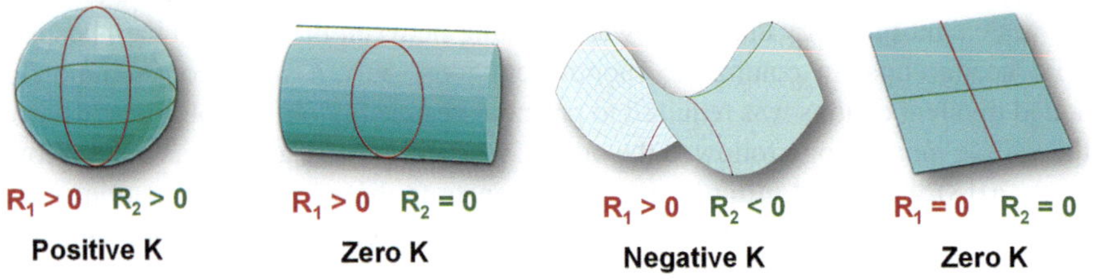

Fig. 2.5 The principal curvatures of a surface are multiplied to create the Gaussian (K) curvature. Positive (convex) or negative (concave) curvatures are defined with respect to the cytoplasm

known as the Gaussian curvature ($K = C_1. C_2$), but the total curvature ($J = C_1 + C_2$) have a deep geometrical significance in practice. The mean curvature (H) is the average of the two major curvature, $H = \frac{(C1+C2)}{2}$, can be nonzero, but the two sides (inner and outer leaflets) will have opposite signs for H. The Gaussian curvature is intrinsic and remains the same on both sides and is essential for the explanation of membrane deformations. The cell organelles can be differentiated by their Gaussian curvature; for example, an endosome is known for its positive curvature, whereas the cytokinetic furrow has the intrinsic negative curvature (Dorn et al. 2016; Nawara et al. 2022). The Gaussian curvature actually illustrates how closely a surface resembles a sphere (for positive values) or a saddle (for negative values) (Cail and Drubin 2023). The Gaussian (K) curvature of the four basic surfaces is shown in Fig. 2.5.

The non-deformed unstressed shape of the membrane is controlled by the membrane structure, which is referred to as the "spontaneous shape." Spontaneous curvature (J_s) refers to the curvature that a lipid monolayer acquires when no forces are working on its surface (Chernomordik and Kozlov 2003; Gruner 1989; Helfrich 1973). The spontaneous shape is determined by the spontaneous curvatures of the inner and outer membrane monolayers, J_s^{in} and J_s^{out}, and is defined by the spontaneous curvature of the membrane bilayer, J_s^B (Zimmerberg and Kozlov 2006). The intrinsic shapes of lipid molecules generate the respective spontaneous (intrinsic) curvature, which plays an instrumental role in the stabilization of the mechanism of membrane fusion (Chernomordik et al. 1995; Cooke and Deserno 2006; Israelachvili et al. 1976). A membrane bilayer leaflet is essentially a thin, two-dimensional surface. When membrane protrusion takes place, if the outer leaflet generates positive (convex) curvature, then the inner leaflet must correspondingly induce negative (concave) curvature to maintain the structure, which is a geometric necessity for thin materials. The ability of lipids to arrange into various phases (such as lamellar, hexagonal, or cubic) in response to environmental factors (temperature, hydration, and lipid composition) is known as lipid polymorphism. The balance between the headgroup repulsion (polar and hydrophilic), acyl chain interactions (van der Waals forces, nonpolar, and hydrophobic), and spontaneous curvature of lipid monolayer controls the molecular packing constraints (Epand 1998; Österberg et al. 1995). Lipids and proteins interact to achieve dynamic membrane remodeling, and high-membrane-curvature regions usually last for a short time; the main way to do this is by altering the morphology of the surrounding proteins (McMahon and Gallop 2005). The major factors influencing the membrane curvature in the process of membrane merger are as follows:

2.7.1 Lipid Composition Affecting Membrane Curvature

The lipid composition of the membrane is crucial for membrane fusion, as their distribution significantly influences the activity of various cell types. Alterations in lipid composition can also affect membrane curvature, its assembly, scaffolding proteins, and the reversible insertion of protein regions (McMahon and Gallop 2005). Molecular packing is an important parameter that is quantitatively correlated to curvature (Marsh 1996). The polar headgroups and the nonpolar acyl chains are subjected to attractive and repulsive forces, which are demonstrated by the way the lipids are packed within the aggregate (Israelachvili 2011). The weighted average of the spontaneous curvatures of the lipids that make up a bilayer determines its spontaneous curvature, which is equal to the difference between the spontaneous curvatures of its inner and outer monolayers. Many lipids, like phosphoinositide, phosphatidic acid, diacylglycerol, and sterols, directly regulate membrane fusion events and are called regulatory lipids (Fratti et al. 2004). The ratio of surface area of headgroup to hydrophobic acyl chain length establishes the concave/convex lipidic shape, which significantly affects membrane curvature (McMahon and Boucrot 2015). Cone-shaped and inverted cone-shaped lipids induce either positive or negative intrinsic curvature to maintain the membrane fusion mechanism. The lipids such as PE, cholesterol, phosphatidic acid (PA), DAG, or cardiolipin generate intrinsic negative curvature, forming reverse hexagonal (H_{II})defined by two major curvatures, and promote stalk intermediate formation because of their smaller polar headgroup, which leads to the bending of monolayer in such a way that the polar headgroups come nearer together, forming a concave structure. On the other hand, the formation of the stalk is typically impeded by lipids that induce intrinsic positive curvature and form micelles or normal hexagonal (H_I) phases. Lysolipids with a single acyl chain like LPC, and LPA, lipids with anionic headgroups like PI, or deprotonated fatty acids due to their larger headgroup, which bends the monolayer away from the polar headgroups, creating a convex structure (McMahon and Gallop 2005; Poojari et al. 2021). Cylindrical lipids like PC and PS create a flat monolayer with zero curvature and form lamellar bilayers (Joardar et al. 2022b). The lipidic shape is also influenced by acyl chain unsaturation, which causes the chain to kink, forcing it to take up more space than its saturated counterpart. Therefore, the overall shape of the lipids is defined by the ratio of the polar headgroup to the hydrophobic acyl chain saturation size (McMahon and Boucrot 2015; Pinot et al. 2014; Vanni et al. 2014). In a membrane bilayer, lipids with negative intrinsic curvature at the outer leaflet assist stalk formation, and lipids with positive intrinsic curvature at the inner leaflet accelerate pore formation, which promotes membrane fusion (Yang et al. 2016a), as shown in Fig. 2.6. This shows the significance of lipid composition

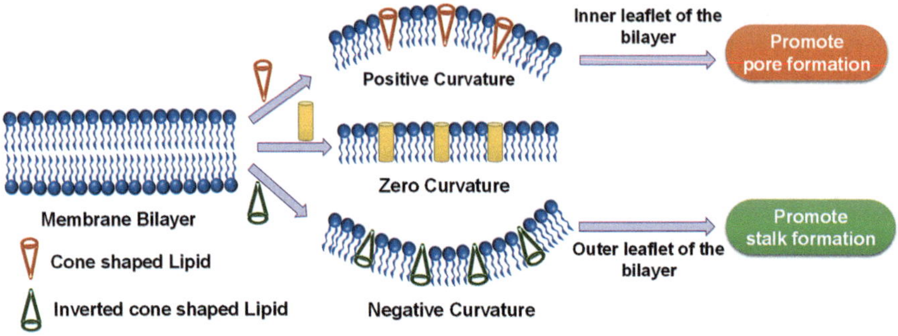

Fig. 2.6 Schematic representation of lipidic shape generating spontaneous (intrinsic) membrane curvature

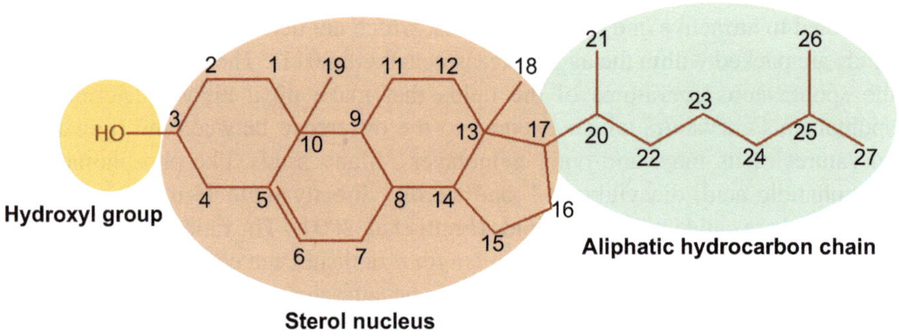

Fig. 2.7 Structure of cholesterol

by reducing the energy needed for stalk intermediate or pore formation (Furber et al. 2009; Meher and Chakraborty 2019).

Another interesting lipid that plays an important role in membrane fusion is cholesterol, which is an essential component of the cell membrane, varying with age and from person to person, and is distributed unevenly in various cell membranes (Lange 1991; Liscum and Underwood 1995; Martin et al. 2010; Pattnaik and Chakraborty 2021a). The structure of cholesterol is unique in terms of its role in driving the fusion process. It comprises 27 carbons having a branched aliphatic hydrocarbon chain at one of the ends, a four-hydrocarbon ring containing a sterol nucleus, and a hydroxyl group that is the only polar part of the cholesterol at the other end (Craig et al. 2018). The polar hydroxyl group of the cholesterol orients close to the headgroup region of phospholipids, while the remaining nonpolar hydrocarbon tail aligns perpendicularly to the bilayer by modifying the bending rigidity and spontaneous curvature of cell membranes, as shown in Fig. 2.7.

The orientation of cholesterol induces a more ordered arrangement of the hydrocarbon chains, hence regulating the physical, structural, and dynamic properties of the membrane. Cholesterol alters membrane fluidity, thickness, compressibility, water penetration, and intrinsic curvature of lipid bilayers. These changes affect the membrane's propensity to undergo fusion by making the bilayer more ordered yet flexible enough to facilitate the merging of two membranes (Moon and Jun 2020; Yang et al. 2016b). It generates intrinsic negative curvature in the membrane because of the smaller headgroup compared to the hydrophobic cross-sectional area of the tail. Consequently, it facilitates membrane fusion and the formation of a highly curved stalk intermediate (Aeffner et al. 2012; Pattnaik and Chakraborty 2021a; Wang et al. 2007; Yang et al. 2016a). It has been reported earlier that cholesterol alters the dynamics and organization of the membrane (Meher and Chakraborty 2019). The interactions between cholesterol, unsaturated phosphatidylcholines, and SM (primarily saturated), the primary constituents of animal cell outer membranes, have drawn significant attention due to the domain formation (Lingwood and Simons 2010; Simons and Ikonen 1997). Furthermore, because SM has a strong affinity for cholesterol, it controls the amount of cholesterol in the plasma membrane (Liscum and Underwood 1995). Cholesterol inhibits spontaneous curvature by increasing the packing of SM and PC in the membrane bilayer (Ikonen 2008). Lipid domains are tiny, dynamic regions comprising cholesterol, saturated phospholipids, and sphingolipids involved in several physiological activities and signaling pathways. The most plausible reason for forming a lipid domain in the presence of SM can be attributed to the formation of H-bonds between cholesterol and SM (Edidin 2003). These microdomains serve as a specialized platform where lipid rearrangement and bilayer merger occur more readily. Lipid domains have been associated with a diverse range of distinct membrane processes, including viral entry (Chazal and Gerlier 2003). A domain is composed of a liquid-ordered (l_o) phase whose boundary is energetically penalized due to structural and chemical variations with the surrounding liquid-disordered (l_d) matrix. The formation of vesicles by domain budding is one of the membrane remodeling mechanisms supported by consequential length minimization of the l_o/l_d phase boundary (Lipowsky 1992).

After PC, PE is the second most abundant phospholipid in mammalian cells (Patel and Witt 2017; van der Veen et al. 2017). PE is essential for membrane fusion, as it induces intrinsic negative curvature because of its smaller headgroup size and lower hydration, as a result of which it stabilizes the stalk-like intermediate, which subsequently moves on to hemifusion diaphragm and fusion pore formation (Kawamoto et al. 2015; McIntosh and Simon 1986; McMahon and Boucrot 2015; Pink et al. 1998; Risselada et al. 2014). It has been reported earlier that PE facilitates membrane fusion by stabilizing the non-lamellar intermediate structures during the fusion process. On the other hand, membrane fusion is inhibited by OA, which generates positive curvature (Joardar et al. 2021). The intrinsic curvature of OA changes in a pH-dependent manner. The deprotonated species exhibits intrinsic positive curvature at high pH because of the greater degree of hydration of the headgroup (Lähdesmäki et al. 2010). At pH 7.4, DOPE and cholesterol both enhance the interfacial and acyl chain order of DOPC membranes and prevent water percolation in both regions. Even at low concentrations (10 mol%), DOPE has a notable impact and a larger influence on the rate of hemifusion formation. On the other hand, above 20 mol% membrane cholesterol, the impact of cholesterol on the rate of hemifusion formation becomes apparent (Mishra and Chakraborty 2023).

Cholesterol has a higher negative intrinsic curvature as compared to DOPE (Kollmitzer et al. 2013). However, higher concentrations of cholesterol are necessary for the intrinsic negative curvature to overcome its ordering ability and encourage hemifusion formation because cholesterol has a greater capacity to generate order at the membrane's interfacial region. On the other hand, at a lesser concentration (10 mol%), the intrinsic negative curvature of DOPE predominates over the ordering ability due to its comparatively modest effect on interfacial ordering (Mishra and Chakraborty 2023). PA, constituting 1–4% of membrane lipids, promotes negative intrinsic curvature in the plasma membrane, thereby facilitating the formation of the highly curved stalk and extended transmembrane contact. The opening of pores between two lipid bilayers is ensured by the stability of these two intermediates (Ashery et al. 2014; Chernomordik et al. 1995; Epand et al. 2015; Meher and Chakraborty 2019). The combination of two monounsaturated acyl chains and a bulky headgroup, with the percentage of negatively charged lipids like PS and PI, determines membrane electrostatics. Lipids with smaller headgroups and/or unsaturated acyl chains are known to increase packing defects. Membrane curvature, which arises due to these lipid-packing defects, can reach a radius of 20 nm, in contrast to the membrane thickness, which is 4 nm (Bigay and Antonny 2012). Lysolipid-containing PE shows negative curvature, whereas LPC induces positive curvature (Fuller and Rand 2001).

2.7.2 Peptide-Induced Change in Membrane Curvature

Membrane proteins are efficient in generating membrane curvature; conversely, the function and distribution of membrane proteins within cells are affected by membrane curvature (McMahon and Gallop 2005; Shibata et al. 2009). For scrutinizing the effect of membrane curvature on membrane proteins, liposomes are suitable model membranes. The size of liposomes plays a decisive role in modulating their membrane curvature. Smaller liposomes exhibit a higher membrane curvature (Pomorski et al. 2014). The impacts of membrane-associated proteins, or "coat proteins," induce curvature, leading to the formation of highly curved vesicles. When the coats separate, the curvature is easily reversible, making the vesicle more fusogenic (since its curvature is not stabilized) and allowing the coat proteins to be recycled in a subsequent vesicle formation cycle, which increases protein utilization efficiency (McMahon and Gallop 2005). Proteins like rhodopsin are activated by using the curvature stress caused by non-lamellar lipids (like PE) (Deese et al. 1981). Peptides, especially fusion peptides that cause the membrane to curve inward, allowing for a larger contact area between the two membranes, promote fusion. In contrast, peptides and other molecules that cause the membrane to curve outward, preventing multiple contacts between fusion proteins and target membranes, will inhibit viral fusion (Colotto et al. 1996).

Fusion peptides are generally a stretch of 20–25 amino acid residues that have unique properties like amphiphilicity and a highly conserved sequence and are located at the N-terminal region of the fusion protein, which are essential for viral entry. They play a crucial role in membrane fusion, as mutations in the peptide sequence can disrupt both viral and host cell membrane fusion, thereby affecting subsequent infection (Freed et al. 1990; Gething et al. 1986; Steinhauer et al. 1995). The fusogenic action of these peptides demonstrates their capacity to actively participate in the fusion process beyond their role as membrane-anchoring units. The viral and cell membranes should come into close contact during the conformational transition if the fusion peptide integrates into the target membrane. As a result, fusion peptides not only cause local deformation of the target membrane but also serve as anchors, allowing fusion proteins to apply mechanical forces to the target membrane (Harrison et al. 2013; Skehel and Wiley 2000). Another interesting property of fusion peptides is their high dynamic nature in the presence of lipids. The interaction between fusion peptides and liposomes offers insights into how viral fusion peptides function (Martin et al. 1999). The capacity of biomolecules to preferentially localize to membrane areas of a particular curvature is known as curvature sensing. It has been demonstrated that helical peptides with both hydrophilic and hydrophobic regions, known as amphipathic helices (AHs), are capable of sensing both positive and negative curvature in lipid membranes. The lipid-packing defects that are developed in regions of positive and negative curvature give rise to these AHs. Using coarse-grained and all-atom simulations with curved POPC bilayers, a correlation between the peptide insertion depth in the membrane and the preferred mean curvature has been established. Deeper peptide insertion takes place with the increasing hydrophobicity of peptides, and increasing hydrophilicity of peptides leads to the shallow insertion of the peptide (Pajtinka and Vácha 2024). However, the lipids that induce positive spontaneous curvature (e.g., lysolipids) promote pore formation and inhibit stalk formation and fusion. On the other hand, a deep insertion of a peptide induces negative spontaneous curvature, creating a bulge (Chernomordik 1996), as shown in Fig. 2.8.

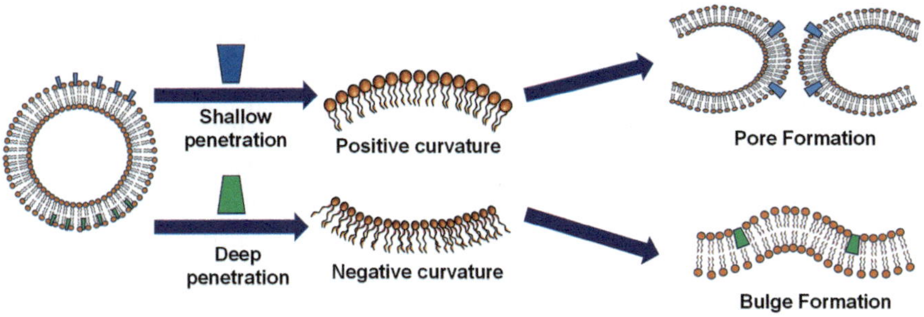

Fig. 2.8 Schematic representation of how peptide binding modulates the membrane shape. Spontaneous positive curvature and pore formation are supported by the shallow insertion of the peptide, whereas deep insertion induces negative spontaneous curvature and facilitates bulge formation

The mechanism by which fusion peptides induce fusion is quite fascinating. According to a well-established model (Epand et al. 1994; Epand and Epand 1994), the insertion of the fusion peptide disrupts the lipid bilayer arrangement, leading to increased fusogenicity. The fusion peptide essentially disrupts the lamellar lipid phase, which is responsible for peptide-induced membrane fusion (Agirre et al. 2000; Colotto and Epand 1997; Colotto et al. 1996; Davies et al. 1998; Epand and Epand 1994; Epand et al. 2001; Pereira et al. 1999; Tenchov et al. 2013). The fusion peptide of the HA of the influenza virus significantly affects the entry of the virus into the host cell. According to small-angle X-ray diffraction studies, at pH 5.0, the HA fusion peptide produces a curved lipid structure, promoting cubic phases by converting the lamellar kinetics to inverted phase transitions. The reason behind the development of these phases of lipids is the intrinsic negative curvature generated by the HA fusion peptide (Colotto and Epand 1997; Epand and Epand 1994; Siegel and Epand 2000; Tenchov et al. 2013). It was suggested that at low pH, stress on the lipid bilayer causes the influenza HA fusion domain to undergo a conformational change, and the deeper insertion of this inverted V-shaped HA fusion peptide raises surface tension in the lipid bilayer's interfacial region and lateral pressure within the acyl chain region. Then, the lipid bilayer induces negative curvature in response to this stress, which contributes to the formation of stalk intermediates and facilitates fusion (Han et al. 2001; Qiao et al. 1999). Fusion peptides having helical structures derived from different enveloped viruses (HIV, HA, and simian immunodeficiency virus (SIV)) can insert freely into the lipid membranes, and such orientation shows a one-to-one correlation with the fusogenic property of these peptides (Epand et al. 2001; Horth et al. 1991). Viral envelopes are highly enriched in cholesterol and are required for the efficient fusion of enveloped viruses with host cell membranes, as they induce negative curvature in the lipid bilayer. Therefore, cholesterol plays a divergent role in the entry of many viruses, such as HIV and influenza virus, and in the oligomerization of fusion proteins and peptides to accept a fusion-proficient conformation (Pattnaik and Chakraborty 2021a). The presence of cholesterol in viral membranes affects the rigidity and fusogenic activity of viral fusion peptides, such as those in HIV gp41, which require cholesterol-rich environments to mediate membrane merger effectively (Apellániz et al. 2014). The presence of cholesterol further modulates the depth of penetration of the fusion peptide to the lipid bilayer (Panda et al. 2025b). The spike (S) protein of severe acute respiratory syndrome coronaviruses (SARS-CoVs) contains multiple putative peptide sequences (the N-terminal fusion peptide (FP), internal fusion peptides (IFPs), and a segment located upstream of the transmembrane domain known as the pre-trans membrane domain (PTM)) that contribute to their entry into the host cell, and the synergistic action of these multiple fusion peptides might be crucial for hemifusion formation, leading to the opening of the pore (Guillén et al. 2008). The propensity of oligomerization of the N-terminal fusion peptide of the S-protein from SARS-CoV enhances with increasing concentration of membrane cholesterol in DOPC/DOPE/DOPG (60/30/10 mol%) membranes, even at a peptide-to-lipid ratio higher than

0.01, whereas, in POPC/POPG (80/20 mol%) membranes, peptide aggregation was noticed at a very high concentration of peptide (peptide-to-lipid ratio: 0.04). This shows the significance of PE and cholesterol in inducing oligomerization of the S2 fusion peptide due to their negative intrinsic curvature (Meher et al. 2019, 2023). Differential scanning calorimetry observations state that FP causes positive curvature on DOPE vesicles, while IFP, depending on the ionic strength, promotes opposing stresses on the intrinsic negative curvature. Additionally, electron spin resonance spectroscopy revealed that fusion peptides decrease the intramembrane water content for anionic membranes while increasing lipid packing and headgroup ordering. This causes the bilayer to produce a bending moment and promote negative curvature to facilitate hemifusion formation (Basso et al. 2016). It is believed that many viruses, including HIV and the influenza virus, exploit lipid domains for assembly and budding (Freed 2015; Scheiffele et al. 1999). Remarkably, membrane cholesterol influences the inhibitory efficacy of the TG-23 peptide derived from coronin 1, a phagosomal protein, as a peptide-based fusion inhibitor. With increasing concentration of membrane cholesterol in two membranes fusing with the same composition, as well as in asymmetric membrane fusion with 0 and 20 mol% of cholesterol, the inhibitory efficiency of TG-23 decreases. This is due to the cholesterol's ability to induce negative curvature, which overtakes the inhibitory action of TG-23 (Pattnaik and Chakraborty 2019; Pattnaik and Chakraborty 2021b). It has been previously reported that the HA2 fusion peptide of class I influenza virus promotes the formation of negative curvature and penetrates deeply into the hydrophobic region of membranes, whereas the class II and III fusion loops of tick-borne encephalitis virus (TBEV) glycoprotein E and vesicular stomatitis virus (VSV) G-protein tend to counteract negative membrane curvature (Surmeier et al. 2022).

An aqueous-membrane partitioning of a voltage-gated channel-forming peptide, alamethicin, was studied in relation to the spontaneous curvature of the DOPC membrane with varied percentages of DOPE, N-methyl tagged DOPE (DOPE-Me), and N, N-dimethyl tagged DOPE (DOPE-Me_2) by using EPR spectroscopic techniques, which demonstrated that the binding of alamethicin decreases and the binding free energy increases with increasing mole fraction of DOPE in the mixtures, whereas in the case of DOPE-Me and DOPE-Me_2, the increase in binding free energy is less compared to DOPE. The binding free energy is found to be linearly dependent on the spontaneous curvature when the free energy changes for alamethicin binding are compared to the previously established spontaneous curvatures for a mixture of DOPC/DOPE and DOPC/DOPE-Me. It is hypothesized that the cooperative behavior and sensitivity to spontaneous curvature observed in the binding curves for alamethicin originate from a localized thinning of the bilayer that is facilitated by this peptide (Lewis and Cafiso 1999). It has been proposed that the β-strand conformation of the paramyxovirus PIV5 fusion peptide induces negative Gaussian curvature and that the dehydration of the DOPE membrane by the fusion peptide might be essential for hemifusion intermediate formation to promote membrane fusion (Yao and Hong 2014).

2.8 Implication of Lipid–Peptide Interaction in Membrane Merger

The process of membrane fusion is not impulsive in biological systems, as either their fusion proteins or their complexes drive it (White 1992; Wickner and Schekman 2008). Most enveloped viruses fuse their membranes to enter their host cells, and this process is mediated by fusion proteins found on the viral surface and can take place at the endosomal membrane or the plasma membrane. The fusion proteins bind the viral envelope and cell membrane together and pull them toward each other through a refolding mechanism (Baquero et al. 2015; Boonstra et al. 2018; Harrison 2008; Harrison 2015; Modis 2014; White and Whittaker 2016). Lipid composition plays a decisive role in modulating membrane fusion by altering the structure and physical properties of the membrane (Meher and Chakraborty 2019). Additionally, some lipids influence the fusogenic properties of fusion proteins and peptides by directly binding to them (Moon and Jun 2020). According to a molecular dynamics simulation study, the vesicles fuse rapidly from the lower PE concentration (e.g., 2:1 POPC: POPE), and the metastable hemifusion state is stabilized at higher PE concentrations (more than 33 mol%) (Kasson and Pande 2007). It has been confirmed from previous experiments that DOPC/DOPE (70/30 mol%) membranes develop pores via a stalk-like intermediate without hemifusion diaphragm formation, which circumvents the classical stalk model because of its intrinsic negative curvature (Joardar et al. 2021). gp41fusion peptide of HIV shifts the fusion mechanism from a nonclassical to a classical stalk model when membranes contain OA but fails to do so for DOPC/DOPE membranes (Joardar et al. 2022a). Guillén et al. had proposed that the insertion of the N-terminal FP of the S-protein of SARS-CoV into the host cell gives rise to the pre-hairpin intermediate, followed by refolding of heptad regions (HR1 and HR2), resulting in the formation of a six-helix bundle. Consequently, IFP and PTM are placed next to each other to facilitate hemifusion formation, which aids in the destabilization of lipids in the last stages of the fusion process, leading to the opening of the fusion pore (Guillén et al. 2008). Though the oligomerization of the S2 fusion peptide of SARS-CoV correlates with an enhancement of the extent and rate of lipid mixing with increasing concentration of membrane cholesterol in DOPC/DOPE/DOPG (60/30/10 mol%) membranes, it is incapable of opening a pore, even in the presence of PEG, in contrast to that of the HA fusion peptide (Chakraborty et al. 2013; Pattnaik et al. 2021). The fusion peptide of class II dengue virus (DENV FP) facilitates hemifusion formation during the fusion of SUVs with DOPC/DOPE/DOPG membrane with varying concentrations of cholesterol, predominantly at pH 5.0, and the amount of membrane cholesterol and fusogenic ability of the peptide are strongly correlated in the fusion process. Both at pH 5.0 and pH 7.4, the highest binding affinity of the peptide was observed in the presence of 10 mol% cholesterol (Mishra and Chakraborty 2025). The DENV FP has an overall positive charge due to the presence of positively charged amino acids, which enhances its preference for anionic phospholipids (Schmidt et al. 2021). Because insect cells contain more negatively charged lipids than mammalian cells, the low pH-triggered fusion of DENV is more effective with insect cells (Zaitseva et al. 2010). Two coiled-coil peptides were derived from the SNARE protein; one is positively charged (named K), and the other is negatively charged (named E). For membrane fusion, the coiled-coil forming peptides, "K," which is enriched in lysine, and "E," which is enriched in glutamic acid, have been employed as a minimum SNARE mimetic system. Detailed biophysical studies of these two peptides in a membrane environment revealed that peptide E is completely available for coiled-coil assembly with the corresponding peptide and does not interact with the membrane. Peptide K, on the other hand, binds to the membrane, penetrates the membrane, and exists in a state of equilibrium between membrane-bound and coiled-coil assembly-ready states. Peptide K brings two membranes into close apposition by forming a coiled-coil assembly with peptide E and destabilizes the membrane to initiate membrane fusion. Depending on the lipid composition, both peptides experience repulsion or attraction by membranes. The E/K peptide ratio is a crucial parameter, as it is highly flexible and adaptable, and has been used for numerous applications. By changing this parameter, it is possible to regulate membrane fusion so that two liposomes can fuse or allow for multiple rounds of fusion, resulting in a single giant liposome (Khvotchev and Soloviev 2022; Wagle et al. 2019). The rate of membrane fusion was accelerated by increasing the number of heptad repeats in E/K peptides, which increased the stability of the coiled-coil assembly (Zheng et al. 2013). Lipid and content mixing in liposome fusion assays is facilitated by E/K peptides fused with native transmembrane domains of SNARE proteins (Meyenberg et al. 2011). Peptides are attached to their liposomes in this model system by means of a lipid anchor and a PEG spacer. The length of the PEG spacer and lipid composition have a significant impact on the efficiency of the membrane fusion mechanism and the function of peptides (Daudey et al. 2019).

Lipid composition acts as a critical parameter that regulates the behavior of the peptide and the peptide-induced change in membrane organization and dynamics. Wu et al. reported earlier that the presence of LPC in the outer leaflet orders the interfacial as well as the hydrophobic chain region of the membrane (Wu et al. 1996). The ability of stalk formation is correlated with the disordering of the interfacial region of the membrane (Joardar et al. 2021; Mishra and Chakraborty 2025; Pattnaik and Chakraborty 2019). The study of the wild type (WT) and mutants (G1S, G1V, W14A) of the N-terminal HA fusion peptide of influenza viruses influences lipid bilayer ordering at different depths. The fluorescence anisotropy and lifetime measurements with anthroyloxy stearic acid (n-AS) probes revealed that fusion is facilitated by the WT fusion peptide, which orders the mid-upper region of the bilayer. Contrarily, fusion is hindered by mutant fusion peptides, as their ordering patterns are different from the WT (Chakraborty et al. 2017). The monolayer properties were measured to examine the ability of the gp41 fusion peptide of HIV to promote fusion among SUVs with DOPC/DOPE/DOPG (60/30/10 mol%) membranes with varying concentrations of cholesterol. This study revealed that in the absence of cholesterol, the peptide is adsorbed on the lipid headgroup in the aqueous subphase; however, in the presence of cholesterol in the lipidic mixture, it is oriented parallel to the lipid molecule. The gp41 fusion peptide disrupts the acyl chain region when the bilayer contains 10–20 mol% cholesterol, but it has a greater effect on the interfacial region in the absence of cholesterol. Since the percentage change in lipid mixing is correlated with the change in the area liftoff parameter in monolayer measurements, the orientation of the peptide in the lipid is considered to be essential for its fusogenic ability (Panda et al. 2025b).

A short chain of amino acids, known as an inhibitor peptide, inhibits or reduces the activity of a certain biological target molecule, typically a protein or an enzyme. A unique amino acid sequence of the inhibitor peptide is designed to have a high affinity for its target, which determines its selectivity. Similar to the secondary structure and physical characteristics of TG-23, GG-21 exhibits inhibitory efficacy in a wide range of lipid compositions (Pandia et al. 2024). Utilizing the sequence of coronin 1 protein as a template, QG-22 and AG-22 are derived and are known to inhibit membrane fusion irrespective of the mole fraction of membrane cholesterol, and they hinder fusion by increasing the acyl chain order of the membrane bilayer, which prevents the opening of the pore (Pandia and Chakraborty 2024). Another derived peptide, mTG-23 of the coronin 1 protein, prevents viral infection by likely inhibiting the viral envelope and endosomal membrane fusion. By altering the biophysical properties of the participating bilayers, mTG-23 inhibits around 30–40% of the fusion of SUVs with varying concentrations of cholesterol (Panda et al. 2025a). The positions of tryptophan (W) and aspartic acid (D) in coronin 1 are conserved across species, and WD sequences are known to prevent membrane fusion. This approach can be used to design broad-spectrum antiviral agents (Eckert et al. 2011). Two new WD-containing peptides (dmTG-23) with a WD repeat at the center of the peptide sequence, using mTG-23 peptide as a template, and a 14-amino acid peptide with seven WD repeats (WD-14) were designed. By enhancing the membrane order at the acyl chain region, the three WD repeat-containing dmTG-23 exhibits the maximum efficacy against the PEG-induced fusion of SUVs as compared to WD-14 (Panda and Chakraborty 2025). By reducing water percolation at the acyl chain region of the membrane, these peptides inhibited pore opening.

2.9 Conclusion

This chapter throws light on the interplay of lipids and peptides in membrane merger, which is a fundamental biological process and plays a pivotal role in the fusion of lipid bilayers through intricate molecular interactions. The roles of lipids and peptides are extremely important in this context. Lipids provide the necessary structural framework and membrane curvature stress, while peptides alter the interaction pattern and catalyze the merging of membranes by destabilizing lipid packing and inducing either intrinsic negative or positive curvature. This coordinated action ensures efficient, selective, and regulated membrane merger, which plays critical roles in various cellular events such as vesicle trafficking, viral entry, and intracellular communication. Understanding this interplay at a molecular level not only educates us about the fundamental membrane biology but will also help us design and develop therapeutic solutions to emerging and reemerging viral infections. Thus, the coordination between lipids and peptides is necessary to understand the language of membrane remodeling in diverse contexts. This integrated perspective highlights the complexity and precision underlying membrane fusion mechanisms.

Acknowledgments This study was supported by the Core Research Grant [grant number CRG/2021/001515] of the Science and Engineering Research Board (SERB) and the SERB Science and Technology Award for Research [grant number STR/2021/000029]. HC acknowledges the University Grants Commission (UGC) for the UGC assistant professor position. MSP and SM thank SERB for the fellowship. The authors extend their gratitude to the members of the Chakraborty laboratory for their constructive comments and discussions.

Conflict of Interest The authors declare that there is no conflict of interest.

References

Aeffner S, Reusch T, Weinhausen B, Salditt T (2012) Energetics of stalk intermediates in membrane fusion are controlled by lipid composition. Proc Natl Acad Sci USA 109:E1609–E1618

Aganna E, Burrin JM, Hitman GA, Turner MD (2006) Involvement of calpain and synaptotagmin Ca2+ sensors in hormone secretion from excitable endocrine cells. J Endocrinol 190:R1–R7

Agirre A, Flach C, Goñi FM, Mendelsohn R, Valpuesta JM, Wu F, Nieva JL (2000) Interactions of the HIV-1 fusion peptide with large unilamellar vesicles and monolayers. A cryo-TEM and spectroscopic study. Biochim Biophys Acta 1467:153–164

Akimov SA, Molotkovsky RJ, Kuzmin PI, Galimzyanov TR, Batishchev OV (2020) Continuum models of membrane fusion: evolution of the theory. Int J Mol Sci 21:3875

Alabi AA, Tsien RW (2013) Perspectives on kiss-and-run: role in exocytosis, endocytosis, and neurotransmission. Annu Rev Physiol 75:393–422

Alés E, Tabares L, Poyato JM, Valero V, Lindau M, Alvarez de Toledo G (1999) High calcium concentrations shift the mode of exocytosis to the kiss-and-run mechanism. Nat Cell Biol 1:40–44

Apellániz B, Rujas E, Carravilla P, Requejo-Isidro J, Huarte N, Domene C, Nieva JL (2014) Cholesterol-dependent membrane fusion induced by the gp41 membrane-proximal external region-transmembrane domain connection suggests a mechanism for broad HIV-1 neutralization. J Virol 88:13367–13377

Ashery U, Bielopolski N, Lavi A, Barak B, Michaeli L, Ben-Simon Y, Sheinin A, Bar-On D, Shapira Z, Gottfried I (2014) The molecular mechanisms underlying synaptic transmission: a view of the presynaptic terminal. In: The synapse. Elsevier, pp 21–109

Baquero E, Albertini AA, Gaudin Y (2015) Recent mechanistic and structural insights on class III viral fusion glycoproteins. Curr Opin Struct Biol 33:52–60

Basso LG, Vicente EF, Crusca E Jr, Cilli EM, Costa-Filho AJ (2016) SARS-CoV fusion peptides induce membrane surface ordering and curvature. Sci Rep 6:37131

Bigay J, Antonny B (2012) Curvature, lipid packing, and electrostatics of membrane organelles: defining cellular territories in determining specificity. Dev Cell 23:886–895

Binder H, Zschörnig O (2002) The effect of metal cations on the phase behavior and hydration characteristics of phospholipid membranes. Chem Phys Lipids 115:39–61

Boonstra S, Blijleven JS, Roos WH, Onck PR, van der Giessen E, van Oijen AM (2018) Hemagglutinin-mediated membrane fusion: a biophysical perspective. Annu Rev Biophys 47: 153–173

Bottacchiari M, Gallo M, Bussoletti M, Casciola CM (2024) The local variation of the Gaussian modulus enables different pathways for fluid lipid vesicle fusion. Sci Rep 14:23
Brown MF (2012) Curvature forces in membrane lipid-protein interactions. Biochemistry 51:9782–9795
Cail RC, Drubin DG (2023) Membrane curvature as a signal to ensure robustness of diverse cellular processes. Trends Cell Biol 33:427–441
Campelo F, Arnarez C, Marrink SJ, Kozlov MM (2014) Helfrich model of membrane bending: from Gibbs theory of liquid interfaces to membranes as thick anisotropic elastic layers. Adv Colloid Interf Sci 208:25–33
Chakraborty H, Tarafdar PK, Bruno MJ, Sengupta T, Lentz BR (2012) Activation thermodynamics of poly(ethylene glycol)-mediated model membrane fusion support mechanistic models of stalk and pore formation. Biophys J 102:2751–2760
Chakraborty H, Tarafdar PK, Klapper DG, Lentz BR (2013) Wild-type and mutant hemagglutinin fusion peptides alter bilayer structure as well as kinetics and activation thermodynamics of stalk and pore formation differently: mechanistic implications. Biophys J 105:2495–2506
Chakraborty H, Lentz BR, Kombrabail M, Krishnamoorthy G, Chattopadhyay A (2017) Depth-dependent membrane ordering by hemagglutinin fusion peptide promotes fusion. J Phys Chem B 121:1640–1648
Chazal N, Gerlier D (2003) Virus entry, assembly, budding, and membrane rafts. Microbiol Mol Biol Rev 67:226–237
Chernomordik L (1996) Non-bilayer lipids and biological fusion intermediates. Chem Phys Lipids 81:203–213
Chernomordik LV, Kozlov MM (2003) Protein-lipid interplay in fusion and fission of biological membranes. Annu Rev Biochem 72:175–207
Chernomordik LV, Kozlov MM (2008) Mechanics of membrane fusion. Nat Struct Mol Biol 15:675–683
Chernomordik L, Kozlov MM, Zimmerberg J (1995) Lipids in biological membrane fusion. J Membr Biol 146:1–14
Chernomordik LV, Frolov VA, Leikina E, Bronk P, Zimmerberg J (1998) The pathway of membrane fusion catalyzed by influenza hemagglutinin: restriction of lipids, hemifusion, and lipidic fusion pore formation. J Cell Biol 140:1369–1382
Colotto A, Epand RM (1997) Structural study of the relationship between the rate of membrane fusion and the ability of the fusion peptide of influenza virus to perturb bilayers. Biochemistry 36:7644–7651
Colotto A, Martin I, Ruysschaert J-M, Sen A, Hui S, Epand R (1996) Structural study of the interaction between the SIV fusion peptide and model membranes. Biochemistry 35:980–989
Cooke IR, Deserno M (2006) Coupling between lipid shape and membrane curvature. Biophys J 91:487–495
Craig M, Yarrarapu SNS, Dimri M (2018) Biochemistry-Cholesterol. StatPearls
Daudey GA, Schwieger C, Rabe M, Kros A (2019) Influence of membrane–fusogen distance on the secondary structure of fusogenic coiled coil peptides. Langmuir 35:5501–5508
Davies SM, Epand RF, Bradshaw JP, Epand RM (1998) Modulation of lipid polymorphism by the feline leukemia virus fusion peptide: implications for the fusion mechanism. Biochemistry 37:5720–5729
Deese AJ, Dratz EA, Brown MF (1981) Retinal rod outer segment lipids form bilayers in the presence and absence of rhodopsin: a 31P NMR study. FEBS Lett 124:93–99
Dorn JF, Zhang L, Phi TT, Lacroix B, Maddox PS, Liu J, Maddox AS (2016) A theoretical model of cytokinesis implicates feedback between membrane curvature and cytoskeletal organization in asymmetric cytokinetic furrowing. Mol Biol Cell 27:1286–1299
Eckert C, Hammesfahr B, Kollmar M (2011) A holistic phylogeny of the coronin gene family reveals an ancient origin of the tandem-coronin, defines a new subfamily, and predicts protein function. BMC Evol Biol 11:268

Edidin M (2003) The state of lipid rafts: from model membranes to cells. Annu Rev Biophys Biomol Struct 32:257–283
Epand RM (1998) Lipid polymorphism and protein–lipid interactions. Biochem Biophys Acta Rev Biomembr 1376:353–368
Epand RM, Epand RF (1994) Relationship between the infectivity of influenza virus and the ability of its fusion peptide to perturb bilayers. Biochem Biophys Res Commun 202:1420–1425
Epand RM, Epand RF (2009) Lipid domains in bacterial membranes and the action of antimicrobial agents. Biochim Biophys Acta 1788:289–294
Epand RF, Martin I, Ruysschaert JM, Epand RM (1994) Membrane orientation of the SIV fusion peptide determines its effect on bilayer stability and ability to promote membrane fusion. Biochem Biophys Res Commun 205:1938–1943
Epand RM, Epand RF, Martin I, Ruysschaert JM (2001) Membrane interactions of mutated forms of the influenza fusion peptide. Biochemistry 40:8800–8807
Epand RM, D'Souza K, Berno B, Schlame M (2015) Membrane curvature modulation of protein activity determined by NMR. Biochem Biophys Acta Biomembr 1848:220–228
Escribá PV, González-Ros JM, Goñi FM, Kinnunen PK, Vigh L, Sánchez-Magraner L, Fernández AM, Busquets X, Horváth I, Barceló-Coblijn G (2008) Membranes: a meeting point for lipids, proteins and therapies. J Cell Mol Med 12:829–875
Fratti RA, Jun Y, Merz AJ, Margolis N, Wickner W (2004) Interdependent assembly of specific regulatory lipids and membrane fusion proteins into the vertex ring domain of docked vacuoles. J Cell Biol 167:1087–1098
Freed EO (2015) HIV-1 assembly, release and maturation. Nat Rev Microbiol 13:484–496
Freed EO, Myers DJ, Risser R (1990) Characterization of the fusion domain of the human immunodeficiency virus type 1 envelope glycoprotein gp41. Proc Natl Acad Sci USA 87: 4650–4654
Fuller N, Rand R (2001) The influence of lysolipids on the spontaneous curvature and bending elasticity of phospholipid membranes. Biophys J 81:243–254
Furber KL, Churchward MA, Rogasevskaia TP, Coorssen JR (2009) Identifying critical components of native Ca2+-triggered membrane fusion. Integrating studies of proteins and lipids. Ann N Y Acad Sci 1152:121–134
Gamon BL, Virden JW, Berg JC (1989) The aggregation kinetics of an electrostatically stabilized dipalmitoyl phosphatidylcholine vesicle system. J Colloid Interface Sci 132:125–138
Gething MJ, Doms RW, York D, White J (1986) Studies on the mechanism of membrane fusion: site-specific mutagenesis of the hemagglutinin of influenza virus. J Cell Biol 102:11–23
Gingell D, Ginsberg I (1978) Membrane fusion ed G Poste and GL Nicolson. Elsevier, New York
Giraldo-Lorza JM, Leidy C, Manrique-Moreno M (2024) The influence of cholesterol on membrane targeted bioactive peptides: modulating peptide activity through changes in bilayer biophysical properties. Membranes 14:220
Gong H, Hu X, Zhang L, Fa K, Liao M, Liu H, Fragneto G, Campana M, Lu JR (2023) How do antimicrobial peptides disrupt the lipopolysaccharide membrane leaflet of gram-negative bacteria? J Colloid Interface Sci 637:182–192
Gruner SM (1989) Stability of lyotropic phases with curved interfaces. J Phys Chem B 93:7562–7570
Guillén J, Kinnunen PK, Villalaín J (2008) Membrane insertion of the three main membranotropic sequences from SARS-CoV S2 glycoprotein. Biochim Biophys Acta 1778:2765–2774
Haldar S, Kombrabail M, Krishnamoorthy G, Chattopadhyay A (2012) Depth-dependent heterogeneity in membranes by fluorescence lifetime distribution analysis. J Phys Chem Lett 3:2676–2681
Hamm M, Kozlov MM (2000) Elastic energy of tilt and bending of fluid membranes. Eur Phys J E 3:323–335
Han X, Bushweller JH, Cafiso DS, Tamm LK (2001) Membrane structure and fusion-triggering conformational change of the fusion domain from influenza hemagglutinin. Nat Struct Biol 8: 715–720

Harata NC, Aravanis AM, Tsien RW (2006) Kiss-and-run and full-collapse fusion as modes of exo-endocytosis in neurosecretion. J Neurochem 97:1546–1570
Harrison SC (2008) Viral membrane fusion. Nat Struct Mol Biol 15:690–698
Harrison SC (2015) Viral membrane fusion. Virology 479-480:498–507
Harrison JS, Higgins CD, O'Meara MJ, Koellhoffer JF, Kuhlman BA, Lai JR (2013) Role of electrostatic repulsion in controlling pH-dependent conformational changes of viral fusion proteins. Structure 21:1085–1096
Helfrich W (1973) Elastic properties of lipid bilayers: theory and possible experiments. Z Naturforsch C 28:693–703
Hong J, Yang H, Pang D, Wei L, Deng C (2018) Effects of mono- and di-valent metal cations on the morphology of lipid vesicles. Chem Phys Lipids 217:19–28
Horth M, Lambrecht B, Khim MC, Bex F, Thiriart C, Ruysschaert JM, Burny A, Brasseur R (1991) Theoretical and functional analysis of the SIV fusion peptide. EMBO J 10:2747–2755
Hui SW, Stewart T, Boni L, Yeagle P (1981) Membrane fusion through point defects in bilayers. Science 212:921–923
Ikonen E (2008) Cellular cholesterol trafficking and compartmentalization. Nat Rev Mol Cell Biol 9:125–138
Inoue T, Minami H, Shimozawa R, Sugihara G (1992) Stability of DLPA/DLPC mixed vesicles against divalent cation-induced aggregation: importance of the hydration force. J Colloid Interface Sci 152:493–506
Israelachvili JN (2011) Intermolecular and surface forces. Academic
Israelachvili JN, McGuiggan PM (1988) Forces between surfaces in liquids. Science 241:795–800
Israelachvili JN, Mitchell DJ, Ninham BW (1976) Theory of self-assembly of hydrocarbon amphiphiles into micelles and bilayers. J Chem Soc Faraday Trans 72:1525–1568
Jahn R, Scheller RH (2006) SNAREs--engines for membrane fusion. Nat Rev Mol Cell Biol 7:631–643
Jahn R, Lang T, Südhof TC (2003) Membrane fusion. Cell 112:519–533
Joardar A, Pattnaik GP, Chakraborty H (2021) Effect of phosphatidylethanolamine and oleic acid on membrane fusion: phosphatidylethanolamine circumvents the classical stalk model. J Phys Chem B 125:13192–13202
Joardar A, Pattnaik GP, Chakraborty H (2022a) Combination of oleic acid and the gp41 fusion peptide switches the phosphatidylethanolamine-induced membrane fusion mechanism from a nonclassical to a classical stalk model. J Phys Chem B 126:3673–3684
Joardar A, Pattnaik GP, Chakraborty H (2022b) Mechanism of membrane fusion: interplay of lipid and peptide. J Membr Biol 255:211–224
Kasson PM, Pande VS (2007) Control of membrane fusion mechanism by lipid composition: predictions from ensemble molecular dynamics. PLoS Comput Biol 3:e220
Katsov K, Müller M, Schick M (2004) Field theoretic study of bilayer membrane fusion. I Hemifusion mechanism. Biophys J 87:3277–3290
Kawamoto S, Klein ML, Shinoda W (2015) Coarse-grained molecular dynamics study of membrane fusion: curvature effects on free energy barriers along the stalk mechanism. J Chem Phys 143:243112
Khvotchev M, Soloviev M (2022) SNARE modulators and SNARE mimetic peptides. Biomolecules 12:1779
Kinnun JJ, Bolmatov D, Lavrentovich MO, Katsaras J (2020) Lateral heterogeneity and domain formation in cellular membranes. Chem Phys Lipids 232:104976
Kollmitzer B, Heftberger P, Rappolt M, Pabst G (2013) Monolayer spontaneous curvature of raft-forming membrane lipids. Soft Matter 9:10877–10884
Kozlov MM, Chernomordik LV (2002) The protein coat in membrane fusion: lessons from fission. Traffic 3:256–267
Kozlov MM, Markin VS (1983) Possible mechanism of membrane fusion. Biofizika 28:242–247
Kozlov MM, Markin VS (1984) On the theory of membrane fusion. The adhesion-condensation mechanism. Gen Physiol Biophys 3:379–402

Kozlovsky Y, Kozlov MM (2002) Stalk model of membrane fusion: solution of energy crisis. Biophys J 82:882–895

Kozlovsky Y, Chernomordik LV, Kozlov MM (2002) Lipid intermediates in membrane fusion: formation, structure, and decay of hemifusion diaphragm. Biophys J 83:2634–2651

Kozlovsky Y, Efrat A, Siegel DP, Kozlov MM (2004) Stalk phase formation: effects of dehydration and saddle splay modulus. Biophys J 87:2508–2521

Kumar GA, Sarkar P, Stepniewski TM, Jafurulla M, Singh SP, Selent J, Chattopadhyay A (2021) A molecular sensor for cholesterol in the human serotonin(1A) receptor. Sci Adv 7

Kuzmin PI, Zimmerberg J, Chizmadzhev YA, Cohen FS (2001) A quantitative model for membrane fusion based on low-energy intermediates. Proc Natl Acad Sci USA 98:7235–7240

Kuzmin PI, Akimov SA, Chizmadzhev YA, Zimmerberg J, Cohen FS (2005) Line tension and interaction energies of membrane rafts calculated from lipid splay and tilt. Biophys J 88:1120–1133

Lähdesmäki K, Ollila OS, Koivuniemi A, Kovanen PT, Hyvönen MT (2010) Membrane simulations mimicking acidic pH reveal increased thickness and negative curvature in a bilayer consisting of lysophosphatidylcholines and free fatty acids. Biochem Biophys Acta Biomembr 1798:938–946

Lange Y (1991) Disposition of intracellular cholesterol in human fibroblasts. J Lipid Res 32:329–339

Leikin SL, Kozlov MM, Chernomordik LV, Markin VS, Chizmadzhev YAJJ (1987) Membrane fusion: overcoming of the hydration barrier and local restructuring. J Theor Biol 129:411–425

Lentz BR (2007) PEG as a tool to gain insight into membrane fusion. Eur Biophys J 36:315–326

Lentz BR, Malinin V, Haque ME, Evans K (2000) Protein machines and lipid assemblies: current views of cell membrane fusion. Curr Opin Struct Biol 10:607–615

Lewis JR, Cafiso DS (1999) Correlation between the free energy of a channel-forming voltage-gated peptide and the spontaneous curvature of bilayer lipids. Biochemistry 38:5932–5938

Lingwood D, Simons K (2010) Lipid rafts as a membrane-organizing principle. Science 327:46–50

Lipowsky R (1992) Budding of membranes induced by intramembrane domains. J Phys II 2:1825–1840

Liscum L, Underwood KW (1995) Intracellular cholesterol transport and compartmentation. J Biol Chem 270:15443–15446

Lundbaek JA, Maer AM, Andersen OS (1997) Lipid bilayer electrostatic energy, curvature stress, and assembly of gramicidin channels. Biochemistry 36:5695–5701

Majumder R, Quinn-Allen MA, Kane WH, Lentz BR (2008) A phosphatidylserine binding site in factor Va C1 domain regulates both assembly and activity of the prothrombinase complex. Blood 112:2795–2802

Malinin VS, Lentz BR (2004) Energetics of vesicle fusion intermediates: comparison of calculations with observed effects of osmotic and curvature stresses. Biophys J 86:2951–2964

Markin VS, Albanesi JP (2002) Membrane fusion: stalk model revisited. Biophys J 82:693–712

Markin VS, Kozlov MM, Borovjagin VL (1984) On the theory of membrane fusion. The stalk mechanism. Gen Physiol Biophys 3:361–377

Marra J, Israelachvili J (1985) Direct measurements of forces between phosphatidylcholine and phosphatidylethanolamine bilayers in aqueous electrolyte solutions. Biochemistry 24:4608–4618

Marsh D (1996) Intrinsic curvature in normal and inverted lipid structures and in membranes. Biophys J 70:2248–2255

Martin II, Ruysschaert J, Epand RM (1999) Role of the N-terminal peptides of viral envelope proteins in membrane fusion. Adv Drug Deliv Rev 38:233–255

Martin M, Dotti CG, Ledesma MD (2010) Brain cholesterol in normal and pathological aging. Biochim Biophys Acta 1801:934–944

McIntosh TJ, Simon SA (1986) Area per molecule and distribution of water in fully hydrated dilauroylphosphatidylethanolamine bilayers. Biochemistry 25:4948–4952

McMahon HT, Boucrot E (2015) Membrane curvature at a glance. J Cell Sci 128:1065–1070

McMahon HT, Gallop JL (2005) Membrane curvature and mechanisms of dynamic cell membrane remodelling. Nature 438:590–596
Meher G, Chakraborty H (2019) Membrane composition modulates fusion by altering membrane properties and fusion peptide structure. J Membr Biol 252:261–272
Meher G, Bhattacharjya S, Chakraborty H (2019) Membrane cholesterol modulates Oligomeric status and peptide-membrane interaction of severe acute respiratory syndrome coronavirus fusion peptide. J Phys Chem B 123:10654–10662
Meher G, Bhattacharjya S, Chakraborty H (2023) Membrane cholesterol regulates the oligomerization and fusogenicity of SARS-CoV fusion peptide: implications in viral entry. Phys Chem Chem Phys 25:7815–7824
Melikyan GB, Niles WD, Ratinov VA, Karhanek M, Zimmerberg J, Cohen FS (1995) Comparison of transient and successful fusion pores connecting influenza hemagglutinin expressing cells to planar membranes. J Gen Physiol 106:803–819
Meyenberg K, Lygina AS, van den Bogaart G, Jahn R, Diederichsen U (2011) SNARE derived peptide mimic inducing membrane fusion. Chem Commun 47:9405–9407
Mishra S, Chakraborty H (2023) Phosphatidylethanolamine and cholesterol promote Hemifusion formation: a tug of war between membrane interfacial order and intrinsic negative curvature of lipids. J Phys Chem B 127:7721–7729
Mishra S, Chakraborty H (2025) Dengue virus fusion peptide promotes hemifusion formation by disordering the interfacial region of the membrane. J Membr Biol 258:161–171
Modis Y (2014) Relating structure to evolution in class II viral membrane fusion proteins. Curr Opin Virol 5:34–41
Mondal Roy S, Sarkar M (2011) Membrane fusion induced by small molecules and ions. J Lipids 2011:528784
Moon Y, Jun Y (2020) The effects of regulatory lipids on intracellular membrane fusion mediated by dynamin-like GTPases. Front Cell Dev Biol 8:518
Mouritsen OG, Bloom M (1984) Mattress model of lipid-protein interactions in membranes. Biophys J 46:141–153
Müller M, Katsov K, Schick M (2003) A new mechanism of model membrane fusion determined from Monte Carlo simulation. Biophys J 85:1611–1623
Nawara TJ, Williams YD 2nd, Rao TC, Hu Y, Sztul E, Salaita K, Mattheyses AL (2022) Imaging vesicle formation dynamics supports the flexible model of clathrin-mediated endocytosis. Nat Commun 13:1732
Noguchi H, Takasu M (2001) Self-assembly of amphiphiles into vesicles: a Brownian dynamics simulation. Phys Rev E Stat Nonlinear Soft Matter Phys 64:041913
Österberg F, Rilfors L, Wieslander Å, Lindblom G, Gruner SM (1995) Lipid extracts from membranes of Acholeplasma laidlawii a grown with different fatty acids have a nearly constant spontaneous curvature. Biochim Biophys Acta Lipids Lipid Metabol 1257:18–24
Pajtinka P, Vácha R (2024) Amphipathic helices can sense both positive and negative curvatures of lipid membranes. J Phys Chem Lett 15:175–179
Panda MS, Chakraborty H (2025) Importance of the number and position of tryptophan–aspartate repeats in designing peptide-based membrane fusion inhibitors. Chem Asian J 16:e00474
Panda MS, Qazi B, Vishwakarma V, Pattnaik GP, Haldar S, Chakraborty HJRMC (2025a) Developing peptide-based fusion inhibitors as an antiviral strategy utilizing coronin 1 as a template. RSC Med Chem 16:125–136
Panda MS, Raghav S, Ghosh SK, Chakraborty HJL (2025b) gp41 fusion peptide alters the properties of lipid monolayer at the air–water interface in a cholesterol-dependent fashion: implications in membrane fusion. Langmuir 41:6103–6112
Pandia S, Chakraborty H (2024) Strategic Design of tryptophan–aspartic acid-containing peptide inhibitors using coronin 1 as a template: inhibition of fusion by enhancing acyl chain order. J Phys Chem B 128:9163–9171
Pandia S, Mahapatra A, Chakraborty H (2024) A Coronin 1-derived peptide inhibits membrane fusion by modulating membrane organization and dynamics. J Phys Chem B 128:4986–4995

Parnas H, Parnas I (1994) Neurotransmitter release at fast synapses. J Membr Biol 142:267–279
Patel D, Witt SN (2017) Ethanolamine and phosphatidylethanolamine: Partners in Health and Disease. Oxidative Med Cell Longev 2017:4829180
Pattnaik GP, Chakraborty H (2019) Cholesterol alters the inhibitory efficiency of peptide-based membrane fusion inhibitor. Biochim Biophys Acta Biomembr 1861:183056
Pattnaik GP, Chakraborty H (2020) Entry inhibitors: efficient means to block viral infection. J Membr Biol 253:425–444
Pattnaik GP, Chakraborty H (2021a) Cholesterol: a key player in membrane fusion that modulates the efficacy of fusion inhibitor peptides. Vitam Horm 117:133–155
Pattnaik GP, Chakraborty H (2021b) Fusogenic effect of cholesterol prevails over the inhibitory effect of a peptide-based membrane fusion inhibitor. Langmuir 37:3477–3489
Pattnaik GP, Bhattacharjya S, Chakraborty H (2021) Enhanced cholesterol-dependent hemifusion by internal fusion peptide 1 of SARS Coronavirus-2 compared to its N-terminal counterpart. Biochemistry 60:559–562
Pereira FB, Valpuesta JM, Basañez G, Goñi FM, Nieva JL (1999) Interbilayer lipid mixing induced by the human immunodeficiency virus type-1 fusion peptide on large unilamellar vesicles: the nature of the nonlamellar intermediates. Chem Phys Lipids 103:11–20
Pink DA, McNeil S, Quinn B, Zuckermann MJ (1998) A model of hydrogen bond formation in phosphatidylethanolamine bilayers. Biochim Biophys Acta 1368:289–305
Pinot M, Vanni S, Pagnotta S, Lacas-Gervais S, Payet LA, Ferreira T, Gautier R, Goud B, Antonny B, Barelli H (2014) Lipid cell biology. Polyunsaturated phospholipids facilitate membrane deformation and fission by endocytic proteins. Science 345:693–697
Pomorski TG, Nylander T, Cárdenas M (2014) Model cell membranes: discerning lipid and protein contributions in shaping the cell. Adv Colloid Interf Sci 205:207–220
Poojari CS, Scherer KC, Hub JS (2021) Free energies of membrane stalk formation from a lipidomics perspective. Nat Commun 12:6594
Qiao H, Armstrong RT, Melikyan GB, Cohen FS, White JM (1999) A specific point mutant at position 1 of the influenza hemagglutinin fusion peptide displays a hemifusion phenotype. Mol Biol Cell 10:2759–2769
Rand R, Parsegian V (1989) Hydration forces between phospholipid bilayers. Biochem Biophys Acta Biomembr 988:351–376
Rawicz W, Olbrich KC, McIntosh T, Needham D, Evans E (2000) Effect of chain length and unsaturation on elasticity of lipid bilayers. Biophys J 79:328–339
Risselada HJ, Bubnis G, Grubmüller H (2014) Expansion of the fusion stalk and its implication for biological membrane fusion. Proc Natl Acad Sci USA 111:11043–11048
Scheiffele P, Rietveld A, Wilk T, Simons K (1999) Influenza viruses select ordered lipid domains during budding from the plasma membrane. J Biol Chem 274:2038–2044
Schmidt TF, Riske KA, Caseli L, Salesse C (2021) Dengue fusion peptide in Langmuir monolayers: a binding parameter study. Biophys J 271:106553
Sengupta T, Chakraborty H, Lentz BR (2014) The transmembrane domain peptide of vesicular stomatitis virus promotes both intermediate and pore formation during PEG-mediated vesicle fusion. Biophys J 107:1318–1326
Shibata Y, Hu J, Kozlov MM, Rapoport TA (2009) Mechanisms shaping the membranes of cellular organelles. Annu Rev Cell Dev Biol 25:329–354
Siegel DP (1993) Energetics of intermediates in membrane fusion: comparison of stalk and inverted micellar intermediate mechanisms. Biophys J 65:2124–2140
Siegel DP, Epand RM (2000) Effect of influenza hemagglutinin fusion peptide on lamellar/inverted phase transitions in dipalmitoleoylphosphatidylethanolamine: implications for membrane fusion mechanisms. Biochim Biophys Acta 1468:87–98
Simons K, Ikonen E (1997) Functional rafts in cell membranes. Nature 387:569–572
Simons K, Sampaio JL (2011) Membrane organization and lipid rafts. Cold Spring Harb Perspect Biol 3:a004697

Skehel JJ, Wiley DC (2000) Receptor binding and membrane fusion in virus entry: the influenza hemagglutinin. Annu Rev Biochem 69:531–569
Söllner T, Rothman JE (1994) Neurotransmission: harnessing fusion machinery at the synapse. Trends Neurosci 17:344–348
Soubias O, Teague WE Jr, Hines KG, Mitchell DC, Gawrisch K (2010) Contribution of membrane elastic energy to rhodopsin function. Biophys J 99:817–824
Steinhauer DA, Wharton SA, Skehel JJ, Wiley DC (1995) Studies of the membrane fusion activities of fusion peptide mutants of influenza virus hemagglutinin. J Virol 69:6643–6651
Südhof TC, Rothman JE (2009) Membrane fusion: grappling with SNARE and SM proteins. Science 323:474–477
Surmeier G, Dogan-Surmeier S, Paulus M, Albers C, Latarius J, Sternemann C, Schneider E, Tolan M, Nase J (2022) The interaction of viral fusion peptides with lipid membranes. Biophys J 121:3811–3825
Swarnakar S, Chaudhury A, Skoda MWA, Chakraborty H, Basu JK (2025) Microscopic insight into HIV fusion peptide-mediated dehydration and packing regulation in membranes. Biophys J 124:2465–2475
Szleifer I, Kramer D, Ben-Shaul A, Gelbart WM (1990) Molecular theory of curvature elasticity in surfactant films. J Chem Phys 92:6800–6817
Tamm LK, Han X, Li Y, Lai AL (2002) Structure and function of membrane fusion peptides. Biopolymers 66:249–260
Templer RH, Khoo BJ, Seddon JM (1998) Gaussian curvature modulus of an amphiphilic monolayer. Langmuir 14:7427–7434
Tenchov BG, MacDonald RC, Lentz BR (2013) Fusion peptides promote formation of bilayer cubic phases in lipid dispersions. An x-ray diffraction study. Biophys J 104:1029–1037
van der Veen JN, Kennelly JP, Wan S, Vance JE, Vance DE, Jacobs RL (2017) The critical role of phosphatidylcholine and phosphatidylethanolamine metabolism in health and disease. Biochim Biophys Acta Biomembr 1859:1558–1572
van Meer G, Voelker DR, Feigenson GW (2008) Membrane lipids: where they are and how they behave. Nat Rev Mol Cell Biol 9:112–124
Vanni S, Hirose H, Barelli H, Antonny B, Gautier R (2014) A sub-nanometre view of how membrane curvature and composition modulate lipid packing and protein recruitment. Nat Commun 5:4916
Wagle S, Georgiev VN, Robinson T, Dimova R, Lipowsky R, Grafmüller A (2019) Interaction of SNARE mimetic peptides with lipid bilayers: effects of secondary structure, bilayer composition and lipid anchoring. Sci Rep 9:7708
Wang W, Yang L, Huang HW (2007) Evidence of cholesterol accumulated in high curvature regions: implication to the curvature elastic energy for lipid mixtures. Biophys J 92:2819–2830
Wassarman PM, Litscher ES (2008) Mammalian fertilization is dependent on multiple membrane fusion events. Methods Mol Biol 475:99–113
Weinreb G, Lentz BR (2007) Analysis of membrane fusion as a two-state sequential process: evaluation of the stalk model. Biophys J 92:4012–4029
Wen X, Saltzgaber GW, Thoreson WB (2017) Kiss-and-run is a significant contributor to synaptic exocytosis and endocytosis in photoreceptors. Front Cell Neurosci 11:286
White JM (1992) Membrane fusion. Science 258:917–924
White JM, Whittaker GR (2016) Fusion of enveloped viruses in endosomes. Traffic 17:593–614
Wickner W, Schekman R (2008) Membrane fusion. Nat Struct Mol Biol 15:658–664
Wiedmann TS, Pates RD, Beach JM, Salmon A, Brown MF (1988) Lipid-protein interactions mediate the photochemical function of rhodopsin. Biochemistry 27:6469–6474
Wu H, Zheng L, Lentz BR (1996) A slight asymmetry in the transbilayer distribution of lysophosphatidylcholine alters the surface properties and poly (ethylene glycol)-mediated fusion of dipalmitoylphosphatidylcholine large unilamellar vesicles. Biochemistry 35:12602–12611
Xiao T, Cai Y, Chen B (2021) HIV-1 entry and membrane fusion inhibitors. Viruses 13:735

Yang ST, Kiessling V, Tamm LK (2016a) Line tension at lipid phase boundaries as driving force for HIV fusion peptide-mediated fusion. Nat Commun 7:11401

Yang ST, Kreutzberger AJB, Lee J, Kiessling V, Tamm LK (2016b) The role of cholesterol in membrane fusion. Chem Phys Lipids 199:136–143

Yao H, Hong M (2014) Conformation and lipid interaction of the fusion peptide of the paramyxovirus PIV5 in anionic and negative-curvature membranes from solid-state NMR. J Am Chem Soc 136:2611–2624

Zaitseva E, Yang S-T, Melikov K, Pourmal S, Chernomordik LV (2010) Dengue virus ensures its fusion in late endosomes using compartment-specific lipids. PLoS Pathog 6:e1001131

Zheng T, Voskuhl J, Versluis F, Zope HR, Tomatsu I, Marsden HR, Kros A (2013) Controlling the rate of coiled coil driven membrane fusion. Chem Commun 49:3649–3651

Zimmerberg J, Kozlov MM (2006) How proteins produce cellular membrane curvature. Nat Rev Mol Cell Biol 7:9–19

Zimmerberg J, Blumenthal R, Sarkar DP, Curran M, Morris SJ (1994) Restricted movement of lipid and aqueous dyes through pores formed by influenza hemagglutinin during cell fusion. J Cell Biol 127:1885–1894

Chapter 3
Membrane Merger Dynamics in Exocytosis and Endocytosis

Xiaoli Guo, Chung Yu Chan, and Ling-Gang Wu

Abstract Exocytosis releases vesicular cargo to the extracellular space and adds membrane to the cell surface, whereas endocytosis retrieves membrane and internalizes extracellular material; together, they maintain membrane homeostasis and enable rapid intercellular signaling. Super-resolution imaging has revealed that these processes are far richer than the classical "kiss-and-run vs. full-collapse" or flat-to-round schemes. Exocytosis now encompasses at least seven fusion modes with reversible pores, vesicle shrinking or enlargement, and compound or sequential compound fusion, while endocytosis proceeds mainly via modular flat→Λ→Ω→O transitions and rapid closure of preformed Ω-profiles. These dynamic transformations are orchestrated by Soluble N-ethylmaleimide-sensitive factor Attachment protein REceptors, Ca^{2+}/synaptotagmin, dynamin, clathrin, and the actomyosin cortex, and their dysregulation underlies neurological, metabolic, and oncogenic disorders.

Keywords Exocytosis · Endocytosis · Super-resolution imaging · Fusion pore dynamics · Hemi-fusion · Vesicle shrinkage · Vesicle enlargement

3.1 Introduction

3.1.1 Biological Functions of Exocytosis and Endocytosis

Cellular life fundamentally depends on maintaining distinct compartments while enabling controlled material exchange with the extracellular environment. Exocytosis and endocytosis represent opposing yet complementary processes for this exchange across the plasma membrane barrier (Wu et al. 2014a; Chernomordik

X. Guo (✉)
School of Life Sciences, Yunnan Normal University, Kunming, Yunnan, China

National Institute of Neurological Disorders and Stroke, Bethesda, MD, USA

C. Y. Chan · L.-G. Wu
National Institute of Neurological Disorders and Stroke, Bethesda, MD, USA

© The Author(s), under exclusive license to Springer Nature Switzerland AG 2026

J. Weiler et al. (eds.), *Cell Fusion*, Advances in Experimental Medicine and Biology 1509, https://doi.org/10.1007/978-3-032-22637-2_3

and Kozlov 2008). Exocytosis releases intracellular contents and delivers membrane proteins to the cell surface by fusing vesicles with the plasma membrane, while endocytosis retrieves membrane, internalizes receptors, and captures extracellular materials by forming vesicles from the plasma membrane (Gundelfinger et al. 2003; Di Fiore and von Zastrow 2014).

Exocytosis and endocytosis must overcome key physical barriers, including repulsive hydration forces; orchestrate lipid rearrangements; and fine-tune membrane curvature, thereby enabling precise and efficient material transfer across the cell membrane (McMahon and Boucrot 2015; Kozlov et al. 2014). In neurons, they orchestrate the synaptic vesicle cycle with millisecond precision (Sudhof 2013; Kononenko and Haucke 2015), while endocrine and exocrine cells utilize similar mechanisms for hormone and enzyme secretion (Dolensek et al. 2011). Beyond specialized secretory cells, these processes underpin intracellular trafficking throughout endosomal and secretory pathways (Bonifacino and Glick 2004) and regulate plasma membrane composition by selectively internalizing surface molecules (Di Fiore and von Zastrow 2014). These processes also facilitate cell division, migration, and fusion events, including fertilization and viral infection (Mostov et al. 2003; Jena 2009). The molecular machinery driving these processes includes highly conserved protein families that mediate specific steps in membrane transformation. Soluble N-ethylmaleimide-sensitive factor Attachment protein REceptor (SNARE) proteins form the core fusion machinery for exocytosis, while dynamin, clathrin, and BAR (Bin/Amphiphysin/Rvs) domain proteins orchestrate membrane deformation and scission during endocytosis. Dysregulation of exocytosis and endocytosis contributes to numerous pathological conditions, including neurodegenerative diseases, metabolic disorders, and cancer.

3.1.2 From Classical Models to Contemporary Understanding

Our understanding of membrane merger dynamics has evolved dramatically over the past half-century. Initial insights emerged from electron microscopy (EM) studies in the 1970s and 1980s, which captured snapshots of apparent membrane intermediates during fusion and budding events (Heuser and Reese 1981; Ceccarelli et al. 1972; Miller and Heuser 1984). These observations established foundational concepts like full-collapse fusion and clathrin-mediated endocytosis. These pioneering observations led to the development of the classical exo–endocytosis framework, which posited relatively simple transformations: vesicle fusion opening a narrow pore that either rapidly closes (kiss-and-run) or dilates until the vesicle flattens completely (full-collapse fusion), followed by recovery through a flat-to-round endocytic transition (Rizzoli and Jahn 2007; He and Wu 2007). Electrophysiological approaches, particularly capacitance measurements and amperometry, subsequently provided indirect evidence of fusion pore dynamics at millisecond resolution (Neher and

Marty 1982; Albillos et al. 1997). These techniques allowed researchers to monitor membrane addition and retrieval in real time, revealing the temporal characteristics of exocytosis and endocytosis. However, these methods lacked direct visualization of membrane rearrangements, leaving uncertainties about the precise nature of the underlying transformations.

The classical view of membrane dynamics remained largely unchallenged until recent advances in super-resolution microscopy enabled direct visualization of membrane transformations in living cells (Zhao et al. 2016; Shin et al. 2018, 2020, 2021, 2022; Chiang et al. 2014). In particular, stimulated emission depletion (STED) microscopy studies have revealed a repertoire of membrane rearrangements far more intricate than previously recognized (Zhao et al. 2016; Shin et al. 2018; Chiang et al. 2014; Willig et al. 2006). The classical view of fusion as a simple binary choice between kiss-and-run and full-collapse has expanded into a rich spectrum of intermediate states and alternative pathways (Chiang et al. 2014; Shin et al. 2020; Wu and Chan 2024), with at least seven fusion modes characterized by distinct vesicle size changes and pore dynamics. This framework redefines exo–endocytosis to include diverse pore behaviors and recognizes sequential compound fusion as an additional pathway. Furthermore, evidence indicates that endocytosis often proceeds through the closure of preformed or fusion-generated Ω-shaped profiles (Ω-profiles) rather than de novo flat-to-round transitions, explaining the previously puzzling speed of certain endocytic modes (Shin et al. 2021; Wu and Chan 2024). This evolving understanding provides multiple mechanisms for regulating content release dynamics and enables fine-tuned membrane retrieval appropriate to specific cellular contexts (Shin et al. 2020; Alabi and Tsien 2013), thereby preserving membrane homeostasis while allowing adaptation to changing physiological demands (Kononenko and Haucke 2015).

3.1.3 Chapter Overview

This chapter reviews recent advances in our understanding of membrane merger dynamics during exocytosis and endocytosis. It summarizes key imaging breakthroughs that have revealed the complexity of membrane transformations, moving beyond classical models. The discussion covers the main biophysical mechanisms and protein machinery involved, with attention to the diversity of fusion and retrieval modes now recognized. The chapter also considers how these processes are coordinated to maintain membrane balance and briefly addresses their relevance to disease.

3.2 Visualization Technologies Driving Conceptual Advances

3.2.1 Evolution of Imaging Techniques: From Electron Microscopy to Super-resolution

Electron microscopy (EM) provided the first nanometer-scale snapshots of fusion and budding intermediates. Heuser and Reese's pioneering quick-freeze EM studies in the 1970s–1980s captured apparent fusion intermediates at neuromuscular junctions milliseconds after stimulation, leading to influential hypotheses about full-collapse fusion (Heuser and Reese 1981; Heuser et al. 1979). Similarly, Ceccarelli's observation of Ω-profiles suggested a potential kiss-and-run mechanism (Ceccarelli et al. 1972), while Miller and Heuser's documentation of clathrin-coated structures established conceptual frameworks for endocytosis (Miller and Heuser 1984).

Despite nanometer resolution, conventional EM has inherent limitations: Chemical fixation may create artifacts, samples represent single timepoints, and live processes cannot be followed in real time (Wyffels 2001). The newer "flash-and-freeze" approach combines optogenetic stimulation with rapid freezing to capture membrane transformations with millisecond resolution, revealing ultrafast endocytosis occurring within 50–100 ms in hippocampal synapses (Watanabe et al. 2013; Watanabe 2016).

The development of fluorescence microscopy techniques enabled live-cell imaging of membrane dynamics. Conventional confocal microscopy and total internal reflection fluorescence microscopy (TIRFM) modalities monitored docking and secretion but, limited by diffraction (~200 nm), could not visualize critical structures such as nascent fusion pores (Stephens and Allan 2003; Burchfield et al. 2010). The resolution revolution began with super-resolution techniques that bypass the diffraction limit (Huang et al. 2009). Stimulated emission depletion (STED) microscopy uses a donut-shaped depletion beam to confine fluorescence to nanoscale volumes, achieving 30–80 nm resolution (Hell and Wichmann 1994; Klar et al. 2000). Single-molecule localization methods like stochastic optical reconstruction microscopy (STORM) and photoactivated localization microscopy (PALM) provide comparable or better resolution through the precise localization of individual fluorophores (Rust et al. 2006; Betzig et al. 2006).

These super-resolution approaches have transformed membrane research by enabling visualization of structures previously accessible only to electron microscopy but with the critical advantage of compatibility with live-cell imaging (Sahl et al. 2017). STED microscopy, in particular, has emerged as a powerful tool for studying membrane dynamics, with temporal resolution now approaching tens of milliseconds while maintaining spatial resolution sufficient to resolve fusion pores and membrane intermediates (Shin et al. 2018; Willig et al. 2006). This combination of spatial and temporal resolution has enabled paradigm-shifting observations of fusion and fission events in living cells, including direct visualization of hemi-fusion intermediates, fusion pore expansion and constriction, and the stepwise transformation of flat membrane to vesicular structures (Zhao et al. 2016; Shin et al. 2018, 2020).

3.2.2 *Real-Time Visualization and Labeling Strategies*

Time-lapse super-resolution microscopy, particularly STED, now achieves frame rates of tens to hundreds of milliseconds while maintaining ~60–80 nm resolution (Guo et al. 2022; Urban et al. 2011). Optimization strategies include restricting imaging to single planes, using minimal excitation power, and employing a range of probes that provide both morphological and functional information (Shin et al. 2021; Wei et al. 2024). However, confining imaging to one focal plane reduces the probability of capturing infrequent events; this drawback can be partly alleviated with synchronized exocytosis protocols (Zhao et al. 2016; Shin et al. 2020, 2021; Wei et al. 2024). Additional insight comes from correlative methods that pair fluorescence imaging with simultaneous electrophysiological recording to link structural changes to functional consequences (Chiang et al. 2014; Shin et al. 2020; Ales et al. 1999).

These advanced imaging approaches rely on specialized fluorescent probes developed in parallel with microscopy advances (Fernandez-Suarez and Ting 2008; Lavis and Raines 2014). Membrane lipid probes targeting the plasma membrane's cytosolic leaflet—including genetically encoded fluorescent proteins fused to lipid-binding domains like PLCδ1-PH or linked to lipid anchors—visualize topological changes during fusion and fission (Chiang et al. 2014; Varnai and Balla 1998; Zacharias et al. 2002) (Fig. 3.1a). These probes diffuse into fusing vesicles or endocytic invaginations, revealing membrane continuity, transformations, or scission in real time (Zhao et al. 2016) (Fig. 3.1b). Fusion pore dynamics can be visualized indirectly through small water-soluble Alexa or Atto fluorophores that penetrate open pores, detecting formation, expansion, and closure events even

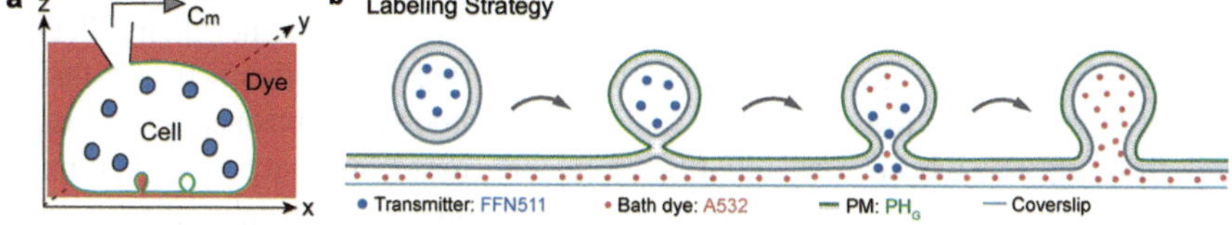

Fig. 3.1 Schematic diagram for setup and labeling strategy. (**a**) Chromaffin cells are overexpressed with PH_G (green) to label the inner leaflet of plasma membrane (PM), extracellular solution is labeled with A532 (red), and vesicles are preloaded with FFN511 (blue). Under whole-cell voltage-clamp configuration, the cell is stimulated with depolarization, and calcium currents (I_{Ca}) and membrane capacitance (C_m) are recorded. (**b**) Visualization of fusion dynamics (hemi-fusion, pore opening, and expansion) with three probes: FFN511 for vesicle content release, PH_G for PM inner leaflet fusion, and A532 for A532-permeable pore. (Reproduced from Wu and Chan 2024, originally published under CC BY 4.0)

below resolution limits (Shin et al. 2018; Chiang et al. 2014; Wu and Chan 2024) (Fig. 3.1a, b). Vesicular content probes—including cargo-fluorescent protein fusions, preloaded fluorescent false neurotransmitter (Fig. 3.1a, b), or pH-sensitive pHluorin—enable tracking of release dynamics (Gubernator et al. 2009; Rodriguez et al. 2013; Miesenbock et al. 1998; Pereira et al. 2016), while labeled protein components of fusion and fission machinery reveal the spatiotemporal organization of the molecular machinery driving membrane transformations (Shin et al. 2018, 2022; Willig et al. 2006; Bademosi et al. 2017; Mund et al. 2018). The most comprehensive insights come from multicolor approaches combining different probe categories to simultaneously visualize membrane topology, content release, and protein machinery dynamics (Zhao et al. 2016; Shin et al. 2018; Wu and Chan 2024; Wei et al. 2024) (Fig. 3.1).

3.3 Exocytotic Membrane Transformations

3.3.1 Hemi-fusion Intermediates and Pathways

Two principal fusion models have been debated for decades: the hemi-fusion pathway (sequential merger of outer then inner leaflets) and the protein-lined pore hypothesis (transmembrane SNARE domains forming the initial pore) (Chernomordik and Kozlov 2008; Jahn et al. 2003; Kozlov and Chernomordik 2015; Jackson 2010; Han et al. 2004; Kozlov and Markin 1983; Hernandez et al. 2012; Diao et al. 2012). This fundamental question remained unresolved until recent super-resolution microscopy studies provided direct evidence for hemi-fusion in living cells. STED microscopy visualized hemi-fused structures as Ω-profiles labeled with inner leaflet markers but impermeable to extracellular molecules, confirming hemi-fusion in physiological membrane fusion (Zhao et al. 2016) (Fig. 3.2a).

Surprisingly, 40–50% of fusion events in chromaffin cells pass through a detectable hemi-fusion state, and about one-third of these intermediates proceed to full fusion within 0.1–26 s (Zhao et al. 2016) (Fig. 3.2b). This significant proportion implies that hemi-fusion functions as a metastable checkpoint rather than a fleeting transition (Zhao et al. 2016; Wu and Chan 2024). Because the hemi-fusion-to-full-fusion transition requires extensive membrane remodeling, it constitutes a major rate-limiting step whose substantial heterogeneity may contribute to asynchronous neurotransmitter release (Shin et al. 2018; Chernomordik and Kozlov 2005).

Calcium influx unexpectedly influences hemi-fusion stability, with reduced calcium accelerating the transition to full fusion (Zhao et al. 2016). This reflects competition between fusion-promoting SNAREs and calcium/dynamin-dependent fission mechanisms (Zhao et al. 2016; Kaeser and Regehr 2014) (Fig. 3.2c). While confirming the hemi-fusion pathway predominance, these findings do not exclude SNARE involvement in pore formation. SNARE proteins may contribute to pore structure after initial hemi-fusion, supporting an emerging model where subsequent

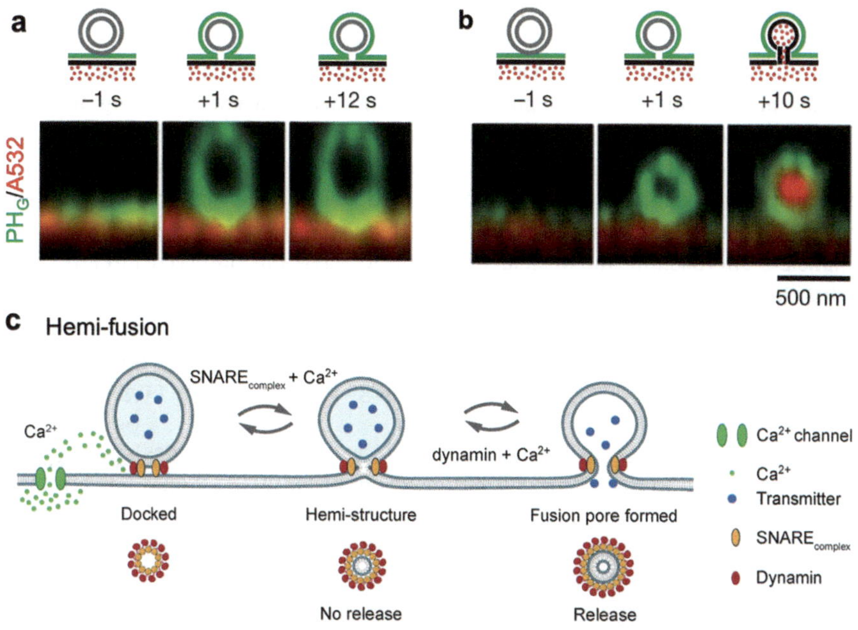

Fig. 3.2 Hemi-fusion of a single vesicle. STED XZ plane images of PH_G (green) and A532 (red) showing hemi-fusion (**a**) and hemi-fusion followed by hemi-to-full fusion (**b**). Images taken at times relative to 1-s depolarization, as labeled (Adapted from Zhao et al. 2016). (**c**) Schematics of the underlying mechanism of hemi-to-full fusion. Fusion undergoes hemi-fusion and hemi-to-full fusion. The dynamics of each of these transitions are the net outcome of competition between the effects of the SNARE complex, plasma membrane tension, and dynamin. Calcium influx triggers vesicle fusion, fusion pore constriction, and closure. There is no release in hemi-fusion. Delayed hemi-to-full fusion causes release. (Reproduced from Wu and Chan 2024, originally published under CC BY 4.0)

pore formation involves both lipidic and proteinaceous components in a hybrid structure (Wu and Chan 2024; Bao et al. 2016).

3.3.2 *Fusion Pore Dynamics: Opening, Expansion, and Constriction*

The fusion pore represents the critical structure connecting vesicular and extracellular compartments, determining the rate and extent of content release (Alabi and Tsien 2013; Alvarez de Toledo et al. 1993). Classical electrophysiological approaches provided indirect evidence for fusion pore formation and expansion by measuring conductance changes and amperometric currents (Albillos et al. 1997; He et al. 2006). These methods suggested initial pores of approximately 1–2 nm that could either close rapidly (kiss-and-run) or expand beyond detection limits (presumed full-collapse fusion) (Lindau and Alvarez de Toledo 2003). However, the limitations of these techniques precluded direct observation of pore dynamics, particularly for larger pores or later stages of expansion (Harata et al. 2006).

Recent super-resolution imaging has revolutionized our understanding of fusion pore behavior by directly visualizing pore dynamics in living cells (Shin et al. 2018; Chiang et al. 2014). STED microscopy reveals that fusion pores in secretory cells exhibit complex dynamic behaviors, including opening, expansion, constriction, and closure, while the Ω-profile persists (Shin et al. 2018). Fusion pores can expand to diameters approaching the vesicle width (up to ~490 nm observed in chromaffin cells), far exceeding the size limits detectable by conductance measurements (Shin et al. 2018; Wu and Chan 2024). Importantly, these expanded pores do not necessarily lead to vesicle collapse but often persist as stable structures for seconds to minutes, challenging the classical view that large pores inevitably progress to full-collapse fusion (Shin et al. 2018).

Pore expansion kinetics display remarkable variability, with rates ranging from barely detectable to exceeding 8.9 nm/ms (Shin et al. 2018). Some pores remain constricted (<60 nm) for several seconds before abruptly expanding, while others rapidly dilate upon formation (Shin et al. 2018). This heterogeneity in expansion kinetics likely contributes to the diversity of content release patterns observed in secretory cells, from slow, partial release to rapid, complete discharge (Shin et al. 2020). The persistence of expanded pores indicates that the classical interpretation of abrupt conductance increases as reflecting full-collapse fusion requires reconsideration; such events may instead represent rapid pore expansion while maintaining the Ω-profile structure (Chiang et al. 2014; Wu and Chan 2024).

Perhaps most significantly, fusion pores exhibit bidirectional dynamics, not only expanding but also constricting and closing while the Ω-profile remains intact (Shin et al. 2018; Anantharam et al. 2011). Pore constriction can proceed from hundreds of nanometers down to complete closure, with the resulting structure resembling the initial Ω-profile but disconnected from the extracellular space (Trexler et al. 2016). This bidirectional pore behavior provides mechanistic insight into kiss-and-run exocytosis, demonstrating that pores of virtually any size can potentially close, rather than being limited to narrow initial pores as classically thought (Wu and Chan 2024; Wu et al. 2019).

Fusion pore behavior is dictated by a tug-of-war between forces that expand the opening, tensions that merge the entire Ω-profile, and forces that constrict the pore (Shin et al. 2018). SNARE proteins and associated fusion machinery generate forces that drive pore expansion, with the specific arrangement of SNARE complexes influencing both the initial pore size and the rate of expansion (Wu et al. 2017a). Cytoskeletal elements, particularly filamentous actin (F-actin), contribute by generating cortical membrane tension that pulls the fusion-generated Ω-profile inward, promoting its shrinkage and facilitating full integration with the plasma membrane (Wen et al. 2016; Gasman et al. 2004). A third major influence comes from dynamin, which assembles into a helical collar around the pore neck and, through GTP-driven conformational changes, constricts or even reseals the pore, leading to classical kiss-and-run fusion events (Shin et al. 2018; Anantharam et al. 2011). Recent research has also identified clathrin as a key pore-constricting factor. By polymerizing around the fusion pore region, clathrin exerts constriction forces due to its intrinsically curved structure, contributing to both pore closure and limiting the release of vesicular contents in various endocytic pathways previously thought to be clathrin-independent (Wei et al. 2024). This interplay of different forces creates a finely tunable system for regulating pore dynamics and controlling content release.

Calcium signaling orchestrates this competitive balance, serving as a master regulator of fusion pore behavior (Zhao et al. 2016; Neher and Sakaba 2008; Wu et al. 2009). While calcium triggers initial fusion through synaptotagmin activation, it also promotes pore constriction and closure by activating dynamin, potentially through calmodulin and calcineurin (Shin et al. 2018; Cousin and Robinson 2001). Consequently, calcium influx has biphasic effects on fusion pore dynamics, initially favoring pore opening but subsequently promoting constriction and closure (Ales et al. 1999). This biphasic action enables precise temporal control of content release and provides a mechanism for coupling exocytosis to subsequent endocytosis (Wu et al. 2009).

3.3.3 Seven Fusion Modes: Beyond Classical Models

Real-time super-resolution imaging has revealed a diverse spectrum of fusion behaviors that can be classified into at least seven distinct modes based on two parameters: changes in Ω-profile size and fusion pore status (Chiang et al. 2014; Shin et al. 2020) (Fig. 3.3a). This expanded classification provides a more nuanced framework for understanding exocytotic membrane dynamics and their functional consequences.

The seven fusion modes identified through live-cell imaging are (Shin et al. 2018, 2020; Chiang et al. 2014) (Fig. 3.3a):

1. *Stay mode*: The Ω-profile remains open after fusion, creating a stable and continuous connection between the vesicle interior and the extracellular space.
2. *Close mode* (classical kiss-and-run): The Ω-profile retains its size, but the fusion pore closes, resulting in the re-formation of a vesicle with dimensions similar to the original vesicle.
3. *Shrink–stay mode*: The Ω-profile partially shrinks while maintaining an open fusion pore, resulting in a smaller but still connected vesicular structure.
4. *Shrink–close mode*: The Ω-profile partially shrinks before the fusion pore closes, generating a smaller recycled vesicle than the original.
5. *Shrink fusion mode*: The Ω-profile continuously shrinks until it becomes undetectable, merging fully into the plasma membrane without undergoing full collapse.
6. *Enlarge–stay mode*: The Ω-profile expands while maintaining an open fusion pore, creating a larger connected vesicular structure.
7. *Enlarge–close mode*: The Ω-profile enlarges before the fusion pore closes, generating a recycled vesicle larger than the original.

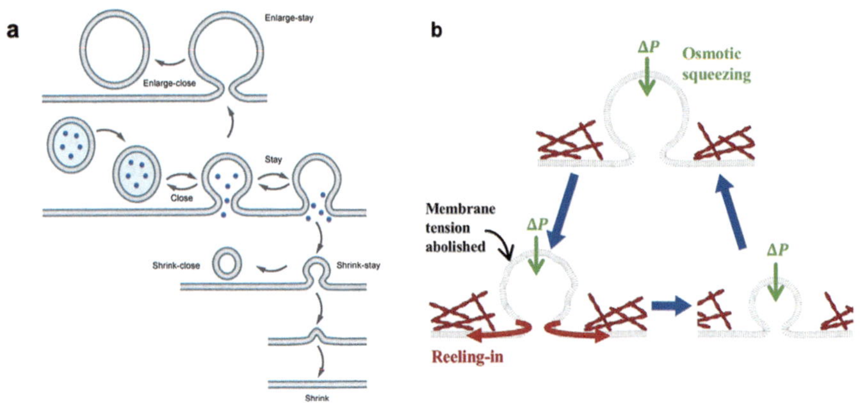

Fig. 3.3 Different fusion modes and shrink fusion. (**a**) Schematic diagram showing seven fusion modes: stay, close, enlarge–stay, enlarge–close, shrink–stay, shrink–close, and shrink fusion. Shrink-related fusion events are associated with a larger pore to generate fast release; enlarge-related fusion events are associated with a smaller pore to generate slow release (Reproduced from Wu and Chan 2024, originally published under CC BY 4.0). (**b**) Schematic diagram depicting the osmotic pressure difference between the intracellular and the extracellular solution (ΔP) that compresses the vesicle, overcoming the vesicular membrane tension and causing the vesicular membrane to be pushed into the plasma membrane by the high plasma membrane tension and the actin cortex. (Schematic diagram taken from the article published in Shin et al. 2020, Copyright Elsevier 2020)

Strikingly, full-collapse fusion as traditionally envisioned—in which the Ω-profile flattens completely into the plasma membrane—was not observed in these super-resolution studies (Shin et al. 2018, 2020). This absence challenges a fundamental tenet of the classical exocytosis model and suggests that alternative mechanisms predominate in physiological settings (Wu and Chan 2024).

These diverse fusion modes have been corroborated by other methodologies, though often without explicit recognition of their distinct nature (Grabner and Moser 2018; Sheng et al. 2012). Capacitance measurements frequently show unequal up- and down-steps, consistent with shrink–close or enlarge–close fusion rather than classical kiss-and-run with identical steps (Monck et al. 1990; Tabares et al. 2003). Carbon fiber amperometry has detected varying release kinetics that align with the different pore dynamics associated with these fusion modes (Chow et al. 1992). Additionally, previous observations of unusually large or small vesicles in electron micrographs likely reflect the products of enlarge–close or shrink–close fusion, respectively (Takei et al. 1995; de Lange et al. 2003).

The functional implications of these diverse fusion modes are profound. Each mode affects content release differently, with shrink-related modes generally associated with larger pores and faster, more complete release, while enlarge-related modes typically involve smaller pores and slower, partial release (Shin et al. 2020).

This relationship between fusion mode and release kinetics provides cells with multiple mechanisms for regulating neurotransmitter or hormone discharge under different physiological conditions. For instance, high-frequency stimulation might favor shrink fusion to ensure rapid and complete content release (Wang and Kaczmarek 1998; Sara et al. 2002), while low-frequency activity could preferentially utilize enlarge–stay or close modes to conserve vesicles and limit release (Wu and Borst 1999; Fernandez-Alfonso and Ryan 2004).

The diversity of fusion modes also significantly impacts the subsequent endocytic pathways available for membrane retrieval. Close, shrink–close, and enlarge–close modes directly generate endocytic vesicles through fusion pore closure, providing immediate recycling without requiring separate endocytic machinery (Shin et al. 2021; Wu and Chan 2024) (Fig. 3.3a). In contrast, stay, shrink–stay, enlarge–stay, and shrink modes necessitate distinct endocytic processes for membrane retrieval, potentially involving different molecular mechanisms and temporal characteristics (Shin et al. 2022) (Fig. 3.3a).

3.3.4 Vesicle Size Changes During Fusion Events

A particularly striking revelation from recent imaging studies is the dynamic nature of vesicle size during and following fusion (Chiang et al. 2014; Shin et al. 2020). Contrary to the classical view of vesicles maintaining fixed dimensions until potential collapse, fused vesicular Ω-profiles can significantly shrink or enlarge while connected to the plasma membrane via fusion pores (Chiang et al. 2014; Shin et al. 2020). These size changes reflect active processes rather than passive consequences of fusion and substantially influence content release and subsequent recycling (Wu and Chan 2024).

Vesicle shrinkage can be partial, as in shrink–stay and shrink–close modes, or complete in shrink fusion mode (Shin et al. 2020) (Fig. 3.3a). Complete shrinkage effectively integrates vesicular membrane into the plasma membrane but through a mechanism fundamentally different from classical full-collapse fusion (Shin et al. 2020). Rather than flattening through progressive pore expansion, the vesicle membrane is gradually pulled into the plasma membrane while maintaining the Ω-profile until dimensions approach the resolution limit (Shin et al. 2018; Chan et al. 2022).

The mechanisms driving vesicle shrinkage involve a combination of osmotic forces and membrane tension (Shin et al. 2020; Wen et al. 2016) (Fig. 3.3b). Mathematical modeling and experimental evidence indicate that cells maintain an outward swelling osmotic pressure. Upon fusion, rapid equilibration between the vesicle lumen and extracellular medium converts this into an inward squeezing pressure on the fused vesicle, effectively abolishing its membrane tension (Shin et al. 2020; Wu and Chan 2024; Finkelstein et al. 1986) (Fig. 3.3b). Simultaneously, the high tension of the plasma membrane, maintained by the F-actin cytoskeleton, creates a tension gradient that reels the tension-free vesicular membrane into the plasma membrane (Wen et al. 2016; Gauthier et al. 2012). This combination of osmotic squeezing force and membrane tension gradient explains why shrink fusion predominates over full-collapse in physiological settings (Shin et al. 2020) (Fig. 3.3b).

At the final stage of shrinkage, mathematical modeling predicts a transformation from Ω-shape to Λ-profile, consistent with experimental observations (Shin et al. 2020). This shape transition may provide the basis for reconciling the conflict between shrink fusion and full-collapse models: Vesicles may undergo a shrink–collapse sequence, in which substantial shrinkage is followed by collapse of the remaining small Ω-profile into a Λ-shape and potentially into the flat membrane (Shin et al. 2020). This unified model accommodates both the predominance of shrinkage for larger vesicles and the potential for collapse-like behavior at the smallest scales (Wu and Chan 2024).

Vesicle enlargement is the opposite transformation; however, the mechanism driving this growth has not yet been resolved (Shin et al. 2020). Enlarged vesicles can subsequently remain connected to the extracellular space (enlarge–stay) or undergo fusion pore closure (enlarge–close), generating unusually large vesicles for future rounds of exocytosis (Shin et al. 2020; Monck et al. 1990). This mechanism potentially explains the origin of oversized vesicles observed in various secretory systems and may contribute to plasticity in quantal size (Xue and Wu 2010; Gong et al. 2003).

3.3.5 Compound and Sequential Fusion Mechanisms

Beyond transformations of individual vesicles, recent research has revealed sophisticated multi-vesicle fusion mechanisms that significantly expand the repertoire of exocytotic behaviors (He et al. 2009; Ge et al. 2022). These mechanisms include compound fusion and sequential compound fusion (Wu and Chan 2024; Pickett and Edwardson 2006) (Fig. 3.4). Both processes create larger or elongated structures that modify release characteristics and increase the secretory capacity (Alvarez de Toledo and Fernandez 1990) (Fig. 3.4c).

Compound fusion describes the homotypic fusion of two or more vesicles prior to their exocytosis at the plasma membrane (He et al. 2009; Hafez et al. 2003) (Fig. 3.4c). This process generates composite vesicles with volumes and membrane areas that represent multiples of the original vesicles. While compound fusion had been inferred from electrophysiological and electron microscopy studies, definitive evidence came from observations of oversized vesicles with capacitance values corresponding to multiples of the standard quantal size, particularly following intense stimulation (He et al. 2009; Scepek and Lindau 1993). Functionally, compound fusion increases quantal size, potentially enhancing synaptic strength during periods of high activity (Xue and Wu 2010; He et al. 2009).

Sequential compound fusion, by contrast, involves the sequential addition of vesicles to an already-fused vesicular Ω-profile at the plasma membrane (Ge et al. 2022; Nemoto et al. 2001) (Fig. 3.4c). This process was initially proposed for large

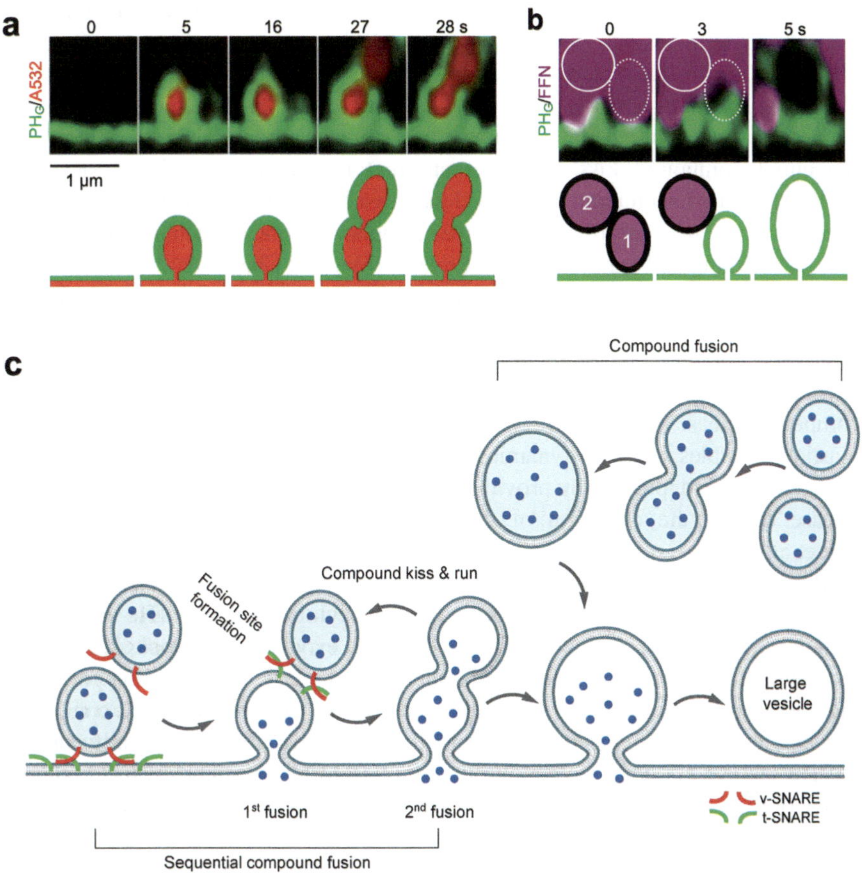

Fig. 3.4 Sequential compound fusion and release observed in chromaffin cells. (**a**) Representative STED XZ plane images of PH_G (green) and A532 (red) showing sequential compound fusion in a chromaffin cell. (**b**) Representative STED XZ plane images of PH_G (green) and FFN511 (magenta) showing sequential compound release (two circles) in a chromaffin cell (Adapted from Ge et al. 2022. © The Authors, some rights reserved; exclusive licensee AAAS. Distributed under a Creative Commons Attribution NonCommercial License 4.0 (CC BY-NC) http://creativecommons.org/licenses/by-nc/4.0/). (**c**) Schematic diagram showing membrane transformations of sequential compound fusion, sequential compound kiss-and-run, and compound fusion. New fusion site formation at the first fused vesicle enables sequential compound fusion, second fusion generates asynchronized release, and sequential compound fusion or compound fusion may generate large Ω-profiles and vesicles. (Reproduced from Wu and Chan 2024, originally published under CC BY 4.0)

exocrine vesicles that release contents very slowly over minutes, where it was suggested to enhance the secretory capacity without requiring multiple plasma membrane fusion sites (Nemoto et al. 2001). Recent super-resolution microscopy has now directly visualized this process in endocrine cells, confirming its existence and revealing its membrane dynamics in unprecedented detail (Wu and Chan 2024; Ge et al. 2022) (Fig. 3.4a, b).

STED imaging demonstrates that sequential compound fusion creates an eight-shaped profile as the second vesicle fuses with the first-fused Ω-profile (Ge et al. 2022) (Fig. 3.4a). This intermediate structure can evolve through several pathways: (1) The intervesicular fusion pore may expand to form a larger, elongated Ω-profile connected to the extracellular space, (2) the fusion pore may close, creating a sequential compound kiss-and-run event that regenerates separate vesicles, or (3) the structure may remain unchanged with two connected vesicular compartments (Ge et al. 2022). These diverse outcomes provide additional mechanisms for regulating the timing and extent of content release (Wu and Chan 2024).

Perhaps most surprisingly, sequential compound fusion reveals the rapid assembly of functional release sites at fused vesicular Ω-profiles (Ge et al. 2022). This discovery challenges the long-held view that exocytosis must occur at preestablished release sites, instead demonstrating that fused vesicles themselves can become competent release sites within seconds to minutes (Sudhof 2012). This remarkable plasticity enables cells to dynamically create new exocytic sites in response to secretory demand, potentially providing a mechanism for sustaining high-frequency release despite the depletion of docked vesicles at conventional release sites (Ge et al. 2022).

Both compound fusion mechanisms significantly impact content release patterns, contributing to the diversity and plasticity of exocytotic behaviors (He et al. 2009; Rudolph et al. 2015). Compound fusion prior to exocytosis increases the quantal size, potentially enhancing the strength of individual release events during periods of high activity (He et al. 2009; Matthews and Sterling 2008). Sequential compound fusion, meanwhile, creates release events that are spatially and temporally coupled but are not synchronous, potentially contributing to asynchronous release components observed at many synapses (Kaeser and Regehr 2014; Ge et al. 2022) (Fig. 3.4a, b).

These diverse exocytotic membrane transformations—hemi-fusion, dynamic fusion pores, multiple fusion modes, vesicle size changes, and compound fusion mechanisms—collectively create a rich repertoire of secretory behaviors far exceeding the simplicity of the classical exocytosis model (Wu and Chan 2024). This expanded framework provides multiple mechanisms for regulating the probability, synchronicity, rate, and extent of content release, enabling fine-tuned control of intercellular communication across diverse physiological contexts (Sudhof 2013; Wu and Chan 2024). Furthermore, the diversity of exocytotic transformations creates various starting points for subsequent endocytic processes, contributing to the coupling between exocytosis and endocytosis that maintains membrane homeostasis (Shin et al. 2021; Chanaday et al. 2019).

3.4 Endocytic Membrane Transformations

3.4.1 *Modular Transitions: Flat→Λ→Ω→O*

Endocytosis is a complex process by which cells internalize membrane and extracellular material, generating intracellular vesicles from the plasma membrane (Doherty and McMahon 2009). Classical models based on electron microscopy suggested a continuous flat-to-round transformation, progressing through increasingly invaginated intermediates (Miller and Heuser 1984; Heuser and Reese 1973). However, the precise sequence and regulation of these membrane transformations remained largely inferential until recent super-resolution imaging studies directly visualized endocytic membrane dynamics in living cells (Shin et al. 2022).

STED microscopy in chromaffin cells has revealed that endocytic membrane transformation proceeds through three distinct modular transitions: flat→Λ-shape, Λ→Ω-shape, and Ω→O-shape (closed vesicle) (Shin et al. 2021; Wu and Chan 2024). Each transition represents a discrete step with its own kinetics, molecular requirements, and probability of completion (Wu and Chan 2024; Chan et al. 2022). This modular framework provides a foundation for understanding the diverse endocytic modes observed across cell types and conditions (Chanaday et al. 2019; Gan and Watanabe 2018).

The flat→Λ transition initiates the endocytic process, generating a shallow, dome-like invagination from the flat plasma membrane. During this transition, the membrane deforms inward while maintaining a broad base at the plasma membrane (Shin et al. 2021, 2022). Imaging reveals that this deformation often involves the formation of a spike-like membrane protrusion attached to the Λ-profile's tip, suggesting an active pulling mechanism rather than passive membrane invagination (Shin et al. 2021, 2022). Notably, the probability of this transition occurring at any particular membrane location is relatively low (Shin et al. 2021; Wu and Chan 2024).

The Λ→Ω transition transforms the open dome-like structure into the characteristic Ω-shaped profile through constriction of the invagination's base (Shin et al. 2022). This transition narrows the connection between the invaginated membrane and the plasma membrane, creating a distinct neck structure that precedes vesicle fission (Shin et al. 2021). Like the initial flat→Λ transition, this step proceeds with variable kinetics and low probability, introducing additional stochasticity into the endocytic timeline (Shin et al. 2021; Wu and Chan 2024).

The final Ω→O transition completes the endocytic process through scission of the narrow neck connecting the Ω-profile to the plasma membrane, generating a free vesicle within the cytoplasm (Ferguson and De Camilli 2012). Super-resolution imaging reveals that this transition involves progressive narrowing of the Ω-profile's connection to the plasma membrane until it disappears from optical detection, indicating complete fission (Shin et al. 2021). Like the preceding transitions, Ω→O proceeds with variable kinetics and probability, contributing to the overall heterogeneity in endocytic timing (Shin et al. 2021; Wu and Chan 2024).

Importantly, each of these modular transitions can occur independently, without necessarily progressing through the complete sequence (Shin et al. 2021, 2022). Due to the modest transition probabilities at each step, many initiated endocytic events arrest as Λ-profiles or Ω-profiles rather than progressing to completed vesicles (Shin et al. 2021). These arrested intermediates may persist at the plasma membrane for extended periods, creating a reservoir of preformed structures that can resume their progression upon appropriate stimulation (Wu and Chan 2024; Hua et al. 2013). This probabilistic, modular framework helps explain the variable efficiency and timing of endocytosis observed across different cellular contexts (Wu et al. 2016; Watanabe and Boucrot 2017).

3.4.2 *Preformed Ω-Profiles as Major Endocytic Intermediates*

Recent super-resolution studies have revealed that preformed Ω-profiles at the plasma membrane, rather than the traditional flat-to-round membrane transformations, are the primary source of endocytic vesicles following stimulation in secretory cells (Shin et al. 2021). Preformed Ω-profiles exist at the plasma membrane prior to stimulation and undergo direct Ω→O transitions (pore closure and fission) upon appropriate signaling (Shin et al. 2021; Wu and Chan 2024). These structures may originate from previously arrested flat→Λ→Ω transitions or from fusion pores that have not fully collapsed or resealed (Shin et al. 2021, 2022). Quantitative imaging shows that vesicle retrieval after stimulation mainly occurs via closure of these preformed Ω-profiles and kiss-and-run fusion pore closure. In contrast, the classical complete flat→Λ→Ω→O progression contributes only minimally to stimulation-induced endocytosis in these cells (Shin et al. 2021). This finding fundamentally revises the classical view that endocytosis primarily proceeds through the complete flat→Λ→Ω→O sequence in response to exocytotic events (Kononenko and Haucke 2015; Saheki and De Camilli 2012).

The prevalence of preformed-Ω→O transitions offers a mechanistic explanation for several longstanding observations in endocytosis research (Watanabe et al. 2013). First, it resolves the paradox of how endocytosis can occur much more rapidly than the time required for clathrin-mediated assembly of endocytic machinery and membrane deformation (Shin et al. 2021; Watanabe and Boucrot 2017). By bypassing the initial steps of membrane deformation and proceeding directly to fission, preformed-Ω→O transitions enable the ultrafast endocytosis observed at many synapses and secretory cells (Wu and Chan 2024; Watanabe et al. 2013). Second, it explains the phenomenon of "readily retrievable" membrane pools that are preferentially internalized upon stimulation, as preformed Ω-profiles represent membranes already configured for rapid endocytosis (Hua et al. 2011; Wienisch and Klingauf 2006).

Functionally, the predominance of preformed-Ω→O transitions reflects a remarkably efficient design for rapid membrane retrieval (Shin et al. 2021; Wu and Chan 2024). Although the machinery driving flat→Λ→Ω transitions operates with relatively low efficiency at each step, the persistence of the resulting intermediates creates a reservoir of preformed Ω-profiles ready for rapid fission (Shin et al. 2021). This arrangement allows cells to overcome the inherent slowness of initiating membrane deformation while maintaining high capacity for rapid endocytosis when needed (Wu et al. 2016; Watanabe and Boucrot 2017). Thus, what seems like an inefficient first step is actually a practical way to achieve both efficiency and speed by accumulating and using preformed intermediates (Shin et al. 2021; Wu and Chan 2024).

The identification of preformed Ω-profiles as major endocytic intermediates necessitates a reinterpretation of many previous studies on endocytosis modes and mechanisms (Chanaday et al. 2019; Saheki and De Camilli 2012). For instance, the distinction between "compensatory" and "excess" endocytosis likely reflects different proportions of preformed-Ω→O transitions rather than fundamentally different molecular mechanisms (Shin et al. 2021; Wu et al. 2009). Similarly, the observation that endocytosis "overshoots" exocytosis in some conditions can be explained by the closure of preformed Ω-profiles that existed prior to stimulation, in addition to the retrieval of recently added membranes (Xue et al. 2012a, b).

3.5 Molecular Machinery Driving Membrane Transformations

3.5.1 Fusion Proteins: SNAREs and Regulatory Factors

Membrane fusion during exocytosis requires the coordinated action of a sophisticated protein machinery that overcomes the energy barriers preventing spontaneous bilayer merger (Sudhof 2013; Jahn and Fasshauer 2012). At the core of this machinery lies the Soluble N-ethylmaleimide-sensitive factor Attachment protein REceptor (SNARE) protein complex, which provides the driving force for bringing opposing membranes into close proximity and initiating fusion (Weber et al. 1998; Sollner et al. 1993). The fundamental fusion apparatus consists of v-SNAREs (primarily synaptobrevin, also known as vesicle-associated membrane protein, VAMP, on vesicles) and t-SNAREs (syntaxin and synaptosomal-associated protein of 25 kDa, SNAP-25, on target membranes), which assemble into a four-helix bundle that progressively zippers from the membrane-distal N-termini toward membrane-proximal C-termini (Fasshauer et al. 1998; Sutton et al. 1998; Fukuda et al. 2000).

This SNARE zippering process generates considerable mechanical force to overcome the repulsive forces between approaching membranes (Li et al. 2007; Min et al. 2013). Single-molecule force spectroscopy reveals that SNARE assembly proceeds through distinct stages with varying energy contributions, allowing for regulated progression through fusion intermediates (Gao et al. 2012). Importantly, the zippering process extends into the transmembrane domains of synaptobrevin and syntaxin, potentially contributing to the destabilization of the lipid bilayers and fusion pore formation (Hernandez et al. 2014; Chang et al. 2015; Stein et al. 2009). Multiple SNARE complexes likely cooperate at fusion sites, with estimates ranging from 3 to 15 complexes required for fast vesicle fusion depending on the specific cellular context (Mohrmann et al. 2010; Sinha et al. 2011; Montecucco et al. 2005).

A diverse array of regulatory proteins modulates SNARE-mediated fusion, creating a tunable system that responds to cellular signals (Rizo and Sudhof 2012; Sudhof and Rothman 2009). Munc18–1 binds to syntaxin-1 in a closed conformation, preventing premature SNARE assembly, and also interacts with synaptobrevin, forming a template for the correct assembly of the SNARE complex (Baker et al. 2015; Ma et al. 2013; Tang et al. 2021; Lee et al. 2020). Munc13 promotes the transition of syntaxin from closed to open conformation and helps orchestrate SNARE assembly in the appropriate orientation (Lai et al. 2013; Liu et al. 2016). These proteins collectively establish a kinetic barrier that prevents spontaneous fusion while enabling rapid release upon appropriate stimulation (Ma et al. 2011; Yang et al. 2015; Rizo 2022).

Synaptotagmin serves as the primary calcium sensor for fast synchronous release, containing C2 domains that bind calcium and penetrate phospholipid membranes, particularly those enriched in acidic phospholipids like phosphatidylinositol 4,5-bisphosphate (Chapman 2008; Fernandez-Chacon et al. 2001). Upon calcium binding, synaptotagmin undergoes conformational changes that displace the fusion clamp complexin from partially zippered SNARE complexes, allowing for rapid completion of assembly and membrane merger (Zhou et al. 2015; Tang et al. 2006). Recent cryo-electron microscopy structures have revealed intricate details of how synaptotagmin interacts with SNARE complexes and membranes to couple calcium binding to fusion (Zhou et al. 2017; Grushin et al. 2019). Different synaptotagmin isoforms exhibit varied calcium affinities and kinetics, potentially contributing to the diversity of release modes observed across secretory systems (Hui et al. 2005; Sugita et al. 2002).

Beyond these principal components, numerous additional proteins fine-tune fusion efficacy and kinetics (Pang and Sudhof 2010; Neher 2006). Synaptophysin interacts with VAMP2, thereby facilitating rapid and cooperative synaptic vesicle fusion and accelerating neurotransmitter release (Chang et al. 2015; Adams et al. 2015). Calcium-activated protein for secretion (CAPS) facilitates SNARE complex formation and interacts with phospholipids to promote fusion site establishment (James et al. 2009; Khodthong et al. 2011). Complexin acts as a fusion clamp by binding to partially assembled SNARE complexes, preventing full zippering until calcium-synaptotagmin displaces it (Tang et al. 2006; Giraudo et al. 2006; Maximov et al. 2009). This clamping function ensures that fusion occurs synchronously with calcium influx rather than spontaneously (Krishnakumar et al. 2011; Kummel et al. 2011). Tomosyn and other SNARE-related proteins provide additional regulatory layers by competing with v-SNAREs for binding to t-SNAREs and, thus, may function in downregulating exocytosis (Hatsuzawa et al. 2003; Yizhar et al. 2004). These emerging roles in fusion pore dynamics reflect the sophisticated control mechanisms that generate diverse exocytic behaviors from a common molecular foundation (Sudhof 2013).

3.5.2 *Calcium Signaling in Fusion and Fission*

Calcium acts as a key regulator, coordinating membrane fusion and fission through complex signaling pathways (Neher and Sakaba 2008; Kochubey and Schneggenburger 2011). The initial influx of calcium triggers exocytosis by binding to synaptotagmin (Schneggenburger and Neher 2000; Walter et al. 2013), which facilitates SNARE protein-mediated fusion pore opening (Jahn and Fasshauer 2012; Brunger et al. 2018). Beyond triggering initial fusion, calcium influx dynamically regulates the subsequent behavior of fusion pores, including their expansion, constriction, and closure (Lindau and Almers 1995). This modulation is achieved via calcium activation of multiple effectors such as dynamin—a guanosine triphosphate (GTP)-hydrolyzing GTPase that constricts and closes fusion pores—and the actin cytoskeleton, which generates membrane tension to promote pore expansion (Shin et al. 2018, 2022; Wu et al. 2019). Indeed, calcium-dependent activation of dynamin facilitates fusion pore constriction and closure, thereby modulating vesicular content release dynamics and recycling through kiss-and-run or other fusion modes (Shin et al. 2018). Furthermore, calcium influx also triggers endocytic membrane transformations by driving sequential transitions from flat to Λ-, Ω-, and eventually to spherical vesicle profiles, endowing cells with diverse endocytic modes differing in speed, quantity, and vesicle size (Chiang et al. 2014; Shin et al. 2021).

Super-resolution imaging has provided further evidence for this calcium-mediated endocytosis, demonstrating that calcium influx drives both the closure of preformed Ω-profiles and fusion pore constriction in kiss-and-run events (Shin et al. 2018, 2021). The primary calcium sensors for endocytosis remain incompletely characterized but appear distinct from those mediating exocytosis (Xue et al. 2012b; Wu et al. 2007). Calmodulin serves as a universal mediator, binding calcium and subsequently activating downstream effectors, including calcineurin phosphatase (Wu et al. 2009). Calcineurin dephosphorylates multiple endocytic proteins collectively termed "dephosphins," including dynamin, amphiphysin, synaptojanin, and epsin (Cousin and Robinson 2001). This dephosphorylation activates these proteins and enhances their interactions, promoting assembly of the endocytic machinery (Slepnev et al. 1998). Importantly, calcium–calmodulin–calcineurin signaling appears necessary for all kinetically distinct forms of endocytosis at synapses (ultrafast, fast, and slow), suggesting a common initial calcium-sensing mechanism with divergent downstream pathways (Wu et al. 2009, 2014b). The calcium dependence of endocytosis creates a natural coupling mechanism between exocytosis and membrane retrieval (Xue et al. 2012a). The same calcium influx that triggers fusion also initiates endocytic processes, ensuring proportional membrane retrieval following secretion (Wu et al. 2009; Xue et al. 2012a).

3.5.3 Molecular Machinery of Membrane Scission

Dynamin GTPases play central roles in membrane fission during endocytosis and in regulating fusion pore dynamics during exocytosis (Ferguson and De Camilli 2012; Antonny et al. 2016). These large, multi-domain proteins (~100 kDa) assemble into helical oligomers around membrane necks, using GTP hydrolysis to generate a mechanical force that promotes membrane constriction and ultimately scission (Takei et al. 1995; Praefcke and McMahon 2004). The three mammalian dynamin isoforms (dynamin-1, 2, and 3) exhibit distinctive expression patterns and regulatory mechanisms but share a conserved domain architecture and fundamental mechanism (Ferguson et al. 2007, 2009).

Dynamin assembly around membrane necks occurs through a series of ordered steps (Sundborger and Hinshaw 2014; Schmid and Frolov 2011). Initially, dynamin is recruited to nascent endocytic sites through interactions with membrane-bound adaptor proteins and direct binding to phosphoinositides (Schmid and Frolov 2011; Meinecke et al. 2013). Dynamin molecules then self-assemble into spiral collars around membrane necks, with adjacent turns stabilized by interactions between the stalks of neighboring molecules (Roux et al. 2006). Recent cryo-electron microscopy studies have revealed the structural details of this assembly process, showing how dynamin forms a helical scaffold with approximately 13 dimers per turn and a diameter that can adapt to different neck dimensions (Kong et al. 2018; Sundborger et al. 2014).

Beyond its classical role in endocytic vesicle scission, dynamin regulates fusion pore dynamics during exocytosis (Anantharam et al. 2011). Super-resolution microscopy has revealed dynamin recruitment to the necks of fusion pores, where it promotes constriction and eventual closure (Zhao et al. 2016). This function explains earlier observations that dynamin inhibition prolongs fusion pore lifetimes and increases quantal size during neurotransmitter release (Graham et al. 2002; Fulop et al. 2008). The ability of dynamin to constrict fusion pores provides a mechanistic basis for kiss-and-run exocytosis and establishes dynamin as a key determinant of release mode.

While dynamin has traditionally been viewed as the primary force generator for membrane fission, emerging evidence highlights clathrin's crucial role in this process. Clathrin polymerization at the base/pore region of Ω-profiles generates forces from its intrinsically curved shape to constrict and close membrane pores in neuroendocrine chromaffin cells (Wei et al. 2024). Mathematical modeling supports this mechanism, confirming that clathrin's intrinsic curvature can generate sufficient elastic forces for pore constriction (Wei et al. 2024). This mechanism operates in both clathrin-coated and non-coated vesicles ranging from 40 to 1500 nm in diameter and underlies diverse endocytic modes previously considered clathrin-independent, including ultrafast, fast, slow, bulk, and overshoot endocytosis. Experimental findings indicate that neither dynamin nor clathrin alone is sufficient for effective fission—both proteins are required to generate the necessary constriction forces (Wei et al. 2024). This insight suggests that clathrin and potentially other intrinsically curved coat proteins constitute an important class of fission proteins working in concert with dynamin to facilitate membrane scission across various cellular contexts.

3.5.4 Cytoskeletal Contributions to Membrane Dynamics

Complementing the scission machinery, the cytoskeleton provides essential mechanical support and force generation for membrane transformations during both exocytosis and endocytosis (Wen et al. 2016; Wu et al. 2016). These cytoskeletal elements work in concert with the membrane remodeling proteins to orchestrate precise spatial and temporal control over membrane dynamics. Actin filaments, in particular, form dynamic networks beneath the plasma membrane that contribute to multiple aspects of membrane remodeling (Rust and Maritzen 2015). Recent super-resolution imaging and molecular perturbation studies reveal that actin and associated proteins generate, sense, and respond to mechanical forces during membrane fusion and fission (Shin et al. 2022).

Traditionally viewed as a passive barrier requiring local disassembly for vesicle docking and fusion (Porat-Shliom et al. 2013; Meunier and Gutierrez 2016), cortical actin in neuroendocrine chromaffin cells is now recognized as an active participant in membrane fusion (Wen et al. 2016). After the fusion pore opens, F-actin assembles within ~1 s at the fusion site, heightening plasma membrane tension and promoting shrinkage of the fusion-generated Ω-profile until it merges into the membrane (Wen et al. 2016; Gasman et al. 2004). This "vesicle-shrinking" mode keeps the fusion pore wide and speeds cargo discharge (Shin et al. 2020; Wen et al. 2016). Inhibiting actin polymerization significantly impairs fusion pore expansion, resulting in prolonged, narrow pores that restrict release (Wen et al. 2016).

Myosin II motor proteins cooperate with actin during exocytosis, exerting tensile pressure on the cortical F-actin network to promote fusion pore expansion (Neco et al. 2004, 2008; Doreian et al. 2008; Berberian et al. 2009). The combined action of actin polymerization and myosin tension creates a mechanical system that actively transforms fusion pores from initial narrow connections to dilated structures that permit efficient content discharge (Wen et al. 2016), working in functional opposition to the constriction machinery.

In endocytosis, actin plays multiple essential roles across all kinetically distinct modes (ultrafast, fast, and slow) (Wu et al. 2016). During clathrin-mediated endocytosis, actin polymerization provides forces that overcome membrane tension and drive invagination (Boulant et al. 2011). This function becomes particularly important under conditions of high membrane tension, where the energetic cost of membrane deformation increases (Wu et al. 2016, 2017b). Actin assembly begins at mid-to-late stages of coated pit formation and intensifies during vesicle scission, suggesting a primarily mechanical role in the final steps of vesicle detachment (Mund et al. 2018), where it works alongside dynamin and clathrin.

Recent super-resolution studies have revealed the precise spatiotemporal dynamics of actin during the flat-to-round transformation of endocytosis (Shin et al. 2022). In chromaffin cells, actin forms the structural framework for generating and advancing membrane invaginations, organizing in distinctive patterns during each step of endocytosis (Shin et al. 2022). During the flat→Λ transition, actin assembles into a striking spike-like structure that pulls the plasma membrane inward, creating the initial invagination (Shin et al. 2022). This observation challenges the traditional view of membrane "pushing" during endocytosis, instead suggesting an active "pulling" mechanism (Shin et al. 2022).

3.6 Pathological Contexts

Impaired membrane transformation dynamics contributes to numerous diseases through the disruption of cellular homeostasis and intercellular communication. These defects manifest across diverse pathological contexts, from neurological disorders to infectious diseases.

Dysregulated membrane fusion and fission dynamics represent a common pathogenic mechanism in neurodegenerative and neurodevelopmental disorders. Alzheimer's disease involves multiple defects in membrane trafficking, including impaired endosomal function and altered synaptic vesicle recycling (Rajendran and Annaert 2012; Marsh and Alifragis 2018); Parkinson's disease pathogenesis similarly involves significant dysfunction in synaptic vesicle dynamics (Nemani et al. 2010). Accumulating evidence shows that many Parkinson's disease genes—such as LRRK2, SNCA, parkin, synaptojanin-1, and auxilin—encode presynaptic proteins that orchestrate synaptic vesicle endocytosis and recycling (Zou et al. 2021; Biskup et al. 2006; Sharma and Burre 2023; Mani et al. 2007). Pathogenic mutations in these genes disrupt vesicle budding, uncoating, or degradation pathways, leading to dopamine terminal loss and ultimately the dopaminergic neurodegeneration characteristic of Parkinson's disease (Soukup et al. 2018); autism spectrum disorders are also closely linked to dysfunctions in synaptic vesicle exocytosis and endocytosis, particularly affecting vesicle priming, fusion, and recycling steps (Yeo et al. 2022). Mutations in presynaptic proteins such as CAPS, Rab3-interacting molecule (RIM), synapsin, synaptotagmin, and DYRK1A disrupt neurotransmitter release and vesicle retrieval, contributing to autism pathogenesis (Corradi et al. 2014; Baker et al. 2018; Imig et al. 2014; Jockusch et al. 2007).

Beyond neurological contexts, dysfunctional membrane transformations play central roles in metabolic disorders (Seino et al. 2011). Diabetes, particularly type 2 diabetes, is closely linked to defects in insulin exocytosis. Pancreatic β-cell dysfunction leads to reduced or dysregulated insulin secretion, often due to abnormalities in the molecular machinery of exocytosis (Rorsman and Ashcroft 2018; Rosengren et al. 2012; Rorsman et al. 2012).

Cancer progression frequently involves dysregulated membrane trafficking that contributes to invasive behavior and therapy resistance (Mosesson et al. 2008;

Mellman and Yarden 2013). Many cancers upregulate exocytosis to release factors promoting angiogenesis, matrix remodeling, and immune evasion (Olejarz et al. 2020; Machado et al. 2015; Morrissey and Yan 2020). Concurrently, altered endocytic trafficking disrupts normal receptor downregulation mechanisms, enhancing signaling from growth factor and adhesion receptors and further promoting malignant phenotypes (Mosesson et al. 2008).

Immune disorders frequently involve defects in the secretory processes essential for immune function (de Saint Basile et al. 2010). Familial hemophagocytic lymphohistiocytosis includes mutations in proteins regulating cytotoxic granule exocytosis, preventing effective target cell elimination (de Saint Basile et al. 2010; Pachlopnik Schmid et al. 2010; Zur Stadt et al. 2006). Similarly, Griscelli syndrome type 2 features mutations in Rab27a that disrupt secretory vesicle docking at the immunological synapse (Stinchcombe et al. 2001), preventing effective cytotoxic responses (Menasche et al. 2000).

Finally, pathogens frequently target or hijack membrane transformation machinery to facilitate infections and evade host defenses. Bacterial toxins such as tetanus and botulinum neurotoxins specifically cleave SNARE proteins to prevent exocytosis, thereby blocking and halting neurotransmitter release (Binz and Rummel 2009). Viral pathogens extensively manipulate membrane dynamics through complementary mechanisms. For instance, HIV utilizes endocytic machinery to achieve productive entry into the cytosol, providing an alternative route to direct fusion at the plasma membrane (Daecke et al. 2005). Understanding these pathogen–host interactions provides potential therapeutic opportunities by protecting critical host membrane transformation machinery.

3.7 Discussion

Advances in super-resolution imaging and molecular tools have significantly refined our understanding of membrane merger dynamics in exocytosis and endocytosis. Rather than a binary distinction between full-collapse and kiss-and-run fusion (Rizzoli and Jahn 2007), current evidence supports a spectrum of fusion pore behaviors and vesicle fate decisions, including various modes of pore expansion, constriction, closure, and changes in vesicle size, such as shrinking, enlargement, and compound fusion (Shin et al. 2018; Chiang et al. 2014). In particular, recent imaging studies suggest that some events previously classified as full-collapse fusion may actually represent shrink fusion, where the vesicle membrane is gradually incorporated into the plasma membrane (Shin et al. 2020). To reconcile the debate between shrink and full-collapse fusion, a "shrink–collapse" model has been proposed, in which vesicles shrink to a critical size before undergoing final merging (Chiang et al. 2014; Wu and Chan 2024).

Endocytosis is now recognized as a modular process, with flat→Λ→Ω→O transitions occurring at low probability (Shin et al. 2022), while preformed and fusion-generated Ω-profile closures emerge as the primary mechanisms underlying rapid and compensatory membrane retrieval (Shin et al. 2021). These transformations are regulated by the coordinated actions of SNARE proteins, dynamin, actin cytoskeleton, and calcium signaling, which together determine the efficiency and kinetics of both content release and membrane recovery. Calcium influx serves as a common trigger, coupling exocytosis and endocytosis to maintain membrane homeostasis (Wu and Chan 2024).

Despite these advances, many questions remain. The molecular mechanisms that govern transitions between different fusion and endocytic modes are not fully understood, particularly how specific proteins and lipids coordinate to control membrane curvature, tension, and scission at the nanoscale. Technical challenges also persist, as even with current super-resolution microscopy, capturing rapid protein dynamics in small vesicles remains difficult.

Future progress will rely on further improvements in imaging resolution and speed, as well as new biochemical and genetic tools to dissect the contributions of individual molecular players. Integrating these approaches with quantitative modeling will help clarify how cells adapt exo–endocytic dynamics to diverse physiological needs. Ultimately, a deeper understanding of these processes may inform new therapeutic strategies for diseases associated with membrane trafficking dysfunction, such as neurodegenerative disorders, diabetes, and cancer.

Acknowledgment This work was supported by the National Institute of Neurological Disorders and Stroke Intramural Research Program (ZIA NS003009-21 and ZIA NS003105-16 to L.G.W.).

References

Adams DJ, Arthur CP, Stowell MH (2015) Architecture of the synaptophysin/synaptobrevin complex: structural evidence for an entropic clustering function at the synapse. Sci Rep 5: 13659. https://doi.org/10.1038/srep13659

Alabi AA, Tsien RW (2013) Perspectives on kiss-and-run: role in exocytosis, endocytosis, and neurotransmission. Annu Rev Physiol 75:393–422. https://doi.org/10.1146/annurev-physiol-020911-153305

Albillos A, Dernick G, Horstmann H, Almers W, Alvarez de Toledo G, Lindau M (1997) The exocytotic event in chromaffin cells revealed by patch amperometry. Nature 389:509–512. https://doi.org/10.1038/39081

Ales E, Tabares L, Poyato JM, Valero V, Lindau M, Alvarez de Toledo G (1999) High calcium concentrations shift the mode of exocytosis to the kiss-and-run mechanism. Nat Cell Biol 1:40–44. https://doi.org/10.1038/9012

Alvarez de Toledo G, Fernandez JM (1990) Compound versus multigranular exocytosis in peritoneal mast cells. J Gen Physiol 95:397–409. https://doi.org/10.1085/jgp.95.3.397

Alvarez de Toledo G, Fernandez-Chacon R, Fernandez JM (1993) Release of secretory products during transient vesicle fusion. Nature 363:554–558. https://doi.org/10.1038/363554a0

Anantharam A, Bittner MA, Aikman RL, Stuenkel EL, Schmid SL, Axelrod D, Holz RW (2011) A new role for the dynamin GTPase in the regulation of fusion pore expansion. Mol Biol Cell 22: 1907–1918. https://doi.org/10.1091/mbc.E11-02-0101

Antonny B, Burd C, De Camilli P, Chen E, Daumke O, Faelber K, Ford M, Frolov VA, Frost A, Hinshaw JE et al (2016) Membrane fission by dynamin: what we know and what we need to know. EMBO J 35:2270–2284. https://doi.org/10.15252/embj.201694613

Bademosi AT, Lauwers E, Padmanabhan P, Odierna L, Chai YJ, Papadopulos A, Goodhill GJ, Verstreken P, van Swinderen B, Meunier FA (2017) In vivo single-molecule imaging of syntaxin1A reveals polyphosphoinositide- and activity-dependent trapping in presynaptic nanoclusters. Nat Commun 8:13660. https://doi.org/10.1038/ncomms13660

Baker RW, Jeffrey PD, Zick M, Phillips BP, Wickner WT, Hughson FM (2015) A direct role for the Sec1/Munc18-family protein Vps33 as a template for SNARE assembly. Science 349:1111–1114. https://doi.org/10.1126/science.aac7906

Baker K, Gordon SL, Melland H, Bumbak F, Scott DJ, Jiang TJ, Owen D, Turner BJ, Boyd SG, Rossi M et al (2018) SYT1-associated neurodevelopmental disorder: a case series. Brain 141: 2576–2591. https://doi.org/10.1093/brain/awy209

Bao H, Goldschen-Ohm M, Jeggle P, Chanda B, Edwardson JM, Chapman ER (2016) Exocytotic fusion pores are composed of both lipids and proteins. Nat Struct Mol Biol 23:67–73. https://doi.org/10.1038/nsmb.3141

Berberian K, Torres AJ, Fang Q, Kisler K, Lindau M (2009) F-actin and myosin II accelerate catecholamine release from chromaffin granules. J Neurosci 29:863–870. https://doi.org/10.1523/JNEUROSCI.2818-08.2009

Betzig E, Patterson GH, Sougrat R, Lindwasser OW, Olenych S, Bonifacino JS, Davidson MW, Lippincott-Schwartz J, Hess HF (2006) Imaging intracellular fluorescent proteins at nanometer resolution. Science 313:1642–1645. https://doi.org/10.1126/science.1127344

Binz T, Rummel A (2009) Cell entry strategy of clostridial neurotoxins. J Neurochem 109:1584–1595. https://doi.org/10.1111/j.1471-4159.2009.06093.x

Biskup S, Moore DJ, Celsi F, Higashi S, West AB, Andrabi SA, Kurkinen K, Yu SW, Savitt JM, Waldvogel HJ et al (2006) Localization of LRRK2 to membranous and vesicular structures in mammalian brain. Ann Neurol 60:557–569. https://doi.org/10.1002/ana.21019

Bonifacino JS, Glick BS (2004) The mechanisms of vesicle budding and fusion. Cell 116:153–166. https://doi.org/10.1016/s0092-8674(03)01079-1

Boulant S, Kural C, Zeeh JC, Ubelmann F, Kirchhausen T (2011) Actin dynamics counteract membrane tension during clathrin-mediated endocytosis. Nat Cell Biol 13:1124–1131. https://doi.org/10.1038/ncb2307

Brunger AT, Choi UB, Lai Y, Leitz J, Zhou Q (2018) Molecular mechanisms of fast neurotransmitter release. Annu Rev Biophys 47:469–497. https://doi.org/10.1146/annurev-biophys-070816-034117

Burchfield JG, Lopez JA, Mele K, Vallotton P, Hughes WE (2010) Exocytotic vesicle behaviour assessed by total internal reflection fluorescence microscopy. Traffic 11:429–439. https://doi.org/10.1111/j.1600-0854.2010.01039.x

Ceccarelli B, Hurlbut WP, Mauro A (1972) Depletion of vesicles from frog neuromuscular junctions by prolonged tetanic stimulation. J Cell Biol 54:30–38. https://doi.org/10.1083/jcb.54.1.30

Chan CY, Faragalla Y, Wu LG (2022) Illuminating membrane structural dynamics of fusion and endocytosis with advanced light imaging techniques. Biochem Soc Trans 50:1157–1167. https://doi.org/10.1042/BST20210263

Chanaday NL, Cousin MA, Milosevic I, Watanabe S, Morgan JR (2019) The synaptic vesicle cycle revisited: new insights into the modes and mechanisms. J Neurosci 39:8209–8216. https://doi.org/10.1523/JNEUROSCI.1158-19.2019

Chang CW, Hui E, Bai J, Bruns D, Chapman ER, Jackson MB (2015) A structural role for the synaptobrevin 2 transmembrane domain in dense-core vesicle fusion pores. J Neurosci 35: 5772–5780. https://doi.org/10.1523/JNEUROSCI.3983-14.2015

Chapman ER (2008) How does synaptotagmin trigger neurotransmitter release? Annu Rev Biochem 77:615–641. https://doi.org/10.1146/annurev.biochem.77.062005.101135

Chernomordik LV, Kozlov MM (2005) Membrane hemifusion: crossing a chasm in two leaps. Cell 123:375–382. https://doi.org/10.1016/j.cell.2005.10.015

Chernomordik LV, Kozlov MM (2008) Mechanics of membrane fusion. Nat Struct Mol Biol 15: 675–683. https://doi.org/10.1038/nsmb.1455

Chiang HC, Shin W, Zhao WD, Hamid E, Sheng J, Baydyuk M, Wen PJ, Jin A, Momboisse F, Wu LG (2014) Post-fusion structural changes and their roles in exocytosis and endocytosis of dense-core vesicles. Nat Commun 5:3356. https://doi.org/10.1038/ncomms4356

Chow RH, von Ruden L, Neher E (1992) Delay in vesicle fusion revealed by electrochemical monitoring of single secretory events in adrenal chromaffin cells. Nature 356:60–63. https://doi.org/10.1038/356060a0

Corradi A, Fadda M, Piton A, Patry L, Marte A, Rossi P, Cadieux-Dion M, Gauthier J, Lapointe L, Mottron L et al (2014) SYN2 is an autism predisposing gene: loss-of-function mutations alter synaptic vesicle cycling and axon outgrowth. Hum Mol Genet 23:90–103. https://doi.org/10.1093/hmg/ddt401

Cousin MA, Robinson PJ (2001) The dephosphins: dephosphorylation by calcineurin triggers synaptic vesicle endocytosis. Trends Neurosci 24:659–665. https://doi.org/10.1016/s0166-2236(00)01930-5

Daecke J, Fackler OT, Dittmar MT, Krausslich HG (2005) Involvement of clathrin-mediated endocytosis in human immunodeficiency virus type 1 entry. J Virol 79:1581–1594. https://doi.org/10.1128/JVI.79.3.1581-1594.2005

de Lange RP, de Roos AD, Borst JG (2003) Two modes of vesicle recycling in the rat calyx of held. J Neurosci 23:10164–10173. https://doi.org/10.1523/JNEUROSCI.23-31-10164.2003

de Saint Basile G, Menasche G, Fischer A (2010) Molecular mechanisms of biogenesis and exocytosis of cytotoxic granules. Nat Rev Immunol 10:568–579. https://doi.org/10.1038/nri2803

Di Fiore PP, von Zastrow M (2014) Endocytosis, signaling, and beyond. Cold Spring Harb Perspect Biol 6. https://doi.org/10.1101/cshperspect.a016865

Diao J, Grob P, Cipriano DJ, Kyoung M, Zhang Y, Shah S, Nguyen A, Padolina M, Srivastava A, Vrljic M et al (2012) Synaptic proteins promote calcium-triggered fast transition from point contact to full fusion. elife 1:e00109. https://doi.org/10.7554/eLife.00109

Doherty GJ, McMahon HT (2009) Mechanisms of endocytosis. Annu Rev Biochem 78:857–902. https://doi.org/10.1146/annurev.biochem.78.081307.110540

Dolensek J, Skelin M, Rupnik MS (2011) Calcium dependencies of regulated exocytosis in different endocrine cells. Physiol Res 60:S29–S38. https://doi.org/10.33549/physiolres.932176

Doreian BW, Fulop TG, Smith CB (2008) Myosin II activation and actin reorganization regulate the mode of quantal exocytosis in mouse adrenal chromaffin cells. J Neurosci 28:4470–4478. https://doi.org/10.1523/JNEUROSCI.0008-08.2008

Fasshauer D, Sutton RB, Brunger AT, Jahn R (1998) Conserved structural features of the synaptic fusion complex: SNARE proteins reclassified as Q- and R-SNAREs. Proc Natl Acad Sci USA 95:15781–15786. https://doi.org/10.1073/pnas.95.26.15781

Ferguson SM, De Camilli P (2012) Dynamin, a membrane-remodelling GTPase. Nat Rev Mol Cell Biol 13:75–88. https://doi.org/10.1038/nrm3266

Ferguson SM, Brasnjo G, Hayashi M, Wolfel M, Collesi C, Giovedi S, Raimondi A, Gong LW, Ariel P, Paradise S et al (2007) A selective activity-dependent requirement for dynamin 1 in synaptic vesicle endocytosis. Science 316:570–574. https://doi.org/10.1126/science.1140621

Ferguson SM, Raimondi A, Paradise S, Shen H, Mesaki K, Ferguson A, Destaing O, Ko G, Takasaki J, Cremona O et al (2009) Coordinated actions of actin and BAR proteins upstream of dynamin at endocytic clathrin-coated pits. Dev Cell 17:811–822. https://doi.org/10.1016/j.devcel.2009.11.005

Fernandez-Alfonso T, Ryan TA (2004) The kinetics of synaptic vesicle pool depletion at CNS synaptic terminals. Neuron 41:943–953. https://doi.org/10.1016/s0896-6273(04)00113-8

Fernandez-Chacon R, Konigstorfer A, Gerber SH, Garcia J, Matos MF, Stevens CF, Brose N, Rizo J, Rosenmund C, Sudhof TC (2001) Synaptotagmin I functions as a calcium regulator of release probability. Nature 410:41–49. https://doi.org/10.1038/35065004

Fernandez-Suarez M, Ting AY (2008) Fluorescent probes for super-resolution imaging in living cells. Nat Rev Mol Cell Biol 9:929–943. https://doi.org/10.1038/nrm2531

Finkelstein A, Zimmerberg J, Cohen FS (1986) Osmotic swelling of vesicles: its role in the fusion of vesicles with planar phospholipid bilayer membranes and its possible role in exocytosis. Annu Rev Physiol 48:163–174. https://doi.org/10.1146/annurev.ph.48.030186.001115

Fukuda R, McNew JA, Weber T, Parlati F, Engel T, Nickel W, Rothman JE, Sollner TH (2000) Functional architecture of an intracellular membrane t-SNARE. Nature 407:198–202. https://doi.org/10.1038/35025084

Fulop T, Doreian B, Smith C (2008) Dynamin I plays dual roles in the activity-dependent shift in exocytic mode in mouse adrenal chromaffin cells. Arch Biochem Biophys 477:146–154. https://doi.org/10.1016/j.abb.2008.04.039

Gan Q, Watanabe S (2018) Synaptic vesicle endocytosis in different model systems. Front Cell Neurosci 12:171. https://doi.org/10.3389/fncel.2018.00171

Gao Y, Zorman S, Gundersen G, Xi Z, Ma L, Sirinakis G, Rothman JE, Zhang Y (2012) Single reconstituted neuronal SNARE complexes zipper in three distinct stages. Science 337:1340–1343. https://doi.org/10.1126/science.1224492

Gasman S, Chasserot-Golaz S, Malacombe M, Way M, Bader MF (2004) Regulated exocytosis in neuroendocrine cells: a role for subplasmalemmal Cdc42/N-WASP-induced actin filaments. Mol Biol Cell 15:520–531. https://doi.org/10.1091/mbc.e03-06-0402

Gauthier NC, Masters TA, Sheetz MP (2012) Mechanical feedback between membrane tension and dynamics. Trends Cell Biol 22:527–535. https://doi.org/10.1016/j.tcb.2012.07.005

Ge L, Shin W, Arpino G, Wei L, Chan CY, Bleck CKE, Zhao W, Wu LG (2022) Sequential compound fusion and kiss-and-run mediate exo- and endocytosis in excitable cells. Sci Adv 8: eabm6049. https://doi.org/10.1126/sciadv.abm6049

Giraudo CG, Eng WS, Melia TJ, Rothman JE (2006) A clamping mechanism involved in SNARE-dependent exocytosis. Science 313:676–680. https://doi.org/10.1126/science.1129450

Gong LW, Hafez I, Alvarez de Toledo G, Lindau M (2003) Secretory vesicles membrane area is regulated in tandem with quantal size in chromaffin cells. J Neurosci 23:7917–7921. https://doi.org/10.1523/JNEUROSCI.23-21-07917.2003

Grabner CP, Moser T (2018) Individual synaptic vesicles mediate stimulated exocytosis from cochlear inner hair cells. Proc Natl Acad Sci USA 115:12811–12816. https://doi.org/10.1073/pnas.1811814115

Graham ME, O'Callaghan DW, McMahon HT, Burgoyne RD (2002) Dynamin-dependent and dynamin-independent processes contribute to the regulation of single vesicle release kinetics and quantal size. Proc Natl Acad Sci USA 99:7124–7129. https://doi.org/10.1073/pnas.102645099

Grushin K, Wang J, Coleman J, Rothman JE, Sindelar CV, Krishnakumar SS (2019) Structural basis for the clamping and Ca(2+) activation of SNARE-mediated fusion by synaptotagmin. Nat Commun 10:2413. https://doi.org/10.1038/s41467-019-10391-x

Gubernator NG, Zhang H, Staal RG, Mosharov EV, Pereira DB, Yue M, Balsanek V, Vadola PA, Mukherjee B, Edwards RH et al (2009) Fluorescent false neurotransmitters visualize dopamine release from individual presynaptic terminals. Science 324:1441–1444. https://doi.org/10.1126/science.1172278

Gundelfinger ED, Kessels MM, Qualmann B (2003) Temporal and spatial coordination of exocytosis and endocytosis. Nat Rev Mol Cell Biol 4:127–139. https://doi.org/10.1038/nrm1016

Guo X, Han S, Wei L, Arpino G, Shin W, Wang X, Wu LG (2022) Real-time visualization of exo- and endocytosis membrane dynamics with confocal and super-resolution microscopy. STAR Protoc 3:101404. https://doi.org/10.1016/j.xpro.2022.101404

Hafez I, Stolpe A, Lindau M (2003) Compound exocytosis and cumulative fusion in eosinophils. J Biol Chem 278:44921–44928. https://doi.org/10.1074/jbc.M306013200

Han X, Wang CT, Bai J, Chapman ER, Jackson MB (2004) Transmembrane segments of syntaxin line the fusion pore of Ca2+-triggered exocytosis. Science 304:289–292. https://doi.org/10.1126/science.1095801

Harata NC, Aravanis AM, Tsien RW (2006) Kiss-and-run and full-collapse fusion as modes of exo-endocytosis in neurosecretion. J Neurochem 97:1546–1570. https://doi.org/10.1111/j.1471-4159.2006.03987.x

Hatsuzawa K, Lang T, Fasshauer D, Bruns D, Jahn R (2003) The R-SNARE motif of tomosyn forms SNARE core complexes with syntaxin 1 and SNAP-25 and down-regulates exocytosis. J Biol Chem 278:31159–31166. https://doi.org/10.1074/jbc.M305500200

He L, Wu LG (2007) The debate on the kiss-and-run fusion at synapses. Trends Neurosci 30:447–455. https://doi.org/10.1016/j.tins.2007.06.012

He L, Wu XS, Mohan R, Wu LG (2006) Two modes of fusion pore opening revealed by cell-attached recordings at a synapse. Nature 444:102–105. https://doi.org/10.1038/nature05250

He L, Xue L, Xu J, McNeil BD, Bai L, Melicoff E, Adachi R, Wu LG (2009) Compound vesicle fusion increases quantal size and potentiates synaptic transmission. Nature 459:93–97. https://doi.org/10.1038/nature07860

Hell SW, Wichmann J (1994) Breaking the diffraction resolution limit by stimulated emission: stimulated-emission-depletion fluorescence microscopy. Opt Lett 19:780–782. https://doi.org/10.1364/ol.19.000780

Hernandez JM, Stein A, Behrmann E, Riedel D, Cypionka A, Farsi Z, Walla PJ, Raunser S, Jahn R (2012) Membrane fusion intermediates via directional and full assembly of the SNARE complex. Science 336:1581–1584. https://doi.org/10.1126/science.1221976

Hernandez JM, Kreutzberger AJ, Kiessling V, Tamm LK, Jahn R (2014) Variable cooperativity in SNARE-mediated membrane fusion. Proc Natl Acad Sci USA 111:12037–12042. https://doi.org/10.1073/pnas.1407435111

Heuser JE, Reese TS (1973) Evidence for recycling of synaptic vesicle membrane during transmitter release at the frog neuromuscular junction. J Cell Biol 57:315–344. https://doi.org/10.1083/jcb.57.2.315

Heuser JE, Reese TS (1981) Structural changes after transmitter release at the frog neuromuscular junction. J Cell Biol 88:564–580. https://doi.org/10.1083/jcb.88.3.564

Heuser JE, Reese TS, Dennis MJ, Jan Y, Jan L, Evans L (1979) Synaptic vesicle exocytosis captured by quick freezing and correlated with quantal transmitter release. J Cell Biol 81: 275–300. https://doi.org/10.1083/jcb.81.2.275

Hua Y, Sinha R, Thiel CS, Schmidt R, Huve J, Martens H, Hell SW, Egner A, Klingauf J (2011) A readily retrievable pool of synaptic vesicles. Nat Neurosci 14:833–839. https://doi.org/10.1038/nn.2838

Hua Y, Woehler A, Kahms M, Haucke V, Neher E, Klingauf J (2013) Blocking endocytosis enhances short-term synaptic depression under conditions of normal availability of vesicles. Neuron 80:343–349. https://doi.org/10.1016/j.neuron.2013.08.010

Huang B, Bates M, Zhuang X (2009) Super-resolution fluorescence microscopy. Annu Rev Biochem 78:993–1016. https://doi.org/10.1146/annurev.biochem.77.061906.092014

Hui E, Bai J, Wang P, Sugimori M, Llinas RR, Chapman ER (2005) Three distinct kinetic groupings of the synaptotagmin family: candidate sensors for rapid and delayed exocytosis. Proc Natl Acad Sci USA 102:5210–5214. https://doi.org/10.1073/pnas.0500941102

Imig C, Min SW, Krinner S, Arancillo M, Rosenmund C, Sudhof TC, Rhee J, Brose N, Cooper BH (2014) The morphological and molecular nature of synaptic vesicle priming at presynaptic active zones. Neuron 84:416–431. https://doi.org/10.1016/j.neuron.2014.10.009

Jackson MB (2010) SNARE complex zipping as a driving force in the dilation of proteinaceous fusion pores. J Membr Biol 235:89–100. https://doi.org/10.1007/s00232-010-9258-1

Jahn R, Fasshauer D (2012) Molecular machines governing exocytosis of synaptic vesicles. Nature 490:201–207. https://doi.org/10.1038/nature11320

Jahn R, Lang T, Sudhof TC (2003) Membrane fusion. Cell 112:519–533. https://doi.org/10.1016/s0092-8674(03)00112-0

James DJ, Kowalchyk J, Daily N, Petrie M, Martin TF (2009) CAPS drives trans-SNARE complex formation and membrane fusion through syntaxin interactions. Proc Natl Acad Sci USA 106: 17308–17313. https://doi.org/10.1073/pnas.0900755106

Jena BP (2009) Membrane fusion: role of SNAREs and calcium. Protein Pept Lett 16:712–717. https://doi.org/10.2174/092986609788681869

Jockusch WJ, Speidel D, Sigler A, Sorensen JB, Varoqueaux F, Rhee JS, Brose N (2007) CAPS-1 and CAPS-2 are essential synaptic vesicle priming proteins. Cell 131:796–808. https://doi.org/10.1016/j.cell.2007.11.002

Kaeser PS, Regehr WG (2014) Molecular mechanisms for synchronous, asynchronous, and spontaneous neurotransmitter release. Annu Rev Physiol 76:333–363. https://doi.org/10.1146/annurev-physiol-021113-170338

Khodthong C, Kabachinski G, James DJ, Martin TF (2011) Munc13 homology domain-1 in CAPS/UNC31 mediates SNARE binding required for priming vesicle exocytosis. Cell Metab 14:254–263. https://doi.org/10.1016/j.cmet.2011.07.002

Klar TA, Jakobs S, Dyba M, Egner A, Hell SW (2000) Fluorescence microscopy with diffraction resolution barrier broken by stimulated emission. Proc Natl Acad Sci USA 97:8206–8210. https://doi.org/10.1073/pnas.97.15.8206

Kochubey O, Schneggenburger R (2011) Synaptotagmin increases the dynamic range of synapses by driving Ca(2)+-evoked release and by clamping a near-linear remaining Ca(2)+ sensor. Neuron 69:736–748. https://doi.org/10.1016/j.neuron.2011.01.013

Kong L, Sochacki KA, Wang H, Fang S, Canagarajah B, Kehr AD, Rice WJ, Strub MP, Taraska JW, Hinshaw JE (2018) Cryo-EM of the dynamin polymer assembled on lipid membrane. Nature 560:258–262. https://doi.org/10.1038/s41586-018-0378-6

Kononenko NL, Haucke V (2015) Molecular mechanisms of presynaptic membrane retrieval and synaptic vesicle reformation. Neuron 85:484–496. https://doi.org/10.1016/j.neuron.2014.12.016

Kozlov MM, Chernomordik LV (2015) Membrane tension and membrane fusion. Curr Opin Struct Biol 33:61–67. https://doi.org/10.1016/j.sbi.2015.07.010

Kozlov MM, Markin VS (1983) Possible mechanism of membrane fusion. Biofizika 28:242–247

Kozlov MM, Campelo F, Liska N, Chernomordik LV, Marrink SJ, McMahon HT (2014) Mechanisms shaping cell membranes. Curr Opin Cell Biol 29:53–60. https://doi.org/10.1016/j.ceb.2014.03.006

Krishnakumar SS, Radoff DT, Kummel D, Giraudo CG, Li F, Khandan L, Baguley SW, Coleman J, Reinisch KM, Pincet F, Rothman JE (2011) A conformational switch in complexin is required for synaptotagmin to trigger synaptic fusion. Nat Struct Mol Biol 18:934–940. https://doi.org/10.1038/nsmb.2103

Kummel D, Krishnakumar SS, Radoff DT, Li F, Giraudo CG, Pincet F, Rothman JE, Reinisch KM (2011) Complexin cross-links prefusion SNAREs into a zigzag array. Nat Struct Mol Biol 18: 927–933. https://doi.org/10.1038/nsmb.2101

Lai Y, Lou X, Jho Y, Yoon TY, Shin YK (2013) The synaptotagmin 1 linker may function as an electrostatic zipper that opens for docking but closes for fusion pore opening. Biochem J 456: 25–33. https://doi.org/10.1042/BJ20130949

Lavis LD, Raines RT (2014) Bright building blocks for chemical biology. ACS Chem Biol 9:855–866. https://doi.org/10.1021/cb500078u

Lee S, Shin J, Jung Y, Son H, Shin J, Jeong C, Kweon DH, Shin YK (2020) Munc18-1 induces conformational changes of syntaxin-1 in multiple intermediates for SNARE assembly. Sci Rep 10:11623. https://doi.org/10.1038/s41598-020-68476-3

Li F, Pincet F, Perez E, Eng WS, Melia TJ, Rothman JE, Tareste D (2007) Energetics and dynamics of SNAREpin folding across lipid bilayers. Nat Struct Mol Biol 14:890–896. https://doi.org/10.1038/nsmb1310

Lindau M, Almers W (1995) Structure and function of fusion pores in exocytosis and ectoplasmic membrane fusion. Curr Opin Cell Biol 7:509–517. https://doi.org/10.1016/0955-0674(95)80007-7

Lindau M, Alvarez de Toledo G (2003) The fusion pore. Biochim Biophys Acta 1641:167–173. https://doi.org/10.1016/s0167-4889(03)00085-5
Liu X, Seven AB, Camacho M, Esser V, Xu J, Trimbuch T, Quade B, Su L, Ma C, Rosenmund C, Rizo J (2016) Functional synergy between the Munc13 C-terminal C1 and C2 domains. elife 5. https://doi.org/10.7554/eLife.13696
Ma C, Li W, Xu Y, Rizo J (2011) Munc13 mediates the transition from the closed syntaxin-Munc18 complex to the SNARE complex. Nat Struct Mol Biol 18:542–549. https://doi.org/10.1038/nsmb.2047
Ma C, Su L, Seven AB, Xu Y, Rizo J (2013) Reconstitution of the vital functions of Munc18 and Munc13 in neurotransmitter release. Science 339:421–425. https://doi.org/10.1126/science.1230473
Machado E, White-Gilbertson S, van de Vlekkert D, Janke L, Moshiach S, Campos Y, Finkelstein D, Gomero E, Mosca R, Qiu X et al (2015) Regulated lysosomal exocytosis mediates cancer progression. Sci Adv 1:e1500603. https://doi.org/10.1126/sciadv.1500603
Mani M, Lee SY, Lucast L, Cremona O, Di Paolo G, De Camilli P, Ryan TA (2007) The dual phosphatase activity of synaptojanin1 is required for both efficient synaptic vesicle endocytosis and reavailability at nerve terminals. Neuron 56:1004–1018. https://doi.org/10.1016/j.neuron.2007.10.032
Marsh J, Alifragis P (2018) Synaptic dysfunction in Alzheimer's disease: the effects of amyloid beta on synaptic vesicle dynamics as a novel target for therapeutic intervention. Neural Regen Res 13:616–623. https://doi.org/10.4103/1673-5374.230276
Matthews G, Sterling P (2008) Evidence that vesicles undergo compound fusion on the synaptic ribbon. J Neurosci 28:5403–5411. https://doi.org/10.1523/JNEUROSCI.0935-08.2008
Maximov A, Tang J, Yang X, Pang ZP, Sudhof TC (2009) Complexin controls the force transfer from SNARE complexes to membranes in fusion. Science 323:516–521. https://doi.org/10.1126/science.1166505
McMahon HT, Boucrot E (2015) Membrane curvature at a glance. J Cell Sci 128:1065–1070. https://doi.org/10.1242/jcs.114454
Meinecke M, Boucrot E, Camdere G, Hon WC, Mittal R, McMahon HT (2013) Cooperative recruitment of dynamin and BIN/amphiphysin/Rvs (BAR) domain-containing proteins leads to GTP-dependent membrane scission. J Biol Chem 288:6651–6661. https://doi.org/10.1074/jbc.M112.444869
Mellman I, Yarden Y (2013) Endocytosis and cancer. Cold Spring Harb Perspect Biol 5:a016949. https://doi.org/10.1101/cshperspect.a016949
Menasche G, Pastural E, Feldmann J, Certain S, Ersoy F, Dupuis S, Wulffraat N, Bianchi D, Fischer A, Le Deist F, de Saint Basile G (2000) Mutations in RAB27A cause Griscelli syndrome associated with haemophagocytic syndrome. Nat Genet 25:173–176. https://doi.org/10.1038/76024
Meunier FA, Gutierrez LM (2016) Captivating new roles of F-actin cortex in exocytosis and bulk endocytosis in neurosecretory cells. Trends Neurosci 39:605–613. https://doi.org/10.1016/j.tins.2016.07.003
Miesenbock G, De Angelis DA, Rothman JE (1998) Visualizing secretion and synaptic transmission with pH-sensitive green fluorescent proteins. Nature 394:192–195. https://doi.org/10.1038/28190
Miller TM, Heuser JE (1984) Endocytosis of synaptic vesicle membrane at the frog neuromuscular junction. J Cell Biol 98:685–698. https://doi.org/10.1083/jcb.98.2.685
Min D, Kim K, Hyeon C, Cho YH, Shin YK, Yoon TY (2013) Mechanical unzipping and rezipping of a single SNARE complex reveals hysteresis as a force-generating mechanism. Nat Commun 4:1705. https://doi.org/10.1038/ncomms2692
Mohrmann R, de Wit H, Verhage M, Neher E, Sorensen JB (2010) Fast vesicle fusion in living cells requires at least three SNARE complexes. Science 330:502–505. https://doi.org/10.1126/science.1193134

Monck JR, Alvarez de Toledo G, Fernandez JM (1990) Tension in secretory granule membranes causes extensive membrane transfer through the exocytotic fusion pore. Proc Natl Acad Sci USA 87:7804–7808. https://doi.org/10.1073/pnas.87.20.7804

Montecucco C, Schiavo G, Pantano S (2005) SNARE complexes and neuroexocytosis: how many, how close? Trends Biochem Sci 30:367–372. https://doi.org/10.1016/j.tibs.2005.05.002

Morrissey SM, Yan J (2020) Exosomal PD-L1: roles in tumor progression and immunotherapy. Trends Cancer 6:550–558. https://doi.org/10.1016/j.trecan.2020.03.002

Mosesson Y, Mills GB, Yarden Y (2008) Derailed endocytosis: an emerging feature of cancer. Nat Rev Cancer 8:835–850. https://doi.org/10.1038/nrc2521

Mostov K, Su T, ter Beest M (2003) Polarized epithelial membrane traffic: conservation and plasticity. Nat Cell Biol 5:287–293. https://doi.org/10.1038/ncb0403-287

Mund M, van der Beek JA, Deschamps J, Dmitrieff S, Hoess P, Monster JL, Picco A, Nedelec F, Kaksonen M, Ries J (2018) Systematic nanoscale analysis of endocytosis links efficient vesicle formation to patterned actin nucleation. Cell 174:884–896. https://doi.org/10.1016/j.cell.2018.06.032

Neco P, Giner D, Viniegra S, Borges R, Villarroel A, Gutierrez LM (2004) New roles of myosin II during vesicle transport and fusion in chromaffin cells. J Biol Chem 279:27450–27457. https://doi.org/10.1074/jbc.M311462200

Neco P, Fernandez-Peruchena C, Navas S, Gutierrez LM, de Toledo GA, Ales E (2008) Myosin II contributes to fusion pore expansion during exocytosis. J Biol Chem 283:10949–10957. https://doi.org/10.1074/jbc.M709058200

Neher E (2006) A comparison between exocytic control mechanisms in adrenal chromaffin cells and a glutamatergic synapse. Pflugers Arch 453:261–268. https://doi.org/10.1007/s00424-006-0143-9

Neher E, Marty A (1982) Discrete changes of cell membrane capacitance observed under conditions of enhanced secretion in bovine adrenal chromaffin cells. Proc Natl Acad Sci USA 79:6712–6716. https://doi.org/10.1073/pnas.79.21.6712

Neher E, Sakaba T (2008) Multiple roles of calcium ions in the regulation of neurotransmitter release. Neuron 59:861–872. https://doi.org/10.1016/j.neuron.2008.08.019

Nemani VM, Lu W, Berge V, Nakamura K, Onoa B, Lee MK, Chaudhry FA, Nicoll RA, Edwards RH (2010) Increased expression of alpha-synuclein reduces neurotransmitter release by inhibiting synaptic vesicle reclustering after endocytosis. Neuron 65:66–79. https://doi.org/10.1016/j.neuron.2009.12.023

Nemoto T, Kimura R, Ito K, Tachikawa A, Miyashita Y, Iino M, Kasai H (2001) Sequential-replenishment mechanism of exocytosis in pancreatic acini. Nat Cell Biol 3:253–258. https://doi.org/10.1038/35060042

Olejarz W, Kubiak-Tomaszewska G, Chrzanowska A, Lorenc T (2020) Exosomes in angiogenesis and anti-angiogenic therapy in cancers. Int J Mol Sci 21. https://doi.org/10.3390/ijms21165840

Pachlopnik Schmid J, Cote M, Menager MM, Burgess A, Nehme N, Menasche G, Fischer A, de Saint Basile G (2010) Inherited defects in lymphocyte cytotoxic activity. Immunol Rev 235:10–23. https://doi.org/10.1111/j.0105-2896.2010.00890.x

Pang ZP, Sudhof TC (2010) Cell biology of Ca2+-triggered exocytosis. Curr Opin Cell Biol 22:496–505. https://doi.org/10.1016/j.ceb.2010.05.001

Pereira DB, Schmitz Y, Meszaros J, Merchant P, Hu G, Li S, Henke A, Lizardi-Ortiz JE, Karpowicz RJ Jr, Morgenstern TJ et al (2016) Fluorescent false neurotransmitter reveals functionally silent dopamine vesicle clusters in the striatum. Nat Neurosci 19:578–586. https://doi.org/10.1038/nn.4252

Pickett JA, Edwardson JM (2006) Compound exocytosis: mechanisms and functional significance. Traffic 7:109–116. https://doi.org/10.1111/j.1600-0854.2005.00372.x

Porat-Shliom N, Milberg O, Masedunskas A, Weigert R (2013) Multiple roles for the actin cytoskeleton during regulated exocytosis. Cell Mol Life Sci 70:2099–2121. https://doi.org/10.1007/s00018-012-1156-5

Praefcke GJ, McMahon HT (2004) The dynamin superfamily: universal membrane tubulation and fission molecules? Nat Rev Mol Cell Biol 5:133–147. https://doi.org/10.1038/nrm1313

Rajendran L, Annaert W (2012) Membrane trafficking pathways in Alzheimer's disease. Traffic 13: 759–770. https://doi.org/10.1111/j.1600-0854.2012.01332.x

Rizo J (2022) Molecular mechanisms underlying neurotransmitter release. Annu Rev Biophys 51: 377–408. https://doi.org/10.1146/annurev-biophys-111821-104732

Rizo J, Sudhof TC (2012) The membrane fusion enigma: SNAREs, Sec1/Munc18 proteins, and their accomplices – guilty as charged? Annu Rev Cell Dev Biol 28:279–308. https://doi.org/10.1146/annurev-cellbio-101011-155818

Rizzoli SO, Jahn R (2007) Kiss-and-run, collapse and 'readily retrievable' vesicles. Traffic 8:1137–1144. https://doi.org/10.1111/j.1600-0854.2007.00614.x

Rodriguez PC, Pereira DB, Borgkvist A, Wong MY, Barnard C, Sonders MS, Zhang H, Sames D, Sulzer D (2013) Fluorescent dopamine tracer resolves individual dopaminergic synapses and their activity in the brain. Proc Natl Acad Sci USA 110:870–875. https://doi.org/10.1073/pnas.1213569110

Rorsman P, Ashcroft FM (2018) Pancreatic beta-cell electrical activity and insulin secretion: of mice and men. Physiol Rev 98:117–214. https://doi.org/10.1152/physrev.00008.2017

Rorsman P, Braun M, Zhang Q (2012) Regulation of calcium in pancreatic alpha- and beta-cells in health and disease. Cell Calcium 51:300–308. https://doi.org/10.1016/j.ceca.2011.11.006

Rosengren AH, Braun M, Mahdi T, Andersson SA, Travers ME, Shigeto M, Zhang E, Almgren P, Ladenvall C, Axelsson AS et al (2012) Reduced insulin exocytosis in human pancreatic beta-cells with gene variants linked to type 2 diabetes. Diabetes 61:1726–1733. https://doi.org/10.2337/db11-1516

Roux A, Uyhazi K, Frost A, De Camilli P (2006) GTP-dependent twisting of dynamin implicates constriction and tension in membrane fission. Nature 441:528–531. https://doi.org/10.1038/nature04718

Rudolph S, Tsai MC, von Gersdorff H, Wadiche JI (2015) The ubiquitous nature of multivesicular release. Trends Neurosci 38:428–438. https://doi.org/10.1016/j.tins.2015.05.008

Rust MB, Maritzen T (2015) Relevance of presynaptic actin dynamics for synapse function and mouse behavior. Exp Cell Res 335:165–171. https://doi.org/10.1016/j.yexcr.2014.12.020

Rust MJ, Bates M, Zhuang X (2006) Sub-diffraction-limit imaging by stochastic optical reconstruction microscopy (STORM). Nat Methods 3:793–795. https://doi.org/10.1038/nmeth929

Saheki Y, De Camilli P (2012) Synaptic vesicle endocytosis. Cold Spring Harb Perspect Biol 4: a005645. https://doi.org/10.1101/cshperspect.a005645

Sahl SJ, Hell SW, Jakobs S (2017) Fluorescence nanoscopy in cell biology. Nat Rev Mol Cell Biol 18:685–701. https://doi.org/10.1038/nrm.2017.71

Sara Y, Mozhayeva MG, Liu X, Kavalali ET (2002) Fast vesicle recycling supports neurotransmission during sustained stimulation at hippocampal synapses. J Neurosci 22:1608–1617. https://doi.org/10.1523/JNEUROSCI.22-05-01608.2002

Scepek S, Lindau M (1993) Focal exocytosis by eosinophils – compound exocytosis and cumulative fusion. EMBO J 12:1811–1817. https://doi.org/10.1002/j.1460-2075.1993.tb05829.x

Schmid SL, Frolov VA (2011) Dynamin: functional design of a membrane fission catalyst. Annu Rev Cell Dev Biol 27:79–105. https://doi.org/10.1146/annurev-cellbio-100109-104016

Schneggenburger R, Neher E (2000) Intracellular calcium dependence of transmitter release rates at a fast central synapse. Nature 406:889–893. https://doi.org/10.1038/35022702

Seino S, Shibasaki T, Minami K (2011) Dynamics of insulin secretion and the clinical implications for obesity and diabetes. J Clin Invest 121:2118–2125. https://doi.org/10.1172/JCI45680

Sharma M, Burre J (2023) Alpha-Synuclein in synaptic function and dysfunction. Trends Neurosci 46:153–166. https://doi.org/10.1016/j.tins.2022.11.007

Sheng J, He L, Zheng H, Xue L, Luo F, Shin W, Sun T, Kuner T, Yue DT, Wu LG (2012) Calcium-channel number critically influences synaptic strength and plasticity at the active zone. Nat Neurosci 15:998–1006. https://doi.org/10.1038/nn.3129

Shin W, Ge L, Arpino G, Villarreal SA, Hamid E, Liu H, Zhao WD, Wen PJ, Chiang HC, Wu LG (2018) Visualization of membrane pore in live cells reveals a dynamic-pore theory governing fusion and endocytosis. Cell 173:934–945. https://doi.org/10.1016/j.cell.2018.02.062

Shin W, Arpino G, Thiyagarajan S, Su R, Ge L, McDargh Z, Guo X, Wei L, Shupliakov O, Jin A et al (2020) Vesicle shrinking and enlargement play opposing roles in the release of exocytotic contents. Cell Rep 30:421–431. https://doi.org/10.1016/j.celrep.2019.12.044

Shin W, Wei L, Arpino G, Ge L, Guo X, Chan CY, Hamid E, Shupliakov O, Bleck CKE, Wu LG (2021) Preformed omega-profile closure and kiss-and-run mediate endocytosis and diverse endocytic modes in neuroendocrine chromaffin cells. Neuron 109:3119–3134. https://doi.org/10.1016/j.neuron.2021.07.019

Shin W, Zucker B, Kundu N, Lee SH, Shi B, Chan CY, Guo X, Harrison JT, Turechek JM, Hinshaw JE et al (2022) Molecular mechanics underlying flat-to-round membrane budding in live secretory cells. Nat Commun 13:3697. https://doi.org/10.1038/s41467-022-31286-4

Sinha R, Ahmed S, Jahn R, Klingauf J (2011) Two synaptobrevin molecules are sufficient for vesicle fusion in central nervous system synapses. Proc Natl Acad Sci USA 108:14318–14323. https://doi.org/10.1073/pnas.1101818108

Slepnev VI, Ochoa GC, Butler MH, Grabs D, De Camilli P (1998) Role of phosphorylation in regulation of the assembly of endocytic coat complexes. Science 281:821–824. https://doi.org/10.1126/science.281.5378.821

Sollner T, Whiteheart SW, Brunner M, Erdjument-Bromage H, Geromanos S, Tempst P, Rothman JE (1993) SNAP receptors implicated in vesicle targeting and fusion. Nature 362:318–324. https://doi.org/10.1038/362318a0

Soukup SF, Vanhauwaert R, Verstreken P (2018) Parkinson's disease: convergence on synaptic homeostasis. EMBO J 37. https://doi.org/10.15252/embj.201898960

Stein A, Weber G, Wahl MC, Jahn R (2009) Helical extension of the neuronal SNARE complex into the membrane. Nature 460:525–528. https://doi.org/10.1038/nature08156

Stephens DJ, Allan VJ (2003) Light microscopy techniques for live cell imaging. Science 300:82–86. https://doi.org/10.1126/science.1082160

Stinchcombe JC, Barral DC, Mules EH, Booth S, Hume AN, Machesky LM, Seabra MC, Griffiths GM (2001) Rab27a is required for regulated secretion in cytotoxic T lymphocytes. J Cell Biol 152:825–834. https://doi.org/10.1083/jcb.152.4.825

Sudhof TC (2012) The presynaptic active zone. Neuron 75:11–25. https://doi.org/10.1016/j.neuron.2012.06.012

Sudhof TC (2013) Neurotransmitter release: the last millisecond in the life of a synaptic vesicle. Neuron 80:675–690. https://doi.org/10.1016/j.neuron.2013.10.022

Sudhof TC, Rothman JE (2009) Membrane fusion: grappling with SNARE and SM proteins. Science 323:474–477. https://doi.org/10.1126/science.1161748

Sugita S, Shin OH, Han W, Lao Y, Sudhof TC (2002) Synaptotagmins form a hierarchy of exocytotic Ca(2+) sensors with distinct Ca(2+) affinities. EMBO J 21:270–280. https://doi.org/10.1093/emboj/21.3.270

Sundborger AC, Hinshaw JE (2014) Regulating dynamin dynamics during endocytosis. F1000Prime Rep 6:85

Sundborger AC, Fang S, Heymann JA, Ray P, Chappie JS, Hinshaw JE (2014) A dynamin mutant defines a superconstricted prefission state. Cell Rep 8:734–742. https://doi.org/10.1016/j.celrep.2014.06.054

Sutton RB, Fasshauer D, Jahn R, Brunger AT (1998) Crystal structure of a SNARE complex involved in synaptic exocytosis at 2.4 a resolution. Nature 395:347–353. https://doi.org/10.1038/26412

Tabares L, Lindau M, Alvarez de Toledo G (2003) Relationship between fusion pore opening and release during mast cell exocytosis studied with patch amperometry. Biochem Soc Trans 31: 837–841. https://doi.org/10.1042/bst0310837

Takei K, McPherson PS, Schmid SL, De Camilli P (1995) Tubular membrane invaginations coated by dynamin rings are induced by GTP-gamma S in nerve terminals. Nature 374:186–190. https://doi.org/10.1038/374186a0

Tang J, Maximov A, Shin OH, Dai H, Rizo J, Sudhof TC (2006) A complexin/synaptotagmin 1 switch controls fast synaptic vesicle exocytosis. Cell 126:1175–1187. https://doi.org/10.1016/j.cell.2006.08.030

Tang F, Xiao D, Chen L, Gao H, Li X (2021) Role of Munc18-1 in the biological functions and pathogenesis of neurological disorders (review). Mol Med Rep 23. https://doi.org/10.3892/mmr.2021.11837

Trexler AJ, Sochacki KA, Taraska JW (2016) Imaging the recruitment and loss of proteins and lipids at single sites of calcium-triggered exocytosis. Mol Biol Cell 27:2423–2434. https://doi.org/10.1091/mbc.E16-01-0057

Urban NT, Willig KI, Hell SW, Nagerl UV (2011) STED nanoscopy of actin dynamics in synapses deep inside living brain slices. Biophys J 101:1277–1284. https://doi.org/10.1016/j.bpj.2011.07.027

Varnai P, Balla T (1998) Visualization of phosphoinositides that bind pleckstrin homology domains: calcium- and agonist-induced dynamic changes and relationship to myo-[3H]inositol-labeled phosphoinositide pools. J Cell Biol 143:501–510. https://doi.org/10.1083/jcb.143.2.501

Walter AM, Pinheiro PS, Verhage M, Sorensen JB (2013) A sequential vesicle pool model with a single release sensor and a Ca(2+)-dependent priming catalyst effectively explains Ca(2+)-dependent properties of neurosecretion. PLoS Comput Biol 9:e1003362. https://doi.org/10.1371/journal.pcbi.1003362

Wang LY, Kaczmarek LK (1998) High-frequency firing helps replenish the readily releasable pool of synaptic vesicles. Nature 394:384–388. https://doi.org/10.1038/28645

Watanabe S (2016) Flash-and-freeze: coordinating optogenetic stimulation with rapid freezing to visualize membrane dynamics at synapses with millisecond resolution. Front Synaptic Neurosci 8:24. https://doi.org/10.3389/fnsyn.2016.00024

Watanabe S, Boucrot E (2017) Fast and ultrafast endocytosis. Curr Opin Cell Biol 47:64–71. https://doi.org/10.1016/j.ceb.2017.02.013

Watanabe S, Rost BR, Camacho-Perez M, Davis MW, Sohl-Kielczynski B, Rosenmund C, Jorgensen EM (2013) Ultrafast endocytosis at mouse hippocampal synapses. Nature 504:242–247. https://doi.org/10.1038/nature12809

Weber T, Zemelman BV, McNew JA, Westermann B, Gmachl M, Parlati F, Sollner TH, Rothman JE (1998) SNAREpins: minimal machinery for membrane fusion. Cell 92:759–772. https://doi.org/10.1016/s0092-8674(00)81404-x

Wei L, Guo X, Haimov E, Obashi K, Lee SH, Shin W, Sun M, Chan CY, Sheng J, Zhang Z et al (2024) Clathrin mediates membrane fission and budding by constricting membrane pores. Cell Discov 10:62. https://doi.org/10.1038/s41421-024-00677-w

Wen PJ, Grenklo S, Arpino G, Tan X, Liao HS, Heureaux J, Peng SY, Chiang HC, Hamid E, Zhao WD et al (2016) Actin dynamics provides membrane tension to merge fusing vesicles into the plasma membrane. Nat Commun 7:12604. https://doi.org/10.1038/ncomms12604

Wienisch M, Klingauf J (2006) Vesicular proteins exocytosed and subsequently retrieved by compensatory endocytosis are nonidentical. Nat Neurosci 9:1019–1027. https://doi.org/10.1038/nn1739

Willig KI, Rizzoli SO, Westphal V, Jahn R, Hell SW (2006) STED microscopy reveals that synaptotagmin remains clustered after synaptic vesicle exocytosis. Nature 440:935–939. https://doi.org/10.1038/nature04592

Wu LG, Borst JG (1999) The reduced release probability of releasable vesicles during recovery from short-term synaptic depression. Neuron 23:821–832. https://doi.org/10.1016/s0896-6273(01)80039-8

Wu LG, Chan CY (2024) Membrane transformations of fusion and budding. Nat Commun 15:21. https://doi.org/10.1038/s41467-023-44539-7

Wu LG, Ryan TA, Lagnado L (2007) Modes of vesicle retrieval at ribbon synapses, calyx-type synapses, and small central synapses. J Neurosci 27:11793–11802. https://doi.org/10.1523/JNEUROSCI.3471-07.2007

Wu XS, McNeil BD, Xu J, Fan J, Xue L, Melicoff E, Adachi R, Bai L, Wu LG (2009) Ca(2+) and calmodulin initiate all forms of endocytosis during depolarization at a nerve terminal. Nat Neurosci 12:1003–1010. https://doi.org/10.1038/nn.2355

Wu LG, Hamid E, Shin W, Chiang HC (2014a) Exocytosis and endocytosis: modes, functions, and coupling mechanisms. Annu Rev Physiol 76:301–331. https://doi.org/10.1146/annurev-physiol-021113-170305

Wu XS, Zhang Z, Zhao WD, Wang D, Luo F, Wu LG (2014b) Calcineurin is universally involved in vesicle endocytosis at neuronal and nonneuronal secretory cells. Cell Rep 7:982–988. https://doi.org/10.1016/j.celrep.2014.04.020

Wu XS, Lee SH, Sheng J, Zhang Z, Zhao WD, Wang D, Jin Y, Charnay P, Ervasti JM, Wu LG (2016) Actin is crucial for all kinetically distinguishable forms of endocytosis at synapses. Neuron 92:1020–1035. https://doi.org/10.1016/j.neuron.2016.10.014

Wu Z, Thiyagarajan S, O'Shaughnessy B, Karatekin E (2017a) Regulation of exocytotic fusion pores by SNARE protein transmembrane domains. Front Mol Neurosci 10:315. https://doi.org/10.3389/fnmol.2017.00315

Wu XS, Elias S, Liu H, Heureaux J, Wen PJ, Liu AP, Kozlov MM, Wu LG (2017b) Membrane tension inhibits rapid and slow endocytosis in secretory cells. Biophys J 113:2406–2414. https://doi.org/10.1016/j.bpj.2017.09.035

Wu Q, Zhang Q, Liu B, Li Y, Wu X, Kuo S, Zheng L, Wang C, Zhu F, Zhou Z (2019) Dynamin 1 restrains vesicular release to a subquantal mode in mammalian adrenal Chromaffin cells. J Neurosci 39:199–211. https://doi.org/10.1523/JNEUROSCI.1255-18.2018

Wyffels JT (2001) Principles and techniques of electron microscopy: biological applications, fourth edition, by M. A. Hayat. Microsc Microanal 7:66

Xue L, Wu LG (2010) Post-tetanic potentiation is caused by two signalling mechanisms affecting quantal size and quantal content. J Physiol 588:4987–4994. https://doi.org/10.1113/jphysiol.2010.196964

Xue L, Zhang Z, McNeil BD, Luo F, Wu XS, Sheng J, Shin W, Wu LG (2012a) Voltage-dependent calcium channels at the plasma membrane, but not vesicular channels, couple exocytosis to endocytosis. Cell Rep 1:632–638. https://doi.org/10.1016/j.celrep.2012.04.011

Xue L, McNeil BD, Wu XS, Luo F, He L, Wu LG (2012b) A membrane pool retrieved via endocytosis overshoot at nerve terminals: a study of its retrieval mechanism and role. J Neurosci 32:3398–3404. https://doi.org/10.1523/JNEUROSCI.5943-11.2012

Yang X, Wang S, Sheng Y, Zhang M, Zou W, Wu L, Kang L, Rizo J, Zhang R, Xu T, Ma C (2015) Syntaxin opening by the MUN domain underlies the function of Munc13 in synaptic-vesicle priming. Nat Struct Mol Biol 22:547–554. https://doi.org/10.1038/nsmb.3038

Yeo XY, Lim YT, Chae WR, Park C, Park H, Jung S (2022) Alterations of presynaptic proteins in autism spectrum disorder. Front Mol Neurosci 15:1062878. https://doi.org/10.3389/fnmol.2022.1062878

Yizhar O, Matti U, Melamed R, Hagalili Y, Bruns D, Rettig J, Ashery U (2004) Tomosyn inhibits priming of large dense-core vesicles in a calcium-dependent manner. Proc Natl Acad Sci USA 101:2578–2583. https://doi.org/10.1073/pnas.0308700100

Zacharias DA, Violin JD, Newton AC, Tsien RY (2002) Partitioning of lipid-modified monomeric GFPs into membrane microdomains of live cells. Science 296:913–916. https://doi.org/10.1126/science.1068539

Zhao WD, Hamid E, Shin W, Wen PJ, Krystofiak ES, Villarreal SA, Chiang HC, Kachar B, Wu LG (2016) Hemi-fused structure mediates and controls fusion and fission in live cells. Nature 534:548–552. https://doi.org/10.1038/nature18598

Zhou Q, Lai Y, Bacaj T, Zhao M, Lyubimov AY, Uervirojnangkoorn M, Zeldin OB, Brewster AS, Sauter NK, Cohen AE et al (2015) Architecture of the synaptotagmin-SNARE machinery for neuronal exocytosis. Nature 525:62–67. https://doi.org/10.1038/nature14975

Zhou Q, Zhou P, Wang AL, Wu D, Zhao M, Sudhof TC, Brunger AT (2017) The primed SNARE-complexin-synaptotagmin complex for neuronal exocytosis. Nature 548:420–425. https://doi.org/10.1038/nature23484

Zou L, Tian Y, Zhang Z (2021) Dysfunction of synaptic vesicle endocytosis in Parkinson's disease. Front Integr Neurosci 15:619160. https://doi.org/10.3389/fnint.2021.619160

Zur Stadt U, Beutel K, Kolberg S, Schneppenheim R, Kabisch H, Janka G, Hennies HC (2006) Mutation spectrum in children with primary hemophagocytic lymphohistiocytosis: molecular and functional analyses of PRF1, UNC13D, STX11, and RAB27A. Hum Mutat 27:62–68. https://doi.org/10.1002/humu.20274

Chapter 4
Membrane Fusions During Mammalian Fertilization

Taylor Gierke, Pulan Liu, Yonggang Lu, and Shaogeng Tang

Abstract Successful mammalian fertilization, resulting in the formation of a diploid zygote from haploid gametes, requires a series of events that enable a sperm to reach and fuse with an oocyte. To become competent for fertilization, the sperm must first undergo capacitation, a process that is followed by the acrosome reaction. During the acrosome reaction, the acrosome, a membrane-bound organelle localized to the sperm head, fuses with the sperm's plasma membrane and releases proteolytic enzymes that allow the sperm to penetrate the zona pellucida, an extracellular matrix surrounding the oocyte. After traveling through the zona pellucida, the sperm recognizes, binds to, and fuses with the oocyte plasma membrane, or the oolemma. This process is mediated by surface receptors on both the sperm and the oocyte. Once fertilization occurs, the oocyte blocks polyspermy by preventing additional sperm from accessing the oocyte. The oocyte undergoes a cortical reaction, in which cortical granules fuse with the oolemma to release enzymes that modify the zona pellucida, rendering it impermeable to the sperm entry. At the oolemma, the sperm–oocyte-binding capacity is reduced, preventing further sperm attachment.

Keywords Fertilization · Zona pellucida · Oolemma · Acrosome reaction · Acrosome proteins · Sperm–oocyte binding · Sperm–oocyte fusion · Cortical reaction · Polyspermy block

---

T. Gierke · P. Liu · S. Tang (✉)
Department of Molecular Biophysics and Biochemistry, Yale University, New Haven, CT, USA
e-mail: shaogeng.tang@yale.edu

Y. Lu (✉)
Premium Research Institute for Human Metaverse Medicine (WPI-PRIMe), The University of Osaka, Osaka, Japan
e-mail: lu.yonggang.prime@osaka-u.ac.jp

© The Author(s), under exclusive license to Springer Nature Switzerland AG 2026
J. Weiler et al. (eds.), *Cell Fusion*, Advances in Experimental Medicine and Biology 1509, https://doi.org/10.1007/978-3-032-22637-2_4

4.1 Introduction

Successful fertilization of an oocyte by a sperm requires a series of membrane fusion events (Fig. 4.1a, b) (Wassarman and Litscher 2008). First, for the sperm to become fertilization-competent, a membrane-bound organelle on the sperm head, known as the acrosome, undergoes exocytosis. During the acrosome reaction, the plasma membrane of sperm fuses with the outer acrosomal membrane to release acrosomal enzymes to digest the zona pellucida (ZP) (Kim et al. 2008). The acrosome reaction

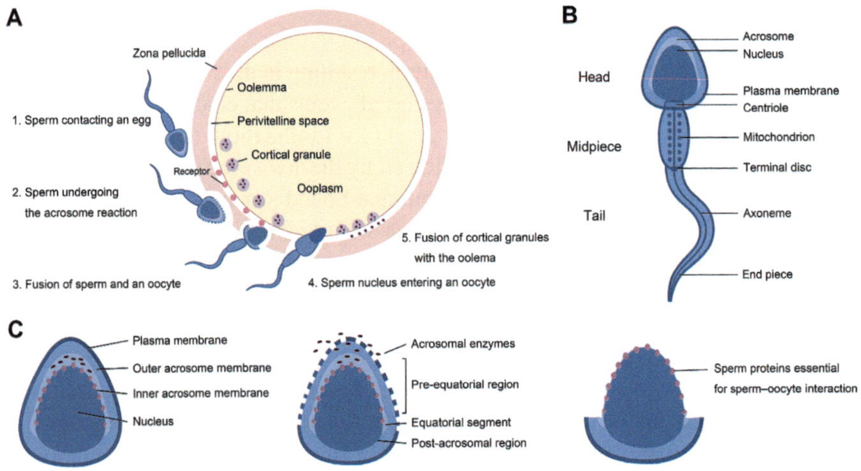

Fig. 4.1 Overview of the three membrane fusion events in mammalian fertilization. (**a**) Schematic of mammalian fertilization. Sperm must interact with and penetrate the ZP of the oocyte for successful fertilization. The primary binding between sperm and the ZP is mediated by sperm-surface proteins and ZP glycoproteins. Primary ZP binding at the sperm head leads to acrosomal exocytosis, exposing intra-acrosomal proteins and allowing secondary sperm–ZP binding. ZP1, ZP3, and ZP4 are involved in primary ZP binding (Chiu et al. 2008; Ganguly et al. 2010), while ZP2 mediates secondary ZP binding (Chiu et al. 2008). While ZP interaction is likely not required for the acrosome reaction (Baibakov et al. 2007), it facilitates zona penetration of the sperm by releasing acrosomal enzymes that facilitate cleavage of glycosylated zona proteins. Upon zona penetration, sperm enter the perivitelline space outside the oolemma to facilitate adhesion and fusion between the equatorial segment of the inner acrosomal membrane on the sperm head and the oolemma. Following membrane fusion, sperm delivers haploid genetic material into the cytoplasm of the oocyte to form a diploid zygote. After fertilization, the zygote prevents polyspermy at the oolemma and the ZP. Single-point exocytotic fusions of cortical granules with the oolemma (Abbott and Ducibella 2001) lead to modification of the ZP structure to prevent penetration of additional sperm. (**b**) Illustration of sperm anatomy. The head contains the nucleus and the acrosome. The midpiece contains mitochondria for the generation of ATP for sperm motility. (**c**) Diagram of the acrosome reaction. Acrosomal exocytosis involves fusion between the outer acrosomal and plasma membranes in the pre-equatorial region of the sperm head, mediated by SNARE interactions. The multipoint fusion during acrosomal exocytosis generates mixed vesicles at the apical side of the sperm head. The equatorial segment of the sperm head remains resistant to exocytotic fusion, with a continuous plasma membrane and outer acrosomal membrane. The resulting acrosome-reacted sperm expose the inner acrosomal membrane, facilitating protein–protein interactions with the oocyte

exposes soluble acrosomal matrix proteins and acrosomal membrane proteins, facilitating interaction with the ZP (Gadella 2010). Sperm passage through the ZP is made possible by hyperactive motility gained during capacitation. Second, the sperm enters the perivitelline space outside the oolemma to facilitate adhesion and fusion between the equatorial segment of the sperm head and the oolemma. After fertilization, the zygote must prevent polyspermy, or the fusion of the zygote by additional sperm. Sperm factors introduced into the oocyte cytoplasm induce the third membrane fusion event: the fusion of cortical granules with the oolemma (Abbott and Ducibella 2001). The secretion of these cortical granules leads to modifications of the ZP structure, which prevent binding and penetration of additional sperm. This chapter covers these three independent membrane fusion events. The acrosome reaction and the cortical reaction involve fusion of cytosolic secretory vesicles with the plasma membrane, whereas the fusion of two plasma membranes involves direct interactions in the extracellular space. Therefore, sperm–oocyte fusion is a topologically inverted process compared to membrane fusion during the acrosome reaction and the cortical reaction.

4.2 The Sperm–Zona Interaction

4.2.1 Sperm Prepares for ZP Binding

For mammalian fertilization, the sperm must move through the female reproductive tract to reach the oviduct, or the fallopian tube. The sperm meets the oocyte for fertilization in the ampulla of the oviduct, where an oocyte is held after ovulation (Coy et al. 2012). Here, a mature egg is ready for fertilization after the developing egg, or oocyte, reaches metaphase II of meiosis during its final stage of development. In the oviduct, the sperm undergoes a series of cellular activation processes called capacitation (Rodriguez-Martinez 2007) to release a protective coat known as the glycocalyx and gain hyperactive motility, enabling the acrosome reaction and facilitating interaction with the ZP and membrane fusion with the oocyte.

4.2.2 Sperm Travels Through the ZP

The oocyte is surrounded by the ZP, a glycoprotein-rich, mesh-like extracellular matrix. To penetrate the zona and move toward the oolemma, the sperm acrosome releases proteolytic enzymes during the acrosome reaction, including hyaluronidase (Kim et al. 2005) and acrosin, allowing the sperm to digest zona proteins. The site and trigger of the acrosome reaction are unclear in vivo. Zona interaction was generally accepted to induce the sperm acrosome reaction (Gadella 2010; Henkel et al. 1998; Hoshi et al. 1993; Schuffner et al. 2002). However, sperm binding to the ZP has been found to be insufficient to trigger the acrosome reaction (Baibakov et al.

2007). Studies have found partially acrosome-reacted sperm in the cumulus outside the ZP during in vitro fertilization (IVF) (Chen and Sathananthan 1986; Jin et al. 2011) and acrosome-reacted sperm that are capable of binding to and penetrating the zona (Inoue et al. 2011; Morales et al. 1989), suggesting the ZP may accelerate rather than induce the acrosome reaction.

The composition and structure of the ZP are important for sperm interaction and fertilization success. The sperm-surface proteins and ZP glycosylation determine the species-specificity of sperm–ZP binding, particularly in mice and humans (Tumova et al. 2021). Structural changes in the ZP can negatively affect fertilization rates. In aging mice, the ZP structure changes from rugged to smooth, and its fiber density increases, impairing sperm–ZP binding (Ishikawa-Yamauchi et al. 2024).

The mechanism for sperm selection by the ZP is unknown, yet the ZP in humans has been shown to interact with sperm having higher fertilization capacity and genetic integrity (Leung et al. 2023). Recent advancements have led to the development of a deep learning algorithm to identify human sperm with ZP-binding capability (Leung et al. 2025). ZP-bound sperm also have higher expression of HSPA2 and SPACA3, which are markers of sperm-binding ability (Leung et al. 2023).

4.2.3 Sperm and ZP Proteins Mediate Sperm–ZP Interactions

The sperm–ZP interaction has been thought to involve the initial binding of a sperm to the ZP, secondary ZP binding via the sperm inner acrosomal membrane after acrosomal exocytosis, and adhesion and fusion of the plasma membranes of a sperm and an oocyte. However, this model may be more complex, as ZP is suggested to accelerate the progress of rather than induce acrosomal exocytosis. The initial binding between sperm and the ZP is mediated by sperm-surface proteins and ZP glycoproteins. This binding mainly involves the recognition of carbohydrate sequences on the ZP by sperm lectin-like oocyte-binding proteins, with the sialyl-Lewis(x) sequence as the major carbohydrate ligand for human sperm–oocyte binding (Clark 2014; Pang et al. 2011). The primary zona binding at the apical region of the sperm head is followed by secondary sperm–ZP binding and digestion and penetration of the ZP. After acrosome exocytosis, the intra-acrosomal proteins are exposed, allowing for the interaction with the ZP and causing cleavage of glycosylated ZP proteins. It remains unclear which sperm proteins mediate binding to the ZP, although high-molecular-weight sperm protein complexes are likely to mediate binding (Tumova et al. 2021).

Three or four ZP glycoproteins are involved in mammalian fertilization (Tumova et al. 2021). The human ZP consists of the four glycoproteins ZP1, ZP2, ZP3, and ZP4. ZP1, ZP3, and ZP4 have been shown to bind to capacitated sperm (Chiu et al. 2008; Ganguly et al. 2010). ZP2 has been suggested to mediate secondary ZP binding to sperm through a sperm receptor, as it is known to bind to acrosome-reacted sperm (Chiu et al. 2008). Mutations of ZP genes have been linked to female infertility and likely disrupt ZP polymerization (Nishio et al. 2024). Supporting this notion, premature ZP2 oligomerization contributes to infertility by blocking sperm penetration of the ZP (Nishio et al. 2024). All ZP glycoproteins have been found to play key roles in female fertility, with a series of mutations identified in recent studies that alter ZP's structures and functions (Cao et al. 2020; Liu et al. 2023; Luo et al. 2020; Okutman et al. 2020; Sa et al. 2025; Sun et al. 2021, 2023; Wei et al. 2022; Xu et al. 2020; Zeng et al. 2023; Zhang et al. 2021).

A collection of sperm proteins is involved in interactions with the ZP (Tumova et al. 2021). Of these proteins, acrosin, an acrosomal serine protease, is essential for sperm penetration through the ZP, as shown by a gene knockout model in hamsters (Hirose et al. 2020). In humans, low acrosin activity is associated with reduced sperm motility and IVF rate. Sperm samples with low acrosin activity have reduced expression of acrosin and PH-20 (Spam1) and a deficiency in glutathione (GSH), which may reduce sperm motility and acrosin activity to affect fertility (Lin et al. 2023). In patients receiving IVF or intracytoplasmic sperm injection (ICSI), low acrosin activity is correlated with low cumulative live birth rates (Zhao et al. 2024). A homozygous nonsense mutation in acrosin has been found that leads to male infertility and total fertilization failure by causing acrosin deficiency and sperm inability to penetrate the ZP (Hua et al. 2023). PH-20 is known to facilitate primary binding of the sperm to the ZP, containing an epidermal growth factor (EGF)-like domain and a catalytic domain that are likely involved in ZP recognition, with a recently solved structure (Im et al. 2025). CALR3 is a testis-specific lectin-deficient chaperone required for ADAM3 maturation and sperm fertility (Ikawa et al. 2011). Mutations in CALR3 cause fertilization failure associated with impaired sperm–ZP binding (Gao et al. 2024). An additional sperm protein, C1orf56, from a capacitated sperm binds to the sialyl-Lewis(x) sequence on human ZP glycoproteins and has been shown to bind to human ZP (Wang et al. 2021b).

4.3 Acrosome Reaction

4.3.1 Calcium Ion Regulates the Acrosome Reaction

Calcium signaling regulates sperm motility, capacitation, and the acrosome reaction. Ca^{2+} influx is regulated primarily by the Ca^{2+} channel complex CatSpermasome (CatSper) localized specifically to the principal piece of the sperm flagellum (Ren et al. 2001), as well as other Ca^{2+} channels (Cohen et al. 2014), such as Hv1 (Lishko et al. 2010, 2011), transient receptor potential channels (Castellano et al. 2003; Gees et al. 2010), and cyclic nucleotide-gated channels (Biel and Michalakis 2009). CatSper mediates hyperactivated motility of the sperm during capacitation, facilitating successful fertilization. Ca^{2+} efflux is maintained by the Ca^{2+} adenosine triphosphatase (ATPase) (Dragileva et al. 1999) and the Na^{+}/Ca^{2+} exchanger (Krasznai et al. 2006).

Elevated and sustained intracellular calcium signaling is required for the induction of human sperm acrosome reaction (Prajapati et al. 2022). Intracellular Ca^{2+} concentrations are regulated by Ca^{2+} pumps, Na^{+}/Ca^{2+} exchangers, and Ca^{2+} channels in the plasma membrane and intracellular organelles, including the acrosome, the redundant nuclear envelope, and the mitochondria. During capacitation, an increase in intracellular Ca^{2+} leads to the activation of Ca^{2+} channels in the plasma membrane and outer acrosomal membrane and the depletion of Ca^{2+} in the acrosome. Ca^{2+} depletion triggers the activation of store-operated Ca^{2+} channels to increase cytosolic Ca^{2+} concentration, enabling completion of the acrosome reaction (Finkelstein et al. 2020).

4.3.2 The Acrosome Reaction Is an Exocytosis Event

In capacitated sperm, the acrosome reaction involves the fusion of the outer acrosomal membrane with the plasma membrane and the release of soluble acrosomal contents (Fig. 4.1c). This fusion occurs at multiple points over the pre-equatorial region of the sperm head and generates vesicles containing a mix of the outer acrosomal membrane and the plasma membrane (Tsai et al. 2010; Zanetti and Mayorga 2009). After the acrosome reaction, the outer acrosomal and plasma membranes undergo vesiculation. Unfused acrosomal membranes remain at the equatorial segment of the acrosome and at the inner acrosomal membrane covering the apical region of the sperm nucleus. Acrosomal exocytosis facilitates penetration of the ZP through the secretion of acrosomal enzymes and allows the sperm to reach the perivitelline space. The acrosome reaction also exposes proteins on the inner acrosomal membrane of the sperm head required for oolemma binding and fusion, rendering the sperm fertilization-competent.

Soluble N-ethylmaleimide-sensitive factor Attachment Protein Receptors (SNAREs) drive fusion between the outer acrosomal membrane and the plasma membrane, involving both acrosome docking and exocytosis steps (Fig. 4.2). During sperm capacitation, the two membranes become docked by a *trans-* ternary protein complex, including syntaxin1, vesicle-associated membrane protein 1 (VAMP1), and synaptosomal-associated protein 23 (SNAP23) (Tomes et al. 2002; Tsai et al. 2010). Complete acrosomal membrane fusion requires additional Ca^{2+} entry (Tsai et al. 2010). Several SNARE complex-interacting proteins are involved, including complexins (Roggero et al. 2007; Tsai et al. 2010; Zhao et al. 2007a), dynamins (Zhao et al. 2007b), Rab3A (Lopez et al. 2007; Yunes et al. 2000), synaptotagmins (Castillo Bennett et al. 2010), multi-PDZ domain protein MUPP1 (Ackermann et al. 2008), calmodulin, CaMKIIalpha (Ackermann et al. 2009), syntaxin-binding proteins (Hutt et al. 2005), and Munc18 (Rodriguez et al. 2012). Complexin and Munc18b stabilize the *trans-* SNARE protein complex and prevent spontaneous acrosomal reaction. Additionally, the scaffolding proteins MUPP1/CaMKIIα have been reported to control the docking of the *trans-* ternary SNARE protein complex. Rab-mediated dephosphorylation of synaptotagmins is important for the acrosome

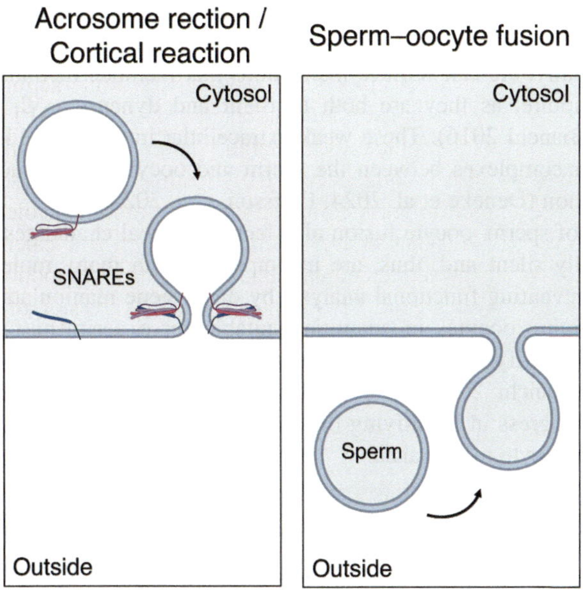

Fig. 4.2 Topology of the three membrane fusion events in mammalian fertilization. Left panel: The acrosome reaction and the cortical reaction represent exocytotic fusion events where secretory vesicles fuse with the plasma membrane. In both processes, the SNARE proteins mediate vesicle docking and fusion. The acrosome reaction involves multipoint fusion of the acrosome with the sperm's plasma membrane, while the cortical reaction involves single-point fusions of multiple cortical granules with the oolemma. Right panel: Sperm–oocyte fusion represents a topologically inverted membrane fusion where two morphologically distinct plasma membranes fuse together, allowing the sperm to enter the oocyte and deliver its genetic material. Both exocytotic events in the left panel are Ca^{2+}-dependent and involve trimeric SNARE complex formation, whereas sperm–oocyte fusion in the right panel represents a cell–cell fusion process involving direct plasma membrane fusion in the extracellular space

reaction. It remains unclear how these proteins interact with the SNAREs, how they stabilize the *trans*- SNARE complex, and how they facilitate the Ca^{2+}-dependent conversion to *cis*- complexes.

4.4 Sperm–Oocyte Interaction at the Fertilization Synapse

4.4.1 Determining the Mechanisms for Sperm–Oocyte Fusion Remains Challenging

Determining the molecular mechanisms behind sperm–oocyte membrane fusion has been met with multiple challenges. Many germline-specific genes are pleiotropic, with multiple functions in reproduction, creating difficulty in identifying their role in membrane fusion. Many others exhibit functional redundancy, performing the same function and preventing an observable loss-of-function phenotype after gene knockouts. Additionally, the cell–surface interactions that facilitate membrane fusion are difficult to capture, as they are both transient and dynamic with weak affinity (Wright and Bianchi 2016). These weak extracellular interactions likely cooperate to form larger complexes between the sperm and oocyte membranes to facilitate membrane fusion (Deneke et al. 2024; Elofsson et al. 2024).

The study of sperm–oocyte fusion also faces technical challenges. Gametes are transcriptionally silent and, thus, are incompatible with many molecular biology techniques, preventing functional analysis by direct gene manipulation. The feasibility of obtaining oocytes in quantities suitable for experiments, such as using proteomics to identify cell–surface receptors on the oocytes, poses another obstacle (Wright and Bianchi 2016). The challenge of capturing these interactions has hindered the progress in identifying the mechanisms for protein assembly between the sperm and oocyte membranes.

4.4.2 Key Sperm Proteins Are Essential for Fertilization

The identification of factors essential for sperm–oocyte fusion has been enabled by targeted gene knockouts in mice. Initial research on the molecular basis for gamete recognition and fusion was advanced by the isolation of the OBF13 antisperm antibody (Okabe et al. 1988). This antibody was found to inhibit fertilization between murine sperm and oocytes. However, for many years, its antigen remained unknown until the identification of IZUMO1 by mass spectrometry (Inoue et al. 2005). Mice lacking *Izumo1* were sterile due to defects in sperm–oocyte interaction. Furthermore, antihuman IZUMO1 polyclonal antibodies were shown to prevent sperm–oocyte binding. This work demonstrated that sperm IZUMO1 is essential for sperm–oocyte fusion and initiated the discovery of several additional sperm factors in fertilization.

As shown in Fig. 4.3, a collection of additional acrosomal proteins has been identified as essential for fertilization through gene disruption. These proteins include SPACA6 (Barbaux et al. 2020; Lorenzetti et al. 2014; Noda et al. 2020), TMEM95 (Fernandez-Fuertes et al. 2017; Lamas-Toranzo et al. 2020; Noda et al. 2020; Pausch et al. 2014; Tang et al. 2022), TMEM81 (Deneke et al. 2024; Elofsson et al. 2024), FIMP (Fujihara et al. 2020), DCST1/2 (Inoue et al. 2021; Noda et al. 2022), and SOF1 (Noda et al. 2020). C11orf94/Frey is implicated in the assembly of IZUMO1 complexes by retaining IZUMO1 in the endoplasmic reticulum of spermatids during spermiogenesis (Contreras et al. 2022; Hao et al. 2022; Lu et al. 2023). All the known factors are transmembrane proteins except for SOF1, which is a putative secreted soluble protein. Other sperm proteins such as members of the ADAM family, CRISPs (Vjugina and Evans 2008), and Tssk6 (Sosnik et al. 2009) are also implicated in sperm–oocyte fusion. The mechanistic roles these proteins play in sperm–oocyte interaction and membrane fusion require further study.

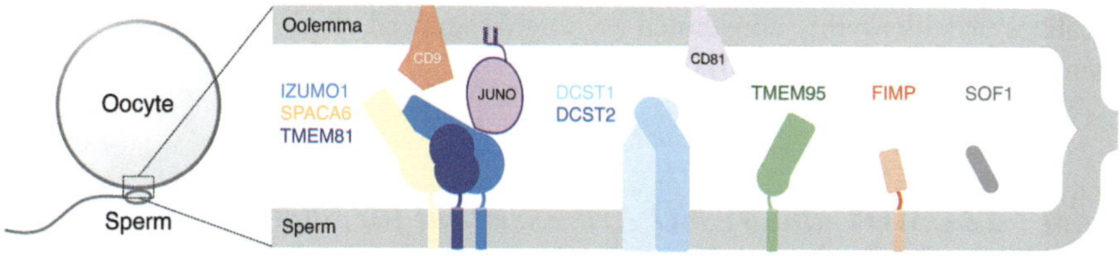

Fig. 4.3 The fertilization synapse and the proteins essential for sperm–oocyte interaction. This diagram illustrates sperm and oocyte proteins involved in membrane binding and fusion. On the oolemma, JUNO (Bianchi et al. 2014) is required for fertilization through interaction with sperm IZUMO1. To date, JUNO remains the only characterized oocyte receptor that forms a receptor–ligand pair with sperm membranes. Two additional oocyte membrane proteins implicated in sperm–oocyte fusion are the tetraspanin proteins CD9 and CD81. *Cd9* knockout female mice exhibit subfertility, while *Cd9/Cd81* double knockout female mice are sterile, indicating that CD9 and CD81 play complementary roles in fertilization (Rubinstein et al. 2006). CD9 is important for sperm–oocyte fusion, as fusion is impaired by *Cd9* knockout oocytes in IVF assays (Kaji et al. 2000; Le Naour et al. 2000; Miyado et al. 2000). On the sperm acrosomal membrane, IZUMO1 (Inoue et al. 2005), SPACA6 (Barbaux et al. 2020; Lorenzetti et al. 2014; Noda et al. 2020), TMEM95 (Fernandez-Fuertes et al. 2017; Lamas-Toranzo et al. 2020; Noda et al. 2020; Pausch et al. 2014; Tang et al. 2022), TMEM81 (Deneke et al. 2024; Elofsson et al. 2024), FIMP (Fujihara et al. 2020), DCST1/2 (Inoue et al. 2021; Noda et al. 2022), and SOF1 (Noda et al. 2020) are essential for sperm–oocyte interaction. All of these are transmembrane proteins except SOF1, which is a putative secreted soluble protein. IZUMO1, SPACA6, and TMEM81 form a heterotrimer on the sperm membrane (Deneke et al. 2024; Elofsson et al. 2024), and DCST1 and DSCT2 form a heterodimer (Inoue et al. 2021; Noda et al. 2022)

4.4.3 Key Oocyte Proteins Are Essential for Fertilization

IZUMO1 was found to bind to the oolemma (Inoue et al. 2013) via its oocyte receptor JUNO (Bianchi et al. 2014). JUNO was discovered using pentameric IZUMO1 against a murine oocyte complementary DNA library, revealing JUNO as a glycosylphosphatidylinositol (GPI)-anchored receptor annotated as folate receptor 4. The binding between IZUMO1 and JUNO was found to be essential for fertilization.

Apart from JUNO, the tetraspanin family proteins CD9 (Kaji et al. 2000; Le Naour et al. 2000; Miyado et al. 2000) and CD81 (Rubinstein et al. 2006) on the oolemma are also essential for fertilization. CD9 and CD81 have complementary roles in sperm–oocyte interaction (Frolikova et al. 2018; Ohnami et al. 2012) with sterility after loss of both genes (Rubinstein et al. 2006). Although experimental evidence is limited, CD9 and CD81 are predicted by AlphaFold to bind to the same region of IZUMO1 (Elofsson et al. 2024). The large extracellular loop of CD9 is important in fertilization, presumably through interactions with unidentified proteins (Umeda et al. 2020). CD9 is important for the activation of the cytosolic protein kinase PTK2B and actin remodeling at the sperm-binding site (Wang et al. 2021a). JUNO and CD9 colocalize on the oolemma (Frolikova et al. 2023). Although the functions of CD9 and CD81 in fertilization remain to be fully understood, CD9 localizes to microvilli, constituted by highly curved, fusion-competent plasma membranes (Runge et al. 2007).

4.4.4 *IZUMO1 and JUNO Are Essential for Sperm–Oocyte Adhesion*

Fertilization involves both sperm–oocyte adhesion and fusion. The molecular architecture of the IZUMO1–JUNO complex has been determined by X-ray crystallography (Aydin et al. 2016; Kato et al. 2016; Ohto et al. 2016). IZUMO1 contains an N-terminal helical bundle domain and an immunoglobulin-like domain, which are connected by a central β-hairpin hinge that provides the receptor-binding site for JUNO. IZUMO1 undergoes a conformational change upon JUNO binding (Aydin et al. 2016; Lu et al. 2025). The IZUMO1–JUNO interaction is important for sperm–oocyte adhesion, but its precise role in membrane fusion is not resolved.

4.4.5 *Sperm Proteins Assemble* **In-Cis** *on the Membrane*

The discovery of a collection of proteins essential for sperm–oocyte interaction, along with the evidence that the IZUMO1–JUNO pairing is insufficient for membrane fusion, has led to a working model of the "fertilization synapse" (Krauchunas et al. 2016). The fertilization synapse is believed to have parallels with cellular synapses, a microenvironment of cell–cell interactions involving adhesion, signaling, secretion, and high-order molecular organization (Krauchunas et al. 2016). This notion is supported by the identification of protein complexes on the sperm membrane. Recognition of monomeric IZUMO1 by JUNO leads to dimerization of IZUMO1, which further interacts with an additional oocyte receptor (Inoue et al. 2015; Inoue and Wada 2018). IZUMO1 is also implicated in *in-cis* interactions with other sperm proteins. IZUMO1 has been shown to form larger multiprotein complexes on the sperm membrane (Ellerman et al. 2009). IZUMO1, TMEM81, and SPACA6 form a trimeric complex on the sperm membrane in zebrafish, and such a complex is predicted in humans and mice as well (Deneke et al. 2024). This trimer is predicted to associate with JUNO and CD9 on the oolemma to form a conserved core fertilization complex (Elofsson et al. 2024).

Another sperm protein complex forms between DCST1 and DCST2, which are involved in gamete recognition and fusion and are well conserved in vertebrates and invertebrates, from humans to nematodes and fruit flies. DCST1 and DCST2 are interdependent but are insufficient to induce sperm–oocyte fusion (Noda et al. 2022) (Inoue et al. 2021).

4.4.6 The Membrane Fusion Machine Driving Sperm–Oocyte Fusion Remains Unknown

A membrane fusion machine is defined by its being both necessary and sufficient to execute membrane fusion. The interaction between IZUMO1 and JUNO is well studied and was previously found to be insufficient for cell–cell fusion (Bianchi et al. 2014). One study suggests that IZUMO1 acts as a membrane fusion machine, as IZUMO1 induced cell–cell fusion unilaterally. Fusion was observed between sperm and Baby Hamster Kidney (BHK) cells overexpressing JUNO (Brukman et al. 2023). This study suggested two functional domains of IZUMO1, one mediating JUNO binding and the other mediating membrane fusion. Additionally, sperm have been found to induce the fusion of JUNO-expressing cells in culture (Brukman et al. 2024). This work suggests that for oocytes to fuse with IZUMO1-expressing cells, additional factors, along with IZUMO1, are required. Furthermore, mammalian cells expressing IZUMO1, FIMP, SPACA6, TMEM95, and SOF1 are unable to fuse with ZP-free oocytes, demonstrating a need for determining other factors contributing to sperm–oocyte fusion (Noda et al. 2020).

A recent study on the mechanism for sperm–oocyte fusion suggests that upon gamete recognition, the oocyte microvilli form lamellipodium-like structures, activate actin polymerization, and engulf sperm through noncanonical phagocytosis for gamete fusion (Inoue et al. 2025). This process involves two phases, with different factors regulating gamete adhesion and sperm engulfment. The first stage involves the IZUMO1–JUNO interaction, triggering JUNO aggregation. The following sperm engulfment activated by the IZUMO1–JUNO linkage and gamete fusion-related factors (SEAL) stage involves the oolemma enveloping the sperm in a phagocytic manner, requiring sperm proteins DCST1, DCST2, SPACA6, TMEM95, FIMP, and TMEM81, and leading to JUNO depletion at the sperm-binding site.

4.5 Oocyte Block to Polyspermy

4.5.1 Modification of the ZP Is Induced After Fertilization

Upon fertilization, polyspermy, or fusion of an oocyte by additional sperm, must be prevented to maintain diploidy of the resulting zygote and ensure proper development. Polyspermy block occurs at both the ZP and oolemma levels. At the ZP level, irreversible modifications to the ZP proteins and their architecture prevent polyspermy. Increased cytosolic Ca^{2+} concentration in the oocyte by the sperm-borne phospholipase C induces the exocytosis of cortical granules to modify the ZP. During exocytosis, the cortical granules fuse with the oolemma and release enzymes into the perivitelline space between the ZP and plasma membrane

(Fahrenkamp et al. 2020). Cortical granule exocytosis involves many single-point fusion events of cortical granules with the oolemma.

The cortical reaction is Ca^{2+}-dependent and, similarly to acrosomal exocytosis, involves molecules that mediate vesicle docking and membrane fusion. These molecules include SNARE proteins, Rab proteins, and Myosin-Va. Members of v-SNARE, such as VAMP and synaptotagmin, are found on vesicle membranes, while members of t-SNARE, such as syntaxin and SNAP-25 (Ikebuchi et al. 1998) are found on target membranes (Liu 2011). Before exocytosis, v-SNARES and Rab bind to t-SNARES for vesicle docking to target membranes. For fusion between the cortical granules and the oolemma, v-SNARE and t-SNARE binding creates a stable protein complex. The SNARE proteins SNAP23, VAMP-1 (t-SNARES), and syntaxin 2 (v-SNARE) likely form a trimeric *trans-* complex prior to fertilization (Tsai et al. 2010). After fertilization, the Ca^{2+} mobilization and related signaling and cytoskeletal rearrangements cause the *trans-* to *cis-* ternary configuration of the trimeric SNARE complex and the dissociation of complexin and clathrin from cortical granules to stimulate endocytosis and membrane recycling (Tsai et al. 2011). Along with the SNAREs, Myosin-Va is involved in the cortical reaction and has been shown to play a role in cortical granule trafficking (Cheeseman et al. 2016).

As a result of cortical granule exocytosis, the ZP proteins are chemically modified, resulting in ZP hardening and more resistance to proteolytic digestion. It is known that the release of zinc ions from cortical granules prevents sperm motility through the zona and precedes complete ZP2 cleavage by ovastacin (Tokuhiro and Dean 2018). Postfertilization cleavage of ZP2 causes ZP2 oligomerization through the N-terminal region to cross-link ZP filaments, hardening the oocyte coat and preventing sperm penetration (Nishio et al. 2024). Moreover, structural information and AlphaFold predictions suggest that protofilaments of ZP proteins form a double helix to create polymers (Nishio et al. 2024). Despite human ZP modification that inhibits sperm penetration, the ZP of fertilized oocytes retains its ability to bind to the sperm and induce intraspermatic calcium influx and the acrosome reaction (Patrat et al. 2006). Additionally, across mammalian species, oocytes are less receptive to fusing with additional sperm found in the perivitelline space after fertilization, suggesting that an additional block at the oolemma may ensure the prevention of polyspermy (Evans 2020).

4.5.2 Molecular Changes at the Oolemma Occur After Fertilization

The mechanisms for the membrane block to polyspermy are still unresolved. It has been observed that JUNO is shed from the oolemma after fertilization, becoming molecular decoys within the perivitelline space (Bianchi et al. 2014). These decoys may neutralize additional acrosome-reacted sperm to block polyspermy. It is unknown whether JUNO shedding is sufficient for the membrane block to polyspermy. At the same time, as JUNO shedding occurs, increases in cytosolic Ca^{2+} concentration facilitate the transition of the oolemma to an unreceptive state for sperm (Evans 2020). Potential mediators to the oolemma block are the oocyte actomyosin cytoskeleton and a receptor–ligand event (Evans 2020).

4.6 Conclusion

This chapter provides an overview of three membrane fusion events in mammalian fertilization. Acrosome exocytosis is the first membrane fusion event, allowing the sperm to reach the oolemma by digesting the ZP. Next, the sperm's plasma membrane at the equatorial region fuses with the oolemma, allowing for delivery of the sperm nucleus to the oocyte. Upon fertilization, the oocyte undergoes blocks to polyspermy in the ZP and on the oolemma, with the last membrane fusion event being the cortical granule exocytosis facilitating the ZP block. Successful fertilization activates the oocyte, triggering the oocyte-to-embryo transition.

The secretions of both the acrosome and the cortical granules involve trimeric SNARE complex formation that causes vesicle docking to the gamete plasma membrane and are Ca^{2+}-dependent exocytosis events. However, the machinery mediating vesicle docking and fusion during these processes requires further clarification.

Between the two exocytotic membrane fusion events lies the sperm–oocyte membrane fusion. Although the precise mechanism is yet to be fully determined, significant progress in identifying the protein factors required for sperm–oocyte interaction has been enabled by CRISPR–Cas9-mediated gene knockouts in mice. This technology, combined with machine learning and artificial intelligence, has provided new insights into the functional organization of the fertilization synapse.

References

Abbott AL, Ducibella T (2001) Calcium and the control of mammalian cortical granule exocytosis. Front Biosci 6:D792–D806. https://doi.org/10.2741/abbott

Ackermann F, Zitranski N, Heydecke D, Wilhelm B, Gudermann T, Boekhoff I (2008) The multi-PDZ domain protein MUPP1 as a lipid raft-associated scaffolding protein controlling the acrosome reaction in mammalian spermatozoa. J Cell Physiol 214(3):757–768. https://doi.org/10.1002/jcp.21272

Ackermann F, Zitranski N, Borth H, Buech T, Gudermann T, Boekhoff I (2009) CaMKIIalpha interacts with multi-PDZ domain protein MUPP1 in spermatozoa and prevents spontaneous acrosomal exocytosis. J Cell Sci 122(Pt 24):4547–4557. https://doi.org/10.1242/jcs.058263

Aydin H, Sultana A, Li S, Thavalingam A, Lee JE (2016) Molecular architecture of the human sperm IZUMO1 and egg JUNO fertilization complex. Nature 534(7608):562–565. https://doi.org/10.1038/nature18595

Baibakov B, Gauthier L, Talbot P, Rankin TL, Dean J (2007) Sperm binding to the zona pellucida is not sufficient to induce acrosome exocytosis. Development 134(5):933–943. https://doi.org/10.1242/dev.02752
Barbaux S, Ialy-Radio C, Chalbi M, Dybal E, Homps-Legrand M, Do Cruzeiro M, Vaiman D, Wolf JP, Ziyyat A (2020) Sperm SPACA6 protein is required for mammalian sperm-egg adhesion/fusion. Sci Rep 10(1):5335. https://doi.org/10.1038/s41598-020-62091-y
Bianchi E, Doe B, Goulding D, Wright GJ (2014) Juno is the egg Izumo receptor and is essential for mammalian fertilization. Nature 508(7497):483–487. https://doi.org/10.1038/nature13203
Biel M, Michalakis S (2009) Cyclic nucleotide-gated channels. Handb Exp Pharmacol 191:111–136. https://doi.org/10.1007/978-3-540-68964-5_7
Brukman NG, Nakajima KP, Valansi C, Flyak K, Li X, Higashiyama T, Podbilewicz B (2023) A novel function for the sperm adhesion protein IZUMO1 in cell-cell fusion. J Cell Biol 222(2). https://doi.org/10.1083/jcb.202207147
Brukman NG, Valansi C, Podbilewicz B (2024) Sperm induction of somatic cell-cell fusion as a novel functional test. elife 13. https://doi.org/10.7554/eLife.94228
Cao Q, Zhao C, Zhang X, Zhang H, Lu Q, Wang C, Hu Y, Ling X, Zhang J, Huo R (2020) Heterozygous mutations in ZP1 and ZP3 cause formation disorder of ZP and female infertility in human. J Cell Mol Med 24(15):8557–8566. https://doi.org/10.1111/jcmm.15482
Castellano LE, Trevino CL, Rodriguez D, Serrano CJ, Pacheco J, Tsutsumi V, Felix R, Darszon A (2003) Transient receptor potential (TRPC) channels in human sperm: expression, cellular localization and involvement in the regulation of flagellar motility. FEBS Lett 541(1–3):69–74. https://doi.org/10.1016/s0014-5793(03)00305-3
Castillo Bennett J, Roggero CM, Mancifesta FE, Mayorga LS (2010) Calcineurin-mediated dephosphorylation of synaptotagmin VI is necessary for acrosomal exocytosis. J Biol Chem 285(34):26269–26278. https://doi.org/10.1074/jbc.M109.095752
Cheeseman LP, Boulanger J, Bond LM, Schuh M (2016) Two pathways regulate cortical granule translocation to prevent polyspermy in mouse oocytes. Nat Commun 7:13726. https://doi.org/10.1038/ncomms13726
Chen C, Sathananthan AH (1986) Early penetration of human sperm through the vestments of human eggs in vitro. Arch Androl 16(3):183–197. https://doi.org/10.3109/01485018608986940
Chiu PC, Wong BS, Lee CL, Pang RT, Lee KF, Sumitro SB, Gupta SK, Yeung WS (2008) Native human zona pellucida glycoproteins: purification and binding properties. Hum Reprod 23(6):1385–1393. https://doi.org/10.1093/humrep/den047
Clark GF (2014) A role for carbohydrate recognition in mammalian sperm-egg binding. Biochem Biophys Res Commun 450(3):1195–1203. https://doi.org/10.1016/j.bbrc.2014.06.051
Cohen R, Buttke DE, Asano A, Mukai C, Nelson JL, Ren D, Miller RJ, Cohen-Kutner M, Atlas D, Travis AJ (2014) Lipid modulation of calcium flux through CaV2.3 regulates acrosome exocytosis and fertilization. Dev Cell 28(3):310–321. https://doi.org/10.1016/j.devcel.2014.01.005
Contreras W, Wiesehofer C, Schreier D, Leinung N, Peche P, Wennemuth G, Gentzel M, Schroder B, Mentrup T (2022) C11orf94/Frey is a key regulator for male fertility by controlling Izumo1 complex assembly. Sci Adv 8(32):eabo6049. https://doi.org/10.1126/sciadv.abo6049
Coy P, Garcia-Vazquez FA, Visconti PE, Aviles M (2012) Roles of the oviduct in mammalian fertilization. Reproduction 144(6):649–660. https://doi.org/10.1530/REP-12-0279
Deneke VE, Blaha A, Lu Y, Suwita JP, Draper JM, Phan CS, Panser K, Schleiffer A, Jacob L, Humer T, Stejskal K, Krssakova G, Roitinger E, Handler D, Kamoshita M, Vance TDR, Wang X, Surm JM, Moran Y, Lee JE, Ikawa M, Pauli A (2024) A conserved fertilization complex bridges sperm and egg in vertebrates. Cell 187(25):7066–7078. https://doi.org/10.1016/j.cell.2024.09.035
Dragileva E, Rubinstein S, Breitbart H (1999) Intracellular Ca(2+)-Mg(2+)-ATPase regulates calcium influx and acrosomal exocytosis in bull and ram spermatozoa. Biol Reprod 61(5):1226–1234. https://doi.org/10.1095/biolreprod61.5.1226

Ellerman DA, Pei J, Gupta S, Snell WJ, Myles D, Primakoff P (2009) Izumo is part of a multiprotein family whose members form large complexes on mammalian sperm. Mol Reprod Dev 76(12): 1188–1199. https://doi.org/10.1002/mrd.21092

Elofsson A, Han L, Bianchi E, Wright GJ, Jovine L (2024) Deep learning insights into the architecture of the mammalian egg-sperm fusion synapse. elife 13. https://doi.org/10.7554/eLife.93131

Evans JP (2020) Preventing polyspermy in mammalian eggs-contributions of the membrane block and other mechanisms. Mol Reprod Dev 87(3):341–349. https://doi.org/10.1002/mrd.23331

Fahrenkamp E, Algarra B, Jovine L (2020) Mammalian egg coat modifications and the block to polyspermy. Mol Reprod Dev 87(3):326–340. https://doi.org/10.1002/mrd.23320

Fernandez-Fuertes B, Laguna-Barraza R, Fernandez-Gonzalez R, Gutierrez-Adan A, Blanco-Fernandez A, O'Doherty AM, Di Fenza M, Kelly AK, Kolle S, Lonergan P (2017) Subfertility in bulls carrying a nonsense mutation in transmembrane protein 95 is due to failure to interact with the oocyte vestments. Biol Reprod 97(1):50–60. https://doi.org/10.1093/biolre/iox065

Finkelstein M, Etkovitz N, Breitbart H (2020) Ca(2+) signaling in mammalian spermatozoa. Mol Cell Endocrinol 516:110953. https://doi.org/10.1016/j.mce.2020.110953

Frolikova M, Manaskova-Postlerova P, Cerny J, Jankovicova J, Simonik O, Pohlova A, Secova P, Antalikova J, Dvorakova-Hortova K (2018) CD9 and CD81 interactions and their structural modelling in sperm prior to fertilization. Int J Mol Sci 19(4). https://doi.org/10.3390/ijms19041236

Frolikova M, Sur VP, Novotny I, Blazikova M, Vondrakova J, Simonik O, Ded L, Valaskova E, Koptasikova L, Benda A, Postlerova P, Horvath O, Komrskova K (2023) Juno and CD9 protein network organization in oolemma of mouse oocyte. Front Cell Dev Biol 11:1110681. https://doi.org/10.3389/fcell.2023.1110681

Fujihara Y, Lu Y, Noda T, Oji A, Larasati T, Kojima-Kita K, Yu Z, Matzuk RM, Matzuk MM, Ikawa M (2020) Spermatozoa lacking fertilization influencing membrane protein (FIMP) fail to fuse with oocytes in mice. Proc Natl Acad Sci USA 117(17):9393–9400. https://doi.org/10.1073/pnas.1917060117

Gadella BM (2010) Interaction of sperm with the zona pellucida during fertilization. Soc Reprod Fertil Suppl 67:267–287. https://doi.org/10.7313/upo9781907284991.023

Ganguly A, Bansal P, Gupta T, Gupta SK (2010) 'ZP domain' of human zona pellucida glycoprotein-1 binds to human spermatozoa and induces acrosomal exocytosis. Reprod Biol Endocrinol 8:110. https://doi.org/10.1186/1477-7827-8-110

Gao Y, Xue R, Guo R, Yang F, Sha X, Li Y, Hua R, Li G, Shen Q, Li K, Liu W, Xu Y, Zhou P, Wei Z, Zhang Z, Cao Y, He X, Wu H (2024) CALR3 defects disrupt sperm-zona pellucida binding in humans: new insights into male factor fertilization failure and relevant clinical therapeutic approaches. Hum Reprod 39(11):2608–2617. https://doi.org/10.1093/humrep/deae205

Gees M, Colsoul B, Nilius B (2010) The role of transient receptor potential cation channels in Ca2+ signaling. Cold Spring Harb Perspect Biol 2(10):a003962. https://doi.org/10.1101/cshperspect.a003962

Hao H, Shi B, Zhang J, Dai A, Li W, Chen H, Ji W, Gong C, Zhang C, Li J, Chen L, Yao B, Hu P, Yang H, Brosius J, Lai S, Shi Q, Deng C (2022) The vertebrate- and testis- specific transmembrane protein C11ORF94 plays a critical role in sperm-oocyte membrane binding. Mol Biomed 3(1):27. https://doi.org/10.1186/s43556-022-00092-1

Henkel R, Franken DR, Habenicht UF (1998) Zona pellucida as physiological trigger for the induction of acrosome reaction. Andrologia 30(4–5):275–280. https://doi.org/10.1111/j.1439-0272.1998.tb01171.x

Hirose M, Honda A, Fulka H, Tamura-Nakano M, Matoba S, Tomishima T, Mochida K, Hasegawa A, Nagashima K, Inoue K, Ohtsuka M, Baba T, Yanagimachi R, Ogura A (2020) Acrosin is essential for sperm penetration through the zona pellucida in hamsters. Proc Natl Acad Sci USA 117(5):2513–2518. https://doi.org/10.1073/pnas.1917595117

Hoshi K, Sugano T, Endo C, Yoshimatsu N, Yanagida K, Sato A (1993) Induction of the acrosome reaction in human spermatozoa by human zona pellucida and effect of cervical mucus on zona-induced acrosome reaction. Fertil Steril 60(1):149–153. https://doi.org/10.1016/s0015-0282(16)56053-1

Hua R, Xue R, Liu Y, Li Y, Sha X, Li K, Gao Y, Shen Q, Lv M, Xu Y, Zhang Z, He X, Cao Y, Wu H (2023) ACROSIN deficiency causes total fertilization failure in humans by preventing the sperm from penetrating the zona pellucida. Hum Reprod 38(6):1213–1223. https://doi.org/10.1093/humrep/dead059

Hutt DM, Baltz JM, Ngsee JK (2005) Synaptotagmin VI and VIII and syntaxin 2 are essential for the mouse sperm acrosome reaction. J Biol Chem 280(21):20197–20203. https://doi.org/10.1074/jbc.M412920200

Ikawa M, Tokuhiro K, Yamaguchi R, Benham AM, Tamura T, Wada I, Satouh Y, Inoue N, Okabe M (2011) Calsperin is a testis-specific chaperone required for sperm fertility. J Biol Chem 286(7):5639–5646. https://doi.org/10.1074/jbc.M110.140152

Ikebuchi Y, Masumoto N, Matsuoka T, Yokoi T, Tahara M, Tasaka K, Miyake A, Murata Y (1998) SNAP-25 is essential for cortical granule exocytosis in mouse eggs. Am J Phys 274(6):C1496–C1500. https://doi.org/10.1152/ajpcell.1998.274.6.C1496

Im SB, Song HN, Jeong TK, Kim N, Kim K, Park SJ, Oh BH (2025) Cryo-EM structure of human hyaluronidase PH-20. Proteins 93(5):1067–1073. https://doi.org/10.1002/prot.26788

Inoue N, Wada I (2018) Monitoring dimeric status of IZUMO1 during the acrosome reaction in living spermatozoon. Cell Cycle 17(11):1279–1285. https://doi.org/10.1080/15384101.2018.1489181

Inoue N, Ikawa M, Isotani A, Okabe M (2005) The immunoglobulin superfamily protein Izumo is required for sperm to fuse with eggs. Nature 434(7030):234–238. https://doi.org/10.1038/nature03362

Inoue N, Satouh Y, Ikawa M, Okabe M, Yanagimachi R (2011) Acrosome-reacted mouse spermatozoa recovered from the perivitelline space can fertilize other eggs. Proc Natl Acad Sci USA 108(50):20008–20011. https://doi.org/10.1073/pnas.1116965108

Inoue N, Hamada D, Kamikubo H, Hirata K, Kataoka M, Yamamoto M, Ikawa M, Okabe M, Hagihara Y (2013) Molecular dissection of IZUMO1, a sperm protein essential for sperm-egg fusion. Development 140(15):3221–3229. https://doi.org/10.1242/dev.094854

Inoue N, Hagihara Y, Wright D, Suzuki T, Wada I (2015) Oocyte-triggered dimerization of sperm IZUMO1 promotes sperm-egg fusion in mice. Nat Commun 6:8858. https://doi.org/10.1038/ncomms9858

Inoue N, Hagihara Y, Wada I (2021) Evolutionarily conserved sperm factors, DCST1 and DCST2, are required for gamete fusion. elife 10. https://doi.org/10.7554/eLife.66313

Inoue N, Saito T, Wada I (2025) Noncanonical phagocytosis-like SEAL establishes mammalian fertilization. Cell Rep 44(4):115463. https://doi.org/10.1016/j.celrep.2025.115463

Ishikawa-Yamauchi Y, Emori C, Mori H, Endo T, Kobayashi K, Watanabe Y, Sagara H, Nagata T, Motooka D, Ninomiya A, Ozawa M, Ikawa M (2024) Age-associated aberrations of the cumulus-oocyte interaction and in the zona pellucida structure reduce fertility in female mice. Commun Biol 7(1):1692. https://doi.org/10.1038/s42003-024-07305-z

Jin M, Fujiwara E, Kakiuchi Y, Okabe M, Satouh Y, Baba SA, Chiba K, Hirohashi N (2011) Most fertilizing mouse spermatozoa begin their acrosome reaction before contact with the zona pellucida during in vitro fertilization. Proc Natl Acad Sci USA 108(12):4892–4896. https://doi.org/10.1073/pnas.1018202108

Kaji K, Oda S, Shikano T, Ohnuki T, Uematsu Y, Sakagami J, Tada N, Miyazaki S, Kudo A (2000) The gamete fusion process is defective in eggs of Cd9-deficient mice. Nat Genet 24(3):279–282. https://doi.org/10.1038/73502

Kato K, Satouh Y, Nishimasu H, Kurabayashi A, Morita J, Fujihara Y, Oji A, Ishitani R, Ikawa M, Nureki O (2016) Structural and functional insights into IZUMO1 recognition by JUNO in mammalian fertilization. Nat Commun 7:12198. https://doi.org/10.1038/ncomms12198

Kim E, Baba D, Kimura M, Yamashita M, Kashiwabara S, Baba T (2005) Identification of a hyaluronidase, Hyal5, involved in penetration of mouse sperm through cumulus mass. Proc Natl Acad Sci USA 102(50):18028–18033. https://doi.org/10.1073/pnas.0506825102

Kim E, Yamashita M, Kimura M, Honda A, Kashiwabara S, Baba T (2008) Sperm penetration through cumulus mass and zona pellucida. Int J Dev Biol 52(5–6):677–682. https://doi.org/10.1387/ijdb.072528ek

Krasznai Z, Krasznai ZT, Morisawa M, Bazsane ZK, Hernadi Z, Fazekas Z, Tron L, Goda K, Marian T (2006) Role of the Na+/Ca2+ exchanger in calcium homeostasis and human sperm motility regulation. Cell Motil Cytoskeleton 63(2):66–76. https://doi.org/10.1002/cm.20108

Krauchunas AR, Marcello MR, Singson A (2016) The molecular complexity of fertilization: introducing the concept of a fertilization synapse. Mol Reprod Dev 83(5):376–386. https://doi.org/10.1002/mrd.22634

Lamas-Toranzo I, Hamze JG, Bianchi E, Fernandez-Fuertes B, Perez-Cerezales S, Laguna-Barraza-R, Fernandez-Gonzalez R, Lonergan P, Gutierrez-Adan A, Wright GJ, Jimenez-Movilla M, Bermejo-Alvarez P (2020) TMEM95 is a sperm membrane protein essential for mammalian fertilization. elife 9. https://doi.org/10.7554/eLife.53913

Le Naour F, Rubinstein E, Jasmin C, Prenant M, Boucheix C (2000) Severely reduced female fertility in CD9-deficient mice. Science 287(5451):319–321. https://doi.org/10.1126/science.287.5451.319

Leung ETY, Lee BKM, Lee CL, Tian X, Lam KKW, Li RHW, Ng EHY, Yeung WSB, Ou JP, Chiu PCN (2023) The role of spermatozoa-zona pellucida interaction in selecting fertilization-competent spermatozoa in humans. Front Endocrinol (Lausanne) 14:1135973. https://doi.org/10.3389/fendo.2023.1135973

Leung ETY, Mei X, Lee BKM, Lam KKW, Lee CL, Li RHW, Ng EHY, Yeung WSB, Yu L, Chiu PCN (2025) Automatic identification of human spermatozoa with zona pellucida-binding capability using deep learning. Hum Reprod Open 2025(3):hoaf024. https://doi.org/10.1093/hropen/hoaf024

Lin M, Ling P, He Q, Chen D, Zheng L, Tang L, Jiang SW (2023) Low acrosin activity is associated with decreased Spam1/acrosin expression and GSH deficiency-caused premature acrosome release of human sperm cells. Cell Tissue Res 394(3):529–545. https://doi.org/10.1007/s00441-023-03826-x

Lishko PV, Botchkina IL, Fedorenko A, Kirichok Y (2010) Acid extrusion from human spermatozoa is mediated by flagellar voltage-gated proton channel. Cell 140(3):327–337. https://doi.org/10.1016/j.cell.2009.12.053

Lishko PV, Botchkina IL, Kirichok Y (2011) Progesterone activates the principal Ca2+ channel of human sperm. Nature 471(7338):387–391. https://doi.org/10.1038/nature09767

Liu M (2011) The biology and dynamics of mammalian cortical granules. Reprod Biol Endocrinol 9:149. https://doi.org/10.1186/1477-7827-9-149

Liu SL, Zuo HY, Zhao BW, Guo JN, Liu WB, Lei WL, Li YY, Ouyang YC, Hou Y, Han ZM, Wang WZ, Sun QY, Wang ZB (2023) A heterozygous ZP2 mutation causes zona pellucida defects and female infertility in mouse and human. iScience 26(10):107828. https://doi.org/10.1016/j.isci.2023.107828

Lopez CI, Belmonte SA, De Blas GA, Mayorga LS (2007) Membrane-permeant Rab3A triggers acrosomal exocytosis in living human sperm. FASEB J 21(14):4121–4130. https://doi.org/10.1096/fj.06-7716com

Lorenzetti D, Poirier C, Zhao M, Overbeek PA, Harrison W, Bishop CE (2014) A transgenic insertion on mouse chromosome 17 inactivates a novel immunoglobulin superfamily gene potentially involved in sperm-egg fusion. Mamm Genome 25(3–4):141–148. https://doi.org/10.1007/s00335-013-9491-x

Lu Y, Shimada K, Tang S, Zhang J, Ogawa Y, Noda T, Shibuya H, Ikawa M (2023) 1700029I15Rik orchestrates the biosynthesis of acrosomal membrane proteins required for sperm-egg interaction. Proc Natl Acad Sci USA 120(8):e2207263120. https://doi.org/10.1073/pnas.2207263120

Lu Y, Ikawa M, Tang S (2025) Allosteric inhibition of the IZUMO1-JUNO fertilization complex by the naturally occurring antisperm antibody OBF13. Proc Natl Acad Sci USA 122(10): e2425952122. https://doi.org/10.1073/pnas.2425952122

Luo G, Zhu L, Liu Z, Yang X, Xi Q, Li Z, Duan J, Jin L, Zhang X (2020) Novel mutations in ZP1 and ZP2 cause primary infertility due to empty follicle syndrome and abnormal zona pellucida. J Assist Reprod Genet 37(11):2853–2860. https://doi.org/10.1007/s10815-020-01926-z

Miyado K, Yamada G, Yamada S, Hasuwa H, Nakamura Y, Ryu F, Suzuki K, Kosai K, Inoue K, Ogura A, Okabe M, Mekada E (2000) Requirement of CD9 on the egg plasma membrane for fertilization. Science 287(5451):321–324. https://doi.org/10.1126/science.287.5451.321

Morales P, Cross NL, Overstreet JW, Hanson FW (1989) Acrosome intact and acrosome-reacted human sperm can initiate binding to the zona pellucida. Dev Biol 133(2):385–392. https://doi.org/10.1016/0012-1606(89)90042-0

Nishio S, Emori C, Wiseman B, Fahrenkamp D, Dioguardi E, Zamora-Caballero S, Bokhove M, Han L, Stsiapanava A, Algarra B, Lu Y, Kodani M, Bainbridge RE, Komondor KM, Carlson AE, Landreh M, de Sanctis D, Yasumasu S, Ikawa M, Jovine L (2024) ZP2 cleavage blocks polyspermy by modulating the architecture of the egg coat. Cell 187(6):1440–1459. https://doi.org/10.1016/j.cell.2024.02.013

Noda T, Lu Y, Fujihara Y, Oura S, Koyano T, Kobayashi S, Matzuk MM, Ikawa M (2020) Sperm proteins SOF1, TMEM95, and SPACA6 are required for sperm-oocyte fusion in mice. Proc Natl Acad Sci USA 117(21):11493–11502. https://doi.org/10.1073/pnas.1922650117

Noda T, Blaha A, Fujihara Y, Gert KR, Emori C, Deneke VE, Oura S, Panser K, Lu Y, Berent S, Kodani M, Cabrera-Quio LE, Pauli A, Ikawa M (2022) Sperm membrane proteins DCST1 and DCST2 are required for sperm-egg interaction in mice and fish. Commun Biol 5(1):332. https://doi.org/10.1038/s42003-022-03289-w

Ohnami N, Nakamura A, Miyado M, Sato M, Kawano N, Yoshida K, Harada Y, Takezawa Y, Kanai S, Ono C, Takahashi Y, Kimura K, Shida T, Miyado K, Umezawa A (2012) CD81 and CD9 work independently as extracellular components upon fusion of sperm and oocyte. Biol Open 1(7):640–647. https://doi.org/10.1242/bio.20121420

Ohto U, Ishida H, Krayukhina E, Uchiyama S, Inoue N, Shimizu T (2016) Structure of IZUMO1-JUNO reveals sperm-oocyte recognition during mammalian fertilization. Nature 534(7608): 566–569. https://doi.org/10.1038/nature18596

Okabe M, Yagasaki M, Oda H, Matzno S, Kohama Y, Mimura T (1988) Effect of a monoclonal anti-mouse sperm antibody (OBF13) on the interaction of mouse sperm with zona-free mouse and hamster eggs. J Reprod Immunol 13:211–219

Okutman O, Demirel C, Tulek F, Pfister V, Buyuk U, Muller J, Charlet-Berguerand N, Viville S (2020) Homozygous splice site mutation in ZP1 causes familial oocyte maturation defect. Genes (Basel) 11(4). https://doi.org/10.3390/genes11040382

Pang PC, Chiu PC, Lee CL, Chang LY, Panico M, Morris HR, Haslam SM, Khoo KH, Clark GF, Yeung WS, Dell A (2011) Human sperm binding is mediated by the sialyl-Lewis(x) oligosaccharide on the zona pellucida. Science 333(6050):1761–1764. https://doi.org/10.1126/science.1207438

Patrat C, Auer J, Fauque P, Leandri RL, Jouannet P, Serres C (2006) Zona pellucida from fertilised human oocytes induces a voltage-dependent calcium influx and the acrosome reaction in spermatozoa, but cannot be penetrated by sperm. BMC Dev Biol 6:59. https://doi.org/10.1186/1471-213X-6-59

Pausch H, Kolle S, Wurmser C, Schwarzenbacher H, Emmerling R, Jansen S, Trottmann M, Fuerst C, Gotz KU, Fries R (2014) A nonsense mutation in TMEM95 encoding a nondescript transmembrane protein causes idiopathic male subfertility in cattle. PLoS Genet 10(1): e1004044. https://doi.org/10.1371/journal.pgen.1004044

Prajapati P, Kane S, McBrinn RC, Dean MS, Martins da Silva SJ, Brown SG (2022) Elevated and sustained intracellular calcium signalling is necessary for efficacious induction of the human sperm acrosome reaction. Int J Mol Sci 23(19). https://doi.org/10.3390/ijms231911253

Ren D, Navarro B, Perez G, Jackson AC, Hsu S, Shi Q, Tilly JL, Clapham DE (2001) A sperm ion channel required for sperm motility and male fertility. Nature 413(6856):603–609. https://doi.org/10.1038/35098027

Rodriguez F, Zanetti MN, Mayorga LS, Tomes CN (2012) Munc18-1 controls SNARE protein complex assembly during human sperm acrosomal exocytosis. J Biol Chem 287(52): 43825–43839. https://doi.org/10.1074/jbc.M112.409649

Rodriguez-Martinez H (2007) Role of the oviduct in sperm capacitation. Theriogenology 68(Suppl 1):S138–S146. https://doi.org/10.1016/j.theriogenology.2007.03.018

Roggero CM, De Blas GA, Dai H, Tomes CN, Rizo J, Mayorga LS (2007) Complexin/synaptotagmin interplay controls acrosomal exocytosis. J Biol Chem 282(36):26335–26343. https://doi.org/10.1074/jbc.M700854200

Rubinstein E, Ziyyat A, Prenant M, Wrobel E, Wolf JP, Levy S, Le Naour F, Boucheix C (2006) Reduced fertility of female mice lacking CD81. Dev Biol 290(2):351–358. https://doi.org/10.1016/j.ydbio.2005.11.031

Runge KE, Evans JE, He ZY, Gupta S, McDonald KL, Stahlberg H, Primakoff P, Myles DG (2007) Oocyte CD9 is enriched on the microvillar membrane and required for normal microvillar shape and distribution. Dev Biol 304(1):317–325. https://doi.org/10.1016/j.ydbio.2006.12.041

Sa SW, Wang LL, Ma QH (2025) Association of Zona Pellucida gene variants with female infertility: a retrospective genetic analysis. BJOG 132(Suppl 2):75–82. https://doi.org/10.1111/1471-0528.18094

Schuffner AA, Bastiaan HS, Duran HE, Lin ZY, Morshedi M, Franken DR, Oehninger S (2002) Zona pellucida-induced acrosome reaction in human sperm: dependency on activation of pertussis toxin-sensitive G(i) protein and extracellular calcium, and priming effect of progesterone and follicular fluid. Mol Hum Reprod 8(8):722–727. https://doi.org/10.1093/molehr/8.8.722

Sosnik J, Miranda PV, Spiridonov NA, Yoon SY, Fissore RA, Johnson GR, Visconti PE (2009) Tssk6 is required for Izumo relocalization and gamete fusion in the mouse. J Cell Sci 122(Pt 15): 2741–2749. https://doi.org/10.1242/jcs.047225

Sun Y, Zeng Y, Chen H, Zhou Z, Fu J, Sang Q, Wang L, Sun X, Chen B, Xu C (2021) A novel homozygous variant in ZP2 causes abnormal zona pellucida formation and female infertility. J Assist Reprod Genet 38(5):1239–1245. https://doi.org/10.1007/s10815-021-02107-2

Sun L, Tong K, Liu W, Tian Y, Liu D, Huang G, Li J (2023) Novel variants in ZP1, ZP2 and ZP3 associated with empty follicle syndrome and abnormal zona pellucida. Reprod Biomed Online 46(5):847–855. https://doi.org/10.1016/j.rbmo.2023.01.010

Tang S, Lu Y, Skinner WM, Sanyal M, Lishko PV, Ikawa M, Kim PS (2022) Human sperm TMEM95 binds eggs and facilitates membrane fusion. Proc Natl Acad Sci USA 119(40): e2207805119. https://doi.org/10.1073/pnas.2207805119

Tokuhiro K, Dean J (2018) Glycan-independent gamete recognition triggers egg zinc sparks and ZP2 cleavage to prevent polyspermy. Dev Cell 46(5):627–640 e625. https://doi.org/10.1016/j.devcel.2018.07.020

Tomes CN, Michaut M, De Blas G, Visconti P, Matti U, Mayorga LS (2002) SNARE complex assembly is required for human sperm acrosome reaction. Dev Biol 243(2):326–338. https://doi.org/10.1006/dbio.2002.0567

Tsai PS, Garcia-Gil N, van Haeften T, Gadella BM (2010) How pig sperm prepares to fertilize: stable acrosome docking to the plasma membrane. PLoS One 5(6):e11204. https://doi.org/10.1371/journal.pone.0011204

Tsai PS, van Haeften T, Gadella BM (2011) Preparation of the cortical reaction: maturation-dependent migration of SNARE proteins, clathrin, and complexin to the porcine oocyte's surface blocks membrane traffic until fertilization. Biol Reprod 84(2):327–335. https://doi.org/10.1095/biolreprod.110.085647

Tumova L, Zigo M, Sutovsky P, Sedmikova M, Postlerova P (2021) Ligands and receptors involved in the sperm-zona Pellucida interactions in mammals. Cells 10(1). https://doi.org/10.3390/cells10010133

Umeda R, Satouh Y, Takemoto M, Nakada-Nakura Y, Liu K, Yokoyama T, Shirouzu M, Iwata S, Nomura N, Sato K, Ikawa M, Nishizawa T, Nureki O (2020) Structural insights into tetraspanin CD9 function. Nat Commun 11(1):1606. https://doi.org/10.1038/s41467-020-15459-7

Vjugina U, Evans JP (2008) New insights into the molecular basis of mammalian sperm-egg membrane interactions. Front Biosci 13:462–476. https://doi.org/10.2741/2693

Wang H, Hong X, Kinsey WH (2021a) Sperm-oocyte signaling: the role of IZUMO1R and CD9 in PTK2B activation and actin remodeling at the sperm binding sitedagger. Biol Reprod 104(6): 1292–1301. https://doi.org/10.1093/biolre/ioab048

Wang Y, Zhao W, Mei S, Chen P, Leung TY, Lee CL, Yeung WSB, Ou JP, Liang X, Chiu PCN (2021b) Identification of Sialyl-Lewis(x)-interacting protein on human spermatozoa. Front Cell Dev Biol 9:700396. https://doi.org/10.3389/fcell.2021.700396

Wassarman PM, Litscher ES (2008) Mammalian fertilization is dependent on multiple membrane fusion events. Methods Mol Biol 475:99–113. https://doi.org/10.1007/978-1-59745-250-2_6

Wei X, Li Y, Liu Q, Liu W, Yan X, Zhu X, Zhou D, Tian Y, Zhang F, Li N, Lu Z (2022) Mutations in ZP4 are associated with abnormal zona pellucida and female infertility. J Clin Pathol 75(3): 201–204. https://doi.org/10.1136/jclinpath-2020-207170

Wright GJ, Bianchi E (2016) The challenges involved in elucidating the molecular basis of sperm-egg recognition in mammals and approaches to overcome them. Cell Tissue Res 363(1): 227–235. https://doi.org/10.1007/s00441-015-2243-3

Xu Q, Zhu X, Maqsood M, Li W, Tong X, Kong S, Wang F, Liu X, Wei Z, Zhang Z, Zhu F, Cao Y, Bao J (2020) A novel homozygous nonsense ZP1 variant causes human female infertility associated with empty follicle syndrome (EFS). Mol Genet Genomic Med 8(7):e1269. https://doi.org/10.1002/mgg3.1269

Yunes R, Michaut M, Tomes C, Mayorga LS (2000) Rab3A triggers the acrosome reaction in permeabilized human spermatozoa. Biol Reprod 62(4):1084–1089. https://doi.org/10.1095/biolreprod62.4.1084

Zanetti N, Mayorga LS (2009) Acrosomal swelling and membrane docking are required for hybrid vesicle formation during the human sperm acrosome reaction. Biol Reprod 81(2):396–405. https://doi.org/10.1095/biolreprod.109.076166

Zeng J, Sun Y, Zhang J, Wu X, Wang Y, Quan R, Song W, Guo D, Wang S, Chen J, Xiao H, Huang HL (2023) Identification of zona pellucida defects revealed a novel loss-of-function mutation in ZP2 in humans and rats. Front Endocrinol (Lausanne) 14:1169378. https://doi.org/10.3389/fendo.2023.1169378

Zhang D, Zhu L, Liu Z, Ren X, Yang X, Li D, Luo Y, Peng X, Zhou X, Jia W, Hou M, Li Z, Jin L, Zhang X (2021) A novel mutation in ZP3 causes empty follicle syndrome and abnormal zona pellucida formation. J Assist Reprod Genet 38(1):251–259. https://doi.org/10.1007/s10815-020-01995-0

Zhao L, Burkin HR, Shi X, Li L, Reim K, Miller DJ (2007a) Complexin I is required for mammalian sperm acrosomal exocytosis. Dev Biol 309(2):236–244. https://doi.org/10.1016/j.ydbio.2007.07.009

Zhao L, Shi X, Li L, Miller DJ (2007b) Dynamin 2 associates with complexins and is found in the acrosomal region of mammalian sperm. Mol Reprod Dev 74(6):750–757. https://doi.org/10.1002/mrd.20660

Zhao S, Zhang R, Wang L, Zhao D, Tang L, Rao M (2024) Acrosin activity negatively influences the cumulative live birth rate in patients undergoing IVF treatment. Reprod Biomed Online 49(3):103993. https://doi.org/10.1016/j.rbmo.2024.103993

Chapter 5
Biophysical Considerations in Cell Fusion

M. Amin Abdolkhani, Alejandro Forigua, and Christopher Moraes

Abstract Physical forces direct cell behaviors such as differentiation, migration, and division. Cell fusion is no exception. In addition to protein-based machinery, mechanical stresses, pressure differentials, and electric fields have been suggested to also regulate fusion. Decades of research have identified numerous fusogenic membrane proteins, yet the precise mechanisms of their action remain unclear. Here, we outline a biophysical framework that links the physics of model membranes to fusion in living cells. We describe how surface free energy, membrane curvature, tension, pressure, and electric fields can play a role in driving fusion, drawing on insights from both model membranes and cellular systems. Finally, we highlight how these biophysical factors contribute to fusion in the context of placental development and how they may inform future work in skeletal muscle, bone tissue, and tumors. By integrating molecular and biophysical perspectives, we provide a starting point to uncover conserved principles of cell fusion and advance a more integrated view of fusion as both a biochemical and biophysical process.

M. A. Abdolkhani
Department of Biological and Biomedical Engineering, McGill University, Montréal, QC, Canada

A. Forigua
Department of Chemical Engineering, McGill University, Montréal, QC, Canada

C. Moraes (✉)
Department of Biological and Biomedical Engineering, McGill University, Montréal, QC, Canada

Department of Chemical Engineering, McGill University, Montréal, QC, Canada

Rosalind and Morris Goodman Cancer Institute, McGill University, Montréal, QC, Canada

Division of Experimental Medicine, McGill University, Montréal, QC, Canada

School of Biomedical Engineering, The University of British Columbia, Vancouver, BC, Canada

International Collaboration on Repair Discoveries (ICORD), Vancouver Coastal Health Research Institute, Vancouver, BC, Canada
e-mail: chris.moraes@ubc.ca

© The Author(s), under exclusive license to Springer Nature Switzerland AG 2026
J. Weiler et al. (eds.), *Cell Fusion*, Advances in Experimental Medicine and Biology 1509, https://doi.org/10.1007/978-3-032-22637-2_5

Keywords Biophysics · Cell fusion · Lipid membranes · Mechanobiology · Membrane tension · Membrane pores · Cytoskeleton dynamics · Trophoblasts · Syncytia

5.1 Introduction

5.1.1 The Role of Biophysical Cues in Cellular Behavior

Biophysical stimuli such as substrate stiffness, electric potential, and fluid forces have significant consequences for the behavior and biology of cells (Zajdel et al. 2020; Marshall et al. 2024; Lim et al. 2010). For example, mesenchymal stem cells differentiate along distinct lineages depending on the rigidity of their substrate (Smith et al. 2018; McBeath et al. 2004): Softer matrices favor neural-like or adipogenic phenotypes, whereas stiffer environments promote osteogenic differentiation (Hu and Bao 2024). In the case of cancer, elevated matrix stiffness actively drives progression by enhancing malignant traits through mechanotransduction pathways such as YAP (Yes-Associated Protein)/TAZ (transcriptional co-activator with PDZ-binding motif) and integrin signaling (Ishihara and Haga 2022; Forigua et al. 2025). Matrix stiffening also influences how cells migrate, as many cancer and stromal cells exhibit durotaxis, preferentially moving from softer toward stiffer regions of the matrix (Lei et al. 2025). Similarly, electromagnetic fields can bias motility through electrotaxis, providing a means of externally directing cell behavior (Zajdel et al. 2020). Mechanical dependence extends across scales, as even isolated actin networks display distinct branching and polymerization behaviors under load, independent of other cellular components (Bieling et al. 2016). Together, these

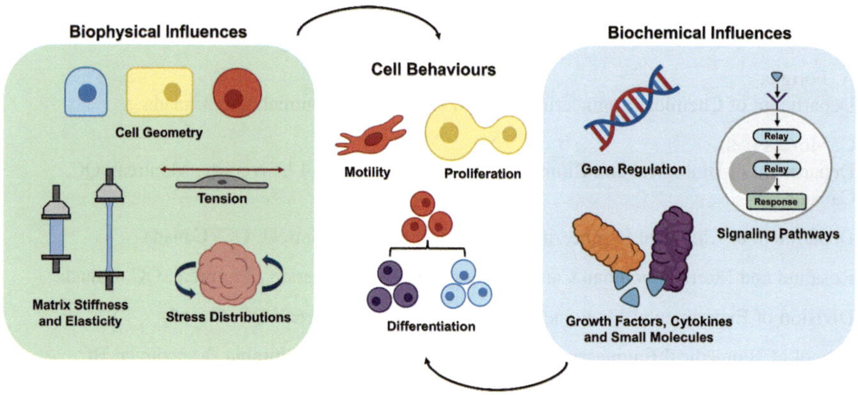

Fig. 5.1 Physicochemical influence of cell behavior. Biophysical influences such as matrix stiffness, elasticity, tension, cell geometry, and stress distributions act alongside biochemical influences such as gene regulation, signaling pathways, and soluble factors to regulate core cell behaviors, including motility, proliferation, and differentiation

examples demonstrate that cellular decision-making is not solely biochemical but emerges from the integration of physical forces with molecular processes (Fig. 5.1).

Early studies in cell biology cataloged proteins and mapped pathways as simplified diagrams of parts and interactions (Marshall et al. 2024). Later work expanded this foundation by demonstrating that cells also sense and respond to their physical surroundings (Hu and Bao 2024; White et al. 2019; Discher et al. 2009; Li et al. 2021). Currently, cell behavior is understood as the integration of biochemical and biophysical processes: Cells interpret molecular cues while simultaneously engaging with forces, geometry, and material properties. This integrated perspective explains phenomena such as stem cell differentiation and cancer progression, with ample evidence supporting its relevance (White et al. 2019; Discher et al. 2009). The same framework offers a natural lens for understanding cell fusion, where proteins and physical forces act together to drive the process.

5.1.2 Molecular Machinery in Cell Fusion

Cell fusion is described as being driven by fusogenic membrane proteins (Brukman et al. 2019; Sieler and Dittmar 2024; Dittmar and Zänker 2011). Protein signaling and expression are considered both necessary and sufficient for the process in many systems (Brukman et al. 2019; Whitlock and Chernomordik 2025; Podbilewicz 2014). This view has shaped how the field approaches fusion, often focusing primarily on the presence and activity of these proteins. A canonical sequence is often described: Soluble factors and signaling pathways first prime cells for fusion, followed by adhesion and close membrane apposition (Fig. 5.2). Fusogenic proteins

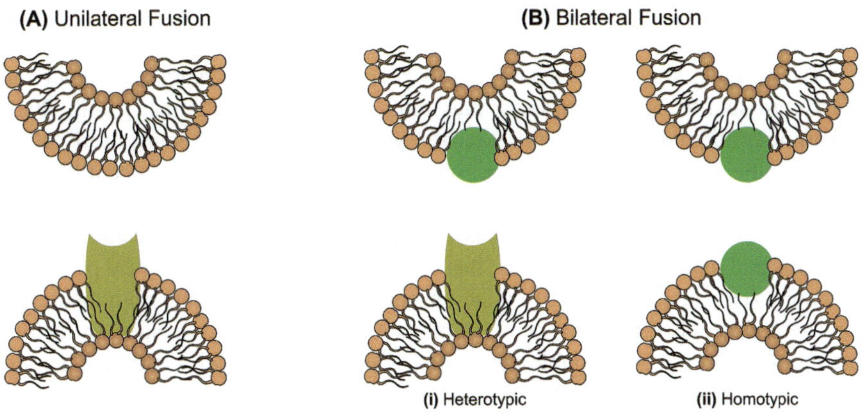

Fig. 5.2 Classifications of cell fusion events. (**a**) Unilateral fusion occurs when only one cell expresses fusogens. (**b**) Bilateral fusion occurs when both cells express fusogens and can be further classified as (*i*) heterotypic, when different fusogens are involved, or (*ii*) homotypic, when the same fusogens are expressed

then remodel membranes into hemifusion intermediates, leading to pore formation and expansion that establish cytoplasmic continuity. This interpretation, reinforced by early studies of fusion phenomena where proteins dominate the process, continues to anchor much of the discussion. At the same time, classifications of fusion events remain centered on fusogenic proteins, distinguishing between unilateral (Fig. 5.2a) and bilateral (Fig. 5.2b) fusion according to fusogen expression on opposing cell membranes (Brukman et al. 2019; Yang and Margam 2021; Ma et al. 2019; Parameshwar et al. 2024). In unilateral fusion, fusogens are present on a single cell, engaging a partner membrane that is not expressing a fusogen (Fig. 5.2a). In contrast, bilateral fusion occurs when both membranes display fusogens and can be further divided into heterotypic fusion (Fig. 5.2bi), where different cell types express complementary fusogens, and homotypic fusion (Fig. 5.2bii), where the same cell type fuses through fusogens shared across like cells.

This protein-centered perspective is reflected across diverse systems. In osteoclastogenesis, macrophages fuse to form multinucleated osteoclasts, and fusogens such as dendritic cell-specific transmembrane protein (DC-STAMP) and osteoclast stimulatory transmembrane protein (OC-STAMP) are crucial for this process (Pereira et al. 2018; Vignery 2005; McNally and Anderson 2011). Numerous cytokines, transcription factors, and signaling pathways have also been identified, but the accompanying biophysical interactions remain less clear. A comparable emphasis on protein regulators is seen in placental trophoblasts, where syncytins, their receptors, and transcriptional regulators drive syncytialization, the fusion of cytotrophoblasts (CTBs) into the multinucleated syncytiotrophoblast (STB) layer (Renaud and Jeyarajah 2022). In skeletal muscle, myoblast fusion depends on membrane proteins such as myomaker and myomerger, which have been studied in detail for how they reshape membrane curvature (Whitlock and Chernomordik 2025; Simionescu and Pavlath 2011; Leikina et al. 2018). However, even in this inherently contractile and mechanically active tissue, the contribution of physical forces to fusion is still not fully understood.

The defining step of cell fusion involves conformational changes in membrane curvature and lipid organization that unify two cells (Brukman et al. 2019; Chernomordik and Kozlov 2008). Across systems, fusogenic proteins catalyze and stabilize this transition, but the same conformational changes can also be influenced by physical factors. Therefore, proteins can be understood to act in tandem with biophysical properties of membranes to achieve fusion. Given the central role of proteins and the emerging evidence that physical factors shape biological processes, a biophysical perspective on cell fusion becomes essential. Insights from theoretical and experimental models of lipid membrane fusion can be extended to cellular membranes to generate testable predictions about how mechanics and the microenvironment influence fusion. At the same time, important gaps remain, including why some cell types lack identifiable genes that directly control fusion (Pereira et al. 2018). These considerations frame the discussion that follows, which explores how biochemical and biophysical cues act together to drive fusion.

5.2 Understanding the Cell Membrane During Fusion

The intrinsic properties of the cell membrane play a key role during fusion, making it important to consider how bilayer properties influence the process. As membranes move through close contact, stalk formation, and pore opening, they are shaped by forces related to curvature, tension, adhesion, pressure, and physical factors. These physical influences act alongside fusogenic proteins, creating the conditions that allow fusion to proceed. To provide a common biophysical framework, we discuss membrane energetics as a basis for comparing parameters such as surface free energy (SFE), curvature, tension, pore stability, pressure, adhesion, and electric fields, and for assessing how each may contribute to cell fusion.

5.2.1 Surface Free Energy of Lipid Membranes

The SFE of a lipid bilayer represents the energy required to create or deform the membrane interface, expressed as energy per unit area. This energy arises from a balance of multiple interactions, including electrostatic forces, van der Waals interactions, and hydrophobic effects between lipid chains and water molecules (Coster 2003). Since lipid bilayers are primarily composed of phospholipids, amphiphilic molecules with a polar headgroup attached to nonpolar fatty acid chains, their molecular structure largely determines membrane energetics. The amphiphilic nature of phospholipids gives rise to three dominant classes of interactions: headgroup–headgroup, headgroup–solvent, and chain–chain (Ben-Shaul 1995). Each of these interactions is sensitive to intramolecular parameters, determined by the molecular structure of the lipid, such as headgroup size and charge, the headgroup-to-chain volume ratio, and chain length, and to intermolecular parameters, arising from interactions with neighboring lipids and the surrounding environment, including lipid spacing and packing, ionic strength of the solution, and local curvature (Coster 2003; Ben-Shaul 1995; Lee 2004).

A membrane displaced from equilibrium therefore stores excess energy, which can be released by changes in geometry such as bending, stretching, or fusion. Because it integrates many of these molecular contributions into a single measurable value, SFE provides a practical parameter for quantifying bilayer stability under different conditions. SFE is experimentally accessible through approaches such as tethered bilayer tensiometry and model lipid membranes, allowing comparisons across systems that differ in lipid composition, curvature, ionic environment, or cytoskeletal coupling (Sun et al. 2011; Barlow et al. 2018). Viewed in this way, SFE serves as an integrative measure that links diverse physical factors to membrane behavior and provides a basis for predicting how biophysical conditions may bias the likelihood of cell fusion.

5.2.2 *The Stalk Hypothesis of Membrane Fusion*

The stalk hypothesis has become the most widely used framework for describing lipid membrane fusion because it offers a physically plausible and broadly applicable pathway for bilayers to merge (Chernomordik and Kozlov 2008; Kozlovsky and Kozlov 2002). As shown in Fig. 5.3, central to the model is the formation of a stalk, a short-lived intermediate in which the outer leaflets of two bilayers reorganize to create the first point of continuity between membranes (Chernomordik and Kozlov 2008; Tamm et al. 2003). This mechanistic description is considered plausible since it requires energy inputs within biologically reasonable limits, and it is quantitatively tractable, as the associated transitions can be analyzed using continuum elastic theory to estimate the energy barriers involved (Kozlovsky and Kozlov 2002; Golani and Schwarz 2023). Intermediates predicted by the stalk hypothesis, including stalk-like structures, have been inferred or observed using multiple methods such as direct sub-nanometer mapping by X-ray diffraction (Aeffner et al. 2012), in situ imaging by cryo-electron microscopy (Fontana et al. 2012), and scattering techniques such as small-angle X-ray scattering (SAXS) in model lipid membranes (Khattari et al. 2015).

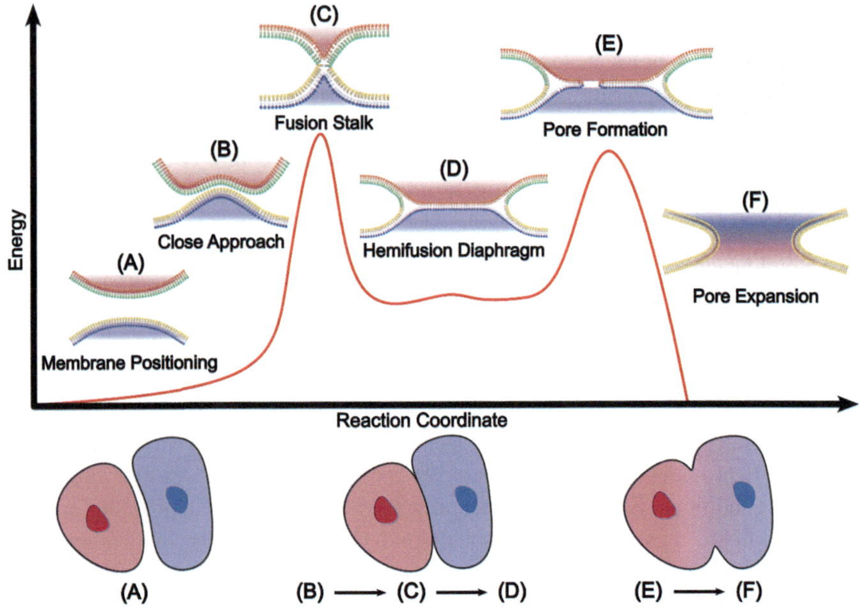

Fig. 5.3 Energy landscape of the stalk model of membrane fusion. Fusion proceeds through a series of intermediates: (**a**) membrane positioning, (**b**) close approach of bilayers, (**c**) formation of a stalk that creates the first continuity between leaflets, (**d**) expansion into a hemifusion diaphragm, (**e**) pore nucleation within the diaphragm, and (**f**) pore expansion that completes fusion. The red line represents the relative energy profile along this pathway, with two major energy barriers dominating the process: stalk formation (**c**) and pore nucleation (**e**)

According to this stalk model, the sequence begins with membrane positioning (Fig. 5.3a) and close approach of the bilayers (Fig. 5.3b) (Chernomordik and Kozlov 2008; Pandia and Chakraborty 2024). At this stage, strong repulsive forces, including hydration repulsion from interfacial water and electrostatic interactions between the charged headgroups (Kozlov et al. 2010), must be overcome to allow the membranes to reach sufficiently close proximity. After this step, lipids reorganize into the stalk (Fig. 5.3c), which represents the first local continuity between the bilayers. The stalk can then expand into a hemifusion diaphragm (Fig. 5.3d), in which the inner leaflets merge to form a continuous bilayer. Fusion is completed during pore formation when a pore nucleates within this diaphragm (Fig. 5.3e) and subsequently expands (Fig. 5.3f), establishing cytoplasmic continuity.

The energy landscape shown in Fig. 5.3 emphasizes two major barriers along the fusion pathway. The first corresponds to the formation of the stalk (Fig. 5.3c) and the second to the nucleation of a critically large pore within the hemifusion diaphragm (Fig. 5.3e). Progress through these transitions requires additional energy input, which cells can provide directly through fusogenic proteins that hydrolyze adenosine triphosphate (ATP) or guanosine triphosphate (GTP) (Yang and Margam 2021; Jahraus et al. 2001) or indirectly by tuning membrane composition, curvature, or tension. Because these steps largely determine whether fusion proceeds to completion, they serve as natural points where biophysical factors can modulate the process. In the following sections, we examine how such parameters influence the energy of the lipid membrane and thereby affect fusion efficiency.

5.2.3 *The Role of Membrane Curvature in Fusion*

Among the physical factors that influence the energy barriers in Fig. 5.3, membrane curvature plays a central role. The geometry and structure of individual lipid monomers determine how bilayers bend, which, in turn, affects the energy required to form intermediates such as the stalk or the hemifusion diaphragm (Coster 2003). Differences in headgroup size, charge, and tail volume give rise to lipids with distinct spontaneous curvatures and associated surface energies (Hamm and Kozlov 2000). As a result, bilayers exhibit characteristic equilibrium curvatures under no-stress conditions, and deviations from this state, whether driven by cellular machinery or experimental manipulation, increase membrane free energy (Cooke and Deserno 2006; Nikolaus et al. 2011). Figure 5.4 illustrates how the balance of intermolecular forces between lipid headgroups and tails can produce negative, positive, or zero spontaneous curvature.

Curvature is closely linked to fusion because the relevant intermediates contain regions of high local curvature. Such conformations store elastic energy that can be released as membranes transition between states. Simulations support this view by showing that increased local curvature raises membrane free energy and can bias transitions that dissipate that energy (Golani and Schwarz 2023). Experimental observations support the same principle. For example, in mast cells, highly curved

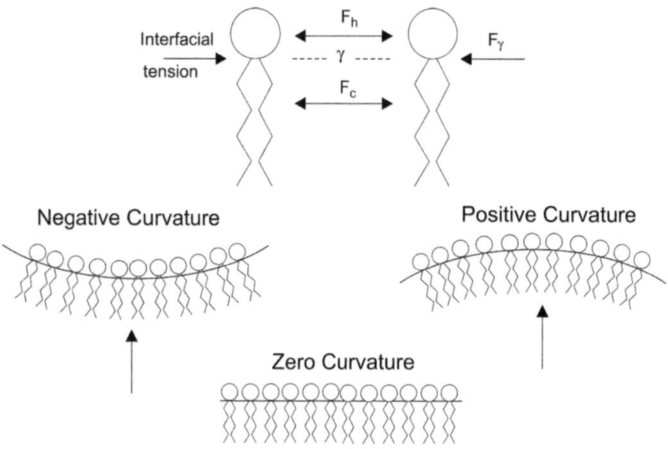

Fig. 5.4 Schematic representation of how intermolecular forces acting on lipid head and tail groups generate different spontaneous curvatures in membranes. Negative curvature arises when lipids favor inward bending, positive curvature when they favor outward bending, and zero curvature when forces are balanced (Lee 2004). (Reproduced from BBA—Biomembranes, 1666 (Zajdel et al. 2020; Marshall et al. 2024))

membrane dimples precede vesicle fusion with the plasma membrane (Monck and Fernandez 1996), and in polyethylene glycol (PEG)-mediated fusion of lipid vesicles, large curvature has also been shown to favor fusion (Lentz 1994). A further example is provided by phosphatidylserine (PS), a naturally occurring lipid that modulates curvature (Hirama et al. 2017; Whitlock and Chernomordik 2021), since its translocation from the inner to the outer leaflet of the plasma membrane is required for trophoblast fusion in placental choriocarcinoma BeWo cells (Das et al. 2004).

Both the curvature imposed by external forces and the spontaneous curvature of membranes play roles in cell fusion. Geometric curvature, imposed externally, describes the actual shape of a membrane under applied forces, while spontaneous curvature refers to the equilibrium shape based on lipid composition. As a result, two membranes may display the same geometric curvature and yet differ in their fusogenic potential if their spontaneous curvatures are different. In biological membranes, enrichment of lipids with negative spontaneous curvature generally facilitates fusion, whereas lipids with positive spontaneous curvature tend to oppose it (Joardar et al. 2022; Chernomordik and Kozlov 2003; Epand and Epand 2000). This principle is evident in several systems. For example, lysophosphatidylcholine (LPC), which imposes positive spontaneous curvature, inhibits fusion in viral entry and other pH-dependent fusion events (Ohki et al. 2006; Dymond 2021). In contrast, asymmetric enrichment of phosphatidylethanolamine (PE), a lipid with negative spontaneous curvature, has been shown to promote fusion in multiple contexts, including osteoclasts (Dymond 2021; Pantazatos and MacDonald 1999; Irie et al. 2017).

Spontaneous curvature varies across the plasma membrane as a function of local composition, leaflet asymmetry, and protein crowding (Kluge et al. 2022). Together with cytoskeletal and other biomechanical stresses, these variations create spatially restricted regions that either favor or resist stalk formation and pore nucleation. Effective fusion, therefore, requires local regulation of curvature in both space and time. While curvature is a central determinant of fusion, it acts alongside other biophysical influences that also shape the energetic landscape. These factors are often more context-dependent, but they can significantly alter the probability of stalk formation or pore nucleation under specific conditions.

5.2.4 *Other Biophysical Factors Influencing Fusion*

In addition to the core determinants of surface free energy, the stalk pathway, and membrane curvature, several other physical parameters contribute to membrane fusion. Their influence is typically modulatory rather than primary, but in specific biological or experimental contexts, they can become critical. These factors bias whether membranes overcome the energy barriers of stalk formation and pore expansion and, in doing so, shape the overall efficiency of fusion.

5.2.4.1 Membrane Tension

Membrane tension reflects the cohesive force that resists deformation of the lipid bilayer, analogous to the stiffness of a spring. It arises from stretching, compression, or curvature stresses imposed on the membrane and represents elastic energy stored in the bilayer. Similar to curvature, membrane surface stress arises from deviations in a bilayer's spontaneous shape, creating a distribution of forces across the plasma membrane (Shearman et al. 2007). The stress profile of a membrane is therefore tightly linked to curvature, since membranes with induced curvature are simultaneously under stress. In addition to curvature, lateral stretching and compression forces (i.e., lateral stress) also contribute to stress within the bilayer (Golani and Schwarz 2023). Through the action of membrane proteins or the cytoskeleton, membranes experience "stored elastic energy" arising from these deformations (Argudo et al. 2016). Membrane surface tension itself is an inherent property defined as the cohesive force that holds lipid monomers together (Mahbubul 2019). The higher this tension, the greater the resistance to deformation and the energy required to form highly curved intermediates.

Tension is therefore a two-sided factor in fusion. High surface tension resists bending into stalks or pores, yet once a membrane is stressed, the release of stored elastic energy can facilitate these transitions. Local stress also provides a biophysical cue, recruiting mechanosensitive proteins and curvature-modifying lipids that can stabilize intermediates (Kozlov et al. 2010; Shearman et al. 2007; Grafmüller et al. 2009; Akimov et al. 2017a; Fuhrmans et al. 2015). In this way, membrane tension contributes directly to the energy demands of the stalk model while also modulating protein organization at fusion sites.

Experimental and biological systems provide support for these roles. Simulations and vesicle experiments suggest that fusion requires induction of a high-stress state that is then relieved by stalk formation (Chernomordik and Kozlov 2008; Shearman et al. 2007; Koshiyama and Wada 2011; Oglęcka et al. 2014; Kim et al. 2015a). Stretching stress above a critical threshold drives vesicle fusion in artificial membranes (Kliesch et al. 2017). In cells, stresses are generated by actin polymerization and cytoskeletal pulling against the plasma membrane (Kozlov et al. 2014) or by interactions with the microenvironment. In myoblasts, myomaker and myomerger proteins induce curvature-based stresses to complete fusion (Leikina et al. 2018). In *Drosophila*, highly curved podosome structures intrude from one cell into another, creating localized stress that drives fusion (Kim et al. 2015a). Stress also influences protein distribution: Myosin-II and spectrin accumulate at high-tension sites during *Drosophila* muscle fusion, where they are thought to stabilize stalk formation and pore expansion (Cong et al. 2019). Together, these findings illustrate how membrane tension shapes both the lipid mechanics and protein dynamics that govern fusion.

5.2.4.2 Membrane Pores

The formation and rupture of a fusion pore in the hemifusion diaphragm is the final step in cell fusion, and it is the second and final energy barrier (Dymond 2021; Fuhrmans et al. 2015; Ryham et al. 2016). In model membranes where conditions are thermodynamically favorable, pores form spontaneously, and membranes maintain a certain porosity (Coster 2003; Akimov et al. 2017b). In biological lipids, however, the flow of material in and out of cells is tightly regulated, and spontaneous pores are scarce. As a result, pore formation during cell fusion must be actively driven by fusogenic proteins or experimentally induced through external stimuli (Ros and García-Sáez 2015).

Pore formation determines whether fusion proceeds to completion. Newly nucleated pores typically exist in transient, unstable "flickering" states, and only when they surpass a critical radius do they irreversibly expand and rupture the bilayer (Coster 2003; Akimov et al. 2017a, b; Glaser et al. 1988; Spencer et al. 2024). Lipid composition strongly influences this barrier: Bilayers enriched in lipids with negative spontaneous curvature lower the activation energy and stabilize early pore states, while positive spontaneous curvature increases the energy required and reduces fusion efficiency (Akimov et al. 2017a, b).

Both artificial systems and biological contexts illustrate these principles. Pores can be experimentally induced by stretching the membrane or by electroporation, which lowers the barrier to nucleation (Akimov et al. 2017a). Biologically, *Drosophila* myoblasts may use tension directly to expand fusion pores by stretching the plasma membrane laterally through cytoskeletal and mechanical components (Koshiyama and Wada 2011; Oglęcka et al. 2014; Cong et al. 2019). These observations emphasize that pore stability is not simply a passive property of lipid bilayers but emerges from the interplay of curvature, stress, and cytoskeletal mechanics, making it a decisive step in the fusion pathway.

5.2.4.3 Hydrostatic and Osmotic Pressure

In biological systems, pressure influences membranes primarily in two forms: hydrostatic pressure, which arises from the confinement and height of fluids, and osmotic pressure, which reflects solute imbalances across an impermeable membrane (Darwish and Lui 2025). Both can alter bilayer organization and, thereby, influence the likelihood of fusion.

Hydrostatic pressure directly compresses lipid bilayers, changing the packing of acyl chains and headgroups (Brooks 2014; Goto et al. 2021). High pressure restricts chain mobility, alters spontaneous curvature, and modifies surface tension (Brooks 2014). In experimental systems, increased hydrostatic pressure accelerates fusion in membranes enriched in negative spontaneous curvature lipids (Dymond 2021; Kim et al. 2024). External applications exploit this principle as well: Ultrasound waves and jet injectors generate transient pressure pulses that create pores in membranes and promote fusion (Deng et al. 2004; Chang et al. 2023). Conversely, viral fusion can be inhibited under high hydrostatic pressure, highlighting the context-dependent role of this parameter (Gaspar et al. 2002). While hydrostatic pressures are relatively stable in most tissues, local variations in vascularized environments may modulate fusion efficiency.

Osmotic pressure, in contrast, does not alter bilayer structure directly but generates cortical tension through water flux across the plasma membrane (Goto et al. 2021; Billah et al. 2022). Osmotic swelling or shrinkage stretches membranes, creating stresses that lower the barrier to pore formation. This effect is well-documented: Osmotic gradients increase fusion yields in electrofusion, promote poration in lipid membranes, and facilitate defect formation that favors fusion (Billah et al. 2022; Przybyło et al. 2021; Rols and Teissie 1990; Polozov et al. 2001). Because osmotic pressures vary substantially in vivo, particularly in the placenta, where syncytiotrophoblast fusion occurs under continuous osmotic flux, they represent a physiologically relevant driver of cell fusion.

5.2.4.4 Adhesion Forces

Adhesion strength refers to the energy and force required to pull two adhered lipid membranes apart (Schneemilch and Quirke 2018). Adhesion can be mediated by external factors such as tissue forces, solution properties, or added adhesive molecules like PEG or through micropatterning and microfluidic technologies (Hu and Bao 2024; Chiu et al. 2016). In cells, however, adhesion is primarily mediated by adhesion proteins and their supporting cytoskeletal connections (Sun et al. 2011; Elineni and Gallant 2011).

Although a direct relationship between adhesion strength and fusion efficiency has not been firmly established, the requirement for specific adhesion molecules in systems such as myoblast fusion suggests that adhesion is a necessary preparatory step (Pizza et al. 2017). Adhesion complexes also respond to the high levels of stress required for stalk formation and pore expansion (Boccafoschi et al. 2010), indicating that sufficiently strong adhesion is likely essential to maintain close membrane apposition during fusion.

5.2.4.5 Electric Fields and Electrostatic Effects

The ionic environment and local electric fields strongly influence membrane properties (Ermakov 2023). Electrostatic forces acting on lipid headgroups alter bilayer organization, stabilizing or destabilizing fusion intermediates (Berkowitz and Vácha 2012; Taghian et al. 2015). For example, cations can stabilize nascent pores during vesicle fusion (Kabaso et al. 2012). Since cells actively maintain transmembrane potentials and rely on action potentials, transient changes in local electrical conditions may provide a physiological means of initiating or stabilizing fusion.

Externally applied electric fields can also drive fusion directly (Zimmermann 1983). In electrofusion, cells exposed to oscillating or pulsed fields experience an increase in membrane surface energy, which is then released through fusion (Rems et al. 2013; Marracino et al. 2022; Li et al. 2019). These fields also lower the barrier for pore formation, making nucleation and expansion more likely. Electrofusion has therefore become a practical experimental tool for inducing cell–cell fusion in both research and biotechnology. Beyond this direct effect, cells can sense electric fields through mechanobiological pathways, altering motility and cytoskeletal protrusions in a process known as electrotaxis (Taghian et al. 2015; Bezanilla 2008; Zhu et al. 2019). Although not yet directly linked to fusion, such behaviors may indirectly promote fusion by organizing cytoskeletal structures at contact sites. Together, these observations highlight how both electrostatic interactions and applied fields can serve as modulators of membrane fusion.

5.3 Trophoblast Fusion During Development

The principles outlined in the previous section, where biochemical signals act in tandem with membrane mechanics, curvature, tension, and external stresses, provide us with a general framework for understanding cell–cell fusion. Trophoblast fusion in the placenta illustrates how these principles operate in vivo. At the maternal–fetal interface, cytotrophoblasts fuse into the multinucleated syncytiotrophoblast (STB), a process essential for fetal development (Kojima et al. 2022). The STB forms the outermost layer of the chorionic villi, simultaneously protecting against maternal immune responses and mediating nutrient and gas exchange. Aberrant syncytialization is linked to complications such as preeclampsia, underscoring the importance of tightly regulated biophysical as well as biochemical control of fusion (Li et al. 2025a; Gauster et al. 2009).

5.3.1 Extracellular Matrix Cues

Substrate stiffness, another factor introduced earlier, also plays a significant role in trophoblast fusion. Both overly soft and overly stiff matrices inhibit fusion, whereas an intermediate stiffness range promotes cytoskeletal reorganization (Fig. 5.5a) (Ma et al. 2020). Within this range, myosin-II redistributes to control membrane tension, consistent with observations in other fusing cell types (Cong et al. 2019; Pelham and Wang 1997; Kim et al. 2015b). In addition to substrate stiffness, the spatial organization of trophoblasts in a three-dimensional (3D) environment makes a dramatic difference in fusion outcomes. When compared to monolayers, trophoblasts seeded in Matrigel, an extracellular matrix (ECM)-based gel that provides

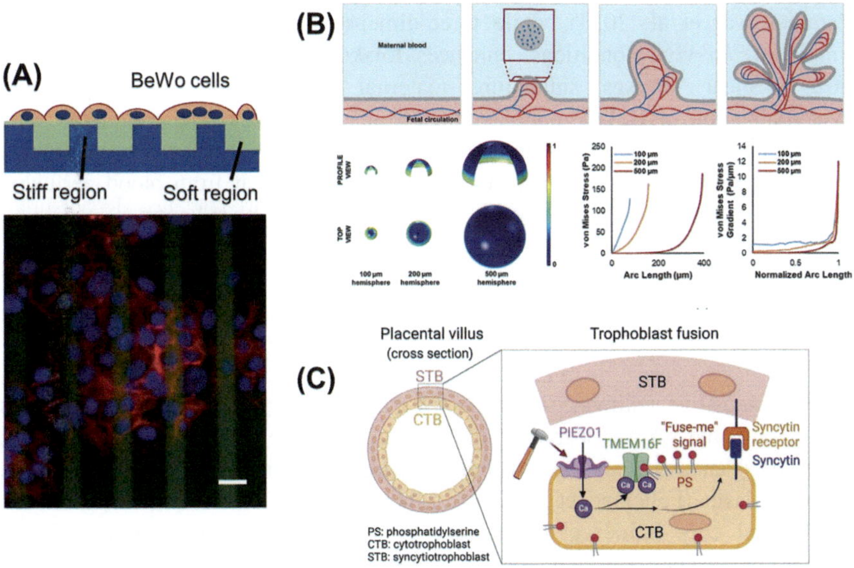

Fig. 5.5 Biophysical regulation of trophoblast fusion. (**a**) Substrate stiffness influences trophoblast fusion efficiency, with an intermediate stiffness range promoting fusion, while both overly soft and stiff conditions inhibit syncytialization (Ma et al. 2020). (Reproduced with permission under a Creative Commons Attribution 4.0 International Public License). (**b**) Micropatterned adhesive islands recreate mechanical stresses that enhance and spatially direct BeWo fusion efficiency compared to unconfined cultures (Ma et al. 2019). (Reproduced with permission from the American Chemical Society through the Copyright Clearance Center). (**c**) Mechanosensitive PIEZO1 channels couple shear and stretch inputs to calcium influx and TMEM16F-mediated phosphatidylserine externalization, priming membranes for fusion (Zhang et al. 2025). (Reproduced with permission under a Creative Commons Attribution 4.0 International Public License)

physiologically relevant microenvironmental cues, spontaneously organize to form highly fusogenic spheroids (Wong et al. 2018). These conditions also facilitate the activation of mechanically relevant cell behaviors relating to migration and intracellular adhesions (Wong et al. 2019). In synthetically derived ECM, trophoblasts and trophoblast-like cell lines exhibit variation in gene expression and metabolic activity across different hydrogel compositions with different concentrations (Slaby et al. 2024a, b). For example, scrambled fibronectin functionalized Arginine (R), Aspartic Acid (D) and Glycine (G) RDG-alginate hydrogels increase the propensity of trophoblasts to secrete hCG beta, a hormone whose expression is used as a marker for fusion (Slaby et al. 2024b).

5.3.2 Mechanical Stresses in the Placental Environment

Compression and shear stresses, described earlier as modulators of fusion, are also evident in trophoblasts. Two-dimensional compression or substrate patterning induces BeWo fusion even in the absence of forskolin (Fig. 5.5b) (Ma et al. 2019; Parameshwar et al. 2024), while three-dimensional compression, more closely mimicking in vivo conditions, enhances forskolin-induced fusion (Delon et al. 2025). Shear stresses simulating maternal blood flow likewise increase syncytialization of BeWo monolayers (Delon et al. 2025). These findings show that the physical forces encountered by chorionic villi within the maternal decidua, including compression from surrounding tissues and shear from blood perfusion, translate into membrane-level events that promote trophoblast fusion.

5.3.3 Mechanosensitive Signaling Pathways

Mechanosensitive pathways integrate the physical environment with trophoblast fusion outcomes. YAP, a transcriptional regulator that responds to stiffness and geometry (Panciera et al. 2017), is inhibited during trophoblast fusion, remaining in the cytoplasm (Feliciano et al. 2021; Lin et al. 2023). This shift suggests that mechanical cues influence the balance between CTB proliferation and fusion. PIEZO1, a mechanosensitive ion channel expressed in trophoblasts, responds to membrane stretch or shear stress by mediating calcium influx and activating the scramblase TMEM16F (Fig. 5.5c) (Zhang et al. 2025; Ranade et al. 2014). Through this mechanism, mechanical inputs are directly coupled to phospholipid externalization and membrane readiness for fusion. RhoE, a mechanical effector that promotes disassembly of actin stress fibers, is also upregulated in fusing BeWo cells, and its inhibition reduces fusion efficiency (Collett et al. 2012; Jie et al. 2015). Even in model systems that induce fusion artificially, induction factors lead to mechanical consequences. Forskolin, a pharmacological activator of BeWo fusion, activates the membrane protein TMEM16F to facilitate PS externalization (Lyden et al. 1993;

Adler et al. 1995; Zhang et al. 2020). As described before, phospholipids that promote negative curvature, such as PS, are essential in fusion, and blocking PS translocation prevents fusion in BeWo cells (Hirama et al. 2017; Das et al. 2004). Together, these pathways illustrate how mechanics, lipid composition, and cytoskeletal organization converge to regulate trophoblast fusion.

Although relatively few studies have examined trophoblast fusion from a biophysical perspective, the available evidence highlights an intricate interplay of lipid remodeling, cytoskeletal stresses, and mechanosensitive signaling. These findings demonstrate how the general principles outlined earlier manifest in a physiologically relevant setting. Trophoblasts, therefore, serve as a useful model to study how mechanical and biochemical cues combine to regulate cell fusion during development while also reminding us that similar biophysical regulation is likely to occur in other fusion systems.

5.4 Avenues for Future Work

Most studies of cell fusion remain centered on identifying fusogenic proteins and signaling pathways, with relatively little emphasis on the accompanying physical context. Building on the framework outlined here, future work should integrate biophysical measurements with molecular studies. Even simple descriptors such as cell circularity or surface area, accessible by light microscopy, can reveal information about contractility and force distributions (Hu and Bao 2024; Bourn et al. 2025). More advanced tools, including embedded force sensors in organoids and tissue models, now allow spatial and temporal mapping of stresses during development (Forigua et al. 2025; Mok et al. 2020; Lee et al. 2019, 2023). Microfluidic and organ-on-chip platforms offer complementary opportunities, enabling precise pairing of cells and the application of defined pressures, adhesion cues, and electrical fields to probe fusion under controlled conditions (Chiu et al. 2016; Li et al. 2025b). Together, these approaches can connect membrane energetics to cell- and tissue-scale mechanics, expanding our ability to dissect how fusogens operate within real physical constraints.

Fusion is a product not only of protein machinery but also of geometry, tension, and the mechanics of the membrane itself. A fuller understanding will emerge once biophysical variables are measured and manipulated alongside biochemical ones, providing a path to resolve outstanding questions across developmental, physiological, and pathological contexts. By embedding biophysical perspectives into fusion research, we can begin to uncover principles that are likely conserved across diverse systems.

5.5 Conclusion

Biophysical factors such as osmotic and hydrostatic pressures, compressive stresses, electric fields, and geometric constraints are fundamental to cell behavior. Cells respond to these influences directly through changes in membrane energetics and indirectly through mechanosensors and signaling pathways. In the context of fusion, these physical effects become particularly important because membrane geometry and tension determine whether stalk formation and pore expansion can occur. Despite this, biophysical contributions to fusion remain relatively underexplored outside of model membranes. The placenta highlights why this perspective matters since abnormal trophoblast fusion is associated with pregnancy disorders such as preeclampsia. A complete understanding of fusion will require treating it as both a biochemical and a biophysical phenomenon, integrating molecular machinery with the mechanics of real cells and tissues.

Acknowledgments This work was supported by funding from the Natural Sciences and Engineering Research Council of Canada Discovery grant program and the Canada Research Chair in Advanced Cellular Microenvironments to CM. AF is partially funded by the Mitacs Accelerate Postdoctoral Fellowship.

Conflict of Interest No conflict of interest is declared by the authors.

Author Contributions MAA and CM conceptualized the chapter. MAA performed the primary investigation. MAA, AF, and CM defined the chapter structure. MAA and AF drafted and revised the manuscript. CM supervised the project. All authors reviewed and approved the final version.

References

Adler RR, Ng AK, Rote NS (1995) Monoclonal Antiphosphatidylserine antibody inhibits intercellular fusion of the Choriocarcinoma line, JAR. Biol Reprod 53(4):905–910. https://doi.org/10.1095/biolreprod53.4.905

Aeffner S, Reusch T, Weinhausen B, Salditt T (2012) Energetics of stalk intermediates in membrane fusion are controlled by lipid composition. Proc Natl Acad Sci 109(25):E1609–E1618. https://doi.org/10.1073/pnas.1119442109

Akimov SA, Volynsky PE, Galimzyanov TR, Kuzmin PI, Pavlov KV, Batishchev OV (2017a) Pore formation in lipid membrane II: energy landscape under external stress. Sci Rep 7(1):12509. https://doi.org/10.1038/s41598-017-12749-x

Akimov SA, Volynsky PE, Galimzyanov TR, Kuzmin PI, Pavlov KV, Batishchev OV (2017b) Pore formation in lipid membrane I: continuous reversible trajectory from intact bilayer through hydrophobic defect to transversal pore. Sci Rep 7(1):12152. https://doi.org/10.1038/s41598-017-12127-7

Argudo D, Bethel NP, Marcoline FV, Grabe M (2016) Continuum descriptions of membranes and their interaction with proteins: towards chemically accurate models. Biochim Biophys Acta Biomembr 1858(7 Part B):1619–1634. https://doi.org/10.1016/j.bbamem.2016.02.003

Barlow NE, Kusumaatmaja H, Salehi-Reyhani A, Brooks N, Barter LMC, Flemming AJ, Ces O (2018) Measuring bilayer surface energy and curvature in asymmetric droplet interface bilayers. J R Soc Interface 15(148):20180610. https://doi.org/10.1098/rsif.2018.0610

Ben-Shaul A (1995) Chapter 7 – molecular theory of chain packing, elasticity and lipid-protein interaction in lipid bilayers. In: Lipowsky R, Sackmann E (eds) Handbook of biological physics, Structure and dynamics of membranes, vol 1, pp 359–401. https://doi.org/10.1016/S1383-8121(06)80024-2

Berkowitz ML, Vácha R (2012) Aqueous solutions at the interface with phospholipid bilayers. Acc Chem Res 45(1):74–82. https://doi.org/10.1021/ar200079x

Bezanilla F (2008) How membrane proteins sense voltage. Nat Rev Mol Cell Biol 9(4):323–332. https://doi.org/10.1038/nrm2376

Bieling P, Li T-D, Weichsel J, McGorty R, Jreij P, Huang B, Fletcher DA, Mullins RD (2016) Force feedback controls motor activity and mechanical properties of self-assembling branched actin networks. Cell 164(1):115–127. https://doi.org/10.1016/j.cell.2015.11.057

Billah MM, Saha SK, Rashid MMO, Hossain F, Yamazaki M (2022) Effect of osmotic pressure on pore formation in lipid bilayers by the antimicrobial peptide Magainin 2. Phys Chem Chem Phys 24(11):6716–6731. https://doi.org/10.1039/D1CP05764B

Boccafoschi F, Bosetti M, Sandra PM, Leigheb M, Cannas M (2010) Effects of mechanical stress on cell adhesion a possible mechanism for morphological changes. Cell Adhes Migr 4(1): 19–25. https://doi.org/10.4161/cam.4.1.9569

Bourn MD, Daly LF, Huggett JF, Braybrook J, Rivera JF (2025) Evaluation of image analysis tools for the measurement of cellular morphology. Front Cell Dev Biol:13. https://doi.org/10.3389/fcell.2025.1572212

Brooks NJ (2014) Pressure effects on lipids and bio-membrane assemblies. IUCrJ 1(Pt 6):470–477. https://doi.org/10.1107/S2052252514019551

Brukman NG, Uygur B, Podbilewicz B, Chernomordik LV (2019) How cells fuse. J Cell Biol 218(5):1436–1451. https://doi.org/10.1083/jcb.201901017

Chang CY, Tai JA, Sakaguchi Y, Nishikawa T, Hirayama Y, Yamashita K (2023) Enhancement of polyethylene glycol-cell fusion efficiency by novel application of transient pressure using a jet injector. FEBS Open Bio 13(3):478–489. https://doi.org/10.1002/2211-5463.13557

Chernomordik LV, Kozlov MM (2003) Protein-lipid interplay in fusion and fission of biological membranes. Annu Rev Biochem 72:175–207. https://doi.org/10.1146/annurev.biochem.72.121801.161504

Chernomordik LV, Kozlov MM (2008) Mechanics of membrane fusion. Nat Struct Mol Biol 15(7): 675–683. https://doi.org/10.1038/nsmb.1455

Chiu FWY, Bagci H, Fisher AG, deMello AJ, Elvira KS (2016) A microfluidic toolbox for cell fusion. J Chem Technol Biotechnol 91(1):16–24. https://doi.org/10.1002/jctb.4803

Collett GP, Goh XF, Linton EA, Redman CWG, Sargent IL (2012) RhoE is regulated by cyclic AMP and promotes fusion of human BeWo choriocarcinoma cells. PLoS One 7(1):e30453. https://doi.org/10.1371/journal.pone.0030453

Cong J, Fang B, Wang Q, Su Y, Gu T, Luo T (2019) The mechanobiology of actin cytoskeletal proteins during cell–cell fusion. J R Soc Interface 16(156):20190022. https://doi.org/10.1098/rsif.2019.0022

Cooke IR, Deserno M (2006) Coupling between lipid shape and membrane curvature. Biophys J 91(2):487–495. https://doi.org/10.1529/biophysj.105.078683

Coster HGL (2003) The physics of cell membranes. J Biol Phys 29(4):363–399. https://doi.org/10.1023/A:1027362704125

Darwish A, Lui F (2025) Physiology, colloid osmotic pressure. In StatPearls. StatPearls Publishing, Treasure Island

Das M, Xu B, Lin L, Chakrabarti S, Shivaswamy V, Rote NS (2004) Phosphatidylserine efflux and intercellular fusion in a BeWo model of human villous cytotrophoblast. Placenta 25(5): 396–407. https://doi.org/10.1016/j.placenta.2003.11.004

Delon L, Busek M, Menezes PD, Gadegaard N, Stokowiec J, Golovin A, Boichuk Y, Combriat T, Aizenshtadt A, Krauss S (2025) Placental Syncytium-on-Chip (PSoC)–comparison of forskolin or mechanical induced-syncytialization. Adv Healthc Mater 14(21):2404462. https://doi.org/10.1002/adhm.202404462

Deng CX, Sieling F, Pan H, Cui J (2004) Ultrasound-induced cell membrane porosity. Ultrasound Med Biol 30(4):519–526. https://doi.org/10.1016/j.ultrasmedbio.2004.01.005

Discher DE, Mooney DJ, Zandstra PW (2009) Growth factors, matrices, and forces combine and control stem cells. Science 324(5935):1673–1677. https://doi.org/10.1126/science.1171643

Dittmar T, Zänker KS (eds) (2011) Cell fusion in health and disease, Advances in experimental medicine and biology, vol 713. Springer, Dordrecht. https://doi.org/10.1007/978-94-007-0763-4

Dymond MK (2021) Lipid monolayer spontaneous curvatures: a collection of published values. Chem Phys Lipids 239:105117. https://doi.org/10.1016/j.chemphyslip.2021.105117

Elineni KK, Gallant ND (2011) Regulation of cell adhesion strength by peripheral focal adhesion distribution. Biophys J 101(12):2903–2911. https://doi.org/10.1016/j.bpj.2011.11.013

Epand RM, Epand RF (2000) Modulation of membrane curvature by peptides. Pept Sci 55(5): 358–363. https://doi.org/10.1002/1097-0282(2000)55:5<358::AID-BIP1009>3.0.CO;2-8

Ermakov Y (2023) Electric fields at the lipid membrane interface. Membranes 13(11):883. https://doi.org/10.3390/membranes13110883

Feliciano D, Ott CM, Espinosa-Medina I, Weigel AV, Benedetti L, Milano KM, Tang Z, Lee T, Kliman HJ, Guller SM et al (2021) YAP1 nuclear efflux and transcriptional reprograming follow membrane diminution upon VSV-G-induced cell fusion. Nat Commun 12(1):4502

Fontana J, Cardone G, Heymann JB, Winkler DC, Steven AC (2012) Structural changes in influenza virus at low pH characterized by Cryo-electron tomography. J Virol 86(6): 2919–2929. https://doi.org/10.1128/jvi.06698-11

Forigua A, Campbell BE, Moraes C (2025) Emerging views of biomechanics via embedded sensors in model tissues: pathways to the clinic. Curr Opin Biomed Eng 34:100587. https://doi.org/10.1016/j.cobme.2025.100587

Fuhrmans M, Marelli G, Smirnova YG, Müller M (2015) Mechanics of membrane fusion/pore formation. Chem Phys Lipids 185:109–128. https://doi.org/10.1016/j.chemphyslip.2014.07.010

Gaspar LP, Silva ACB, Gomes AMO, Freitas MS, Ano Bom APD, Schwarcz WD, Mestecky J, Novak MJ, Foguel D, Silva JL (2002) Hydrostatic pressure induces the fusion-active state of enveloped viruses*. J Biol Chem 277(10):8433–8439. https://doi.org/10.1074/jbc.M106096200

Gauster M, Moser G, Orendi K, Huppertz B (2009) Factors involved in regulating trophoblast fusion: potential role in the development of preeclampsia. Placenta 30:49–54. https://doi.org/10.1016/j.placenta.2008.10.011

Glaser RW, Leikin SL, Chernomordik LV, Pastushenko VF, Sokirko AI (1988) Reversible electrical breakdown of lipid bilayers: formation and evolution of pores. Biochim Biophys Acta Biomembr 940(2):275–287. https://doi.org/10.1016/0005-2736(88)90202-7

Golani G, Schwarz US (2023) High curvature promotes fusion of lipid membranes: predictions from continuum elastic theory. Biophys J 122(10):1868–1882. https://doi.org/10.1016/j.bpj.2023.04.018

Goto M, Kazama A, Fukuhara K, Sato H, Tamai N, Ito H-O, Matsuki H (2021) Membrane fusion of phospholipid bilayers under high pressure: spherical and irreversible growth of giant vesicles. Biophys Chem 277:106639. https://doi.org/10.1016/j.bpc.2021.106639

Grafmüller A, Shillcock J, Lipowsky R (2009) The fusion of membranes and vesicles: pathway and energy barriers from dissipative particle dynamics. Biophys J 96(7):2658–2675. https://doi.org/10.1016/j.bpj.2008.11.073

Hamm M, Kozlov MM (2000) Elastic energy of tilt and bending of fluid membranes. Eur Phys J E 3(4):323–335. https://doi.org/10.1007/s101890070003

Hirama T, Lu SM, Kay JG, Maekawa M, Kozlov MM, Grinstein S, Fairn GD (2017) Membrane curvature induced by proximity of anionic phospholipids can initiate endocytosis. Nat Commun 8(1):1393. https://doi.org/10.1038/s41467-017-01554-9

Hu X, Bao M (2024) Advances in micropatterning technology for mechanotransduction research. Mechanobiol Med 2(3):100066. https://doi.org/10.1016/j.mbm.2024.100066

Irie A, Yamamoto K, Miki Y, Murakami M (2017) Phosphatidylethanolamine dynamics are required for osteoclast fusion. Sci Rep 7(1):46715. https://doi.org/10.1038/srep46715

Ishihara S, Haga H (2022) Matrix stiffness contributes to cancer progression by regulating transcription factors. Cancer 14(4):1049. https://doi.org/10.3390/cancers14041049

Jahraus A, Egeberg M, Hinner B, Habermann A, Sackman E, Pralle A, Faulstich H, Rybin V, Defacque H, Griffiths G (2001) ATP-dependent membrane assembly of F-actin facilitates membrane fusion. Mol Biol Cell 12(1):155–170. https://doi.org/10.1091/mbc.12.1.155

Jie W, Andrade KC, Lin X, Yang X, Yue X, Chang J (2015) Pathophysiological functions of Rnd3/RhoE. Compr Physiol 6(1):169–186. https://doi.org/10.1002/cphy.c150018

Joardar A, Pattnaik GP, Chakraborty H (2022) Mechanism of membrane fusion: interplay of lipid and peptide. J Membr Biol 255(2–3):211–224. https://doi.org/10.1007/s00232-022-00233-1

Kabaso D, Calejo AI, Jorgačevski J, Kreft M, Zorec R, Iglič A (2012) Fusion pore diameter regulation by cations modulating local membrane anisotropy. Sci World J 2012(1):983138. https://doi.org/10.1100/2012/983138

Khattari Z, Köhler S, Xu Y, Aeffner S, Salditt T (2015) Stalk formation as a function of lipid composition studied by X-Ray reflectivity. Biochim Biophys Acta Biomembr 1848(1, Part A):41–50. https://doi.org/10.1016/j.bbamem.2014.08.010

Kim JH, Ren Y, Ng WP, Li S, Son S, Kee Y-S, Zhang S, Zhang G, Fletcher DA, Robinson DN, Chen EH (2015a) Mechanical tension drives cell membrane fusion. Dev Cell 32(5):561–573. https://doi.org/10.1016/j.devcel.2015.01.005

Kim JH, Jin P, Duan R, Chen EH (2015b) Mechanisms of myoblast fusion during muscle development. Curr Opin Genet Dev 32:162–170. https://doi.org/10.1016/j.gde.2015.03.006

Kim BK, Kang D-H, Woo J, Yoon W, Ryu H, Han K, Chung S, Kim TS (2024) Control of artificial membrane fusion in physiological ionic solutions beyond the limits of electroformation. Nat Commun 15(1):4524. https://doi.org/10.1038/s41467-024-48875-0

Kliesch T-T, Dietz J, Turco L, Halder P, Polo E, Tarantola M, Jahn R, Janshoff A (2017) Membrane tension increases fusion efficiency of model membranes in the presence of SNAREs. Sci Rep 7(1):12070. https://doi.org/10.1038/s41598-017-12348-w

Kluge C, Pöhnl M, Böckmann RA (2022) Spontaneous local membrane curvature induced by transmembrane proteins. Biophys J 121(5):671–683. https://doi.org/10.1016/j.bpj.2022.01.029

Kojima J, Ono M, Kuji N, Nishi H (2022) Human chorionic villous differentiation and placental development. Int J Mol Sci 23(14):8003. https://doi.org/10.3390/ijms23148003

Koshiyama K, Wada S (2011) Molecular dynamics simulations of pore formation dynamics during the rupture process of a phospholipid bilayer caused by high-speed equibiaxial stretching. J Biomech 44(11):2053–2058. https://doi.org/10.1016/j.jbiomech.2011.05.014

Kozlov MM, McMahon HT, Chernomordik LV (2010) Protein-driven membrane stresses in fusion and fission. Trends Biochem Sci 35(12):699–706. https://doi.org/10.1016/j.tibs.2010.06.003

Kozlov MM, Campelo F, Liska N, Chernomordik LV, Marrink SJ, McMahon HT (2014) Mechanisms shaping cell membranes. Curr Opin Cell Biol 29:53–60. https://doi.org/10.1016/j.ceb.2014.03.006

Kozlovsky Y, Kozlov MM (2002) Stalk model of membrane fusion: solution of energy crisis. Biophys J 82(2):882–895. https://doi.org/10.1016/S0006-3495(02)75450-7

Lee AG (2004) How lipids affect the activities of integral membrane proteins. Biochim Biophys Acta Biomembr 1666(1):62–87. https://doi.org/10.1016/j.bbamem.2004.05.012

Lee W, Kalashnikov N, Mok S, Halaoui R, Kuzmin E, Putnam AJ, Takayama S, Park M, McCaffrey L, Zhao R, Leask RL, Moraes C (2019) Dispersible hydrogel force sensors reveal patterns of solid mechanical stress in multicellular spheroid cultures. Nat Commun 10(1):144. https://doi.org/10.1038/s41467-018-07967-4

Lee W, Boghdady C-M, Lelarge V, Leask RL, McCaffrey L, Moraes C (2023) Ultrasoft edge-labelled hydrogel sensors reveal internal tissue stress patterns in invasive engineered tumors. Biomaterials 296:122073. https://doi.org/10.1016/j.biomaterials.2023.122073

Lei KF, Bai K-C, Pai P-C (2025) Study of cell migration trajectory on two-dimensional continuous stiffness gradient surface edited by grayscale photopolymerization. Talanta 281:126899. https://doi.org/10.1016/j.talanta.2024.126899

Leikina E, Gamage DG, Prasad V, Goykhberg J, Crowe M, Diao J, Kozlov MM, Chernomordik LV, Millay DP (2018) Myomaker and Myomerger work independently to control distinct steps of membrane remodeling during myoblast fusion. Dev Cell 46(6):767–780.e7. https://doi.org/10.1016/j.devcel.2018.08.006

Lentz BR (1994) Polymer-induced membrane fusion: potential mechanism and relation to cell fusion events. Chem Phys Lipids 73(1):91–106. https://doi.org/10.1016/0009-3084(94)90176-7

Li C, Ke Q, Yao C, Yao C, Mi Y, Wu M, Ge L (2019) Comparison of bipolar and unipolar pulses in cell electrofusion: simulation and experimental research. IEEE Trans Biomed Eng 66(5): 1353–1360. https://doi.org/10.1109/TBME.2018.2872909

Li J, Liu Y, Zhang Y, Yao B, Enhejirigala, Li Z, Song W, Wang Y, Duan X, Yuan X, Fu X, Huang S (2021) Biophysical and biochemical cues of biomaterials guide mesenchymal stem cell behaviors. Front Cell Dev Biol:9. https://doi.org/10.3389/fcell.2021.640388

Li Q, Zhu Y, Yu X, Shao X, Wang Y-L (2025a) Molecular regulation and functional benefits of trophoblast syncytialization in optimizing maternal-fetal nutrient allocation. Placenta. https://doi.org/10.1016/j.placenta.2025.06.003

Li C, Kalashnikov N, Moraes C (2025b) Internal delivery and transport within cellular aggregates via perfusable glass-sheathed hydrogel microtubes. Langmuir. https://doi.org/10.1021/acs.langmuir.5c02197

Lim CT, Bershadsky A, Sheetz MP (2010) Mechanobiology. J R Soc Interface 7(suppl_3):S291–S293. https://doi.org/10.1098/rsif.2010.0150.focus

Lin Q, Cao J, Yu J, Zhu Y, Shen Y, Wang S, Wang Y, Liu Z, Chang Y (2023) YAP-mediated trophoblast dysfunction: the common pathway underlying pregnancy complications. Cell Commun Signal 21(1):353. https://doi.org/10.1186/s12964-023-01371-2

Lyden TW, Ah-Kau NG, Rote NS (1993) Modulation of phosphatidylserine epitope expression by BeWo cells during Forskolin treatment. Placenta 14(2):177–186. https://doi.org/10.1016/S0143-4004(05)80259-0

Ma Z, Sagrillo-Fagundes L, Tran R, Parameshwar PK, Kalashnikov N, Vaillancourt C, Moraes C (2019) Biomimetic micropatterned adhesive surfaces to mechanobiologically regulate placental trophoblast fusion. ACS Appl Mater Interfaces 11(51):47810–47821. https://doi.org/10.1021/acsami.9b19906

Ma Z, Sagrillo-Fagundes L, Mok S, Vaillancourt C, Moraes C (2020) Mechanobiological regulation of placental trophoblast fusion and function through extracellular matrix rigidity. Sci Rep 10(1):5837. https://doi.org/10.1038/s41598-020-62659-8

Mahbubul IM (2019) 4 - Thermophysical properties of nanofluids. In: Mahbubul IM (ed) Preparation, characterization, properties and application of nanofluid, Micro and nano technologies. William Andrew Publishing, pp 113–196. https://doi.org/10.1016/B978-0-12-813245-6.00004-6

Marracino P, Caramazza L, Montagna M, Ghahri R, D'Abramo M, Liberti M, Apollonio F (2022) Electric-driven membrane poration: a rationale for water role in the kinetics of pore formation. Bioelectrochemistry 143:107987. https://doi.org/10.1016/j.bioelechem.2021.107987

Marshall W, Baum B, Fairhall A, Heisenberg C-P, Koslover E, Liu A, Mao Y, Mogilner A, Nelson CM, Paluch EK, Trepat X, Yap A (2024) Where physics and biology meet. Curr Biol 34(20): R950–R960. https://doi.org/10.1016/j.cub.2024.08.022

McBeath R, Pirone DM, Nelson CM, Bhadriraju K, Chen CS (2004) Cell shape, cytoskeletal tension, and RhoA regulate stem cell lineage commitment. Dev Cell 6(4):483–495

McNally AK, Anderson JM (2011) Macrophage fusion and multinucleated giant cells of inflammation. In: Dittmar T, Zänker KS (eds) Cell fusion in health and disease. Springer, Dordrecht, pp 97–111. https://doi.org/10.1007/978-94-007-0763-4_7

Mok S, Al Habyan S, Ledoux C, Lee W, MacDonald KN, McCaffrey L, Moraes C (2020) Mapping cellular-scale internal mechanics in 3D tissues with thermally responsive hydrogel probes. Nat Commun 11(1):4757. https://doi.org/10.1038/s41467-020-18469-7

Monck JR, Fernandez JM (1996) The fusion pore and mechanisms of biological membrane fusion. Curr Opin Cell Biol 8(4):524–533. https://doi.org/10.1016/S0955-0674(96)80031-7

Nikolaus J, Warner JM, O'Shaughnessy B, Herrmann A (2011) The pathway to membrane fusion through hemifusion*. In: Chernomordik LV, Kozlov MM (eds) Current topics in membranes, vol 68. Academic Press, pp 1–32. https://doi.org/10.1016/B978-0-12-385891-7.00001-5

Oglęcka K, Rangamani P, Liedberg B, Kraut RS, Parikh AN (2014) Oscillatory phase separation in giant lipid vesicles induced by transmembrane osmotic differentials. elife 3:e03695. https://doi.org/10.7554/eLife.03695

Ohki S, Baker GA, Page PM, McCarty TA, Epand RM, Bright FV (2006) Interaction of influenza virus fusion peptide with lipid membranes: effect of Lysolipid. J Membr Biol 211(3):191–200. https://doi.org/10.1007/s00232-006-0862-z

Panciera T, Azzolin L, Cordenonsi M, Piccolo S (2017) Mechanobiology of YAP and TAZ in physiology and disease. Nat Rev Mol Cell Biol 18(12):758–770. https://doi.org/10.1038/nrm.2017.87

Pandia S, Chakraborty H (2024) Navigating the mechanistic pathways of membrane fusion: the lipid perspective. Eur Phys J Spec Top 233(21):2965–2979. https://doi.org/10.1140/epjs/s11734-024-01106-5

Pantazatos DP, MacDonald RC (1999) Directly observed membrane fusion between oppositely charged phospholipid bilayers. J Membr Biol 170(1):27–38. https://doi.org/10.1007/s002329900535

Parameshwar PK, Li C, Arnauts K, Jiang J, Rostami S, Campbell BE, Lu H, Rosenzweig DH, Vaillancourt C, Moraes C (2024) Directed biomechanical compressive forces enhance fusion efficiency in model placental trophoblast cultures. Sci Rep 14(1):11312. https://doi.org/10.1038/s41598-024-61747-3

Pelham RJ, Wang Y (1997) Cell locomotion and focal adhesions are regulated by substrate flexibility. Proc Natl Acad Sci USA 94(25):13661–13665. https://doi.org/10.1073/pnas.94.25.13661

Pereira M, Petretto E, Gordon S, Bassett JHD, Williams GR, Behmoaras J (2018) Common signalling pathways in macrophage and osteoclast multinucleation. J Cell Sci 131(11): jcs216267. https://doi.org/10.1242/jcs.216267

Pizza FX, Martin RA, Springer EM, Leffler MS, Woelmer BR, Recker IJ, Leaman DW (2017) Intercellular adhesion molecule-1 augments myoblast adhesion and fusion through homophilic trans-interactions. Sci Rep 7(1):5094. https://doi.org/10.1038/s41598-017-05283-3

Podbilewicz B (2014) Virus and cell fusion mechanisms. Annu Rev Cell Dev Biol 30(1):111–139. https://doi.org/10.1146/annurev-cellbio-101512-122422

Polozov IV, Anantharamaiah GM, Segrest JP, Epand RM (2001) Osmotically induced membrane tension modulates membrane Permeabilization by class 1 amphipathic helical peptides: nucleation model of defect formation. Biophys J 81(2):949–959. https://doi.org/10.1016/S0006-3495(01)75753-0

Przybyło M, Drabik D, Doskocz J, Iglič A, Langner M (2021) The effect of the osmotically active compound concentration difference on the passive water and proton fluxes across a lipid bilayer. Int J Mol Sci 22(20):11099. https://doi.org/10.3390/ijms222011099

Ranade SS, Qiu Z, Woo S-H, Hur SS, Murthy SE, Cahalan SM, Xu J, Mathur J, Bandell M, Coste B, Li Y-SJ, Chien S, Patapoutian A (2014) Piezo1, a mechanically activated ion channel, is required for vascular development in mice. Proc Natl Acad Sci 111(28):10347–10352. https://doi.org/10.1073/pnas.1409233111

Rems L, Ušaj M, Kandušer M, Reberšek M, Miklavčič D, Pucihar G (2013) Cell electrofusion using nanosecond electric pulses. Sci Rep 3(1):3382. https://doi.org/10.1038/srep03382

Renaud SJ, Jeyarajah MJ (2022) How trophoblasts fuse: an in-depth look into placental syncytiotrophoblast formation. Cell Mol Life Sci 79(8):433. https://doi.org/10.1007/s00018-022-04475-z

Rols M, Teissie J (1990) Modulation of electrically induced permeabilization and fusion of Chinese hamster ovary cells by osmotic-pressure. Biochemistry 29(19):4561–4567. https://doi.org/10.1021/bi00471a009

Ros U, García-Sáez AJ (2015) More than a pore: the interplay of pore-forming proteins and lipid membranes. J Membr Biol 248(3):545–561. https://doi.org/10.1007/s00232-015-9820-y

Ryham RJ, Klotz TS, Yao L, Cohen FS (2016) Calculating transition energy barriers and characterizing activation states for steps of fusion. Biophys J 110(5):1110–1124. https://doi.org/10.1016/j.bpj.2016.01.013

Schneemilch M, Quirke N (2018) Free energy of adhesion of lipid bilayers on silica surfaces. J Chem Phys 148(19):194704. https://doi.org/10.1063/1.5028557

Shearman GC, Attard GS, Hunt AN, Jackowski S, Baciu M, Sebai SC, Mulet X, Clarke JA, Law RV, Plisson C, Parker CA, Gee A, Ces O, Templer RH (2007) Using membrane stress to our advantage. Biochem Soc Trans 35(3):498–501. https://doi.org/10.1042/BST0350498

Sieler M, Dittmar T (2024) Cell fusion and syncytia formation in cancer. In: Kloc M, Uosef A (eds) Syncytia: origin, structure, and functions. Springer, Cham, pp 433–465. https://doi.org/10.1007/978-3-031-37936-9_20

Simionescu A, Pavlath GK (2011) Molecular mechanisms of myoblast fusion across species. In: Dittmar T, Zänker KS (eds) Cell fusion in health and disease. Springer, Dordrecht, pp 113–135. https://doi.org/10.1007/978-94-007-0763-4_8

Slaby EM, Hansen N, Sharma R, Pirrotte P, Weaver JD (2024a) Engineered 3D hydrogel matrices to modulate trophoblast stem cell-derived placental organoid phenotype. bioRxiv May 16, p 2024.05.13.594007. https://doi.org/10.1101/2024.05.13.594007

Slaby EM, Plaisier SB, Brady SR, Hiremath SC, Weaver JD (2024b) Controlling placental spheroid growth and phenotype using engineered synthetic hydrogel matrices. Biomater Sci 12(4):933–948. https://doi.org/10.1039/D3BM01393F

Smith LR, Cho S, Discher DE (2018) Stem cell differentiation is regulated by extracellular matrix mechanics. Physiology (Bethesda) 33(1):16–25. https://doi.org/10.1152/physiol.00026.2017

Spencer RKW, Smirnova YG, Soleimani A, Müller M (2024) Transient pores in hemifusion diaphragms. Biophys J 123(16):2455–2475. https://doi.org/10.1016/j.bpj.2024.06.009

Sun Y, Lee C-C, Huang HW (2011) Adhesion and merging of lipid bilayers: a method for measuring the free energy of adhesion and hemifusion. Biophys J 100(4):987–995. https://doi.org/10.1016/j.bpj.2011.01.013

Taghian T, Narmoneva DA, Kogan AB (2015) Modulation of cell function by electric field: a high-resolution analysis. J R Soc Interface 12(107):20150153. https://doi.org/10.1098/rsif.2015.0153

Tamm LK, Crane J, Kiessling V (2003) Membrane fusion: a structural perspective on the interplay of lipids and proteins. Curr Opin Struct Biol 13(4):453–466. https://doi.org/10.1016/S0959-440X(03)00107-6

Vignery A (2005) Macrophage fusion. J Exp Med 202(3):337–340. https://doi.org/10.1084/jem.20051123

White FM, Gatenby RA, Fischbach C (2019) The physics of cancer. Cancer Res 79(9):2107–2110. https://doi.org/10.1158/0008-5472.CAN-18-3937

Whitlock JM, Chernomordik LV (2021) Flagging fusion: phosphatidylserine signaling in cell–cell fusion. J Biol Chem 296:100411. https://doi.org/10.1016/j.jbc.2021.100411

Whitlock JM, Chernomordik LV (2025) Cell–cell fusion: to lose one life and begin another. BioEssays 47(2):2400206. https://doi.org/10.1002/bies.202400206

Wong MK, Shawky SA, Aryasomayajula A, Green MA, Ewart T, Selvaganapathy PR, Raha S (2018) Extracellular matrix surface regulates self-assembly of three-dimensional placental trophoblast spheroids. PLoS One 13(6):e0199632. https://doi.org/10.1371/journal.pone.0199632

Wong MK, Wahed M, Shawky SA, Dvorkin-Gheva A, Raha S (2019) Transcriptomic and functional analyses of 3D placental extravillous trophoblast spheroids. Sci Rep 9(1):12607. https://doi.org/10.1038/s41598-019-48816-8

Yang Y, Margam NN (2021) Structural insights into membrane fusion mediated by convergent small Fusogens. Cells 10(1):160. https://doi.org/10.3390/cells10010160

Zajdel TJ, Shim G, Wang L, Rossello-Martinez A, Cohen DJ (2020) Scheepdog: programming electric cues to dynamically herd large-scale cell migration. Cell Syst 10(6):506–514.e3. https://doi.org/10.1016/j.cels.2020.05.009

Zhang Y, Le T, Grabau R, Mohseni Z, Kim H, Natale DR, Feng L, Pan H, Yang H (2020) TMEM16F phospholipid scramblase mediates trophoblast fusion and placental development. Sci Adv 6(19):eaba0310. https://doi.org/10.1126/sciadv.aba0310

Zhang Y, Shan KZ, Liang P, Lowry AJ, Feng L, Yang H (2025) PIEZO1 drives trophoblast fusion and placental development. Nat Commun 16(1):6895. https://doi.org/10.1038/s41467-025-62254-3

Zhu K, Takada Y, Nakajima K, Sun Y, Jiang J, Zhang Y, Zeng Q, Takada Y, Zhao M (2019) Expression of integrins to control migration direction of electrotaxis. FASEB J 33(8):9131–9141. https://doi.org/10.1096/fj.201802657R

Zimmermann U (1983) Electrofusion of cells: principles and industrial potential. Trends Biotechnol 1(5):149–155. https://doi.org/10.1016/0167-7799(83)90006-9

Chapter 6
Trophoblast Fusion Deficiency in Placenta-Related Pregnancy Disorders

Yan Xiong and Yu-Bin Ding

Abstract Trophoblast cell fusion is a vital developmental process that enables the formation of the multinucleated syncytiotrophoblast (STB), which plays a central role in placental function and maternal–fetal exchange. Fusion defects in this lineage are closely associated with pregnancy complications such as preeclampsia (PE) and fetal growth restriction (FGR). Although the fusion of trophoblasts is a highly coordinated event, it involves multiple interdependent steps, including transcriptional programming, membrane remodeling, cytoskeletal rearrangement, metabolic adaptation, and immune regulation. Recent studies have uncovered critical molecular mediators at each of these levels. Transcription factors such as Glial cells missing transcription factor 1(GCM1), Krüppel-like factor 6(KLF6), and Transcription factor EB(TFEB) govern differentiation timing; fusion proteins, including Syncytin-1/2 and Mfn2, facilitate membrane merger; polarity regulators and actin-associated proteins like Par6 and CNN3 organize cytoskeletal architecture; metabolic reprogramming, particularly a shift from oxidative phosphorylation to glycolysis, supplies energy and biosynthetic precursors; and immune modulators such as pregnancy-induced factor 1 and Interleukin-10(IL-10) ensure a permissive environment for fusion at the maternal–fetal interface. Epigenetic mechanisms, including DNA methylation and histone modifications, further fine-tune the expression of fusion-related genes. Alongside mechanistic discoveries, a wide range of experimental models has been developed to investigate trophoblast fusion in vitro. These include traditional monolayer cell lines (e.g., BeWo), primary human trophoblasts, placental explants, trophoblast stem cells, and trophoblast organoids. Each model system provides distinct advantages in recapitulating aspects of syncytialization and placental physiology. Moreover, the integration of multi-omics technologies—such as single-cell and spatial transcriptomics, proteomics, metabolomics, and epigenomics—has expanded our understanding of the spatiotemporal dynamics and molecular complexity underlying trophoblast fusion. Despite these advances,

Y. Xiong · Y.-B. Ding (✉)
Department of Toxicology, Joint International Research Laboratory of Reproduction and Development of the Ministry of Education of China, School of Public Health, Chongqing Medical University, Chongqing, China
e-mail: dingyb@cqmu.edu.cn

© The Author(s), under exclusive license to Springer Nature Switzerland AG 2026
J. Weiler et al. (eds.), *Cell Fusion*, Advances in Experimental Medicine and Biology 1509, https://doi.org/10.1007/978-3-032-22637-2_6

several key challenges remain unresolved, including the lack of models that fully recapitulate the structural and functional features of the human maternal–fetal interface and the limited understanding of posttranslational and spatiotemporal regulatory mechanisms. Addressing these gaps will be essential for translating basic insights into diagnostic and therapeutic innovations for placental diseases.

Keywords Placenta · Syncytiotrophoblast · Trophoblast fusion · Transcription factors · Multi-omics · Placenta-related pregnancy disorders

6.1 Introduction

Cell–cell fusion is a fundamental biological process critical for sexual reproduction, development, and homeostasis across a broad spectrum of organisms, from fungi to humans (Chen and Olson 2005; Hernandez and Podbilewicz 2017; Brukman et al. 2019). During pregnancy, this process is particularly critical for embryo implantation and placental development, where trophoblast cells undergo fusion to form the syncytiotrophoblast (STB)—a specialized multinucleated structure essential for establishing maternal–fetal exchange and maintaining pregnancy. During human embryogenesis, two distinct forms of STB develop at different stages: the primitive STB and the definitive STB. The primitive STB emerges during the second week postfertilization, playing a central role in implantation and decidual invasion, while the definitive STB arises from the third week onward and lines the chorionic villi throughout gestation. The formation of the primitive STB coincides with implantation, as the blastocyst penetrates the uterine surface epithelium. This structure likely originates from the fusion of cytotrophoblast (CTB) cells at the embryonic pole of the blastocyst. Once established, the primitive STB rapidly expands into the decidua, invading the uterine stroma, glands, and capillaries. Within the primitive STB, fluid-filled cavities known as lacunae form, which subsequently fill with maternal blood and glandular secretions from the eroded decidual tissues. These lacunar contents serve as an early nutrient source for the developing embryo. As development progresses, CTB cells proliferate into the primitive STB, forming primary villi that traverse its entire thickness. These villi interconnect to form the cytotrophoblast shell, anchoring the conceptus to the decidua basalis. Subsequently, the definitive STB develops, covering the outer surfaces of the chorionic villi. Positioned directly in contact with maternal blood, the definitive STB constitutes the principal interface for maternal–fetal exchange throughout pregnancy (Jaremek et al. 2021).

The fusion of CTBs to form the STB is a highly regulated process, orchestrated by both intrinsic factors and the microenvironment. This process occurs in three distinct stages: competence, commitment, and cell fusion (Li et al. 2023; Gerbaud and Pidoux 2015; Renaud and Jeyarajah 2022). In the competence stage, CTBs exit the mitotic cell cycle and differentiate into fusion-competent cells, with the suppression of progenitor maintenance factors (e.g., TEA domain transcription factor 4 (TEAD4) (Saha et al. 2020), Yes-associated protein (YAP) (Mizutani et al. 2022), and GATA binding protein 3(GATA3) (Krendl et al. 2017)) and the activation of pathways that promote STB formation, including Wnt and activin/transforming growth factor beta(TGF-β) signaling. Key transcription factors such as Glial cells missing 1 (GCM1) (Wang et al. 2022a) and OVO-like 1 (OVOL1) (Renaud et al. 2015), along with epigenetic regulators (histone deacetylases, HDACs) and signaling pathways like cyclic adenosine monophosphate (cAMP)/PKA and mitogen-activated protein kinase(MAPK), drive this differentiation process. During the commitment stage, intercellular adhesion is facilitated by junctional complexes (E-cadherin and ZO-1) (Pidoux et al. 2010; Aghababaei et al. 2015) and gap junctions (connexin-43), enabling synchronization and signaling between adjacent cells (Coutifaris et al. 1991). Finally, in the cell fusion stage, the fusogenic proteins Syncytin-1 and Syncytin-2 mediate the opening of fusion pores, leading to the merging of cell membranes and the formation of the multinucleated STB (Borges et al. 2003; Dupressoir et al. 2009). This tightly coordinated process is essential for proper placental development and function.

Disruptions in the trophoblast cell fusion process can lead to a variety of placenta-related pregnancy complications. The STB, a multinucleated layer formed by the continuous fusion of mononucleated CTBs, serves as the primary interface between maternal and fetal circulations, facilitating critical functions such as nutrient and gas exchange (Costa 2016). The proper development of STB is a key indicator of successful placentation, as it ensures the efficient establishment of the maternal–fetal barrier. However, defective syncytialization, coupled with altered expression or dysregulation of fusion modulators, has been implicated in a range of pregnancy-related disorders. These include preeclampsia (PE), fetal growth restriction (FGR), and placental abruption, where the inability to form a functional STB layer impairs placental development and function. In these conditions, the failure of trophoblast fusion results in compromised placental exchange capacity, contributing to maternal and fetal morbidity (Kosinska-Kaczynska 2022; Heazell et al. 2007). Thus, understanding the molecular mechanisms underlying trophoblast fusion and its regulation is essential for the development of therapeutic strategies aimed at preventing or treating placental dysfunction in pregnancy.

6.2 Study Models

One major challenge in studying the mechanisms of trophoblast fusion lies in the lack of ideal experimental models. Currently, human placental samples obtained after delivery and chorionic villi collected from early pregnancy terminations are commonly used in placental research due to their relative accessibility. However, obtaining placental samples from the second trimester remains difficult. In addition, primary trophoblast cells isolated from placental tissues are limited in quantity and exhibit low proliferative capacity, which restrict their widespread application in in vitro studies. Although animal models have provided valuable insights into placental development, significant interspecies differences in placental structure and physiology limit the translational relevance of these findings. As a result, current

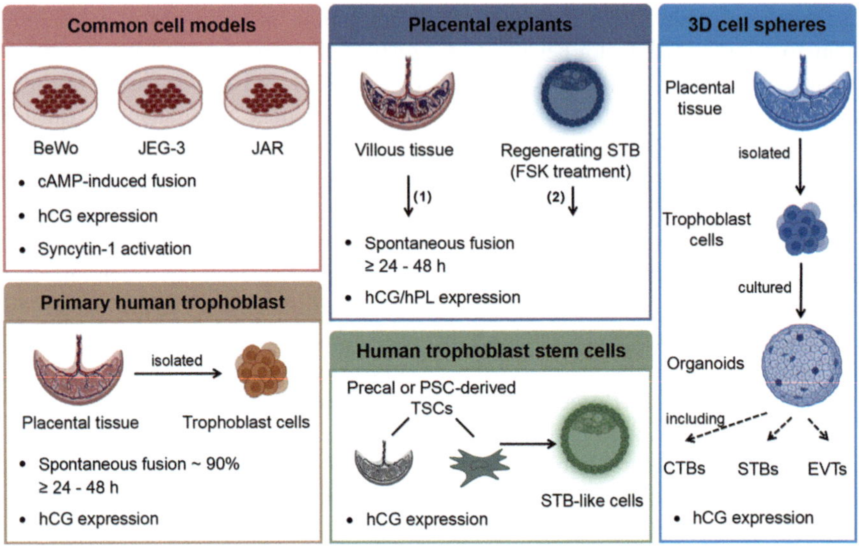

Fig. 6.1 Experimental models for studying human trophoblast fusion and differentiation. This schematic summarizes the current in vitro and ex vivo models employed to study trophoblast fusion

investigations into trophoblast biology primarily rely on various in vitro models, including traditional monolayer cultures and more recently developed three-dimensional (3D) culture systems (Li et al. 2023). Despite notable advancements, existing models still fall short of fully recapitulating the complex structure and function of the human placenta. This chapter systematically reviews the currently available models for studying trophoblast fusion (Fig. 6.1), aiming to inform model selection and application in future research.

6.2.1 Monolayer Cell Lines

6.2.1.1 Common Cell Models

Monolayer culture models have long been the preferred in vitro system for studying trophoblast fusion due to their ease of establishment, maintenance, and high stability. Among these, choriocarcinoma-derived cell lines, such as BeWo, JEG-3, and JAR, have been widely employed to investigate the mechanisms underlying placentation and trophoblast fusion.

The BeWo cell line was originally isolated in 1959 from cerebral metastases of choriocarcinoma in three female patients (Hertz 1959). After multiple passages, a stable cell line capable of producing human chorionic gonadotropin (hCG) was established in 1968 (Pattillo et al. 1968; Pattillo and Gey 1968). Studies have demonstrated that the treatment of BeWo cells with cyclic adenosine monophosphate (cAMP) analogs, such as forskolin, induces cell fusion and the formation of multinucleated STBs, accompanied by the secretion of placenta-specific proteins, including hCG, placental alkaline phosphatase, and SP1 (Wice et al. 1990). As such, the BeWo cell line has become a widely used model for investigating trophoblast fusion mechanisms, drug transport, STB-derived extracellular vesicle release, and placental metabolism (Alfaifi et al. 2020; Li et al. 2020; Abdelkhaliq et al. 2020).

The JEG-3 and JAR cell lines are also derived from human choriocarcinoma, but they differ in their capacity for cell fusion. JEG-3 cells, a subclone of BeWo cells, were originally obtained by reinjecting BeWo cells into hamster cheek pouches (Kohler and Bridson 1971). In contrast, JAR cells were isolated from a placental trophoblastic tumor and are known to secrete various placental hormones. Although both cell lines are capable of producing multiple placental hormones in response to treatment with cAMP analogs, their ability to undergo membrane fusion is limited, and prominent syncytiotrophoblast structures fail to form.

In recent years, studies utilizing BeWo and JEG-3 cell lines have identified a variety of transcription factors (e.g., BCL6 (Jasmer et al. 2017), OVOL1 (Renaud et al. 2015), HDACs (Jaju Bhattad et al. 2020), CREB (Finney 1965), DLX3 (Gilles et al. 1994), GATAs (Cheng and Handwerger 2005), and TBX3 (Lv et al. 2019)) and signaling pathways (e.g., cAMP/PKA (Knerr et al. 2005), cAMP/Epac1/CaMK1 (Chang et al. 2011), PI3K/mTOR (Vatish et al. 2012), JAK/STAT (Leduc et al. 2012), Wnt (Matsuura et al. 2011), and MAPK (Daoud et al. 2005)) as critical regulators of STB formation. In addition, factors associated with membrane remodeling—such as E-cadherin (Ng et al. 2011), calmodulin-3 (Shibukawa et al. 2010), Syncytin-1 (Huppertz et al. 2006; Huang et al. 2013), and Syncytin-2 (Soygur and Moore 2016)—have been extensively investigated. However, it should be noted that these cell lines exhibit heterogeneity and have undergone extensive passaging, leading to various chromosomal abnormalities. Furthermore, their genome-wide DNA methylation profiles differ markedly from those of primary trophoblasts (Novakovic et al. 2011). Therefore, findings obtained from these cell lines should be interpreted in conjunction with data from primary trophoblast models to ensure more comprehensive and reliable conclusions.

6.2.1.2 Primary Human Trophoblast

The primary human trophoblast (PHT) model, established by the direct isolation of trophoblast cells from human placental tissue, is widely employed to study trophoblast fusion mechanisms. This model closely recapitulates the biological characteristics of trophoblasts in vivo, particularly their spontaneous fusion into multinucleated STBs. Studies have demonstrated that fusion rates of approximately 90% can be achieved within 24–48 h of culture and that fused cells secrete hormones, such as hCG and human placental lactogen (hPL), reflecting the functional properties of STBs (Kliman et al. 1986).

However, the PHT model has certain limitations. First, isolated cells may be contaminated with stromal and blood cells, which can affect the accuracy of experimental results. Second, PHT cells lack proliferative capacity, and fused cells typically undergo apoptosis approximately 5 days after in vitro culture, necessitating isolation from fresh tissue for each experiment, thereby increasing complexity and cost. Additionally, due to differences between the in vitro culture environment and the in vivo microenvironment, the PHT model may not fully recapitulate the complex processes of placental development. Nonetheless, the PHT model remains valuable for studying key regulatory genes involved in trophoblast fusion (Renaud et al. 2015), such as YAP (Meinhardt et al. 2020), ID2 (Janatpour et al. 2000), HDACs (Jaju Bhattad et al. 2020), and TFAP2 (Richardson et al. 2001), as well as for investigating placental barrier anti-infective functions, mechanisms of STB exosome release, and drug transport.

6.2.2 Organization-Based Model

6.2.2.1 Placental Explants

The placental explant model, established by culturing human placental villous tissue in vitro, preserves the three-dimensional structure of the placenta and the interactions among various cell types. It has been widely used to study trophoblast proliferation, differentiation, and fusion. During culture, this model supports the regeneration of the STB layer and the secretion of hormones, such as hCG and hPL, reflecting the functional characteristics of STBs. However, the culture duration of placental explants is limited. Cellular swelling, fragmentation, and disruption of STB junctions are typically observed after 24 h of culture, which constrains the model's experimental utility (Xiao et al. 2020).

In recent years, researchers have sought to extend the viability and structural integrity of placental explants by optimizing culture conditions. For example, in 2021, Nadja et al. developed a closed perfusion system to simulate the uterine environment and found that it better preserved tissue viability and structural integrity over a 48-h culture period (Kupper et al. 2021). In addition, the placental explant model has been employed to investigate the placental barrier's antimicrobial functions, the release of extracellular vesicles from STBs, and drug transport mechanisms.

6.2.2.2 Human Trophoblast Stem Cells

Human trophoblast stem cells (TSCs) have emerged as a novel in vitro model for studying trophoblast fusion, characterized by a normal karyotype and the capacity for unlimited self-renewal and multilineage differentiation under defined conditions. TSCs can be derived from two primary sources: (i) direct isolation from human placental trophoblast tissues, for example, proliferative CTBs isolated from early gestation placentas, which can be induced to form TSCs under specific culture conditions, and (ii) human pluripotent stem cells (PSCs), including embryonic stem cells (ESCs), induced pluripotent stem cells (iPSCs), and extended potential stem cells (ePSCs), which can be differentiated into TSCs using defined induction protocols (Tan et al. 2022; Baker and Pera 2018). These TSCs can further differentiate into multinucleated STB-like cells upon stimulation, expressing characteristic trophoblast markers and secreting hCG.

The TSC model offers a highly controllable and homogeneous cell population, making it well-suited for investigating trophoblast differentiation, fusion processes, and their roles in placental development. However, several limitations remain. For instance, compared to primary trophoblast cells, TSCs may exhibit differences in gene expression and epigenetic modifications. Additionally, the culture systems required for TSCs are technically demanding and relatively expensive. In recent years, significant progress has been made using TSC models, including RNA sequencing studies that have revealed transcriptional similarities between TSCs and primary trophoblasts, as well as investigations into the roles of specific transcription factors in trophoblast lineage differentiation.

6.2.3 3D Cell Spheres

Organoid models have recently emerged as advanced 3D culture systems for studying trophoblast cell fusion. By mimicking the 3D architecture of placental villi, these models offer a physiologically relevant platform for investigating trophoblast biology. TSCs are isolated from human placental tissue and cultured under defined conditions to self-organize into organoids that comprise multiple trophoblast subtypes, including CTBs, STBs, and extravillous trophoblast (EVT)s. These organoids can be maintained in long-term culture, undergo spontaneous cell fusion to form functional multinucleated STBs, and secrete hCG, thereby providing a powerful tool for elucidating the mechanisms underlying trophoblast differentiation and fusion.

However, organoid models also present certain limitations. For instance, in conventional culture systems, STBs are predominantly localized on the interior of the organoid, failing to fully recapitulate the structural feature of direct contact with maternal blood. To address this limitation, researchers have developed suspension culture methods that reposition STBs to the outer surface of the organoid, thereby better mimicking the physiological polarity of the placenta. Despite the significant advances made using organoid models to study trophoblast fusion, further refinement of culture conditions is required to more comprehensively replicate the complex placental microenvironment.

In summary, various experimental models have played a pivotal role in advancing the study of trophoblast fusion. Monolayer cell culture systems offer a simple and accessible platform that, despite their limited ability to replicate the structural complexity of the placenta, remains valuable for investigating fundamental cellular behaviors. PHT cells and placental explant models, by preserving tissue architecture and intercellular interactions, have facilitated research into trophoblast proliferation, differentiation, and their interactions with the immune microenvironment. TSC models provide new opportunities for studying long-term proliferation and differentiation of trophoblast lineages and enable the generation of distinct trophoblast subtypes in vitro. Three-dimensional trophoblast spheroids and placental organoids (such as CTB-ORGs (Haider et al. 2018) and SC-TOs (Turco et al. 2018)), as more physiologically relevant systems, better recapitulate trophoblast fusion and function, offering powerful platforms for studies on placental development, viral infection, and drug transport. The combined application of these complementary models offers a multifaceted perspective and robust toolkit for deepening our understanding of trophoblast differentiation and fusion.

6.3 Molecular Mechanisms of Trophoblast Cell Differentiation and Fusion

The differentiation and fusion of trophoblast cells represent a highly intricate and precisely regulated biological process essential for placental development. This transition is orchestrated by a coordinated network of molecular mechanisms, including transcriptional regulation, intercellular interactions, polarity establishment, metabolic remodeling, and immunological support (Fig. 6.2). With the advent of multi-omics technologies such as single-cell and single-nucleus RNA sequencing (sc/snRNA-seq), proteomics, metabolomics, spatial transcriptomics, and epigenomic profiling, our understanding of this complex process has deepened significantly. These tools provide unprecedented resolution in mapping cell states, signaling dynamics, and intercellular communication in both healthy and pathological placentas.

6.3.1 Transcription Factors

The differentiation and fusion of trophoblast stem cells are governed by fine-tuned transcriptional networks. Classical transcription factors such as KLF6, Twist1, and TFEB are central to this regulation. KLF6 promotes trophoblast proliferation and primes cells for differentiation (Miranda et al. 2022), while Twist1 facilitates cytoskeletal remodeling essential for fusion (Lu et al. 2016). As a key regulator of lysosomal biogenesis and autophagy, TFEB may participate in trophoblast fusion by modulating intracellular metabolism and membrane dynamics (Cesana et al. 2024; Esbin et al. 2024).

Recent single-nucleus RNA sequencing (snRNA-seq) studies have reconstructed trophoblast lineage trajectories, revealing stage-specific expression peaks of key

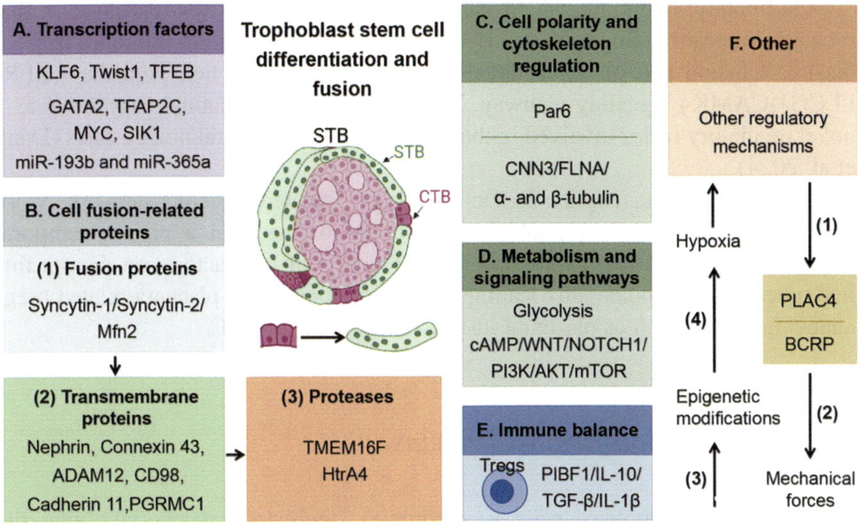

Fig. 6.2 Multilayered regulatory mechanisms governing trophoblast stem cell differentiation and fusion. Trophoblast stem cell differentiation and fusion into multinucleated syncytiotrophoblasts involve a tightly coordinated and multistep biological process. This process is regulated by a complex interplay of transcription factors, dynamic gene expression, cell polarity establishment, cytoskeletal remodeling, and membrane fusion proteins. Concurrently, cellular metabolic pathways and signaling cascades provide both energy and regulatory input essential for driving cellular transition. Intercellular interactions and modulation by the maternal–fetal immune microenvironment further support successful trophoblast differentiation, fusion, and placental function. Collectively, these interdependent regulatory layers ensure precise structural and functional development of the placenta

transcription factors. For instance, factors such as GATA2, TFAP2C, and MYC show dynamic expression patterns during the differentiation of uninucleate trophoblast cells (UNCs) into binucleate trophoblast cells (BNCs), as identified by pseudotime analysis. Upstream transcription factor-binding site enrichment analyses further support their roles in synchronizing differentiation timing (Davenport et al. 2023).

In addition to these classical transcription factors, an increasing number of novel regulatory factors have been identified as influencing the trophoblast syncytialization process. For instance, SIK1, identified through RNA sequencing analysis, has been shown to play a regulatory role in trophoblast stem cell differentiation and fusion, potentially by modulating cellular signaling pathways to facilitate fusion capacity (Msheik et al. 2019). MicroRNAs (miRNAs) also play a critical role in this process. miR-193b and miR-365a promote STB formation by regulating the expression of fusion-related genes, whereas miR-181d-5p is abnormally elevated in FGR placentas and inhibits BeWo cell fusion by repressing CREB3 recruitment factor (CREBRF), highlighting its pivotal role in the regulation of trophoblast fusion (Apicella et al. 2021; Wu et al. 2023).

Additionally, nontraditional transcriptional regulators are also involved in this process. Nuclear binding protein 2 (Nesfatin-1) has been found to influence trophoblast cell fusion and placental development through interaction with the EGFR-PLCG1-CAMK4 signaling pathway. This suggests that Nesfatin-1 may serve as a novel regulatory factor involved in the maintenance of placental homeostasis (Dang et al. 2024).

Overall, these transcription factors and their regulated signaling networks work together to ensure that trophoblast cells undergo fusion in a precise temporal sequence. Any abnormal expression or dysfunction of key factors can disrupt this fine-tuned process, potentially leading to the development of placenta-related pregnancy disorders, such as placental insufficiency, FGR, and PE.

6.3.2 *Cell Fusion-Related Proteins*

The fusion of trophoblast cells is a complex, multistep process involving the interaction and precise regulation of various proteins. These proteins can be broadly classified into three categories: fusion proteins, transmembrane proteins, and proteases. They collaborate at different levels to ensure the successful progression of cell fusion.

Fusion proteins are the direct mediators of cell fusion, primarily by binding to receptors on the cell membrane, activating downstream signaling pathways, or directly promoting membrane fusion. Syncytin-1 and Syncytin-2 are the most typical fusion proteins and play crucial roles in the trophoblast cell fusion process (Huppertz et al. 2006; Huang et al. 2013; Roberts et al. 2021; Toufaily et al. 2013; Xu and Sun 2022). Mfn2, a mitochondrial fusion protein, has also been found to participate in trophoblast cell fusion, particularly in the coordination of intercellular inner membrane fusion and mitochondrial dynamics (Chen et al. 2003). Proteomic analyses have identified dynamic changes in protein expression during BeWo cell syncytialization, highlighting the upregulation of proteins associated with trophoblast fusion (Gehl et al. 2025). Immunofluorescence studies have localized fusion-related proteins, such as Syncytin-2, to specific trophoblast subtypes in the placenta, providing insights into their spatial expression patterns (Gehl et al. 2025; Liu et al. 2018).

Transmembrane proteins are also indispensable in the trophoblast cell fusion process. They promote fusion by mediating intercellular adhesion and signal transduction, forming tight junctions on the cell membrane. Additionally, they influence fusion efficiency indirectly by regulating amino acid transport and membrane fluidity. Transmembrane proteins such as nephrin (Li et al. 2015), connexin 43 (Dunk et al. 2012), ADAM12 (Aghababaei et al. 2015), CD98 (Mori et al. 2004), and cadherin 11 (Getsios and MacCalman 2003) have been shown to play key roles in trophoblast syncytialization, while PGRMC1 affects trophoblast cell fusion capacity to some extent by regulating membrane properties (Mori et al. 2004).

Proteases play a crucial role in trophoblast cell fusion by cleaving cell surface receptors and extracellular matrix components, facilitating membrane remodeling and providing the molecular foundation necessary for cell fusion. TMEM16F, a Ca^{2+}-activated phospholipid scramblase (CaPLSase), enhances membrane fusion capacity by promoting the exposure of phosphatidylserine (PS) (Zhang et al. 2020; Zhang et al. 2022). HtrA4, a serine protease, degrades various extracellular matrix proteins and receptor proteins, assisting in cell shape remodeling and promoting trophoblast syncytialization (Yuan et al. 2023). Therefore, abnormal expression of TMEM16F and HtrA4 may lead to impaired trophoblast cell fusion, which can adversely affect placental development and function.

Generally, these proteins interact with each other to exert a synergistic effect at different stages of trophoblast cell fusion, ensuring that trophoblast stem cells successfully differentiate and ultimately form a multinucleated syncytial structure.

6.3.3 Cell Polarity and Cytoskeleton Regulation

The regulation of cell polarity and the cytoskeleton is also crucial for trophoblast cell differentiation and fusion. These factors not only determine cell morphology and migratory capacity but also provide structural support for cell fusion by modulating intercellular interactions.

Cell polarity is a fundamental basis for cell functional differentiation and tissue development. It ensures effective intercellular signaling and precise fusion by regulating cell membrane regionalization, cell shape remodeling, and molecular transport. Par6, as one of the key regulators of cell polarity, negatively modulates the fusion process of trophoblast cells by controlling tight junctions and cytoskeletal dynamics (Sivasubramaniyam et al. 2013). Therefore, other cell polarity-related factors may also play regulatory roles in trophoblast cell fusion. Abnormal expression of cell polarity factors significantly impacts the fusion process of trophoblast cells.

The cytoskeletal system not only provides support during trophoblast cell differentiation and fusion but also regulates cell morphology and dynamic remodeling. Calponin-3 (CNN3), a widely expressed F-actin-binding protein, interacts with the actin cytoskeleton to regulate cell shape and aggregation, enhancing intercellular tight junctions and supporting the morphological changes required for cell fusion (Liu and Jin 2016). In addition to serving as a mechanical support component, CNN3 dynamically regulates cytoskeletal rearrangement, providing the necessary structural foundation for trophoblast cell fusion. Furthermore, other cytoskeleton-related proteins may also play roles in the fusion process. For example, α- and β-tubulin, components of microtubules, are crucial for cell migration, polarity establishment, and post-division remodeling, and they may influence trophoblast cell fusion efficiency by modulating the redistribution of intracellular structures (McKenna et al. 2023). Similarly, filamin A (FLNA), an actin-cross-linking protein, has been shown to play a role in morphological changes and signal transduction in various cell types (Rosa et al. 2019). It may impact trophoblast fusion by stabilizing the cytoskeleton and coordinating transmembrane protein signaling. Additionally, septins, a family of Guanosine triphosphate (GTP)-binding proteins, are important for membrane curvature, cytoskeletal organization, and cell division, with their potential role in trophoblast cell fusion warranting further investigation (Kinoshita 2003). Therefore, various cytoskeletal components, including actin filaments, microtubules, cross-linking proteins, and septins, cooperate to promote trophoblast cell fusion through multiple mechanisms such as morphological shaping, mechanical support, and signal regulation, thereby ensuring proper placental development and function.

Comprehensively, the regulation of cell polarity and the cytoskeleton, in collaboration with the actions of cell fusion-related proteins, works synergistically at multiple levels to promote trophoblast cell differentiation and fusion, thereby ensuring proper placental development and function.

6.3.4 Metabolism and Signaling Pathways

The differentiation and fusion of trophoblast cells is a highly coordinated process that not only relies on precise transcriptional regulation and protein interactions but is also profoundly influenced by metabolic status and multiple key signaling pathways. Energy metabolism, particularly the activity of glycolysis, plays a crucial role in this process. Metabolomic analyses indicate that differentiating trophoblasts exhibit a metabolic shift toward glycolysis, reducing reliance on oxidative phosphorylation. This glycolytic activity supplies Adenosine Triphosphate (ATP) and biosynthetic precursors essential for membrane fusion during trophoblast differentiation (Brugger et al. 2023; Yu et al. 2024). In addition to glycolysis, several signaling pathways also regulate intracellular energy metabolism. For example, the cAMP signaling pathway activates cAMP-dependent PKA, leading to the phosphorylation of downstream target proteins, which further influence cellular metabolic dynamics and morphological shaping (Omata et al. 2013). This, in turn, plays a synergistic regulatory role in trophoblast cell fusion.

Several signaling pathways play key roles in regulating the proliferation, differentiation, and fusion of trophoblast cells. For example, the activation of the Wnt signaling pathway influences the proliferation of trophoblast stem cells and cytotrophoblasts (Haider et al. 2018; Dietrich et al. 2022). The Notch receptor 1 (NOTCH1) signaling pathway promotes the activation of specific differentiation pathways through intercellular interactions (Haider et al. 2018; Dietrich et al. 2022; Tang et al. 2023). The Phosphatidylinositol 3-kinase / Protein kinase B / Mechanistic target of rapamycin signaling pathway (PI3K/AKT/mTOR pathway), a core regulator of cell growth and metabolism, not only promotes cell survival but also indirectly affects the efficiency of cell fusion by regulating nutrient utilization (Zhao et al. 2024). These signaling pathways are interconnected, collectively maintaining trophoblast cell homeostasis and ensuring proper placental development and function.

In conclusion, trophoblast cell fusion is finely regulated by metabolic dynamics and multiple signaling pathways, with both interacting to maintain the homeostasis of the fusion process. Any abnormalities in metabolism or signal transduction can lead to placental dysfunction, potentially resulting in pregnancy-related disorders. Therefore, further research into their roles in trophoblast cell fusion will not only help reveal the molecular mechanisms of placental development but also provide new theoretical foundations and potential targets for the prevention and treatment of pregnancy-related diseases.

6.3.5 Immune Balance

The differentiation and fusion of trophoblast cells are critical processes in placental development. These processes are not only tightly regulated by intracellular molecular mechanisms but are also profoundly influenced by the immune microenvironment. Single-cell and spatial transcriptomic analyses have provided insights into the distribution of immune cells at the maternal–fetal interface, revealing alterations in trophoblasts at invasion sites and disturbances in the myometrial immune microenvironment. The immune homeostasis at the maternal–fetal interface is essential for proper trophoblast fusion, as immune regulatory factors modulate maternal–fetal immune tolerance, maintaining a suitable microenvironment that ensures the successful transition of CTBs into multinucleated STBs. Among these factors, pregnancy-induced factor 1 (PIBF1) plays a pivotal role in sustaining immune balance and promoting trophoblast fusion (Yu et al. 2024). However, the precise mechanisms by which PIBF1 interacts with specific receptors or signaling pathways to regulate immune cell infiltration, trophoblast fusion, and functional maintenance remain to be further elucidated.

Additionally, various immune cells and inflammatory factors play crucial roles in trophoblast fusion. Regulatory T cells (Tregs) promote proper trophoblast differentiation and fusion by secreting anti-inflammatory cytokines, such as IL-10 and TGF-β, which suppress excessive inflammatory responses and maintain maternal–fetal immune tolerance. In contrast, M1 macrophages, upon stimulation by Interferon-gamma (IFN-γ) or lipopolysaccharides (LPSs), secrete pro-inflammatory cytokines, including Tumor necrosis factor-alpha (TNF-α) and IL-1β, thereby amplifying inflammation. Excessive release of these pro-inflammatory factors may impair trophoblast fusion and contribute to placental dysfunction. Moreover, uterine natural killer (uNK) cells facilitate trophoblast invasion and angiogenesis by secreting factors such as IL-8 and hepatocyte growth factor (HGF). However, when cultured in vitro, uNK cells can be activated by IL-2, acquiring cytotoxic properties against target cells, including trophoblasts (Schust et al. 2021). This suggests that uNK cell function is tightly regulated during normal pregnancy to prevent placental tissue damage.

In future studies, multi-omics analyses could be employed to further elucidate the association between immune dysregulation and the downregulation of fusion-related genes (such as Syncytins), thereby reinforcing the critical role of immune modulation in trophoblast fusion.

6.3.6 Other

In addition to transcriptional regulation, fusion-related proteins, cell polarity, metabolic signaling, and immune modulation, other regulatory mechanisms also play crucial roles in trophoblast stem cell differentiation and fusion. For instance, key proteins such as placenta-specific protein 4 (PLAC4) and breast cancer resistance protein (BCRP), as well as factors such as hypoxic microenvironments, mechanical forces, and epigenetic modifications, may significantly influence the trophoblast fusion process.

PLAC4 is a placenta-specific protein whose mRNA is primarily localized in trophoblast cells and can be released into maternal circulation, making it a key biomarker for noninvasive prenatal testing (NIPT). Studies have shown that PLAC4 expression is upregulated in placentas from severe early-onset preeclampsia cases, with specific localization to STBs (Tuohey et al. 2013). Additionally, both mRNA and protein levels of PLAC4 increase significantly during the syncytialization of BeWo cells, suggesting its potential role in trophoblast fusion, while its aberrant expression may contribute to fusion defects.

Breast cancer resistance protein (BCRP), also known as ATP-binding cassette subfamily G member 2 (ABCG2), is an ATP-binding cassette (ABC) transporter highly expressed in the placental barrier, predominantly localized at the brush border of STB cells (Lye et al. 2019). Its primary function is to utilize ATP hydrolysis to export various xenobiotics (e.g., drugs and toxins) and endogenous metabolites from the placenta, thereby protecting the fetus from potentially harmful substances. Beyond its role in placental barrier function, BCRP is increasingly recognized as a regulator of trophoblast biology. Its expression is closely associated with placental growth and development, and aberrant expression (either upregulation or downregulation) has been observed in pathological conditions such as FGR. Moreover, BCRP has been implicated in the transmembrane transport of cholesterol and steroid hormones, which may indirectly influence trophoblast differentiation and functional maturation. Consequently, abnormal BCRP expression may serve as an indicator of fusion defects.

Hypoxia is a crucial regulatory factor in placental development. In early pregnancy, the placenta develops under a low-oxygen environment, and as gestation progresses, increased placental perfusion leads to a gradual rise in oxygen levels (Kudo et al. 2003; Jaremek et al. 2023). Hypoxia modulates trophoblast proliferation, invasion, and differentiation, primarily through hypoxia-inducible factor-1α (HIF-1α)-mediated gene expression. On one hand, moderate hypoxic stimulation promotes trophoblast stem cell proliferation and maintains their stemness. On the other hand, excessive or prolonged hypoxia may impair trophoblast fusion, resulting in placental dysfunction. For instance, HIF-1α is upregulated in the placenta during early pregnancy and in cases of FGR (Jaremek et al. 2023). It has been suggested that HIF-1α may interact with the NOTCH signaling pathway to maintain cells in an undifferentiated state (Jaremek et al. 2023). In future studies, direct evidence of HIF-1α binding to the promoters of NOTCH and Wnt pathway genes in trophoblasts could be obtained through ATAC-seq and ChIP-seq analyses, thereby further elucidating the mechanisms by which hypoxia regulates trophoblast cell fusion.

During pregnancy, the placenta is subjected to mechanical forces, including uterine hemodynamic changes and tension generated by fetal growth. These mechanical stimuli regulate trophoblast differentiation and fusion by influencing cytoskeletal remodeling, cell adhesion, and signal transduction. For instance, shear stress and stretch forces within the placenta modulate the expression of integrins and E-cadherin, thereby affecting cell–cell adhesion and fusion. Additionally (Groten et al. 2010; Parameshwar et al. 2024), mechanical stimuli may activate the Yes-associated protein/Transcriptional co-activator with PDZ-binding motif (YAP/TAZ) signaling pathway, promoting trophoblast proliferation and differentiation, ultimately impacting placental development (Parameshwar et al. 2024). Although direct proteomic evidence supporting the regulatory role of mechanical forces in trophoblast cell fusion and differentiation is currently lacking, preliminary studies have suggested that mechanical stimulation may exert important effects in this process.

Epigenetic modifications, such as DNA methylation and histone modifications, play a crucial role in placental development and may influence trophoblast fusion. Studies have shown that the promoter region of the Syncytin-1 gene contains multiple CpG islands, whose methylation status directly affects gene expression. Aberrant DNA methylation may lead to reduced Syncytin-1 expression, thereby inhibiting trophoblast fusion. In addition to DNA methylation, histone modifications are also key epigenetic mechanisms regulating gene expression. In trophoblast cells, distinct histone modification patterns can modulate the transcriptional activity of fusion-related genes. For instance, H3K9me3, a well-known transcriptional repression marker, is enriched at the Syncytin-1 promoter, thereby inhibiting its transcription and reducing Syncytin-1 protein expression, ultimately impairing trophoblast fusion (Shankar et al. 2015). Conversely, H3K4me3, typically associated with active transcription, may enhance Syncytin-1 expression when enriched at its promoter, facilitating cell fusion (Shankar et al. 2015; Hahn et al. 2022). Additionally, H3K27ac is linked to enhancer activity and may promote the transcription of fusion-related genes in trophoblasts (Shankar et al. 2015; Hahn et al. 2022). In recent years, multi-omics studies have further revealed a profound link between epigenetic regulation and trophoblast cell fusion. For example, integrated analyses of the DNA methylome and transcriptome have shown that the downregulation of DNA methyltransferases (DNMTs) during the fusion process is closely associated with the upregulation of key fusion factors such as Syncytin-1. The use of DNA methylation inhibitors (e.g., 5-aza-CdR) has been shown to promote fusion, suggesting that the methylation status of DNA plays a functional regulatory role in this process (Yang et al. 2024). In addition, histone acetylation levels are influenced by the cellular metabolic state, particularly through fluctuations in acetyl-CoA availability, which can modulate modifications such as H3K27ac. This, in turn, indirectly regulates the expression of fusion-related genes by altering chromatin accessibility (Shankar et al. 2015). This metabolism–epigenetics interaction mechanism has been supported by combined metabolomic and proteomic analyses. It has been demonstrated in recent studies that aberrant DNA methylation and histone modifications may disrupt trophoblast fusion, impair normal placental development, and be associated with pregnancy-related disorders such as PE and FGR. Further investigation into the epigenetic regulation of trophoblast fusion could provide novel insights into the mechanisms underlying placental dysfunction and offer potential therapeutic targets for related diseases.

Aberrations in these factors have been associated with impaired placental development and an increased risk of pregnancy complications, including preeclampsia and fetal growth restriction. Elucidating the underlying molecular mechanisms is essential for advancing our understanding of the complex regulatory networks that govern placental development and may provide a theoretical foundation and novel targets for the prevention and management of pregnancy-related disorders.

In recent years, the rapid advancement of multi-omics technologies—particularly single-cell and single-nucleus transcriptomics, spatial transcriptomics, proteomics, metabolomics, and epigenomics—has enabled unprecedented resolution in characterizing the molecular features of trophoblast fusion. Single-cell transcriptomic analyses have delineated the differentiation trajectories of distinct trophoblast subtypes and revealed dynamic expression patterns of key transcription factors and fusion-associated genes. Spatial omics approaches have further clarified the anatomical localization of these factors and their intercellular interactions. Proteomic and phosphoproteomic analyses have uncovered expression changes and posttranslational modifications in membrane fusion-associated adhesion molecules and signaling proteins. Metabolomic studies have highlighted the critical roles of energy metabolism and lipid remodeling in the regulation of trophoblast fusion. Meanwhile, epigenomic analyses, such as DNA methylation and histone modification profiling, have demonstrated that chromatin states are tightly linked to the transcriptional regulation of fusion-related genes. The integration of these multi-omics datasets offers a comprehensive framework for identifying core regulatory factors involved in trophoblast fusion defects and provides valuable insights for the development of early diagnostic markers and precision therapeutic strategies for placental dysfunction.

6.4 Trophoblast Cell Fusion and Placenta-Related Pregnancy Diseases

The normal fusion of trophoblast cells is crucial for placental development and the maintenance of maternal–fetal interface stability. Disruption of this process can lead to abnormalities in placental structure and function, subsequently triggering a range

Table 6.1 Summary of reported factors associated with trophoblast dysfunction in placenta-related pregnancy disorders

Disease	Key molecular factors	References
PE	Transcription factors: HPIP, TNFAIP3, ARID1A, miR-193b, miR-365a, PPARG, BCL6	Penugurti et al. (2025), Yang et al. (2024), Fantone et al. (2020), Apicella et al. (2021), Shoaito et al. (2020), Louwen et al. (2022)
	Cell fusion-related proteins: Syncytin-1/2, THBS1, P-glycoprotein, Gal-2	Knerr et al. (2008), Vargas et al. (2011), Zhuang et al. (2014), Duan et al. (2021), Dunk et al. (2018), Hahn et al. (2022)
	Cell polarity and cytoskeleton regulation: RITA, cullin 1, VE-cadherin	Wildner et al. (2019), Zhang et al. (2013), Groten et al. (2010)
	Metabolism and signaling pathways: AMPK, cAMP, mTOR	Drewlo et al. (2020), Johnstone et al. (2005), Aiko et al. (2014)
	Immune balance: -	
	Other: -	
FGR	Transcription factors: miR-181d-5p, ARID1A	Wu et al. (2023), Fantone et al. (2020)
	Cell fusion-related proteins: Syncytin-1/2, Syncytin-a, MT1-MMP	Ruebner et al. (2010), Knerr et al. (2008), Wang et al. (2022b), Hiden et al. (2013)
	Cell polarity and cytoskeleton regulation:	
	Metabolism and signaling pathways: mTOR, TauT/SLC3A2, PPARγ/RXRa	Wang et al. (2024), Aiko et al. (2014), Desforges et al. (2013), Ruebner et al. (2012)
	Immune balance: Immunomodulatory factor	
	Other: -	
Other	NK cells, AMPK, DAPK3	Drewlo et al. (2020), Nakashima et al. (2008), Wang et al. (2020)

of pregnancy-related disorders, including PE, FGR/Intrauterine growth restriction (IUGR), and other placental dysfunction-associated disorders. Growing evidence suggests that impaired trophoblast fusion is often accompanied by the dysregulation of various molecular factors involved in transcriptional control, membrane fusion, cytoskeletal dynamics, metabolism, and immune balance. A summary of the key factors implicated in trophoblast dysfunction and their reported associations with placental pregnancy disorders is provided in Table 6.1. This overview highlights the molecular complexity underlying these diseases and underscores the importance of targeting trophoblast fusion pathways for future diagnostic and therapeutic strategies.

6.4.1 PE

PE is a pregnancy-related disease closely associated with abnormal placental development, typically presenting with symptoms such as hypertension and proteinuria in the later stages of pregnancy. Normal trophoblast cell fusion is essential for placental development and function. When trophoblast cell fusion is impaired, it can lead to abnormal placental vasculature, impaired maternal–fetal blood flow exchange, and subsequent pathological conditions, such as hypoxia and oxidative stress, ultimately resulting in the development of PE.

The mechanisms underlying fusion defects in PE include the abnormal expression of transcription factors and fusion-related proteins, immune microenvironment dysregulation, metabolic disturbances, and aberrant signaling pathway regulation. Studies on PE have revealed that transcription factors such as HPIP (Penugurti et al. 2025), TNFAIP3 (Yang et al. 2024), and ARID1A (Fantone et al. 2020), as well as the imbalance of miRNAs like miR-193b and miR-365a, affect the differentiation of trophoblast cells. Additionally, the dysregulation of PPARG and BCL6 weakens the invasive ability of trophoblast cells, impairing placental function (Shoaito et al. 2020; Louwen et al. 2022). Insufficient expression of fusion-related proteins such as Syncytin-1/2 hinders syncytialization, while abnormalities in THBS1 (Duan et al. 2021), P-glycoprotein (Dunk et al. 2018), and Gal-2 (Hahn et al. 2022) further exacerbate disruptions in intercellular interactions. RBPJ-interacting and tubulin-associated protein(RITA), a microtubule-associated protein (MAP) that interferes with tubulin, regulates microtubule (MT) stability and dynamics, influencing mitosis, migration, and invasion (Wildner et al. 2019). Cullin1 (CUL1), a scaffold protein in the cullin-based ubiquitin ligase complex, plays an important role in cell differentiation and invasion (Zhang et al. 2013). VE-cadherin, a calcium-dependent adhesion molecule, mediates intercellular adhesion, controlling monolayer integrity and barrier function (Groten et al. 2010). The abnormal expression of these proteins contributes to the development of PE. Moreover, abnormalities in signaling pathways such as AMP-activated protein kinase(AMPK) (Drewlo et al. 2020), cAMP (Johnstone et al. 2005), and mTOR (Aiko et al. 2014) weaken trophoblast cell energy supply, hindering their differentiation and fusion, thereby leading to the onset of PE.

Universally, the multilayered regulatory network governing trophoblast cell fusion, when disrupted, can lead to placental dysfunction, thereby promoting the development of pregnancy-related diseases such as PE. Therefore, a deeper understanding of these regulatory mechanisms not only aids in comprehending the pathogenesis of PE but also provides potential targets for early diagnosis and intervention strategies.

6.4.2 FGR

Defects in trophoblast cell fusion may lead to another condition—FGR/IUGR. Specifically, when the fusion of mononuclear CTBs to form multinucleated STBs is impaired, the efficiency of placental nutrient transfer decreases, thereby restricting fetal growth and development.

FGR is a pregnancy complication closely associated with placental dysfunction, involving molecular mechanisms at multiple levels, including transcriptional regulation, cell fusion, cell polarity, metabolic signaling, and immune homeostasis. At the transcriptional level, abnormal expression of miR-181d-5p may disrupt the proliferation and differentiation of trophoblast cells (Wu et al. 2023), while ARID1A inactivation affects chromatin remodeling, ultimately impairing placental development (Fantone et al. 2020). In FGR placentas, the expression of key fusion proteins Syncytin-1/2 is reduced, and abnormalities in Membrane-type 1 matrix metalloproteinase (MT1-MMP) and Syncytin-a further inhibit the formation of multinucleated STBs, affecting placental barrier function (Ruebner et al. 2010; Knerr et al. 2008; Hiden et al. 2013). The stability of cell polarity and the cytoskeleton is crucial for placental structure, and their disruption may lead to abnormal arrangement of trophoblast cells, weakening cell fusion capacity and exacerbating placental dysfunction. Metabolic and signaling pathway imbalances also play a critical role in FGR. For instance, inhibition of the mTOR signaling pathway affects trophoblast cell fusion (Wang et al. 2024), while abnormalities in TauT and SLC3A2 lead to amino acid metabolism disorders (Aiko et al. 2014; Desforges et al. 2013). Additionally, dysregulation of the PPARγ/RXRα signaling pathway impairs lipid metabolism and inflammatory regulation (Ruebner et al. 2012). Disruption of maternal–fetal immune homeostasis is also a key factor, as abnormal expression of placental immune regulatory factors may trigger excessive inflammation, impair placental blood flow, and further worsen placental dysfunction, ultimately limiting fetal growth.

Broadly, the occurrence of FGR is the result of the interplay of various molecular mechanisms, including abnormalities in gene expression regulation, metabolic adaptation, cytoskeletal remodeling, and immune microenvironment regulation. These changes collectively impair the function of trophoblast cells, affecting normal placental development and fetal growth. In the future, further exploration of the interactions between these molecular mechanisms will help clarify the underlying mechanisms of FGR and provide new strategies for its early diagnosis and intervention.

6.4.3 Other Pregnancy-Related Diseases

The normal maintenance of placental function depends on the proliferation, differentiation, and fusion of trophoblast cells, a process critical for maternal–fetal substance exchange, the formation of the placental barrier, and the success of pregnancy. When trophoblast cell fusion is impaired, placental structure and function may be affected to varying degrees, potentially leading to other pregnancy-related diseases. For example, placental abruption and placental infarction are often associated with trophoblast cell dysfunction, leading to reduced placental blood supply, localized hypoxia, and weakened integrity of the placental barrier (Liu et al. 2016). Recurrent spontaneous abortion (RSA) is also closely related to abnormal trophoblast cell fusion. Studies suggest that excessive activation of uterine NK (uNK) cells may trigger the abnormal elimination or dysfunction of trophoblast cells, affecting stable placental implantation (Nakashima et al. 2008). In addition, gestational diabetes mellitus (GDM) not only impacts maternal metabolism but may also alter trophoblast cell metabolic state and invasive ability, impairing placental function (Drewlo et al. 2020). Specifically, irisin (a fat-derived factor) may promote trophoblast cell differentiation by regulating the AMPK signaling pathway, while abnormal expression of death-associated protein kinase 3 (DAPK3) in GDM may inhibit trophoblast cell invasion, further affecting placental development (Wang et al. 2020). These studies reveal the close connection between placental insufficiency and trophoblast cell fusion defects, highlighting that various pathological factors may influence this process through multiple molecular mechanisms, thereby increasing the risk of pregnancy complications.

6.5 Potential Therapeutic Strategies

In exploring strategies to improve trophoblast cell fusion defects and placental insufficiency, various cutting-edge technologies and innovative therapies are offering new opportunities for potential intervention. Gene editing technologies, such as CRISPR–Cas9, have been applied in mouse trophoblast stem cells (mTSCs), providing the possibility for precise repair of genetic mutations and the restoration of disrupted signaling pathways (Perez-Garcia and Perez-Garcia 2024). At the same time, epigenetic regulation plays a key role in trophoblast cell differentiation and fusion. Changes in oxidative status may influence protein glycosylation, thereby regulating cell function (Pidoux et al. 2012). Furthermore, dynamic changes in histone modifications, especially the deacetylation mediated by HDAC1/2 and the increased acetylation at H3K9ac and H3K27ac sites, have been shown to be closely related to syncytialization (Zhuang et al. 2014). Intervening in these epigenetic mechanisms holds promise for improving placental function and lays a foundation for understanding the pathogenesis of placenta-related diseases.

In addition to gene and epigenetic regulation, pharmacological and physical interventions also show significant potential in placental development regulation. Small molecules, such as valproic acid (Ohyama et al. 2023) and sodium butyrate (Kumar et al. 2015), can promote trophoblast cell differentiation by modulating metabolism and signaling pathways. Moreover, mechanical stimuli, as a noninvasive intervention, are also considered effective activators to promote cell fusion

(Parameshwar et al. 2024). In the future, by integrating gene editing, epigenetic regulation, targeted pharmacological interventions, and biophysical modulation, it may be possible to create a systematic, multidimensional placental function regulation framework. This approach could enable precise molecular-level interventions in placenta-related diseases, not only deepening the mechanistic understanding of placental development abnormalities but also opening new possibilities for personalized diagnosis and treatment of pregnancy complications. This would provide broader prospects for the development of maternal–fetal medicine and regenerative medicine.

Despite significant progress in elucidating the mechanisms of trophoblast cell fusion and exploring potential therapeutic strategies in recent years, several critical scientific challenges remain unresolved, thereby limiting the clinical translation of related findings. First, current multi-omics studies on fusion defects have predominantly focused on transcriptional-level regulation, whereas the broader network encompassing posttranslational protein modifications, metabolic reprogramming, and intercellular signaling integration has yet to be systematically delineated. Second, experimental models capable of faithfully recapitulating the complex architecture of human placental development and the maternal–fetal interface remain inadequate. In particular, within organoid and 3D culture systems, technical barriers persist in achieving functionalization and polarity reconstruction of STBs. Third, although fusogenic proteins (e.g., Syncytin-1) and key signaling pathways (e.g., cAMP/PKA) have been identified as critical mediators of trophoblast fusion, their disease-specific expression dynamics and temporal correlation with disease progression in conditions such as PE and FGR remain insufficiently characterized. This knowledge gap hinders the determination of optimal intervention windows for targeted therapy. In addition, the clinical translation of multi-omics approaches for disease diagnosis and precision treatment is challenged by data integration complexity, limited access to high-quality human placental samples, and ethical considerations. Future research should therefore adopt an integrated strategy encompassing mechanistic investigation, model optimization, and translational linkage to bridge the gap between basic research and clinical application.

References

Abdelkhaliq A, van der Zande M, Peters RJB, Bouwmeester H (2020) Combination of the BeWo b30 placental transport model and the embryonic stem cell test to assess the potential developmental toxicity of silver nanoparticles. Part Fibre Toxicol 17:11. https://doi.org/10.1186/s12989-020-00342-6

Aghababaei M, Hogg K, Perdu S, Robinson WP, Beristain AG (2015) ADAM12-directed ectodomain shedding of E-cadherin potentiates trophoblast fusion. Cell Death Differ 22: 1970–1984. https://doi.org/10.1038/cdd.2015.44

Aiko Y et al (2014) Differential levels of amino acid transporters system L and ASCT2, and the mTOR protein in placenta of preeclampsia and IUGR. BMC Pregnancy Childbirth 14:181. https://doi.org/10.1186/1471-2393-14-181

Alfaifi AA et al (2020) Megalin-targeting liposomes for placental drug delivery. J Control Release 324:366–378. https://doi.org/10.1016/j.jconrel.2020.05.033
Apicella C et al (2021) Urothelial cancer associated 1 (UCA1) and miR-193 are two non-coding RNAs involved in trophoblast fusion and placental diseases. Front Cell Dev Biol 9:633937. https://doi.org/10.3389/fcell.2021.633937
Baker CL, Pera MF (2018) Capturing totipotent stem cells. Cell Stem Cell 22:25–34. https://doi.org/10.1016/j.stem.2017.12.011
Borges M, Bose P, Frank HG, Kaufmann P, Potgens AJ (2003) A two-colour fluorescence assay for the measurement of syncytial fusion between trophoblast-derived cell lines. Placenta 24:959–964. https://doi.org/10.1016/s0143-4004(03)00173-5
Brugger BA et al (2023) Fluid shear stress induces a shift from glycolytic to amino acid pathway in human trophoblasts. Cell Biosci 13:163. https://doi.org/10.1186/s13578-023-01114-3
Brukman NG, Uygur B, Podbilewicz B, Chernomordik LV (2019) How cells fuse. J Cell Biol 218: 1436–1451. https://doi.org/10.1083/jcb.201901017
Cesana M et al (2024) TFEB controls syncytiotrophoblast formation and hormone production in placenta. Cell Death Differ 31:1439–1451. https://doi.org/10.1038/s41418-024-01337-y
Chang CW, Chang GD, Chen H (2011) A novel cyclic AMP/Epac1/CaMKI signaling cascade promotes GCM1 desumoylation and placental cell fusion. Mol Cell Biol 31:3820–3831. https://doi.org/10.1128/MCB.05582-11
Chen EH, Olson EN (2005) Unveiling the mechanisms of cell-cell fusion. Science 308:369–373. https://doi.org/10.1126/science.1104799
Chen H et al (2003) Mitofusins Mfn1 and Mfn2 coordinately regulate mitochondrial fusion and are essential for embryonic development. J Cell Biol 160:189–200. https://doi.org/10.1083/jcb.200211046
Cheng YH, Handwerger S (2005) A placenta-specific enhancer of the human syncytin gene. Biol Reprod 73:500–509. https://doi.org/10.1095/biolreprod.105.039941
Costa MA (2016) Scrutinising the regulators of syncytialization and their expression in pregnancy-related conditions. Mol Cell Endocrinol 420:180–193. https://doi.org/10.1016/j.mce.2015.11.010
Coutifaris C et al (1991) E-cadherin expression during the differentiation of human trophoblasts. Development 113:767–777. https://doi.org/10.1242/dev.113.3.767
Dang Q et al (2024) Nuclear binding protein 2/Nesfatin-1 affects trophoblast cell fusion during placental development via the EGFR-PLCG1-CAMK4 pathway. Int J Mol Sci 25. https://doi.org/10.3390/ijms25031925
Daoud G et al (2005) ERK1/2 and p38 regulate trophoblasts differentiation in human term placenta. J Physiol 566:409–423. https://doi.org/10.1113/jphysiol.2005.089326
Davenport KM et al (2023) Single-nuclei RNA sequencing (snRNA-seq) uncovers trophoblast cell types and lineages in the mature bovine placenta. Proc Natl Acad Sci USA 120:e2221526120. https://doi.org/10.1073/pnas.2221526120
Desforges M, Parsons L, Westwood M, Sibley CP, Greenwood SL (2013) Taurine transport in human placental trophoblast is important for regulation of cell differentiation and survival. Cell Death Dis 4:e559. https://doi.org/10.1038/cddis.2013.81
Dietrich B, Haider S, Meinhardt G, Pollheimer J, Knofler M (2022) WNT and NOTCH signaling in human trophoblast development and differentiation. Cell Mol Life Sci 79:292. https://doi.org/10.1007/s00018-022-04285-3
Drewlo S et al (2020) Irisin induces trophoblast differentiation via AMPK activation in the human placenta. J Cell Physiol 235:7146–7158. https://doi.org/10.1002/jcp.29613
Duan FM et al (2021) THBS1 regulates trophoblast fusion through a CD36-dependent inhibition of cAMP, and its upregulation participates in preeclampsia. Genes Dis 8:353–363. https://doi.org/10.1016/j.gendis.2020.05.007
Dunk CE et al (2012) The molecular role of connexin 43 in human trophoblast cell fusion. Biol Reprod 86:115. https://doi.org/10.1095/biolreprod.111.096925
Dunk CE et al (2018) P-glycoprotein (P-gp)/ABCB1 plays a functional role in extravillous trophoblast (EVT) invasion and is decreased in the pre-eclamptic placenta. J Cell Mol Med 22:5378–5393. https://doi.org/10.1111/jcmm.13810

Dupressoir A et al (2009) Syncytin-a knockout mice demonstrate the critical role in placentation of a fusogenic, endogenous retrovirus-derived, envelope gene. Proc Natl Acad Sci USA 106: 12127–12132. https://doi.org/10.1073/pnas.0902925106

Esbin MN et al (2024) TFEB controls expression of human syncytins during cell-cell fusion. Genes Dev 38:718–737. https://doi.org/10.1101/gad.351633.124

Fantone S et al (2020) AT-rich interactive domain 1A protein expression in normal and pathological pregnancies complicated by preeclampsia. Histochem Cell Biol 154:339–346. https://doi.org/10.1007/s00418-020-01892-8

Finney DJ (1965) Statistics in the university and in the community. Biom Z 7:78–89. https://doi.org/10.1002/bimj.19650070203

Gehl AL et al (2025) The proteomic landscape of trophoblasts unravels calcium-dependent syncytialization processes and beta-chorionic gonadotropin (ss-hCG) production. Reprod Biol Endocrinol 23:33. https://doi.org/10.1186/s12958-025-01362-7

Gerbaud P, Pidoux G (2015) Review: an overview of molecular events occurring in human trophoblast fusion. Placenta 36(Suppl 1):S35–S42. https://doi.org/10.1016/j.placenta.2014.12.015

Getsios S, MacCalman CD (2003) Cadherin-11 modulates the terminal differentiation and fusion of human trophoblastic cells in vitro. Dev Biol 257:41–54. https://doi.org/10.1016/s0012-1606(03)00041-1

Gilles JA, Carnes DL, Windeler AS (1994) Development of an in vitro culture system for the study of osteoclast activity and function. J Endodont 20:327–331. https://doi.org/10.1016/s0099-2399(06)80094-4

Groten T et al (2010) Differential expression of VE-cadherin and VEGFR2 in placental syncytiotrophoblast during preeclampsia - new perspectives to explain the pathophysiology. Placenta 31:339–343. https://doi.org/10.1016/j.placenta.2010.01.014

Hahn L et al (2022) Gal-2 increases H3K4me(3) and H3K9ac in trophoblasts and preeclampsia. Biomolecules 12. https://doi.org/10.3390/biom12050707

Haider S et al (2018) Self-renewing trophoblast organoids recapitulate the developmental program of the early human placenta. Stem Cell Rep 11:537–551. https://doi.org/10.1016/j.stemcr.2018.07.004

Heazell AE, Moll SJ, Jones CJ, Baker PN, Crocker IP (2007) Formation of syncytial knots is increased by hyperoxia, hypoxia and reactive oxygen species. Placenta 28(Suppl A):S33–S40. https://doi.org/10.1016/j.placenta.2006.10.007

Hernandez JM, Podbilewicz B (2017) The hallmarks of cell-cell fusion. Development 144:4481–4495. https://doi.org/10.1242/dev.155523

Hertz R (1959) Choriocarcinoma of women maintained in serial passage in hamster and rat. Proc Soc Exp Biol Med 102:77–81. https://doi.org/10.3181/00379727-102-25149

Hiden U et al (2013) Membrane-type matrix metalloproteinase 1 regulates trophoblast functions and is reduced in fetal growth restriction. Am J Pathol 182:1563–1571. https://doi.org/10.1016/j.ajpath.2013.01.011

Huang Q et al (2013) Syncytin-1 modulates placental trophoblast cell proliferation by promoting G1/S transition. Cell Signal 25:1027–1035. https://doi.org/10.1016/j.cellsig.2013.01.008

Huppertz B, Bartz C, Kokozidou M (2006) Trophoblast fusion: fusogenic proteins, syncytins and ADAMs, and other prerequisites for syncytial fusion. Micron 37:509–517. https://doi.org/10.1016/j.micron.2005.12.011

Jaju Bhattad G et al (2020) Histone deacetylase 1 and 2 drive differentiation and fusion of progenitor cells in human placental trophoblasts. Cell Death Dis 11:311. https://doi.org/10.1038/s41419-020-2500-6

Janatpour MJ et al (2000) Id-2 regulates critical aspects of human cytotrophoblast differentiation, invasion and migration. Development 127:549–558. https://doi.org/10.1242/dev.127.3.549

Jaremek A, Jeyarajah MJ, Jaju Bhattad G, Renaud SJ (2021) Omics approaches to study formation and function of human placental Syncytiotrophoblast. Front Cell Dev Biol 9:674162. https://doi.org/10.3389/fcell.2021.674162

Jaremek A et al (2023) Genome-wide analysis of hypoxia-inducible factor binding reveals targets implicated in impaired human placental Syncytiotrophoblast formation under low oxygen. Am J Pathol 193:846–865. https://doi.org/10.1016/j.ajpath.2023.03.006

Jasmer B et al (2017) Involvement of the oncogene B-cell lymphoma 6 in the fusion and differentiation process of trophoblastic cells of the placenta. Oncotarget 8:108643–108654. https://doi.org/10.18632/oncotarget.20586

Johnstone ED et al (2005) Sphingosine-1-phosphate inhibition of placental trophoblast differentiation through a G(i)-coupled receptor response. J Lipid Res 46:1833–1839. https://doi.org/10.1194/jlr.M500095-JLR200

Kinoshita M (2003) The septins. Genome Biol 4:236. https://doi.org/10.1186/gb-2003-4-11-236

Kliman HJ, Nestler JE, Sermasi E, Sanger JM, Strauss JF 3rd. (1986) Purification, characterization, and in vitro differentiation of cytotrophoblasts from human term placentae. Endocrinology 118: 1567–1582. https://doi.org/10.1210/endo-118-4-1567

Knerr I et al (2005) Stimulation of GCMa and syncytin via cAMP mediated PKA signaling in human trophoblastic cells under normoxic and hypoxic conditions. FEBS Lett 579:3991–3998. https://doi.org/10.1016/j.febslet.2005.06.029

Knerr I, Soder S, Licha E, Aigner T, Rascher W (2008) Response of HEK293 and CHO cells overexpressing fusiogenic syncytin-1 to mitochondrion-mediated apoptosis induced by antimycin a. J Cell Biochem 105:766–775. https://doi.org/10.1002/jcb.21874

Kohler PO, Bridson WE (1971) Isolation of hormone-producing clonal lines of human choriocarcinoma. J Clin Endocrinol Metab 32:683–687. https://doi.org/10.1210/jcem-32-5-683

Kosinska-Kaczynska K (2022) Placental syndromes-a new paradigm in perinatology. Int J Environ Res Public Health 19. https://doi.org/10.3390/ijerph19127392

Krendl C et al (2017) GATA2/3-TFAP2A/C transcription factor network couples human pluripotent stem cell differentiation to trophectoderm with repression of pluripotency. Proc Natl Acad Sci USA 114:E9579–E9588. https://doi.org/10.1073/pnas.1708341114

Kudo Y, Boyd CA, Sargent IL, Redman CW (2003) Hypoxia alters expression and function of syncytin and its receptor during trophoblast cell fusion of human placental BeWo cells: implications for impaired trophoblast syncytialisation in pre-eclampsia. Biochim Biophys Acta 1638:63–71. https://doi.org/10.1016/s0925-4439(03)00043-7

Kumar P, Thirkill TL, Ji J, Monte LH, Douglas GC (2015) Differential effects of sodium butyrate and lithium chloride on rhesus monkey trophoblast differentiation. PLoS One 10:e0135089. https://doi.org/10.1371/journal.pone.0135089

Kupper N, Pritz E, Siwetz M, Guettler J, Huppertz B (2021) Placental villous explant culture 2.0: flow culture allows studies closer to the in vivo situation. Int J Mol Sci 22. https://doi.org/10.3390/ijms22147464

Leduc K et al (2012) Leukemia inhibitory factor regulates differentiation of trophoblastlike BeWo cells through the activation of JAK/STAT and MAPK3/1 MAP kinase-signaling pathways. Biol Reprod 86:54. https://doi.org/10.1095/biolreprod.111.094334

Li Y et al (2015) Involvement of nephrin in human placental trophoblast syncytialization. Reproduction 149:339–346. https://doi.org/10.1530/REP-14-0424

Li H et al (2020) Internalization of trophoblastic small extracellular vesicles and detection of their miRNA cargo in P-bodies. J Extracell Vesicles 9:1812261. https://doi.org/10.1080/20013078.2020.1812261

Li X, Li ZH, Wang YX, Liu TH (2023) A comprehensive review of human trophoblast fusion models: recent developments and challenges. Cell Death Discov 9:372. https://doi.org/10.1038/s41420-023-01670-0

Liu R, Jin JP (2016) Calponin isoforms CNN1, CNN2 and CNN3: regulators for actin cytoskeleton functions in smooth muscle and non-muscle cells. Gene 585:143–153. https://doi.org/10.1016/j.gene.2016.02.040

Liu M, Hassana S, Stiles JK (2016) Heme-mediated apoptosis and fusion damage in BeWo trophoblast cells. Sci Rep 6:36193. https://doi.org/10.1038/srep36193

Liu Y et al (2018) Single-cell RNA-seq reveals the diversity of trophoblast subtypes and patterns of differentiation in the human placenta. Cell Res 28:819–832. https://doi.org/10.1038/s41422-018-0066-y

Louwen F et al (2022) BCL6, a key oncogene, in the placenta, pre-eclampsia and endometriosis. Hum Reprod Update 28:890–909. https://doi.org/10.1093/humupd/dmac027

Lu X et al (2016) Twist1 is involved in trophoblast syncytialization by regulating GCM1. Placenta 39:45–54. https://doi.org/10.1016/j.placenta.2016.01.008

Lv B et al (2019) Single-cell RNA sequencing reveals regulatory mechanism for trophoblast cell-fate divergence in human peri-implantation conceptuses. PLoS Biol 17:e3000187. https://doi.org/10.1371/journal.pbio.3000187

Lye P et al (2019) Breast cancer resistance protein (BCRP/ABCG2) inhibits extra villous trophoblast migration: the impact of bacterial and viral infection. Cells 8. https://doi.org/10.3390/cells8101150

Matsuura K et al (2011) Identification of a link between Wnt/beta-catenin signalling and the cell fusion pathway. Nat Commun 2:548. https://doi.org/10.1038/ncomms1551

McKenna ED, Sarbanes SL, Cummings SW, Roll-Mecak A (2023) The tubulin code, from molecules to health and disease. Annu Rev Cell Dev Biol 39:331–361. https://doi.org/10.1146/annurev-cellbio-030123-032748

Meinhardt G et al (2020) Pivotal role of the transcriptional co-activator YAP in trophoblast stemness of the developing human placenta. Proc Natl Acad Sci USA 117:13562–13570. https://doi.org/10.1073/pnas.2002630117

Miranda AL et al (2022) Kruppel-like factor 6 (KLF6) requires its amino terminal domain to promote villous trophoblast cell fusion. Placenta 117:139–149. https://doi.org/10.1016/j.placenta.2021.12.006

Mizutani T et al (2022) Inhibition of YAP/TAZ-TEAD activity induces cytotrophoblast differentiation into syncytiotrophoblast in human trophoblast. Mol Hum Reprod 28. https://doi.org/10.1093/molehr/gaac032

Mori K et al (2004) The functional interaction between CD98 and CD147 in regulation of virus-induced cell fusion and osteoclast formation. Med Microbiol Immunol 193:155–162. https://doi.org/10.1007/s00430-003-0191-0

Msheik H et al (2019) Transcriptomic profiling of trophoblast fusion using BeWo and JEG-3 cell lines. Mol Hum Reprod 25:811–824. https://doi.org/10.1093/molehr/gaz061

Nakashima A et al (2008) Granulysin produced by uterine natural killer cells induces apoptosis of extravillous trophoblasts in spontaneous abortion. Am J Pathol 173:653–664. https://doi.org/10.2353/ajpath.2008.071169

Ng YH, Zhu H, Leung PC (2011) Twist regulates cadherin-mediated differentiation and fusion of human trophoblastic cells. J Clin Endocrinol Metab 96:3881–3890. https://doi.org/10.1210/jc.2010-2714

Novakovic B et al (2011) Wide-ranging DNA methylation differences of primary trophoblast cell populations and derived cell lines: implications and opportunities for understanding trophoblast function. Mol Hum Reprod 17:344–353. https://doi.org/10.1093/molehr/gar005

Ohyama N et al (2023) Effects of valproic acid on syncytialization in human placental trophoblast cell lines. Toxicol Appl Pharmacol 474:116611. https://doi.org/10.1016/j.taap.2023.116611

Omata W, Ackerman WE, t., Vandre, D. D. & Robinson, J. M. (2013) Trophoblast cell fusion and differentiation are mediated by both the protein kinase C and a pathways. PLoS One 8:e81003. https://doi.org/10.1371/journal.pone.0081003

Parameshwar PK et al (2024) Directed biomechanical compressive forces enhance fusion efficiency in model placental trophoblast cultures. Sci Rep 14:11312. https://doi.org/10.1038/s41598-024-61747-3

Pattillo RA, Gey GO (1968) The establishment of a cell line of human hormone-synthesizing trophoblastic cells in vitro. Cancer Res 28:1231–1236

Pattillo RA, Gey GO, Delfs E, Mattingly RF (1968) Human hormone production in vitro. Science 159:1467–1469. https://doi.org/10.1126/science.159.3822.1467

Penugurti V et al (2025) Potential role of hematopoietic PBX-interacting protein (HPIP) in trophoblast fusion and invasion: implications in pre-eclampsia pathogenesis. Cell Signal 128: 111633. https://doi.org/10.1016/j.cellsig.2025.111633

Perez-Garcia I, Perez-Garcia V (2024) CRISPR activation in mouse trophoblast stem cells. Methods Mol Biol 2781:93–103. https://doi.org/10.1007/978-1-0716-3746-3_9

Pidoux G et al (2010) ZO-1 is involved in trophoblastic cell differentiation in human placenta. Am J Physiol Cell Physiol 298:C1517–C1526. https://doi.org/10.1152/ajpcell.00484.2008

Pidoux G et al (2012) Review: human trophoblast fusion and differentiation: lessons from trisomy 21 placenta. Placenta 33(Suppl):S81–S86. https://doi.org/10.1016/j.placenta.2011.11.007

Renaud SJ, Jeyarajah MJ (2022) How trophoblasts fuse: an in-depth look into placental syncytiotrophoblast formation. Cell Mol Life Sci 79:433. https://doi.org/10.1007/s00018-022-04475-z

Renaud SJ et al (2015) OVO-like 1 regulates progenitor cell fate in human trophoblast development. Proc Natl Acad Sci USA 112:E6175–E6184. https://doi.org/10.1073/pnas.1507397112

Richardson BD, Cheng YH, Langland RA, Handwerger S (2001) Differential expression of AP-2gamma and AP-2alpha during human trophoblast differentiation. Life Sci 69:2157–2165. https://doi.org/10.1016/s0024-3205(01)01299-1

Roberts RM et al (2021) Syncytins expressed in human placental trophoblast. Placenta 113:8–14. https://doi.org/10.1016/j.placenta.2021.01.006

Rosa JP, Raslova H, Bryckaert M (2019) Filamin a: key actor in platelet biology. Blood 134:1279–1288. https://doi.org/10.1182/blood.2019000014

Ruebner M et al (2010) Impaired cell fusion and differentiation in placentae from patients with intrauterine growth restriction correlate with reduced levels of HERV envelope genes. J Mol Med (Berl) 88:1143–1156. https://doi.org/10.1007/s00109-010-0656-8

Ruebner M et al (2012) Regulation of the human endogenous retroviral Syncytin-1 and cell-cell fusion by the nuclear hormone receptors PPARgamma/RXRalpha in placentogenesis. J Cell Biochem 113:2383–2396. https://doi.org/10.1002/jcb.24110

Saha B et al (2020) TEAD4 ensures postimplantation development by promoting trophoblast self-renewal: an implication in early human pregnancy loss. Proc Natl Acad Sci USA 117:17864–17875. https://doi.org/10.1073/pnas.2002449117

Schust DJ et al (2021) The immunology of Syncytialized trophoblast. Int J Mol Sci 22. https://doi.org/10.3390/ijms22041767

Shankar K et al (2015) Transcriptomic and epigenomic landscapes during cell fusion in BeWo trophoblast cells. Placenta 36:1342–1351. https://doi.org/10.1016/j.placenta.2015.10.010

Shibukawa Y et al (2010) Calponin 3 regulates actin cytoskeleton rearrangement in trophoblastic cell fusion. Mol Biol Cell 21:3973–3984. https://doi.org/10.1091/mbc.E10-03-0261

Shoaito H et al (2020) Peroxisome proliferator-activated receptor gamma-ligand-binding domain mutations associated with familial partial lipodystrophy type 3 disrupt human trophoblast fusion and fibroblast migration. J Cell Mol Med 24:7660–7669. https://doi.org/10.1111/jcmm.15401

Sivasubramaniyam T et al (2013) Where polarity meets fusion: role of Par6 in trophoblast differentiation during placental development and preeclampsia. Endocrinology 154:1296–1309. https://doi.org/10.1210/en.2012-1823

Soygur B, Moore H (2016) Expression of Syncytin 1 (HERV-W), in the preimplantation human blastocyst, embryonic stem cells and trophoblast cells derived in vitro. Hum Reprod 31:1455–1461. https://doi.org/10.1093/humrep/dew097

Tan JP, Liu X, Polo JM (2022) Establishment of human induced trophoblast stem cells via reprogramming of fibroblasts. Nat Protoc 17:2739–2759. https://doi.org/10.1038/s41596-022-00742-2

Tang HY et al (2023) Tspan5 promotes the EMT process to regulate the syncytialization of trophoblast cells by activating Notch signalling. Zygote 31:498–506. https://doi.org/10.1017/S0967199423000369

Toufaily C et al (2013) MFSD2a, the Syncytin-2 receptor, is important for trophoblast fusion. Placenta 34:85–88. https://doi.org/10.1016/j.placenta.2012.10.012

Tuohey L et al (2013) PLAC4 is upregulated in severe early onset preeclampsia and upregulated with syncytialisation but not hypoxia. Placenta 34:256–260. https://doi.org/10.1016/j.placenta.2012.12.009

Turco MY et al (2018) Trophoblast organoids as a model for maternal-fetal interactions during human placentation. Nature 564:263–267. https://doi.org/10.1038/s41586-018-0753-3
Vargas A et al (2011) Reduced expression of both syncytin 1 and syncytin 2 correlates with severity of preeclampsia. Reprod Sci 18:1085–1091. https://doi.org/10.1177/1933719111404608
Vatish M et al (2012) Inhibition of Akt activity and calcium channel function coordinately drive cell-cell fusion in the BeWO choriocarcinoma placental cell line. PLoS One 7:e29353. https://doi.org/10.1371/journal.pone.0029353
Wang Y et al (2020) Silencing DAPK3 blocks the autophagosome-lysosome fusion by mediating SNAP29 in trophoblast cells under high glucose treatment. Mol Cell Endocrinol 502:110674. https://doi.org/10.1016/j.mce.2019.110674
Wang LJ et al (2022a) Functional antagonism between DeltaNp63alpha and GCM1 regulates human trophoblast stemness and differentiation. Nat Commun 13:1626. https://doi.org/10.1038/s41467-022-29312-6
Wang YN et al (2022b) Inducible knockout of syncytin-a leads to poor placental glucose transport in mice. Placenta 121:155–163. https://doi.org/10.1016/j.placenta.2022.03.016
Wang Y et al (2024) Gestational exposure to PM(2.5) disrupts fetal development by suppressing placental trophoblast syncytialization via progranulin/mTOR signaling. Sci Total Environ **921**:171101. https://doi.org/10.1016/j.scitotenv.2024.171101
Wice B, Menton D, Geuze H, Schwartz AL (1990) Modulators of cyclic AMP metabolism induce syncytiotrophoblast formation in vitro. Exp Cell Res 186:306–316. https://doi.org/10.1016/0014-4827(90)90310-7
Wildner JM et al (2019) RITA is expressed in trophoblastic cells and is involved in differentiation processes of the placenta. Cells 8. https://doi.org/10.3390/cells8121484
Wu ZH et al (2023) miR-181d-5p, which is upregulated in fetal growth restriction placentas, inhibits trophoblast fusion via CREBRF. J Assist Reprod Genet 40:2725–2737. https://doi.org/10.1007/s10815-023-02917-6
Xiao X et al (2020) Upregulation of pannexin-1 hemichannels explains the apparent death of the syncytiotrophoblast during human placental explant culture. Placenta 94:1–12. https://doi.org/10.1016/j.placenta.2020.03.003
Xu L, Sun S (2022) Reconstitution of fusion-competent human placental Fusogen Syncytin-2. J Membr Biol 255:723–732. https://doi.org/10.1007/s00232-022-00242-0
Yang YF et al (2024) Evaluating the role of DNA methyltransferases in trophoblast fusion and preeclampsia development: insights from methylation-regulated genes. FASEB J 38:e23706. https://doi.org/10.1096/fj.202302705R
Yu X et al (2024) Acetyl-CoA metabolism maintains histone acetylation for syncytialization of human placental trophoblast stem cells. Cell Stem Cell 31:1280–1297 e1287. https://doi.org/10.1016/j.stem.2024.07.003
Yuan X et al (2023) Endoplasmic reticulum stress impairs trophoblast syncytialization through upregulation of HtrA4 and causes early-onset preeclampsia. J Hypertens 41:2095–2106. https://doi.org/10.1097/HJH.0000000000003541
Zhang Q et al (2013) CUL1 promotes trophoblast cell invasion at the maternal-fetal interface. Cell Death Dis 4:e502. https://doi.org/10.1038/cddis.2013.1
Zhang Y et al (2020) TMEM16F phospholipid scramblase mediates trophoblast fusion and placental development. Sci Adv **6**:eaba0310. https://doi.org/10.1126/sciadv.aba0310
Zhang Y et al (2022) Functional coupling between TRPV4 channel and TMEM16F modulates human trophoblast fusion. elife 11. https://doi.org/10.7554/eLife.78840
Zhao X et al (2024) 27-hydroxycholesterol inhibits trophoblast fusion during placenta development by activating PI3K/AKT/mTOR signaling pathway. Arch Toxicol 98:849–863. https://doi.org/10.1007/s00204-023-03664-4
Zhuang XW et al (2014) Decreased expression and altered methylation of syncytin-1 gene in human placentas associated with preeclampsia. Curr Pharm Des 20:1796–1802. https://doi.org/10.2174/13816128113199990541

Chapter 7
Pathogen-Mediated Cell Fusion and Actin Cytoskeleton

Malgorzata Kloc

Abstract Under physiological conditions, programmed cell fusion occurs during fertilization, embryonic development, tissue and organ formation, and colony formation. Cell fusion also occurs under pathologic conditions such as inflammation, cancer, and pathogen infection. Various animal and plant pathogens, such as viruses, bacteria, and parasitic nematodes, induce cell fusion to facilitate spred, evade the host immune response, or serve as nutrient source. In this short review, we discuss various examples and underlying mechanisms of pathogen-mediated cell fusion, emphasizing the role of the actin cytoskeleton and the actin nucleation pathway in the pathogen life cycle and the fusion processes.

Keywords Virus · Bacterium · Parasitic nematode · Cell fusion · Syncytium · Fusogen · Actin

7.1 Introduction

The fusion of individual cells into a single entity, called a syncytium, is either natural or pathological. Depending on the type (identical or different) of fusing partners, there are homotypic or heterotypic syncytia. There are also examples of natural interspecific (sometimes called xenogeneic) fusion between different species of bacteria. Additionally, the xenogeneic syncytia derived from fusion between cells from different species have been created in several laboratories. The fusion of cell cytoplasms without the fusion of nuclei creates a homotypic or heterotypic heterokaryotic syncytium with multiple nuclei of the same or different origin, respectively. The fusion of the cytoplasms and nuclei creates a homotypic or heterotypic synkaryotic syncytium (Fig. 7.1). Usually, cell fusion requires the presence of a fusing protein called fusogen, which is either present on one fusing partner only

M. Kloc (✉)
The Houston Methodist Research Institute, Transplant Immunology, Houston, TX, USA

Department of Surgery, The Houston Methodist Hospital, Houston, TX, USA
e-mail: mkloc@houstonmethodist.org

© The Author(s), under exclusive license to Springer Nature Switzerland AG 2026

J. Weiler et al. (eds.), *Cell Fusion*, Advances in Experimental Medicine and Biology 1509, https://doi.org/10.1007/978-3-032-22637-2_7

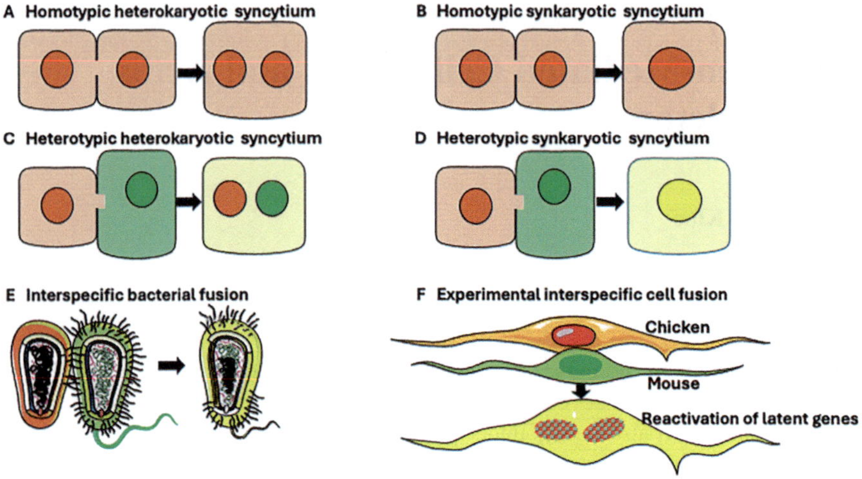

Fig. 7.1 Types of cell fusion. (**a**) Fusion of identical cell cytoplasms without the fusion of nuclei creates homotypic heterokaryotic syncytia, such as the fusion of macrophages to form the osteoclasts or SARS-CoV-2-induced fusion of pneumocytes (Chaudhary et al. 2023; Kloc et al. 2022a, b). (**b**) Fusion of cytoplasms and nuclei of identical cells creates homotypic synkaryotic syncytia, such as in varicella–zoster virus–induced syncytia of ARPE-19 cells (Wang et al. 2017). (**c, d**) Fusion of different cell types produces either heterokaryotic or synkaryotic heterotypic syncytia, such as SARS-CoV-2-induced fusion of pneumocytes with monocytes or neutrophils (Chaudhary et al. 2023) and fusion of the gametes during fertilization. (**e**) The interspecific (xenogeneic) fusion occurs, for example, between different species of *Clostridium* bacteria, and (**f**) fusion between chicken and murine myoblasts was induced in the laboratory (see the main text)

(unilateral fusion) or both (bilateral fusion) (Fig. 7.2). One of the roles of fusogens is to overcome thermodynamic repulsion of cell membranes consisting of proximal and distal lipid monolayers and allow direct contact between the proximal lipid monolayers and their merger. This is followed by the merger of distal lipid monolayers and the formation of a fusion pore. The fusogen is also involved in the opening and widening of the fusion pore (Brukman et al. 2019; Chernomordik and Kozlov 2008; Kloc et al. 2022a, b). A common feature of viral and cell–cell fusogens is their long ectodomain, which extends ~10 nm or more from the membrane. Because the cell surfaces have a glycoprotein and glycolipid coating, which blocks close membrane contact between the cells, such a long ectodomain allows penetration through the surface coating and anchoring of neighboring membranes (Chan et al. 2020).

Programmed cell fusion occurs naturally during the fertilization and the development of organisms, organs, and tissues, while pathological fusion occurs in cancer or pathogen infection (Brukman et al. 2019; Chen and Olson 2005; Dittmar and Hass 2022; Iosilevskii and Podbilewicz 2021; Kloc et al. 2022a, b; Manjunath et al. 2020). Cell fusion is involved not only in the physiologic and pathologic processes within individual organisms but also in the formation of fungal and animal colonies

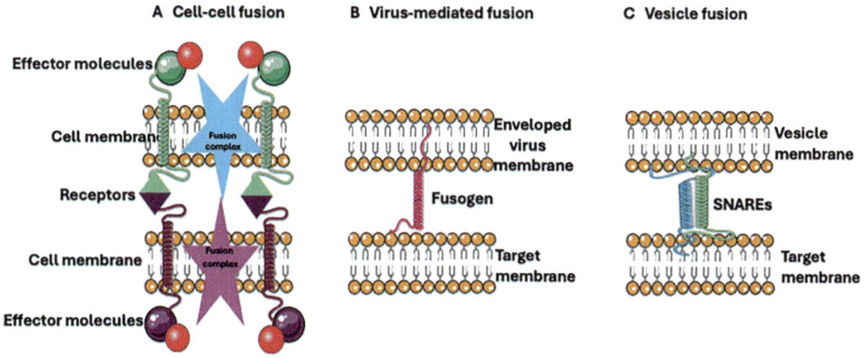

Fig. 7.2 Basic mechanisms of membrane fusion. Although the fusion of cells, organelles, vesicles, and pathogen membranes differs depending on the organism, membrane type, receptor, fusogen, and cooperating effector molecules, it follows one of the basic mechanisms depicted in the figure. (**a**) Fusion of cell membranes relies on the transmembrane receptors mediating close apposition of the membranes and recruitment of cooperating effector molecules and on the fusion protein complexes, which promote membrane destabilization and the formation of fusion pore that interconnects partnering cytoplasms. (**b**) In a fusion of enveloped viruses with the cell or vesicle membrane, viral fusogen anchors the target membrane, brings the viral and target membranes together, and destabilizes membranes to form a fusion pore. (**c**) Fusion between intracellular vesicles and target membranes is mediated by the SNARE proteins in the vesicle and target membranes. Like other fusogens, they bring membranes together, destabilize them, and promote fusion (see the main text and Chen and Olson 2005)

(Clark-Cotton et al. 2022; Fleißner et al. 2022; Herzog et al. 2015; Manjunath et al. 2020; Nawrocki and Cartwright 2012). It can also occur between different species (heterologous fusion; Charubin et al. 2020, 2024). We focus on pathogen-mediated cell fusion and the role of the actin cytoskeleton in these processes.

7.2 Role of the Actin Cytoskeleton in Pathogen Life Cycle and Pathogen-Induced Cell Fusion

Many pathogens exploit the cytoskeleton, including the actin cytoskeleton of the host cells, for entry, movement within the cell, establishment of a replication niche, spreading/dissemination between cells, and cell–cell fusion (Figs. 7.3 and 7.4) (Bradburne et al. 2013; Bugalhão et al. 2015; Choe and Welch 2016; Haglund and Welch 2011; Taylor et al. 2011; Wesolowski and Paumet 2017).

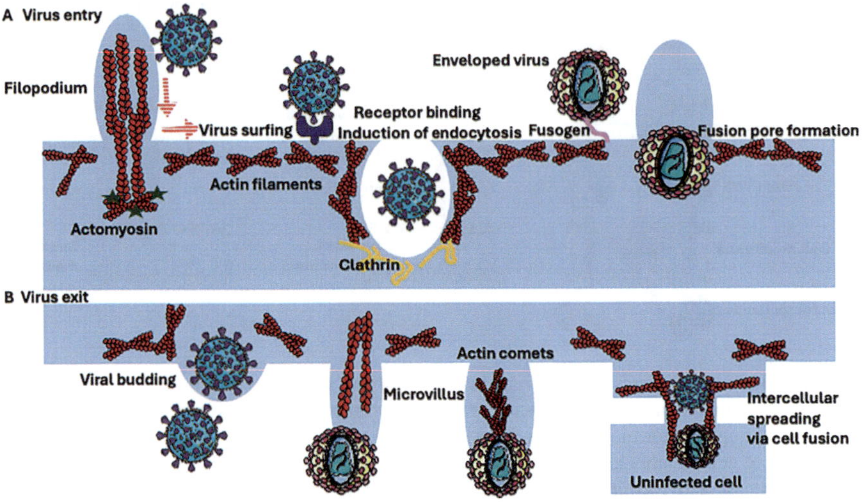

Fig. 7.3 Actin cytoskeleton's role in viral entry, exit, and intercellular spreading. (**a**) Many viruses surf cell surfaces to find suitable spots for entry. Viral surfing is facilitated by the mechanical force created by actomyosin complexes within the filopodia. After locating the suitable receptor, the virus internalizes by actin/clathrin-dependent endocytosis. The enveloped viruses use fusogen-based mechanisms for the destabilization of host cell membranes, fusion pore formation, and entry. (**b**) Depending on the type, viruses exit through viral budding from the cell surface, microvilli, or other cellular projections using actin-polymerization-based motility and actin comets or, lethal for the host, cell lysis. The spreading of the viruses between cells, allowing the avoidance of the immune defense, may occur through the virus-mediated fusion between infected and noninfected cells

7.2.1 Viruses

There are two main categories of viruses: non-enveloped and enveloped. Non-enveloped (naked) viruses (e.g., norovirus, enterovirus, adenovirus, and rhinovirus) consist of a genome (RNA or DNA) and a virus-encoded protein capsid. In enveloped viruses (e.g., influenza, human cytomegalovirus, human immunodeficiency virus (HIV), respiratory syncytial virus, vaccinia virus, and human coronaviruses, such as severe acute respiratory syndrome coronavirus 2 (SARS-CoV-2)), the capsid is surrounded by the lipid bilayer membrane. The membrane of enveloped viruses is co-opted from the cellular membrane. As the newly produced virus exits from an infected host cell, it is wrapped by a fragment of the cell membrane and is ready to infect more cells.

Enveloped virus entry into the cell occurs by fusing a viral membrane with a cell membrane or through endocytosis. First, the virus moves over the cell membrane surface, and after recognizing the appropriate cellular receptor, the viral surface protein binds to it. For example, HIV recognizes the cluster of differentiation 4 (CD4) receptor on immune cells, while influenza viruses recognize sialic acid on respiratory and enteric tract epithelial cells (Yang et al. 2022; Zhao and Pu 2022).

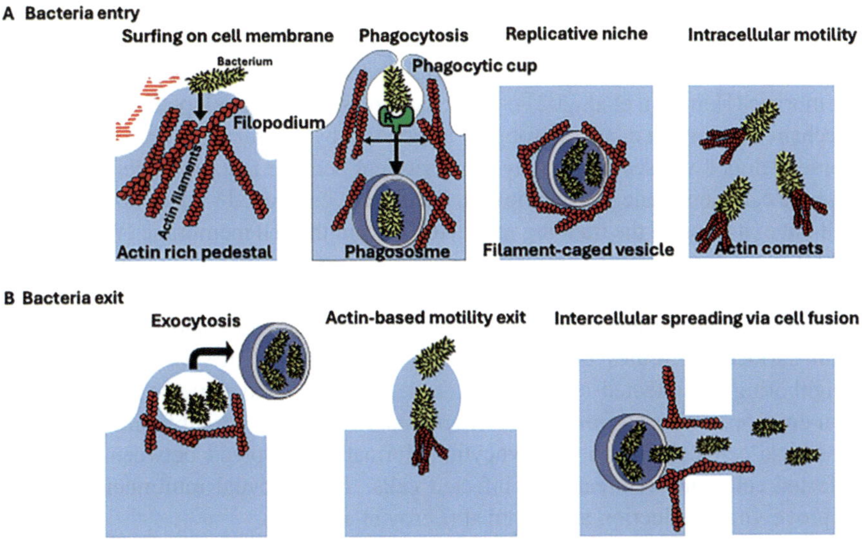

Fig. 7.4 Actin cytoskeleton's role in bacterial entry, exit, and intercellular spreading. (**a**) After binding to the cell surface, bacteria may induce cell protrusion containing an actin filament-rich pedestal whose undulation and flexing promote bacteria surfing on the cell membrane. Binding bacterial surface proteins to the cellular receptors rearranges the actin cytoskeleton, inducing the phagocytic cup and bacteria-containing vesicle (phagosome). Internalized bacteria may modify actin and cytokeratin filaments to form replicative niches, the filamentous cages around bacteria-containing vesicles to prevent destruction by host defense mechanisms. After exiting vesicles, bacteria can move within the host cell cytoplasm by forming actin comets, which use the mechanical force created by actin polymerization for propelling bacteria. (**b**) Bacteria exit the infected cell either by cell lysis (which kills the host cell) or by actin-dependent nonlethal means, such as exocytosis, actin-based motility within the cellular projections, or the induction of cell fusion and intercellular spreading

Such specificity explains why different viruses have different affinity for different cells/tissues/organs and cause different diseases. Because the cell membrane is very stable, the binding itself is not enough to initiate membrane fusion; thus, the role of fusogens, activated by a direct or indirect receptor binding, is to destabilize the membrane bilayer configuration, allowing the fusion (Más and Melero 2013).

If the virus enters through endocytosis, the whole virus with its envelope becomes engulfed. Subsequently, inside the cell, the viral membrane fuses with the endosome's membrane, and the virus is released to the cytoplasm (Cohen 2016).

Non-enveloped viruses enter the cell either by directly penetrating the cell membrane or through endocytosis and penetrating the endosomal membrane (Pletan and Tsai 2022; Thorley et al. 2010). Thorough studies using cryo-electron microscopy (EM) tomography and three-dimensional reconstruction of membrane penetration by bluetongue virus (BTV) and rhesus rotavirus (RRV) (Hermann et al. 2021; Xia et al. 2021) showed that the BVT acidifies the endosome and the low pH modifies viral capsid proteins VP2, VP 5, and VP7. The "dagger domain" in VP5, consisting of histidine residues, senses the low pH and unfolds into a long stylet, penetrating the membrane (Xia et al. 2021). During RRV entry, the VP5 protein spontaneously changes conformation, unfolding the foot domain that embeds in the membrane (Hermann et al. 2021). Other examples of non-enveloped virus entry and mechanisms involved are thoroughly described by Pletan and Tsai (2022). The non-enveloped viruses are usually more virulent because to exit the cell, they must lyse the cell membrane, killing the cell. In contrast, enveloped viruses do not lyse the cell but exit through the budding (exocytosis) from the cell membrane (Welsch et al. 2007).

The ability of viruses to induce cell–cell fusion depends on virus-encoded fusogens expressed at the surface of the virus-infected cells and their interaction with surface molecules or receptors involved in virus entry and expressed on neighboring noninfected cells (Leroy et al. 2020). All enveloped viruses and the non-enveloped viruses from *Orthoreovirus* and *Aquareovirus* subfamilies of reoviruses use cell–cell fusion and syncytium formation to spread between individual infected cells and uninfected or infected cells. The syncytial multinucleated cells improve viral production and survival (Leroy et al. 2020).

Enveloped, double-stranded DNA viruses from the *Herpesviridae* family induce the formation of multinuclear giant cells (MGCs) in eye keratitis, lungs, and skin lesions. Enveloped, negative-stranded RNA viruses from *Paramyxoviruses* family, such as the measles virus and the respiratory syncytial virus, induce MGCs in lymphocytes, thymocytes, epithelial cells, dendritic cells, and alveolar and bronchial epithelial cells. Enveloped, positive, single-stranded RNA viruses from the *Coronaviridae* family, such as SARS-CoV-2, induce MGCs in alveolar epithelial cells, monocytes, and macrophages. Enveloped viruses containing two linear copies of single-stranded, positive RNA from the *Retroviridae* family, such as HIV-1, induce MGCs in T cells and dendritic cells. The examples of other viruses, inducing syncytia, are described in detail by Leroy et al. (2020).

One of the best-studied virus-induced cell fusion is syncytia formation between cells infected by the SARS-CoV-2 virus. The SARS-CoV-2 virus can cause severe coronavirus disease 2019 (COVID-19) associated with the formation of multinuclear pneumocyte syncytia in the lungs. Virus entry or the fusion of infected cells depends on the interaction between the viral spike glycoprotein (S) and its receptor, the angiotensin-converting enzyme 2 (ACE2) expressed on the cell surface, followed by S protein and ACE2 cleavage by cellular proteases, such as transmembrane serine protease 2 (TMPRSS2), and conformational changes of three S1–S2 dimers and the receptor. Also, in vitro cultured cells that express ACE2 receptor U2OS-ACE2, 293TACE2, and A549-ACE2 form large syncytia when infected with SARS-CoV-2 (Buchrieser et al. 2020; Kloc et al. 2022a, b; Rajah et al. 2022).

7.2.1.1 Viral Fusogens

Most viral fusogens belong to one of the structurally and evolutionarily distinct classes (I, II, III, or IV), which are defined on the basis of the fusogen's tertiary structure and the number of monomers. Classes I, II, and III are produced by enveloped viruses, allowing virus–host membrane fusion, virus entry to the host cell, and cell–cell fusion (Leroy et al. 2020; Moi et al. 2022). Class IV fusogenic (fusion-associated small transmembrane (FAST) proteins) are produced by non-enveloped reoviruses to fuse multiple host cells into a syncytium (Podbilewicz 2014; Vance and Lee 2020). Class I fusogens, characteristic of influenza virus, HIV, coronavirus, and Ebola virus, are trimers of coiled coils (two or more alpha helices are wrapped around each other, forming a superhelix). Class II fusogens, characteristic of West Nile, Semliki Forest, Sindbis, dengue fever, Zika, and tick-borne encephalitis viruses, have a three-domain structure. Domain I is an eight-stranded β barrel, and domain II contains 12 β strands and two α helices. Domain III is similar to the immunoglobulin IgC domain (two layers of antiparallel β-strands arranged in two β-sheets), with ten β strands. Recently, an archaeal protein, fusexin1, homologous to fusexins, a superfamily of fusogens mediating eukaryotic gamete and somatic cell fusion, and similar to viral class II and possibly class I fusogens, was discovered (Moi et al. 2022). Class III fusogens are trimers with both α-helices and β-sheets that dissociate into monomers, insert fusion loops (FLs) into membranes, and, after fusion, oligomerize into trimers (Modis 2013; Podbilewicz 2014; Vance and Lee 2020). Despite the differences in structure, the class I–III fusogens use the same principle to drive membrane fusion. First, fusogen proteins are presented on the virus surface with their hydrophobic fusion peptide (FP) or fusion loop concealed within the fusogen. The receptor binding or another external trigger changes fusogen conformation, causing the release of the FP or FL, which then embeds in the membrane of the host cell and forms a temporary anchor between the virus and host. The subsequent collapse of the anchor results in a fusion of two outer lipid layers (state of hemifusion). Next, the inner lipid layers of the membranes merge, the fusion pore expands, and the fusogen changes conformation to the stable post-fusion form (Podbilewicz 2014; Vance and Lee 2020). Class IV fusogens (FAST proteins) encoded by fusogenic reoviruses (*Orthoreovirus* and *Aquareovirus*—the only non-enveloped viruses inducing syncytium formation) are very different in structure and function from class I to III. Because non-enveloped viruses do not have a lipid bilayer, they do not use membrane fusion for cell entry. Thus, the FAST proteins are not essential for the virus entry, but when present on the surface of the infected cell, they induce cell–cell fusion instead of virus–cell fusion (Ciechonska and Duncan 2014; Podbilewicz 2014). The FAST proteins consist of three domains: the transmembrane (TM) domain, the ectodomain, and the cytoplasmic endodomain. They are synthesized by the virus-infected cell and gradually traffic through the endoplasmic reticulum (ER) and Golgi pathway to the cell membrane, where they promote cell–cell fusion. In FAST-mediated fusion, an initial membrane apposition and the binding of fusing partners are mediated by calcium-dependent cadherin interactions and actin polymerization, with FAST proteins assembling in lipid rafts of the membrane. The TM domain induces membrane curvature and final close membrane apposition. During the membrane fusion, the ectodomain, TM domain, and endodomain induce hemifusion and the transition to stable pore formation. Pore expansion after fusion depends on calcium and annexin A1 (AX1). The endodomain of FAST proteins interacts with AX1 and other cofactors to disrupt cortical actin and promote pore expansion and syncytium formation (Podbilewicz 2014). It seems that efficient replication of the virus requires the formation of a syncytium. Studies of mammalian cells infected with wild-type and recombinant viruses showed that cell–cell fusion induced by FAST proteins dramatically improved replication of non-enveloped double-stranded RNA (dsRNA) viruses that did not encode FAST proteins (Kanai et al. 2019).

7.2.1.2 Role of Actin in Cell- and Virus-Induced Cell Fusion

The actin cytoskeleton is involved in cell–cell fusion in eukaryotic organisms (Martin 2016) and participates in many phases of the virus life cycle, such as entry and exit from the cell, replication, transcription, movement at the cell surface and between cells, and cell fusion (Fig. 7.3) (Kloc et al. 2022a, b; Molho et al. 2022; Ospina Stella and Turville 2018; Spear and Wu 2014; Taylor et al. 2011).

Many viruses, such as human immunodeficiency virus, vesicular stomatitis virus, papillomavirus, and murine leukemia virus, surf the surface of the cell to find the most permissible place for entry, which is very often located at the surface of actin-based cellular filopodia. The contraction (retrograde F-actin flow) of the filopodium's actin filaments by myosin II motors located within the cortical actin at the base of the filopodium creates a dragging force that powers the surfing and moves virus–receptor complexes to the base of the filopodium, where it becomes endocytosed. Endocytosis and the egress of the virus from the cell by exocytosis also depend on actin filaments. Some viruses (e.g., baculovirus, Ebola, Marburg, and vaccinia) use actin comets to move to the cell periphery and exit. Nucleation and propelling force of actin comets depend on the actin-related protein 2/3 (Arp2/3) complex and myosin and kinesin motors (Dibsy et al. 2023; Frischknecht and Way 2001; Kloc et al. 2022a, b; Lehmann et al. 2005; Mueller et al. 2014; Serrano et al. 2023).

Also, the fusion machinery during natural and virus-mediated cell fusion often consists, besides fusogens, of an actin cytoskeleton juxtaposing the membranes of fusing partners (Chen and Olson 2005; Martin 2016). Studies of cell–cell fusion in the yeast *Schizosaccharomyces pombe* discovered specialized actin fusion foci (asters). Asters, with actin-barbed ends localized under the cell membrane, are organized by the Fus1 formin, which nucleates actin filaments, and the type V myosin motor, which moves and concentrates vesicles carrying cell wall hydrolases. The aster focuses cell wall dissolution to a restricted region within the broader zone of wall synthesis, prompting localized cell wall degradation and cell fusion instead of cell growth (Dudin et al. 2015). Similar but not identical actin foci facilitate cell fusion during the development of myotubes. Studies in *Drosophila* showed that the fusion of fusion-competent myoblast (FCM) with the founder cell (FC) depends on the temporary formation of an actin focus in the FCM, which brings membranes together and supplies mechanical force for defeating the mechanical barrier separating individual cells (Martin 2016). The formation of actin focus depends on the interacting Arp2/3 complex and its coactivators: Wiskott–Aldrich syndrome protein (WASP) and WASP-family Verprolin homolog or suppressor of cyclic AMP receptor (WAVE/SCAR), which regulate actin branching, while actin polymerization delivers mechanical (pushing) force (Rout et al. 2022). Similar actin-related molecular pathways and the actin cytoskeleton are also involved in the virus-induced cell fusion.

The reptilian orthoreovirus fusion protein p14 from the FAST family of fusogens has a very short (0.7–1.5 nm) ectodomain, which is, theoretically, too short for bringing together fusing membranes. Despite that, the expression of p14 in noninfected cultured cells efficiently drives their fusion (Corcoran and Duncan 2004; Duncan et al. 2004). Studies of cell fusion in HEK293T cells transiently expressing p14 showed that p14 induces, through the Neural Wiskott–Aldrich syndrome protein (N-WASP) pathway, the assembly of branched actin (Chan et al. 2020). The N-WASP and interacting actin-related protein 2/3 complex (Arp 2/3) regulate actin polymerization (Kovacs et al. 2011; Yamaguchi et al. 2005). The energy created during actin assembly physically forces a short fusogenic ectodomain to penetrate the cell surface coating, overcoming membrane repulsion and allowing cell fusion. This indicates that viral fusogens with short ectodomains use alternative, actin-based mechanisms for cell fusion (Chan et al. 2020, 2021).

The human immunodeficiency virus type 1 (HIV-1)-induced fusion with a target cell and the fusion of an infected cell with a noninfected cell also depend on the actin cytoskeleton. The fusion initiation starts with the binding of HIV envelope glycoprotein (Env) to the CD4 receptor and its co-receptors (CCR5 or CXCR4) expressed by CD4+ immune cells. The CCR5 transduces the signal from the Env expressed by HIV or HIV-infected cells to the RhoA-family GTPase Rac1, which is one of the main regulators of the actin cytoskeleton. This leads to the restructuring of actin cytoskeleton, cell membrane extension, and virus–cell or cell–cell fusion (Pontow et al. 2004).

7.2.2 *Bacteria*

Bacteria, like viruses, also use the actin cytoskeleton in various phases of infection (Fig. 7.4). They surf the surface of the host cell to establish the best entry point. The movement (surfing) of bacteria across the cell membranes often involves the actin cytoskeleton. Enteropathogenic *Escherichia coli* triggers (via the Arp2/3 pathway) the formation of a moving, actin-rich pedestal beneath the bacteria that, by flexing and undulating, promotes its movement (Miner and Rauch 2024; Smith et al. 2010).

The entry of bacteria into the host cells occurs through phagocytosis or macropinocytosis, which are actin-dependent processes. Some bacteria, such as *Listeria monocytogenes* and *Yersinia* spp., enter through the so-called zipper mechanism, in which binding of bacterial surface proteins to the host cell receptors induces actin-dependent membrane rearrangement and phagocytosis. *Salmonella enterica* and *Shigella flexneri* use the "trigger" mechanism. They inject, via the type 3 secretion system (T3SS), effector proteins into the host cell, inducing actin-rich membrane ruffles that mediate macropinocytotic engulfment of bacteria. The phagocytic cup and ruffle formation depend on Rho GTPases and a (Arp2/3) complex that nucleates branched actin networks (Sumida and Yamada 2015). In contrast, *Clostridium difficile* produces a transferase Cytolethal distending toxin (CDT) toxin that, after being endocytosed by intestinal epithelial cells, Adenosine Diphosphate (ADP)-ribosylates actin, inducing actin filament disassembly and promoting stabilization and growth of microtubule-containing cell extensions, which wrap around bacteria (Bugalhão et al. 2015; Haglund and Welch 2011). After entering the host cell, some bacteria, e.g., *Chlamydia trachomatis,* modify actin and cytokeratin filaments to form replicative niches, the filamentous cages around bacteria-containing vesicles to prevent destruction by host defense mechanisms (Kumar and Valdivia 2008). Many species of bacteria, such as *Listeria* or *Shigella*, after exiting the phagosome, replicate freely in the host cytoplasm and induce Arp2/3-dependent actin polymerization (actin comets) at their surface to propel them within the cytoplasm and spread between cells (Haglund and Welch 2011). After successful replication, bacteria exit the host cell and spread to new cells. The exit occurs through inclusion/extrusion (exocytosis), a rapture of the cell membrane, which exposes bacteria to the host immune defense, or a direct spread between cells, which prevents immunodetection (Colonne et al. 2016). For example, *Chlamydia* exits either by cell membrane rupture and cell lysis or extrusion. The extrusion is actin/N-WASP/myosin II/Rho GTPase-dependent and consists of the formation of bacteria-containing vesicles at the cell periphery, which pinch off the cell surface (Hybiske and Stephens 2007). *Ehrlichia chaffeensis*, which infects macrophages, uses a dual exit strategy. In the early stage of infection, it spreads between cells through actin-dependent protrusions, which extend to the neighboring cells, while in the late stage of infection, it lyses the host cell (Thomas et al. 2010). Studies on intercellular spreading of *Shigella, Listeria, Rickettsia,* and *Burkholderia* showed that they modify the actin cytoskeleton and the host cell membrane to form the protrusions extending to neighboring cells (Raab et al. 2024). The protrusions resolve into vesicles from which bacteria escape to the cytosol. Formation of the protrusions at the cell periphery depends on actin bundle formation, regulated by the actin-polymerizing formins (mDia1 and mDia2) (Raab et al. 2024).

7.2.2.1 Bacteria-Induced Cell Fusion

So far, the only known bacterial species inducing fusion of host cells are *Burkholderia pseudomallei* and *Burkholderia mallei* of the pseudomallei group, which cause the human disease melioidosis and the equine disease glanders, respectively. The third species from this group, *Burkholderia thailandensis*, is nonpathogenic and is commonly used as a model organism (Kostow and Welch 2022). Live-cell imaging studies showed that these bacteria use an actin-based motility to push on the cell membrane and create protrusions, which extend to the neighboring cell and fuse with its membrane. This fusion also requires bacterial type

VI secretion system-5 (T6SS-5) and the expression within the protrusion of the TGSS-5 effector proteins VgrG5 and TagD5 (Kostow and Welch 2022). Another study on *B. mallei*-induced cell–cell fusion found that the fusion also requires a host cell tetraspanin CD81 (TSPAN28) (Sangsri et al. 2020) from the transmembrane 4 superfamily (TM4SF) proteins, which participate in non-pathologic cell–cell fusion events such as muscle fiber and bone formation (Fanaei et al. 2011).

7.2.2.2 Interspecific Bacterial Fusion

Heterologous cell fusion between different species of bacteria may occur in syntrophy (Charubin et al. 2020). Syntrophy is a nutritional codependence of two or more species, which cooperate in degrading a single substrate and/or exchange metabolites or metabolic intermediates. In evolution, syntrophy has probably been involved in the origin of the eukaryotic cell (López-García and Moreira 2019, 2020). Syntrophy is not limited to the exchange of metabolites per se but may also involve the exchange of proteins or RNAs, interspecific transfer of electrons, or cross-protection (Charubin et al. 2020). Direct interspecies electron transfer (DIET) is common for *Geobacter* and *Methanothrix* species (Holmes et al. 2017; Lovley 2017), while *Acinetobacter baumannii* cross-protects *Klebsiella pneumoniae* from certain antibiotics (Semenec et al. 2023). Recently, a stable syntrophy was experimentally established between tryptophan auxotrophic *Saccharomyces cerevisiae* deletion mutants that spontaneously formed syntrophic, synergistic communities (Aulakh et al. 2023). Electron microscopy and electron tomography combined with correlative fluorescence transmission electron microscopy studies of syntrophy between *Clostridium ljungdahlii* and *Clostridium acetobutylicum* showed that they fuse cell walls and membranes and that the resulting hybrid cells contain proteins and RNAs from both fusion partners. The authors suggest that heterologous interspecific cell fusions may also widely occur in nature in non-syntrophic organisms but have remained undetected (Charubin et al. 2020). Interspecific (xenogeneic) cell fusion, i.e., fusion between cells from different species, was achieved in several laboratories. For example, chick skeletal myoblasts were fused with murine myoblasts and formed interspecific multinuclear myotubes (Takaya et al. 2017). Interestingly, some studies indicate that in xenogeneic fusion, the genes latent in both fusing partners are activated and expressed in the xenogeneic hybrid (Yuan et al. 2019).

7.3 Parasite-Induced Syncytia in Vascular Plants

Many agriculturally important plants, such as potato, tomato, sugar beet, eggplant, soybean, cereals, rubber plants, figs, various cacti, and ornamental plants, are parasitized by the endoparasitic cyst nematodes, mainly from the *Heterodera* and *Globodera* genera (Agrios 2005; Jones and Northcote 1972; Jones et al. 2013). They hatch from the egg as second-stage juveniles [(J2s); the first molt and first-stage juvenile (J1) occur while still in the egg], which is the only infective and migratory stage in these nematodes' life cycle (Mimee et al. 2015). The chemicals released from the plant root attract J2s to infect the host and form, within the root differentiation and elongation zone, a syncytial cyst that will be their sole source of nutrients (Wang et al. 2018; Zhang et al. 2017). The J2 injects, using the hollow oral stylet, the secretion of its esophageal glands into a chosen single initial syncytial cell (ISC), initiating the formation of a syncytial cyst. Stylet-injected secretion contains many different compounds (effectors), including a group of those mimicking plant CLAVATA3/Embryo Surrounding Region-Related (CLE) peptide hormones crucial for plant cell signaling and fate determination (Liu and Mitchum 2024). The CLE-like effector proteins injected through the stylet mimic the ligand for plant A-type and B-type CLE peptides. The B-type CLE peptide TDR is a receptor for (TDIF) tracheary element differentiation inhibitorty factor participate in the CLE–receptor kinase–WOX pathway, which promotes cyst formation. Treatment with exogenous A-type and B-type CLE peptides induces massive cell proliferation in uninfected roots, suggesting that the nematode-derived CLEs may regulate cell proliferation during cyst formation (Guo et al. 2017). The development of a syncytial cyst has several consecutive steps. First, the ISC fuses with the neighboring cells by widening existing plasmodesmata openings. Next, the cyst wall and adjacent cells' walls disassemble locally, leading to the fusion of their protoplasts and hypertrophy of participating cells. Subsequently, the cyst expands by incorporating hundreds of neighboring cells to become a metabolically highly active nutritional depot (Ohtsu et al. 2017; Rodiuc et al. 2014; Zhang et al. 2017).

The formation of syncytial cysts involves degradation and extensive restructuring of the cell wall. A study of tomato (*Lycopersicon esculentum* L.) cysts induced by potato cyst nematode *Globodera rostochiensis* suggests the role of LeEXPA4 and LeEXPA5 alpha-expansins in these processes. LeEXPA4 is possibly involved in cell wall relaxation or disassembly and LeEXPA5 in hypertrophy of cells included in the syncytium (Fudali et al. 2008). Screening of cDNA library, RT-PCR, and GeneChip analysis of *Arabidopsis thaliana* root cysts induced by *Heterodera schachtii* nematode showed syncytia-specific upregulation of AtEXPA3, AtEXPA6, AtEXPA8, AtEXPA10, and AtEXPA16 expansins (Wieczorek and Grundler 2006). Expansins are present in all plants, with expansin-like proteins present in bacteria, fungi, and animal cells (Cosgrove 2015). Expansins loosen plant cell walls in a pH-dependent manner by acidification and probably without lytic activity. They bind to cell walls and somehow disrupt noncovalent binding between cellulose microfibrils (Sampedro and Cosgrove 2005). Interestingly, under physiological conditions, expansins facilitate the assembly of an actin filament network during the formation of plant root hairs. Studies of root hair formation in *Arabidopsis* showed that the assembly of actin/profilin-based elongation machinery at the tip of the growing root requires the formation of expansin-enriched bulges and expansin-dependent local loosening of the cell wall (Baluska et al. 2000). Formation of the syncytial cyst also involves a profound change in the actin cytoskeleton. In plants, positioning and fusion of cell wall-forming vesicles depend on microtubules and actin filaments (Gu and Rasmussen 2022; Sinclair et al. 2022). Ultrastructural studies in *Arabidopsis* showed that actin filaments are disorganized and misaligned in the cysts. Here, the organization of the actin cytoskeleton depends on formin and actin depolymerization factor 2 (ADF2). The authors propose that a high concentration of ADF2 and other actin-binding proteins induced by the parasite destabilizes the actin cytoskeleton in the cyst, diminishing the density of the actin network and facilitating nematode feeding (Engler Jde et al. 2010). Proteomic studies of nematode-induced cysts in tomato roots identified 48 proteins specific for cysts, related to protein turnover, stress responses, and intracellular trafficking (Filipecki et al. 2021), indicating that the formation of parasite-induced feeding syncytia is a multifaceted process requiring further study.

7.4 Eukaryotic Fusogens and Their Evolutionary Origin

The striking similarities between viral and eukaryotic fusogens raise the question of whether, early in evolution, the eukaryotic fusogens were co-opted from the viruses or whether the viruses acquired eukaryotic fusogens (Doms 2017; Vance and Lee 2020). For example, the fusion protein hapless 2/generative cell-specific 1(HAP2) that promotes gamete fusion in sexually reproducing eukaryotes shares structural similarity with class II viral fusogens, while syncytin 1 and 2, involved in the formation of syncytiotrophoblasts of the mammalian placenta, and the soluble N-ethylmaleimide-sensitive factor attachment protein receptor (SNARE) proteins, mediating membrane fusion between organelles and cell membranes, are similar to viral class I fusogens (Brukman et al. 2022; Jahn et al. 2024; Roberts et al. 2021). Studies of the syncytin-Opo1 in the South American marsupial opossum, which has a primitive short-lived placenta and is the closest living relative to eutherian mammals from which it diverged $\sim$190 Mya, recapitulate an ancient origin of syncytin from the retroviral envelope (env) genes to the mammalian placenta (Cornelis et al. 2015).

Recent discovery of the archaeal fusion protein, homologous to HAP2, sheds new light on the evolution of eukaryotic fusogens. *Archaea* have unique features that separate them from *Bacteria* and *Eukaryota* and some common to both (Adam et al. 2017). Although the archaeal fusexin1 and 2 have a unique structure with a six-helix bundle and a globular domain, fusexin1 can fuse mammalian cells when expressed ectopically. In the archaeal genome, fusexin (fsx) genes are located within the integrated mobile elements (IMEs) (Krupovic et al. 2019). The association of *fsx1* with IMEs indicates its lateral mobility. It suggests that fusexin1 mediates a cell fusion-dependent genetic exchange in archaea and that these ancient organisms evolved conjugative cell fusion to support the horizontal gene transfer (HGT) mechanism (Moi et al. 2022).

Acknowledgments The author acknowledge JC Walter Jr.'s support to the Houston Methodist Transplantation Center. Some drawings used in the figures were adapted from images at Servier Medical Art (https://smart.servier.com/), licensed under CC BY 4.0 (https://creativecommons.org/licenses/by/4.0/).

References

Adam PS, Borrel G, Brochier-Armanet C, Gribaldo S (2017) The growing tree of archaea: new perspectives on their diversity, evolution and ecology. ISME J 11(11):2407–2425. https://doi.org/10.1038/ismej.2017.122

Agrios GN (2005) Chapter 15: plant diseases caused by nematodes. In: Agrios GN (ed) Plant pathology, 5th edn. Academic Press, pp 825–874. https://doi.org/10.1016/B978-0-08-047378-9.50021-X

Aulakh SK, Sellés Vidal L, South EJ, Peng H, Varma SJ, Herrera-Dominguez L, Ralser M, Ledesma-Amaro R (2023) Spontaneously established syntrophic yeast communities improve bioproduction. Nat Chem Biol 19(8):951–961. https://doi.org/10.1038/s41589-023-01341-2

Baluska F, Salaj J, Mathur J, Braun M, Jasper F, Samaj J, Chua NH, Barlow PW, Volkmann D (2000) Root hair formation: F-actin-dependent tip growth is initiated by local assembly of profilin-supported F-actin meshworks accumulated within expansin-enriched bulges. Dev Biol 227(2):618–632. https://doi.org/10.1006/dbio.2000.9908

Bradburne CE, Verhoeven AB, Manyam GC, Chaudhry SA, Chang EL, Thach DC, Bailey CL, van Hoek ML (2013) Temporal transcriptional response during infection of type II alveolar epithelial cells with Francisella tularensis live vaccine strain (LVS) supports a general host suppression and bacterial uptake by macropinocytosis. J Biol Chem 288(15):10780–10791. https://doi.org/10.1074/jbc.M112.362178

Brukman NG, Uygur B, Podbilewicz B, Chernomordik LV (2019) How cells fuse. J Cell Biol 218(5):1436–1451. https://doi.org/10.1083/jcb.201901017

Brukman NG, Li X, Podbilewicz B (2022) Fusexins, HAP2/GCS1 and evolution of gamete fusion. Front Cell Dev Biol 9:824024. https://doi.org/10.3389/fcell.2021.824024

Buchrieser J, Dufloo J, Hubert M, Monel B, Planas D, Rajah MM, Planchais C, Porrot F, Guivel-Benhassine F, Van der Werf S, Casartelli N, Mouquet H, Bruel T, Schwartz O (2020) Syncytia formation by SARS-CoV-2-infected cells. EMBO J 39(23):e106267. https://doi.org/10.15252/embj.2020106267

Bugalhão JN, Mota LJ, Franco IS (2015) Bacterial nucleators: actin' on actin. Pathog Dis 73(9): ftv078. https://doi.org/10.1093/femspd/ftv078

Chan KMC, Son S, Schmid EM, Fletcher DA (2020) A viral fusogen hijacks the actin cytoskeleton to drive cell-cell fusion. elife 9:e51358. https://doi.org/10.7554/eLife.51358

Chan KMC, Arthur AL, Morstein J, Jin M, Bhat A, Schlesinger D, Son S, Stevens DA, Drubin DG, Fletcher DA (2021) Evolutionarily related small viral fusogens hijack distinct but modular actin nucleation pathways to drive cell-cell fusion. Proc Natl Acad Sci USA 118(1):e2007526118. https://doi.org/10.1073/pnas.2007526118

Charubin K, Modla S, Caplan JL, Papoutsakis ET (2020) Interspecies microbial fusion and large-scale exchange of cytoplasmic proteins and RNA in a syntrophic clostridium Coculture. MBio 11(5):e02030–e02020. https://doi.org/10.1128/mBio.02030-20

Charubin K, Hill JD, Papoutsakis ET (2024) DNA transfer between two different species mediated by heterologous cell fusion in clostridium coculture. MBio 15(2):e0313323. https://doi.org/10.1128/mbio.03133-23

Chaudhary S, Yadav RP, Kumar S, Yadav SC (2023) Ultrastructural study confirms the formation of single and heterotypic syncytial cells in bronchoalveolar fluids of COVID-19 patients. Virol J 20(1):97. https://doi.org/10.1186/s12985-023-02062-7

Chen EH, Olson EN (2005) Unveiling the mechanisms of cell-cell fusion. Science 308(5720): 369–373. https://doi.org/10.1126/science.1104799

Chernomordik LV, Kozlov MM (2008) Mechanics of membrane fusion. Nat Struct Mol Biol 15(7): 675–683. https://doi.org/10.1038/nsmb.1455

Choe JE, Welch MD (2016) Actin-based motility of bacterial pathogens: mechanistic diversity and its impact on virulence. Pathog Dis. 74(8):ftw099. https://doi.org/10.1093/femspd/ftw099

Ciechonska M, Duncan R (2014) Reovirus FAST proteins: virus-encoded cellular fusogens. Trends Microbiol 22(12):715–724. https://doi.org/10.1016/j.tim.2014.08.005

Clark-Cotton MR, Jacobs KC, Lew DJ (2022) Chemotropism and cell-cell fusion in fungi. Microbiol Mol Biol Rev 86(1):e0016521. https://doi.org/10.1128/mmbr.00165-21

Cohen FS (2016) How viruses invade cells. Biophys J 110(5):1028–1032. https://doi.org/10.1016/j.bpj.2016.02.006

Colonne PM, Winchell CG, Voth DE (2016) Hijacking host cell highways: manipulation of the host actin cytoskeleton by obligate intracellular bacterial pathogens. Front Cell Infect Microbiol 6: 107. https://doi.org/10.3389/fcimb.2016.00107

Corcoran JA, Duncan R (2004) Reptilian reovirus utilizes a small type III protein with an external myristylated amino terminus to mediate cell-cell fusion. J Virol 78:4342–4351. https://doi.org/10.1128/JVI.78.8.4342-4351.2004

Cornelis G, Vernochet C, Carradec Q, Souquere S, Mulot B, Catzeflis F, Nilsson MA, Menzies BR, Renfree MB, Pierron G, Zeller U, Heidmann O, Dupressoir A, Heidmann T (2015) Retroviral envelope gene captures and syncytin exaptation for placentation in marsupials. Proc Natl Acad Sci USA 112(5):E487–E496. https://doi.org/10.1073/pnas.1417000112

Cosgrove DJ (2015) Plant expansins: diversity and interactions with plant cell walls. Curr Opin Plant Biol 25:162–172. https://doi.org/10.1016/j.pbi.2015.05.014

Dibsy R, Bremaud E, Mak J, Favard C, Muriaux D (2023) HIV-1 diverts cortical actin for particle assembly and release. Nat Commun 14(1):6945. https://doi.org/10.1038/s41467-023-41940-0

Dittmar T, Hass R (2022) Extracellular events involved in cancer cell-cell fusion. Int J Mol Sci 23(24):16071. https://doi.org/10.3390/ijms232416071

Doms RW (2017) What came first-the virus or the egg? Cell 168(5):755–757. https://doi.org/10.1016/j.cell.2017.02.012

Dudin O, Bendezú FO, Groux R, Laroche T, Seitz A, Martin SG (2015) A formin-nucleated actin aster concentrates cell wall hydrolases for cell fusion in fission yeast. J Cell Biol 208(7): 897–911. https://doi.org/10.1083/jcb.201411124

Duncan R, Corcoran J, Shou J, Stoltz D (2004) Reptilian reovirus: a new fusogenic Orthoreovirus species. Virology 319:131–140. https://doi.org/10.1016/j.virol.2003.10.025

Engler Jde A, Rodiuc N, Smertenko A, Abad P (2010) Plant actin cytoskeleton re-modeling by plant parasitic nematodes. Plant Signal Behav 5(3):213–217. https://doi.org/10.4161/psb.5.3.10741

Fanaei M, Monk PN, Partridge LJ (2011) The role of tetraspanins in fusion. Biochem Soc Trans 39(2):524–528. https://doi.org/10.1042/BST0390524

Filipecki M, Żurczak M, Matuszkiewicz M, Święcicka M, Kurek W, Olszewski J, Koter MD, Lamont D, Sobczak M (2021) Profiling the proteome of cyst nematode-induced syncytia on tomato roots. Int J Mol Sci 22(22):12147. https://doi.org/10.3390/ijms222212147

Fleißner A, Oostlander AG, Well L (2022) Highly conserved, but highly specific: somatic cell-cell fusion in filamentous fungi. Curr Opin Cell Biol 79:102140. https://doi.org/10.1016/j.ceb.2022.102140

Frischknecht F, Way M (2001) Surfing pathogens and the lessons learned for actin polymerization. Trends Cell Biol 11(1):30–38. https://doi.org/10.1016/s0962-8924(00)01871-7

Fudali S, Janakowski S, Sobczak M, Griesser M, Grundler FM, Golinowski W (2008) Two tomato alpha-expansins show distinct spatial and temporal expression patterns during development of nematode-induced syncytia. Physiol Plant 132(3):370–383. https://doi.org/10.1111/j.1399-3054.2007.01017.x

Gu Y, Rasmussen CG (2022) Cell biology of primary cell wall synthesis in plants. Plant Cell 34(1): 103–128. https://doi.org/10.1093/plcell/koab249

Guo X, Wang J, Gardner M, Fukuda H, Kondo Y, Etchells JP, Wang X, Mitchum MG (2017) Identification of cyst nematode B-type CLE peptides and modulation of the vascular stem cell pathway for feeding cell formation. PLoS Pathog 13(2):e1006142. https://doi.org/10.1371/journal.ppat.1006142

Haglund CM, Welch MD (2011) Pathogens and polymers: microbe-host interactions illuminate the cytoskeleton. J Cell Biol 195(1):7–17. https://doi.org/10.1083/jcb.201103148

Herzog S, Schumann MR, Fleißner A (2015) Cell fusion in Neurospora crassa. Curr Opin Microbiol 28:53–59. https://doi.org/10.1016/j.mib.2015.08.002

Holmes DE, Shrestha PM, Walker DJF, Dang Y, Nevin KP, Woodard TL, Lovley DR (2017) Metatranscriptomic evidence for direct interspecies electron transfer between Geobacter and Methanothrix species in methanogenic Rice Paddy soils. Appl Environ Microbiol 83(9): e00223–e00217. https://doi.org/10.1128/AEM.00223-17

Hermann T, Torres R, Salgado EN et al. (2021) Functional refolding of the penetration protein on a non-enveloped virus. Nature 590:666–670. https://doi.org/10.1038/s41586-020-03124-4

Hybiske K and Stephens RS (2007) Mechanisms of Chalmydia trachomatis entry into nonphagocytic cells.Unfet, Immun. 75(8):3925–4. https://doi.org/10.1128/IA.00106-07

Iosilevskii Y, Podbilewicz B (2021) Programmed cell fusion in development and homeostasis. Curr Top Dev Biol 144:215–244. https://doi.org/10.1016/bs.ctdb.2020.12.013

Jahn R, Cafiso DC, Tamm LK (2024) Mechanisms of SNARE proteins in membrane fusion. Nat Rev Mol Cell Biol 25(2):101–118. https://doi.org/10.1038/s41580-023-00668-x

Jones MGK, Northcote DH (1972) Nematode-induced syncytium-a multinucleate transfer cell. J Cell Sci 10(3):789–809

Jones JT, Haegeman A, Danchin EG, Gaur HS, Helder J, Jones MG, Kikuchi T, Manzanilla-López R, Palomares-Rius JE, Wesemael WM, Perry RN (2013) Top 10 plant-parasitic nematodes in molecular plant pathology. Mol Plant Pathol 14(9):946–961. https://doi.org/10.1111/mpp.12057

Kanai Y, Kawagishi T, Sakai Y, Nouda R, Shimojima M, Saijo M, Matsuura Y, Kobayashi T (2019) Cell-cell fusion induced by reovirus FAST proteins enhances replication and pathogenicity of non-enveloped dsRNA viruses. PLoS Pathog 15(4):e1007675. https://doi.org/10.1371/journal.ppat.1007675

Kloc M, Subuddhi A, Uosef A, Kubiak JZ, Ghobrial RM (2022a) Monocyte–macrophage lineage cell fusion. Int J Mol Sci 23:6553. https://doi.org/10.3390/ijms23126553

Kloc M, Uosef A, Wosik J, Kubiak JZ, Ghobrial RM (2022b) Virus interactions with the actin cytoskeleton-what we know and do not know about SARS-CoV-2. Arch Virol 167(3):737–749. https://doi.org/10.1007/s00705-022-05366-1

Kostow N, Welch MD (2022) Plasma membrane protrusions mediate host cell-cell fusion induced by *Burkholderia thailandensis.* Mol Biol Cell 33(8):ar70. https://doi.org/10.1091/mbc.E22-02-0056

Kovacs EM, Verma S, Ali RG, Ratheesh A, Hamilton NA, Akhmanova A, Yap AS (2011) N-WASP regulates the epithelial junctional actin cytoskeleton through a non-canonical post-nucleation pathway. Nat Cell Biol 13(8):934–943. https://doi.org/10.1038/ncb2290

Krupovic M, Makarova KS, Wolf YI, Medvedeva S, Prangishvili D, Forterre P, Koonin EV (2019) Integrated mobile genetic elements in Thaumarchaeota. Environ Microbiol 21(6):2056–2078. https://doi.org/10.1111/1462-2920.14564

Kumar Y, Valdivia RH (2008) Actin and intermediate filaments stabilize the chlamydia trachomatis vacuole by forming dynamic structural scaffolds. Cell Host Microbe 4(2):159–169. https://doi.org/10.1016/j.chom.2008.05.018

Lehmann MJ, Sherer NM, Marks CB, Pypaert M, Mothes W (2005) Actin- and myosin-driven movement of viruses along filopodia precedes their entry into cells. J Cell Biol 170(2):317–325. https://doi.org/10.1083/jcb.200503059

Leroy H, Han M, Woottum M, Bracq L, Bouchet J, Xie M, Benichou S (2020) Virus-mediated cell-cell fusion. Int J Mol Sci 21(24):9644. https://doi.org/10.3390/ijms21249644

Liu X, Mitchum MG (2024) A major role of class III HD-ZIPs in promoting sugar beet cyst nematode parasitism in Arabidopsis. PLoS Pathog 20(11):e1012610. https://doi.org/10.1371/journal.ppat.1012610

López-García P, Moreira D (2019) Eukaryogenesis, a syntrophy affair. Nat Microbiol 4(7): 1068–1070. https://doi.org/10.1038/s41564-019-0495-5

López-García P, Moreira D (2020) The Syntrophy hypothesis for the origin of eukaryotes revisited. Nat Microbiol 5:655–667. https://doi.org/10.1038/s41564-020-0710-4

Lovley DR (2017) Syntrophy goes electric: direct interspecies electron transfer. Ann Rev Microbiol 71:643–664. https://doi.org/10.1146/annurev-micro-030117-020420

Manjunath Y, Porciani D, Mitchem JB, Suvilesh KN, Avella DM, Kimchi ET, Staveley-O'Carroll KF, Burke DH, Li G, Kaifi JT (2020) Tumor-cell-macrophage fusion cells as liquid biomarkers and tumor enhancers in cancer. Int J Mol Sci 21(5):1872. https://doi.org/10.3390/ijms21051872

Martin SG (2016) Role and organization of the actin cytoskeleton during cell-cell fusion. Semin Cell Dev Biol 60:121–126. https://doi.org/10.1016/j.semcdb.2016.07.025

Más V, Melero JA (2013) Entry of enveloped viruses into host cells: membrane fusion. Subcell Biochem 68:467–487. https://doi.org/10.1007/978-94-007-6552-8_16

Mimee B, Dauphinais N, Bélair G (2015) Life cycle of the Golden cyst nematode, Globodera rostochiensis, in Quebec, Canada. J Nematol 47(4):290–295

Miner MV, Rauch I (2024) Why put yourself on a pedestal? The pathogenic role of the A/E pedestal. Infect Immun 92(9):e0048923. https://doi.org/10.1128/iai.00489-23

Modis Y (2013) Class II fusion proteins. In: Madame Curie bioscience database. Landes Bioscience, Austin

Moi D, Nishio S, Li X, Valansi C, Langleib M, Brukman NG, Flyak K, Dessimoz C, de Sanctis D, Tunyasuvunakool K, Jumper J, Graña M, Romero H, Aguilar PS, Jovine L, Podbilewicz B (2022) Discovery of archaeal fusexins homologous to eukaryotic HAP2/GCS1 gamete fusion proteins. Nat Commun 13(1):3880. https://doi.org/10.1038/s41467-022-31564-1

Molho M, Zhu S, Nagy PD (2022) Race against time between the virus and host: actin-assisted rapid biogenesis of replication organelles is used by TBSV to limit the recruitment of cellular restriction factors. J Virol 96(12):e0016821. https://doi.org/10.1128/jvi.00168-21

Mueller J, Pfanzelter J, Winkler C, Narita A, Le Clainche C, Nemethova M, Carlier MF, Maeda Y, Welch MD, Ohkawa T, Schmeiser C, Resch GP, Small JV (2014) Electron tomography and simulation of baculovirus actin comet tails support a tethered filament model of pathogen propulsion. PLoS Biol 12(1):e1001765. https://doi.org/10.1371/journal.pbio.1001765

Nawrocki AM, Cartwright P (2012) A novel mode of colony formation in a hydrozoan through fusion of sexually generated individuals. Curr Biol 22(9):825–829. https://doi.org/10.1016/j.cub.2012.03.026

Ohtsu M, Sato Y, Kurihara D, Suzaki T, Kawaguchi M, Maruyama D, Higashiyama T (2017) Spatiotemporal deep imaging of syncytium induced by the soybean cyst nematode Heterodera glycines. Protoplasma 254(6):2107–2115. https://doi.org/10.1007/s00709-017-1105-0

Ospina Stella A, Turville S (2018) All-round manipulation of the actin cytoskeleton by HIV. Viruses 10(2):63. https://doi.org/10.3390/v10020063

Pletan ML, Tsai B (2022) Non-enveloped virus membrane penetration: new advances leading to new insights. PLoS Pathog 18(12):e1010948. https://doi.org/10.1371/journal.ppat.1010948

Podbilewicz B (2014) Virus and cell fusion mechanisms. Annu Rev Cell Dev Biol 30:111–139. https://doi.org/10.1146/annurev-cellbio-101512-122422

Pontow SE, Heyden NV, Wei S, Ratner L (2004) Actin cytoskeletal reorganizations and coreceptor-mediated activation of rac during human immunodeficiency virus-induced cell fusion. J Virol 78(13):7138–7147. https://doi.org/10.1128/JVI.78.13.7138-7147.2004

Raab JE, Hamilton DJ, Harju TB, Huynh TN, Russo BC (2024) Pushing boundaries: mechanisms enabling bacterial pathogens to spread between cells. Infect Immun 92(9):e0052423. https://doi.org/10.1128/iai.00524-23

Rajah MM, Bernier A, Buchrieser J, Schwartz O (2022) The mechanism and consequences of SARS-CoV-2 spike-mediated fusion and syncytia formation. J Mol Biol 434(6):167280. https://doi.org/10.1016/j.jmb.2021.167280
Roberts RM, Ezashi T, Schulz LC, Sugimoto J, Schust DJ, Khan T, Zhou J (2021) Syncytins expressed in human placental trophoblast. Placenta 113:8–14. https://doi.org/10.1016/j.placenta.2021.01.006
Rodiuc N, Vieira P, Banora MY, de Almeida EJ (2014) On the track of transfer cell formation by specialized plant-parasitic nematodes. Front Plant Sci 5:160. https://doi.org/10.3389/fpls.2014.00160
Rout P, Preußner M, Önel SF (2022) Drosophila melanogaster: a model system to study distinct genetic programs in myoblast fusion. Cells 11(3):321. https://doi.org/10.3390/cells11030321
Sampedro J, Cosgrove DJ (2005) The expansin superfamily. Genome Biol 6(12):242. https://doi.org/10.1186/gb-2005-6-12-242
Sangsri T, Saiprom N, Tubsuwan A, Monk P, Partridge LJ, Chantratita N (2020) Tetraspanins are involved in *Burkholderia pseudomallei*-induced cell-to-cell fusion of phagocytic and non-phagocytic cells. Sci Rep 10(1):17972. https://doi.org/10.1038/s41598-020-74737-y
Semenec L, Cain AK, Dawson CJ, Liu Q, Dinh H, Lott H, Penesyan A, Maharjan R, Short FL, Hassan KA, Paulsen IT (2023) Cross-protection and cross-feeding between Klebsiella pneumoniae and Acinetobacter baumannii promotes their co-existence. Nat Commun 14(1):702. https://doi.org/10.1038/s41467-023-36252-2
Serrano T, Frémont S, Echard A (2023) Get in and get out: remodeling of the cellular actin cytoskeleton upon HIV-1 infection. Biol Cell 115(4):e2200085. https://doi.org/10.1111/boc.202200085
Sinclair R, Hsu G, Davis D, Chang M, Rosquete M, Iwasa JH, Drakakaki G (2022) Plant cytokinesis and the construction of new cell wall. FEBS Lett 596(17):2243–2255. https://doi.org/10.1002/1873-3468.14426
Smith K, Humphreys D, Hume PJ, Koronakis V (2010) Enteropathogenic Escherichia coli recruits the cellular inositol phosphatase SHIP2 to regulate actin-pedestal formation. Cell Host Microbe 7(1):13–24. https://doi.org/10.1016/j.chom.2009.12.004
Spear M, Wu Y (2014) Viral exploitation of actin: force-generation and scaffolding functions in viral infection. Virol Sin 29(3):139–147. https://doi.org/10.1007/s12250-014-3476-0
Sumida GM, Yamada S (2015) Rho GTPases and the downstream effectors actin-related protein 2/3 (Arp2/3) complex and myosin II induce membrane fusion at self-contacts. J Biol Chem 290(6):3238–3247. https://doi.org/10.1074/jbc.M114.612168
Takaya T, Nihashi Y, Kojima S, Ono T, Kagami H (2017) Autonomous xenogenic cell fusion of murine and chick skeletal muscle myoblasts. Anim Sci J 88(11):1880–1885. https://doi.org/10.1111/asj.12884
Taylor MP, Koyuncu OO, Enquist LW (2011) Subversion of the actin cytoskeleton during viral infection. Nat Rev Microbiol 9(6):427–439. https://doi.org/10.1038/nrmicro2574
Thomas S, Popov VL, Walker DH (2010) Exit mechanisms of the intracellular bacterium Ehrlichia. PLoS One 5(12):e15775. https://doi.org/10.1371/journal.pone.0015775
Thorley JA, McKeating JA, Rappoport JZ (2010) Mechanisms of viral entry: sneaking in the front door. Protoplasma 244(1–4):15–24. https://doi.org/10.1007/s00709-010-0152-6
Vance TDR, Lee JE (2020) Virus and eukaryote fusogen superfamilies. Curr Biol 30(13):R750–R754. https://doi.org/10.1016/j.cub.2020.05.029
Wang W, Yang L, Huang X, Fu W, Pan D, Cai L, Ye J, Liu J, Xia N, Cheng T, Zhu H (2017) Outer nuclear membrane fusion of adjacent nuclei in varicella-zoster virus-induced syncytia. Virology 512:34–38. https://doi.org/10.1016/j.virol.2017.09.002
Wang C, Masler EP, Rogers ST (2018) Responses of Heterodera glycines and Meloidogyne incognita infective juveniles to root tissues, root exudates, and root extracts from three plant species. Plant Dis 102(9):1733–1740. https://doi.org/10.1094/PDIS-09-17-1445-RE

Welsch S, Müller B, Kräusslich HG (2007) More than one door – budding of enveloped viruses through cellular membranes. FEBS Lett 581(11):2089–2097. https://doi.org/10.1016/j.febslet.2007.03.060

Wesolowski J, Paumet F (2017) Taking control: reorganization of the host cytoskeleton by chlamydia. F1000Res 6:2058. https://doi.org/10.12688/f1000research.12316.1

Wieczorek K, Grundler FM (2006) Expanding nematode-induced syncytia: the role of expansins. Plant Signal Behav 1(5):223–224. https://doi.org/10.4161/psb.1.5.3426

Xia X, Wu W, Cui Y, Roy P, Zhou ZH (2021) Bluetongue virus capsid protein VP5 perforates membranes at low endosomal pH during viral entry. Nat Microbiol 6(11):1424–1432. https://doi.org/10.1038/s41564-021-00988-8

Yamaguchi H, Lorenz M, Kempiak S, Sarmiento C, Coniglio S, Symons M, Segall J, Eddy R, Miki H, Takenawa T, Condeelis J (2005) Molecular mechanisms of invadopodium formation: the role of the N-WASP-Arp2/3 complex pathway and cofilin. J Cell Biol 168(3):441–452. https://doi.org/10.1083/jcb.200407076

Yang Z, Dam KA, Gershoni JM, Zolla-Pazner S, Bjorkman PJ (2022) Antibody recognition of CD4-induced open HIV-1 Env trimers. J Virol 96(24):e0108222. https://doi.org/10.1128/jvi.01082-22

Yuan C, Freeman BT, McArdle TJ, Jung JP, Ogle BM (2019) Conserved pathway activation following xenogeneic, heterotypic fusion. FASEB J 33(6):6767–6777. https://doi.org/10.1096/fj.201801700R

Zhang L, Lilley CJ, Imren M, Knox JP, Urwin PE (2017) The complex Cell Wall composition of syncytia induced by plant parasitic cyst nematodes reflects both function and host plant. Front Plant Sci 8:1087. https://doi.org/10.3389/fpls.2017.01087

Zhao C, Pu J (2022) Influence of host sialic acid receptors structure on the host specificity of influenza viruses. Viruses 14(10):2141. https://doi.org/10.3390/v14102141

Chapter 8
Cell Fusion in Reprogramming and Regeneration

Marta Cadevall Angles, Eimear M. Byrne, and Maria Pia Cosma

Abstract Cell fusion has evolved at an exciting pace over the past 50 years. From experimental studies in vitro, later applications in vivo, and various discoveries of innate in vivo cell fusion, this field continues to expand, particularly as biotechnological strategies advance. Initially developed as an experimental approach to study nuclear reprogramming, cell fusion provided some of the first and most convincing evidence that differentiated somatic nuclei are still reprogrammable. Fusion-based systems changed long-held beliefs about the stability of cell fate by demonstrating that significant transcriptional, epigenetic, and chromatin changes can occur within hours. Subsequent in vivo studies demonstrated that similar fusion events occur naturally in adult tissues, principally in response to injury, inflammation, or degenerative stress. In the context of regeneration, cell fusion has emerged as an important and previously under-recognized mechanism underlying cellular plasticity, tissue repair, and functional recovery.

Authors Marta Cadevall Angles and Eimear M. Byrne have equally contributed to this chapter.

M. C. Angles
Center for Genomic Regulation (CRG), The Barcelona Institute of Science and Technology, Barcelona, Spain

Universitat Pompeu Fabra (UPF), Barcelona, Spain

E. M. Byrne
Center for Genomic Regulation (CRG), The Barcelona Institute of Science and Technology, Barcelona, Spain

M. P. Cosma (✉)
Center for Genomic Regulation (CRG), The Barcelona Institute of Science and Technology, Barcelona, Spain

Universitat Pompeu Fabra (UPF), Barcelona, Spain

ICREA, Barcelona, Spain

Medical Research Institute, Guangdong Provincial People's Hospital (Guangdong Academy of Medical Sciences), Southern Medical University, Guangzhou, China
e-mail: pia.cosma@crg.eu

© The Author(s), under exclusive license to Springer Nature Switzerland AG 2026
J. Weiler et al. (eds.), *Cell Fusion*, Advances in Experimental Medicine and Biology 1509, https://doi.org/10.1007/978-3-032-22637-2_8

In this chapter, we discuss the molecular and cellular mechanisms that control cell fusion, the experimental tools used to study it, and the insights gained from fusion-based models of reprogramming. We further examine the growing evidence supporting the role of cell fusion in tissue regeneration across multiple organs, including the liver, muscle, nervous system, and heart. Finally, we address the challenges, limitations, and safety considerations that currently restrict therapeutic translation and outline future directions for understanding and applying cell fusion in regenerative biology.

Keywords Cell fusion · Reprogramming · Regeneration · Polyploidy

8.1 Introduction

Cell fusion is a highly regulated process essential for many biological processes, such as embryonic development and the formation of multinucleated cells like myofibers and osteoclasts. It also has well-characterized roles in cell and tissue reprogramming and regeneration (Willkomm and Bloch 2015; Alvarez-Dolado et al. 2003; Aguilar et al. 2013; Larsson et al. 2008). The propensity to undergo cell fusion is highly dependent on the cell type. In most adult cell types, cell membranes are composed of lipid bilayers that show strong repulsive forces and high energy barriers, making spontaneous fusion a rare phenomenon. In this chapter, we will discuss the current state of the art regarding cell fusion, with a focus on reprogramming and regeneration.

8.1.1 Mechanisms of Cell Fusion

To initiate cell fusion, membrane bending, dehydration, and close apposition of bilayers are required. Beyond these biophysical requirements, fusion is coordinated through genetic programs, specialized fusogens, and signaling pathways. No single fusogenic pathway alone is sufficient for the induction of effective fusion. Thus, the coordinated action of multiple fusogens, including membrane, cytoskeletal, and signaling regulators, is required (Willkomm and Bloch 2015; Brukman et al. 2019; Martens and HT 2008; Dörnen and Dittmar 2021).

In mammals, membrane glycoproteins derived from endogenous retroviral envelope genes, known as syncytins, are key fusogens (Brukman et al. 2019). These proteins mediate trophoblast fusion, prior to syncytiotrophoblast formation. Syncytins also act in osteoclasts, myoblasts, and breast cancer cells, indicating broader roles in regeneration and pathology. Growth factors such as epidermal growth factor (EGF) and transforming growth factor beta (TGF-β) further promote cell fusion through syncytins (Willkomm and Bloch 2015; Chen and Cheong 2011; Gupta et al. 2015).

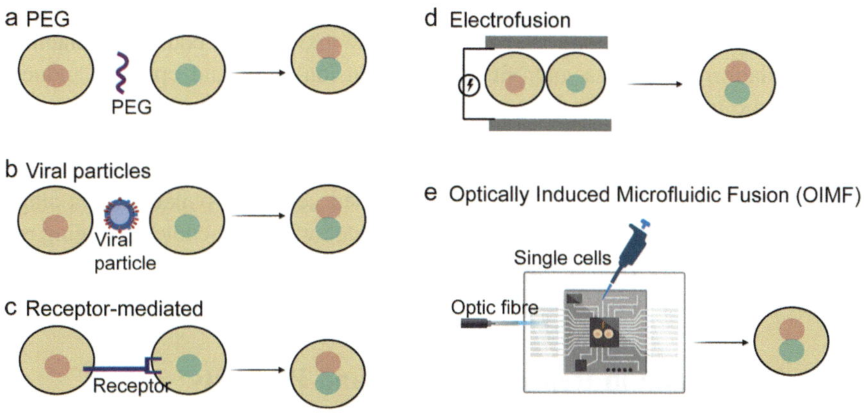

Fig. 8.1 Mechanisms of induction of cell fusion. Fusion events, triggered by (**a**) PEG, (**b**) virus-mediated fusion through naïve mammalian cell receptors, e.g., syncytins, (**c**) receptor–ligand-mediated, e.g., cell-adhesion receptor–ligands, (**d**) electrofusion, or (**e**) optically induced microfluidic fusion merging two or more cells into a single cytoplasm, producing bi- or multinucleated cells

Several experimental approaches have been used to induce cell fusion, facilitating studies of the underlying mechanisms and enabling reprogramming studies (Fig. 8.1). Polyethylene glycol (PEG) has been extensively used. PEG promotes fusion through a volume-exclusion and dehydration effect that forces close contact between cell membranes (Ahkong et al. 1975; Arnold et al. 1985). Another common method is electrofusion, which uses pulsed electric fields to transiently permeabilize membranes, increasing hybrid formation. Electrofusion is considered the standard method for somatic cell hybrid generation (Chiu 2001; Ramos and Teissie 2000).

Other methods used to promote cell fusion include inactivated Sendai virus envelopes and related viral fusogens. Vesicular stomatitis virus glycoprotein (VSV-G)-based fusion mediates membrane binding and merging through the virus glycoprotein, enabling efficient fusion of targeted cells. Inactivated Sendai viruses directly bind sialic acid residues on the cell surface of mammalian cells and drive immediate membrane–membrane fusion between cells at physiological pH. Whereas VSV-G-mediated fusion relies on low-pH activation following endocytic uptake, inactivated Sendai virus envelopes fuse immediately at neutral pH through their native receptors, enabling faster and often more efficient cell–cell fusion (Harris 1965).

A more recently developed strategy involves microfluidic electrofusion chips, which align pairs of cells in microscale channels and apply a brief electric pulse that destabilizes their membranes, causing them to fuse (Wu et al. 2015). Similarly, optically induced microfluidic fusion focuses laser energy within a microfluidic device to locally disrupt cell membranes, triggering fusion with high spatial precision (Yang et al. 2016).

Although classical fusogens such as PEG, electrofusion, and inactivated Sendai virus (Fig. 8.1) are effective, they are limited by toxicity and poor controllability. Newer platforms, such as microfluidic electrofusion chips and optically induced microfluidic fusion, enable more precise pairing of defined cell types, higher efficiency, and better viability, making them well-suited for mechanistic dissection of fusion-mediated reprogramming. These systems allow controlled generation of heterokaryons (two separated nuclei in one fused cell) and synkaryons (a single fused nucleus, polyploid cell) and can be integrated with high-content imaging and single-cell genomics (Wu et al. 2015; Yang et al. 2016).

8.2 Cell Fusion in Reprogramming and Plasticity

8.2.1 Cell Fusion-Enabled Investigation of Cellular Reprogramming

Since the initial studies of cell fusion in the 1970s, the possibility of nuclear reprogramming has been under investigation. Fusing mouse embryonic carcinoma cells (ECCs) with primary thymocytes yielded hybrids that acquired ECC-like pluripotent properties (Miller and Ruddle 1976). Since then, a wide variety of somatic cells have been fused with mouse or human embryonic stem cells (ESCs), demonstrating that cell–cell fusion can reset the state of a somatic nucleus to that of a pluripotent stem cell when it merges with a pluripotent partner (Cowan et al. 2005; Tada et al. 2001). Fusion between differentiated cells can also generate hybrids that adopt one parental identity, as seen when murine muscle cells impose a myogenic program on human primary cells from multiple germ layers (Blau et al. 1983; Blau et al. 1985; Baron and Maniatis 1986).

Cloning of Dolly the sheep by somatic cell nuclear transfer (SCNT) in 1997 provided the proof that fully differentiated mammalian cells can be reprogrammed to totipotency (Wilmut et al. 1997). Earlier work by Briggs and King had already shown that nuclei from differentiated amphibian cells retain developmental potential (Briggs and King 1952). Subsequent SCNT in mammals, including cows, mice, pigs, and primates, generated cloned animals and pluripotent ESCs with broad differentiation potential (Wilmut et al. 1997; Wakayama et al. 1998; Wakayama and Yanagimachi 1998; Onishi et al. 2000; Byrne et al. 2007). However, SCNT is inefficient (3–5% success), often associated with developmental abnormalities such as large offspring syndrome (Han and Sidhu 2008), and provides limited mechanistic insight into reprogramming, largely due to incomplete DNA methylation reprogramming in donor nuclei (Morgan et al. 2005; Carlson et al. 1992; Bourc'his et al. 2001; Dean et al. 2001; Kang et al. 2001). These limitations motivated the development of fusion-based reprogramming models using stem cells.

The use of pluripotent cells in vitro began in 1981 with the derivation of mouse embryonic stem cells (ESCs) (Evans and Kaufman 1981; Reubinoff et al. 2000;

Thomson et al. 1998). Soon after, embryonic germ cells (EGCs) were established from primordial germ cells (PGCs), both capable of long-term culture and maintaining pluripotency (Donovan 1994). However, before these cell lines were established, the initial studies of fusion-based reprogramming were performed using embryonic carcinoma cells (ECCs), which originate from the stem cells of teratocarcinomas and are closely related to ESCs. ECCs were preferred at the time because they were easier to manipulate (Han and Sidhu 2008; Evans and Kaufman 1981; Reubinoff et al. 2000; Thomson et al. 1998; Donovan 1994; Martin 1981; Jacob 1978).

8.2.1.1 Embryonic Carcinoma Cell (ECC)–Somatic Cell Hybrids

Miller and Ruddle showed in 1976 that the hybrids from teratocarcinoma and thymus cells retain pluripotency and can differentiate into multiple tissue types (Miller and Ruddle 1976). These hybrid cells resembled ECCs morphologically, reactivated specific pluripotency genes, and reactivated the inactive somatic X chromosome (Miller and Ruddle 1976; Andrews and Goodfellow 1980; Rousset et al. 1983). Cross-species fusion further underscored ECC reprogramming capacity: Murine ECC–human T-lymphoma hybrids reactivated human OCT4 and SOX2. Later studies showed that ECCs can reprogram neural stem cells (NSCs) and erase their epigenetic memory, leading hybrids to converge toward the ECC state (Do et al. 2007). These ECC–somatic hybrids can form carcinomas and embryoid bodies (Andrews and Goodfellow 1980; Rousset et al. 1983; Atsumi et al. 1982; Takahashi et al. 2007). However, ECCs are not always sufficient to reprogram all the somatic lineages; for example, ECCs fused with fibroblasts sometimes retain fibroblast-like morphology (Rousset et al. 1979).

Although ECCs were very useful in the early stages of studying reprogramming, they have several limitations: They exhibit lower developmental potential than ESCs or EGCs, rarely form chimeras, and fail to contribute to the germ line (Rousset et al. 1983). Therefore, ESC– or EGC–somatic hybrids are preferred models for reprogramming studies.

8.2.1.2 Embryonic Stem Cell (ESC)–Somatic Cell Hybrids

Early work debated whether reprogramming determinants reside in the cytoplasm or the nucleus. Nuclear transfer and egg extracts indicated the role of cytoplasmic factors (Eggan et al. 2004; Jaenisch et al. 2004; Taranger et al. 2005), but fusing NSCs separately with ESC karyoplasts or cytoplasts demonstrated that only karyoplasts activate Oct4-Green Fluorescent Protein(GFP), providing strong evidence that nuclear factors from ESCs have a strong capacity to reprogram somatic nuclei (Do and Scholer 2004).

Initial studies on animal cells indicated that fusion of somatic cells with ESCs is sufficient for the acquisition of embryonic antigens, X chromosome activation, and the formation of neural differentiating teratomas. An early study found that male $Hprt^-$ mouse Embryonic Stem Cells(mESCs) fused with female spleen cells, producing HAT-resistant hybrids that expressed embryonic antigens, lacked somatic markers, displayed alkaline phosphatase activity, formed embryoid bodies, and generated chimeric mice (Matveeva et al. 1998, 2001). Hybrids carried chromosomes from both partners, some of which were tetraploid although unstable, and were lost when the selection was removed. Later, an Oct4-GFP reporter system showed that thymocytes fused with ESCs reactivated GFP and the thymocyte X chromosome within ~48 h, indicating rapid reprogramming. However, parental imprints at H19 and Igf2r were not reset, even though hybrids contributed to all three germ layers in E7.5 chimeras (Tada et al. 2001). Fusion of $Hprt^-$ ESCs with thymocytes carrying a ROSA26 βgeo transgene yielded hybrids capable of teratoma formation and differentiation into dopaminergic neurons in vitro and in vivo (Tada et al. 2003).

Functional assays show that somatic genomes remain transcriptionally competent. For example, ESC–mesenchymal stem cell (MSC) hybrids generated from $Rag2^-/^-$ ESCs and wild-type MSCs restored Rag2 expression upon hematopoietic differentiation and produced myeloid and lymphoid progenitors in vitro, though their tetraploid state prevented hematopoietic reconstitution in vivo (Sumer et al. 2014). This highlights gene dosage and ploidy as constraints on the therapeutic use of fusion-derived pluripotent cells.

Recent work has extended these findings beyond conventional ESCs by showing that mouse extended pluripotent stem cells (EPSCs) can also reprogram somatic cells. PEG-mediated fusion of EPSCs with mouse neural stem cells (NSCs) produced stable tetraploid hybrids that activated Oct4-GFP within 48 hours, adopted EPSC-specific transcriptional signatures, and acquired dual embryonic and extra-embryonic developmental potential in vitro and in chimeric embryos. These results indicate that both pluripotent and extended pluripotent nuclei can dominantly reset somatic chromatin states and transfer their broader lineage competence through cell fusion (Song et al. 2022).

Similarly, fusion of human somatic cells with human ESCs (hESCs) was found to achieve reprogramming (Cowan et al. 2005; Yu et al. 2006). PEG-mediated fusion of hygromycin-resistant hESCs with puromycin-resistant fibroblasts produced ~92-chromosome hybrids with demethylated OCT4 promoters, pluripotency marker expression, and fibroblast gene repression, able to form embryoid bodies and teratomas (Cowan et al. 2005).

Cross-species fusion has been widely used to probe conservation of pluripotency networks. Mouse ESCs fused with porcine pluripotent stem cells or fibroblasts induce porcine OCT4, NANOG, and SOX2 and support embryoid body formation, indicating partial activation of porcine pluripotency genes (Fang et al. 2020; Bou et al. 2021). Similar approaches reprogram rabbit, bovine, and zebra fibroblasts into ESC-like hybrids (Imai et al. 2020). These cross-species models are valuable but often limited by chromosomal instability of the non-mouse genome (Bou et al. 2021; Imai et al. 2020).

8.2.1.3 Embryonic Germ Cell (EGC)–Somatic Cell Hybrids

EGCs, derived from primordial germ cells (PGCs), are globally demethylated at imprinted and non-imprinted loci (Hajkova et al. 2002; Monk et al. 1987). Fusion of female EGCs carrying a Rosa26 βgeo transgene with thymic lymphocytes produced hybrids phenotypically similar to EGCs, with extensive demethylation of the thymocyte genome, X chromosome reactivation, and erasure of parental imprints at H19 and Igf2r (features not observed with ESCs). Hybrids contributed to chimeric embryos at E9.5–E10.5 (Tada et al. 1997). Male germ line-derived multipotent germ line stem (mGS) cells show comparable reprogramming power: Fusion with thymocytes generates pluripotent hybrids that form teratomas, erase imprints, and reactivate the X chromosome. Fusion of GS cells with ESCs or mGS cells likewise produces pluripotent hybrids, reinforcing the idea that the pluripotent state of one partner is the key determinant of fusion-mediated reprogramming efficiency (Takehashi et al. 2012).

8.2.2 Mechanisms Controlling Fusion-Mediated Reprogramming

Cell fusion activates a set of transcription factors, signaling pathways, and epigenetic regulators that determine whether a somatic nucleus can regain pluripotency. Below, we summarize the main molecular mechanisms involved in this reprogramming process.

8.2.2.1 Key Regulators: Nanog and Sall4

Nanog, a homeodomain transcription factor, is central for ESC pluripotency and self-renewal (Chambers et al. 2003; Mitsui et al. 2003; Silva et al. 2009; Yang et al. 2008; Andrews and Gokhale 2024). Overexpressing Nanog in ESCs increases reprogrammed NSC colonies ~200-fold and enhances thymocyte and fibroblast hybrid yields (Silva et al. 2006). While dispensable for early reprogramming steps, Nanog is essential for the completion of dedifferentiation (Silva et al. 2009). Sall4 is another essential factor for ESC pluripotency that regulates numerous genes, including Oct4 (Frei et al. 1988; Elling et al. 2006; Zhang et al. 2006). When expressed in Mouse Embryonic Fibroblasts (MEFs) carrying an Oct4-GFP reporter, only Sall4 strongly increased GFP^+ hybrids, while Nanog and Sox2 had moderate effects, and Oct4 had little impact (Wong et al. 2008).

Heterokaryon-based screens have identified new reprogramming factors, such as NKX3-1 (downstream of IL6–STAT3) capable of substituting OCT4 in fibroblast reprogramming and Pou3f2 as a driver of epithelial differentiation in ESC–endothelial heterokaryons (Wong et al. 2017; Mai et al. 2018).

8.2.2.2 Signaling Pathways: Wnt/β-Catenin and AKT

Wnt/β-catenin signaling promotes ESC self-renewal and pluripotency gene expression (Sato et al. 2004; Cole et al. 2008; Lluis et al. 2008; de Jaime-Soguero et al. 2018). Activation of this pathway in ESCs with Wnt3a or GSK-3 inhibition, both of which lead to nuclear accumulation of β-catenin, boosts fusion-mediated reprogramming of NSCs, thymocytes, and MEFs (Lluis et al. 2008). Downstream effectors of Wnt may include Tcf3, which co-localizes with other factors regulating the pluripotency, like OCT4, SOX2, and NANOG (Cole et al. 2008; Tam et al. 2008; Yi et al. 2008). Interestingly, it has been proven that the activation of Nanog and Wnt can act synergistically (Lluis and Cosma 2009).

Another important pathway is PI3K/Akt, which regulates pluripotency and the tumorigenic potential. Akt, activated downstream of PI3K, phosphorylates multiple targets to control diverse cellular responses. This pathway supports PGC dedifferentiation into EGCs and maintains pluripotency of mouse and primate ESCs without Leukemia inhibitory factor (LIF) or feeders (Nakamura et al. 2008; Watanabe et al. 2006). Akt activation increases pluripotent hybrid colony formation after ESC–somatic cell fusion but paradoxically reduces nuclear transfer reprogramming efficiency, likely due to differential activation of target genes in the two methods (Nakamura et al. 2008). Beyond these classical pluripotency pathways, cell fusion also activates intrinsic signaling events specific to the fusion process itself. In particular, membrane remodeling and transient metabolic stress following fusion can stimulate the AMP-activated protein kinase (AMPK) pathway, which, in turn, inhibits Yes-associated protein-1 (YAP1) and contribute to the transcriptional reprogramming of fused cells (Feliciano et al. 2021).

8.2.2.3 Epigenetic Control: PRC2, AID, and G9a

Genome-wide analyses in mouse–rat hybrids distinguish genes silenced in *trans* (rapidly reactivated without DNA replication) from those silenced in *cis* (requiring replication and chromatin remodeling). Somatic–somatic cell fusion relieves trans-silencing, but only ESC–somatic fusions efficiently erase *cis*-silencing and reestablish pluripotency circuits such as OCT4 and NANOG. This supports a biphasic model: early trans-reprogramming, followed by slower, replication-dependent *cis*-reprogramming (Foshay et al. 2012).

DNA methylation is one of the major barriers to reprogramming (Meissner et al. 2008; MVC 2020; Bhutani et al. 2010). In mESC–human fibroblast heterokaryons, DNA demethylase Auxin-Inducible Degron (AID) is required for rapid demethylation and activation of OCT4 and NANOG promoters in somatic genomes; AID knockdown blocks the initiation of reprogramming and selectively binds methylated promoters in fibroblasts (Bhutani et al. 2010).

Polycomb complexes PRC1 and PRC2 maintain repression of developmental regulators via H3K27me3 (Ringrose and Paro 2004; Chamberlain et al. 2008; Pasini et al. 2007). In ESCs, PRCs mark "bivalent" promoters of developmental transcription factors, including pluripotency-associated genes (Boyer et al. 2005; Lee et al. 2006; Bernstein et al. 2006). Loss of PRC function leads to spontaneous differentiation and reduced developmental potential (Chamberlain et al. 2008; Boyer et al. 2006). Strikingly, deletion of PRC components (Eed, Suz12, Ezh2, and Ring1a/b) abolishes the ability of ESCs to reprogram human lymphocytes in heterokaryons, even though PRC-deficient ESCs remain pluripotent, demonstrating that pluripotency and reprogramming capacity are separable properties (Pereira et al. 2010).

H3K9 methylation also restricts reprogramming: Knockdown of histone H3 lysine 9 (H3K9) methyltransferase G9a or overexpression of the H3K9 demethylase Jhdm2a enhances Oct4-Enhanced Green Fluorescent Protein(EGFP) reactivation in NSCs after ESC fusion; co-expression of Nanog and Jhdm2a further boosts efficiency and reduces DNA methylation at the Oct4 promoter (Ma et al. 2008).

Fusion systems also provide a fast model for X chromosome reactivation (XCR): Fusion of female fibroblasts with mESCs rapidly silences X-inactive specific transcript (XIST), removes H3K27me3, and induces biallelic X-linked gene expression, with potential applications such as MECP2 reactivation in Rett syndrome (Cantone et al. 2016; Cantone and Fisher 2017; Janiszewski et al. 2019; Carrette et al. 2018).

8.2.2.4 Replication, Cell Cycle, and Mitochondria

Replication is tightly linked to epigenetic resetting. In ESC–human fibroblast heterokaryons, DNA synthesis is required for the stable activation of OCT4, SOX2, and NANOG (Tsubouchi et al. 2013). Reprogramming is most efficient when somatic nuclei fuse with ESCs in S phase, coinciding with redistribution of H3K9me3 and H3K27me3 and increased accessibility at pluripotency loci (Tsubouchi et al. 2013). Reprogramming efficiency also depends on the somatic cell type, developmental history, cell cycle state, and replicative age, with early passage fibroblasts and MSCs outperforming NSCs or senescent cells (Tat et al. 2011).

The cell cycle in pluripotent cells (ESCs, EGCs, and ECCs) is distinctive from that of somatic cells, with a short G1 and extended S phase, enabling them to drive G1 somatic nuclei into early S phase and promote DNA replication-coupled chromatin remodeling (Tsubouchi et al. 2013; Piccolo et al. 2013; Lowe et al. 2016; Aligianni and Varga-Weisz 2006). Pluripotent cell-mediated reprogramming via fusion is extremely efficient (~70% of somatic cells) and rapid (5–7 days) (Yamanaka and Blau 2010; Matveeva et al. 2017; Serov et al. 2011; Hasegawa et al. 2010).

Mitochondrial dynamics are also critical to reprogramming efficiency. Fusion mixes parental cytoplasm, allowing analysis of mitochondrial inheritance (Munoz and Zorzano 2017; Choi et al. 2015). Mitochondrial fission/fusion remodeling and mitophagy are required for efficient reprogramming (Acquistapace et al. 2011; Prieto et al. 2016; Naik et al. 2019; Spurlock et al. 2025). Importantly, cytoplasmic mixing also reveals mitochondrial–nuclear incompatibilities similar to those observed in somatic cell nuclear transfer, where the absence of paternal mitochondria and mtDNA–nuclear mismatches can impair metabolic remodeling and compromise developmental competence. Such mitonuclear discordance underscores the need for coordinated nucleus–mitochondria communication to support stable and efficient reprogramming (Takeda 2019).

8.2.2.5 Chromatin Architecture

Chromatin organization and phase separation form additional barriers for reprogramming (Wang et al. 2020; Larson et al. 2017; Brown and Fisher 2021; Cossec et al. 2018). Pluripotent cells such as ESC have open chromatin conformations, while somatic cells such as fibroblasts have closed chromatin conformations (Fig. 8.2). ESC–somatic fusions trigger extensive nuclear reorganization within

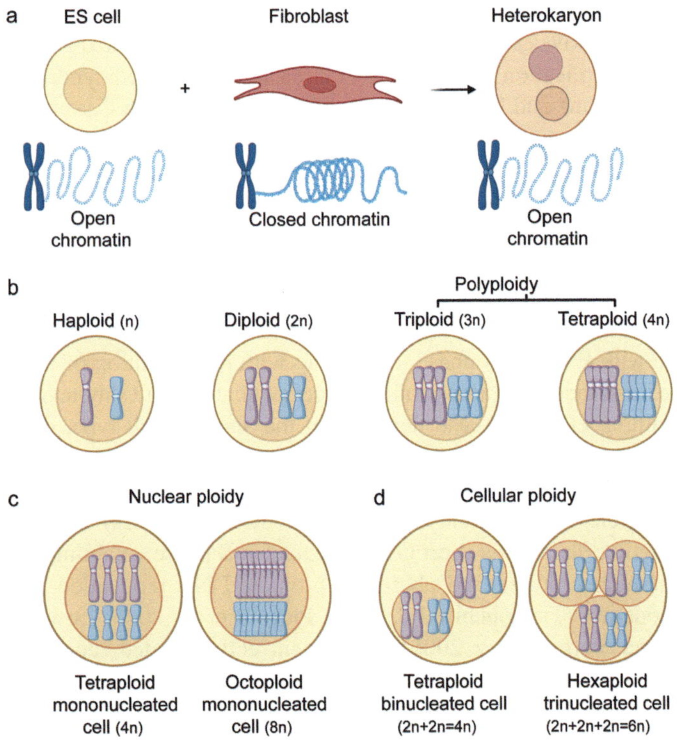

Fig. 8.2 Chromosome conformation and number in polyploid cells. (**a**) Somatic fibroblast cells have a closed chromatin configuration until fusion with pluripotent embryonic stem cells, which have an open chromatin configuration. This fusion-mediated reprogramming leads to heterokaryon formation, which has an open chromatin conformation. (**b**) Ploidy is determined by the number of complete sets of chromosomes present in the cell. Haploid cells have one set, diploid cells have two sets, triploid cells have three sets, and tetraploid cells have four sets of chromosomes, where n = chromosome number. Polyploidy is defined by nuclear ploidy (**c**) and cellular ploidy (**d**), corresponding to DNA content per nucleus and the number of nuclei per cell, respectively

12–24 hours (Cantone et al. 2016; Piccolo et al. 2013; Brown et al. 2013) (Fig. 8.2). ESC contains chromatin-opening factors that are rapidly engaged during fusion-driven reprogramming, and OCT4 is essential for this process; depletion of OCT4 severely impairs reprogramming efficiency (Pereira et al. 2008; Zaret and Mango 2016; Lopes Novo et al. 2016; Hörnblad and Remeseiro 2022; Peñalosa-Ruiz et al. 2019). Phase separation proteins such as HP1α and Small Ubiquitin-related MOdifier (SUMO)-modified factors help stabilize cell identity, and their modulation can enhance reprogramming (Wang et al. 2020; Larson et al. 2017; Cossec et al. 2018). Recent single-cell imaging further shows that somatic chromatin undergoes global decondensation only after 36–48 hours of fusion, with late removal of H3K9me3 and H3K27me3, while OCT4 transcription is reactivated much earlier (6–12 hours) even without local chromatin opening, underscoring how chromatin structure imposes gene-specific constraints on reprogramming trajectories (Martinez-Sarmiento et al. 2024).

8.2.3 Lineage Reprogramming by Cell Fusion

Beyond dedifferentiation into pluripotency, fusion can directly convert one differentiated cell type into another. ESCs can impose epigenetic and transcriptional changes on specialized cells (Tada et al. 2001; Silva et al. 2006), but differentiated cells can also reprogram other somatic nuclei (Blau et al. 1983; Baron and Maniatis 1986; Wright 1984).

In early synkaryon studies, fusing murine hepatoma cells with human leukocytes produced hybrids secreting both mouse and human albumin, indicating reactivation of human albumin by murine trans-acting factors (Darlington et al. 1974). Later, transcription factors such as MyoD, C/EBPα/β, or Ascl1/Brn2/Myt1l were shown to drive muscle, macrophage, or neuronal identities, respectively, highlighting parallels between direct Tissue Factor (TF)-based and fusion-mediated lineage conversion (Aurade et al. 1994; Choi et al. 1990; Weintraub et al. 1989; Xie et al. 2004; Bussmann et al. 2009; Vierbuchen et al. 2010).

Not all cell types are equally permissive: BNL liver cells, but not HepG2 hepatocytes, can be converted to muscle by MyoD (Choi et al. 1990; Tapscott et al. 1988). However, fusion of refractory HepG2 cells with myotubes can still activate human muscle genes (Pomerantz et al. 2009). Phenotypic differences between MyoD-transduced hepatocytes and hepatocyte-derived heterokaryons illustrate how hybrids integrate complex regulatory inputs from both parents, beyond what a single TF can achieve (Blau 1989).

8.2.3.1 Nuances of Reprogramming Directionality: Synkaryons Versus Heterokaryons, Gene Dosage, and Chromatin

Stable synkaryons, generated by PEG-mediated fusion of erythroid cells with fibroblasts, revealed that adult human globin genes can be reactivated in fibroblast

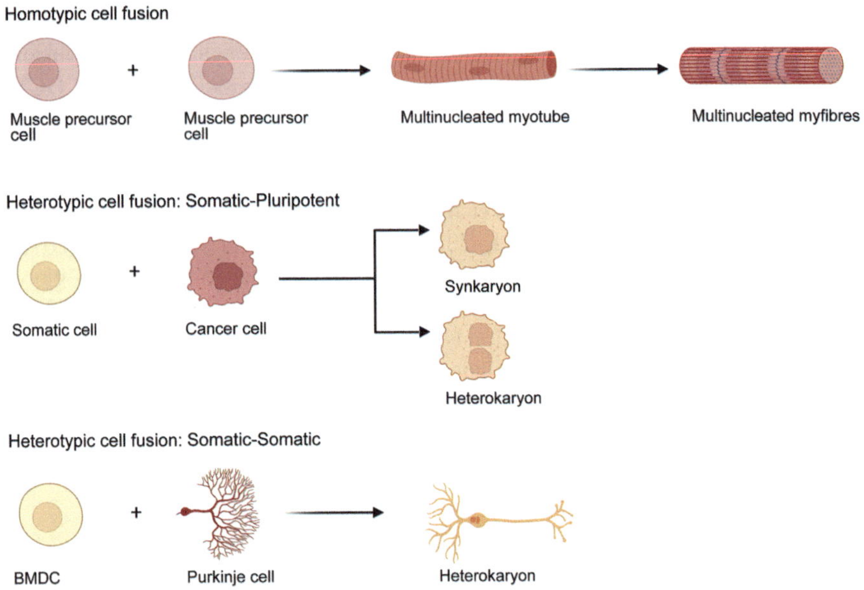

Fig. 8.3 Types of cell fusion that undergo lineage reprogramming. The major types of cell fusion investigated for regeneration purposes are homotypic and heterotypic cell fusion, between somatic–somatic, somatic–pluripotent, and pluripotent–pluripotent cells

genomes (Deisseroth and Hendrick 1979; Willing et al. 1979; Papayannopoulou et al. 1988), but long-term culturing and chromosome loss complicated mechanistic interpretation. Transient heterokaryons overcame these issues: Fusion of erythroid cells with non-erythroid cells and analysis of RNA at 24 hours revealed rapid reactivation of globin genes in previously silent nuclei (Blau et al. 1983). Because heterokaryons preserve separate nuclei and prevent mitosis, they allow species-specific transcriptomic analyses of early reprogramming events.

In typical pluripotent–somatic fusions, synkaryons usually form a single fused nucleus (Fig. 8.3), which impedes isolation of genome-specific responses. Heterokaryon systems, therefore, allow dissecting directional responses without nuclear fusion. Nondividing mESC–human heterokaryons show human OCT4 and NANOG expression within 24 hours, with initial activation driven by mouse trans-acting factors and later stabilization by human *cis*-regulatory elements (Bhutani et al. 2010; Pereira et al. 2008; Blau and Blakely 1999). Not all hybrids form via classical membrane fusion: Coculture of ESCs with MSCs yields both fusion-derived heterokaryons and entosis-derived structures (cell-in-cell without cytoplasmic mixing). Only fusion-derived heterokaryons give rise to ESC-like colonies; entosis-derived structures behave like MSCs, underscoring that different cell–cell interaction modes produce distinct outcomes (Sottile et al. 2016).

Muscle–keratinocyte heterokaryons showed that keratinocyte nuclei can fully adopt a muscle gene program while silencing keratinocyte genes, indicating global transcriptional conversion and chromatin remodeling (Blau et al. 1983; Baron and

Maniatis 1986; Pomerantz et al. 2009; Chiu and Blau 1984; Chiu and Blau 1985). Reprogramming efficiency depends on tissue origin and developmental history, with mesoderm-derived fibroblasts reprogramming faster than keratinocytes or hepatocytes (Pomerantz et al. 2009).

Lineage direction can be bidirectional: Keratinocyte nuclei can acquire muscle identity, and muscle nuclei can shift toward keratinocyte phenotypes, with outcomes dictated by nuclear stoichiometry and cytoplasmic factor balance rather than a single master regulator. Varying myoblast:keratinocyte ratios bias lineage choice, emphasizing the role of gene dosage (Palermo et al. 2009).

Histone Deacetylases (HDACs) and other epigenetic regulators also shape lineage reprogramming. During lymphocyte-to-muscle conversion in heterokaryons, lymphocyte nuclei enlarge, heterochromatin is reorganized, and muscle genes are activated sequentially, while lymphocyte genes are silenced. HDAC inhibition induces co-expression of both lineages, showing that activation and silencing can be uncoupled. Pretreatment with HDAC inhibitors enhances reprogramming, likely by relaxing chromatin at muscle loci. Muscle cells can also impose tissue-specific DNA methylation patterns on epidermal progenitor nuclei, demonstrating active epigenetic rewriting without cell division (Terranova et al. 2006).

8.2.4 Cell Fusion as a Mechanism of Stem Cell Plasticity

Stem cells maintain tissue homeostasis by differentiation, paracrine signaling and, in some cases, by cell fusion (Marshak et al. 2001; Wagers and Weissman 2004). Bone-marrow derived Stem Cells (BMSCs), including both hematopoietic and mesenchymal populations, show an unexpected broad plasticity. Early transplantation studies reported that they can derive microglia, neurons, hepatocytes, pancreatic cells, skeletal muscle, endothelium, and myocardium (Wagers and Weissman 2004; Meir and Li 2021). Although initially, it was deduced as a result of transdifferentiation, many of these observations could now be explained, at least in part, by fusion (Álvarez-Dolado and Martínez-Losa 2011).

According to the fusion hypothesis, bone marrow-derived cells (BMDCs) merge with local progenitors or mature cells, forming heterokaryons, which go through genetic and epigenetic reprogramming toward the phenotype of the host cell (Terada et al. 2002; Ying et al. 2002)(Fig. 8.3). In vitro fusion between ESCs and bone marrow (BM) or adult neural stem cells, as well as in vivo studies in muscles, provided early proof of principle. Later, genetic lineage-tracing strategies (e.g., Cre–lox reporters) offered strong evidence that cells containing both donor and host genomes can be found in multiple adult organs, where they can contribute to functional repair (Terada et al. 2002; Ying et al. 2002; Ferrari et al. 1998; Gibson et al. 1995; Gussoni et al. 2002; Gussoni et al. 1999).

Notably, fusion-derived hybrids have been shown to correct genetic defects in vivo, for example, in fumarylacetoacetate hydrolase (FAH)-deficient liver models and certain neurodegenerative or metabolic disorders (Vassilopoulos and Russell

2003; Wang et al. 2003). These results suggest that cell fusion is more than just a normal regenerative process and may also play a role in restoring tissue function. Studies using bone marrow transplantation and lineage tracing have shown that fusion occurs in several organs during repair and can even correct genetic defects in vivo, pointing to its potential use in future therapies.

8.3 Cell Fusion in Tissue Regeneration and Disease

Direct conversion of one adult cell type into another, bypassing a stem cell stage, has major implications for regeneration. If differentiated cells can dedifferentiate or switch identity in vivo, adult tissues are far more plastic than previously assumed. Although somatic nuclear reprogramming by cell fusion was demonstrated in the 1980s, later, it became clear that similar fusion events occur naturally in adult vertebrate tissues. Cell fusion is now recognized beyond an experimental tool as a physiological mechanism contributing to homeostasis and repair.

BMDCs, including hematopoietic stem/progenitor cells (HSPCs) and MSCs, are particularly plastic. After bone marrow transplantation, BMDCs can contribute to both hematopoietic and non-hematopoietic tissues, giving rise to myocytes (Ferrari et al. 1998), hepatocytes (Petersen et al. 1999), neurons (Mezey et al. 2000), and other cell types (Krause et al. 2001). These observations challenged rigid fate models and sparked a debate regarding (i) transdifferentiation, in which stem/progenitor cells directly adopt new lineages under microenvironmental cues, and (ii) cell fusion-mediated reprogramming, in which BMDCs fuse with differentiated cells and their nuclei are reprogrammed by host factors. Terada et al. showed that BMDCs spontaneously fuse with ESCs in vitro and adopt an ESC-like phenotype (Terada et al. 2002). Together with in vivo data, these findings support a model in which adult cell plasticity can arise from both direct differentiation and fusion events, with a growing body of evidence favoring fusion as a major contributor in many contexts.

8.3.1 Experimental Evidence for Cell Fusion-Mediated Reprogramming In Vivo

8.3.1.1 Brain, Retina, and Neurodegeneration

One of the most compelling examples of fusion in vivo involves cerebellar Purkinje neurons. In a brain biopsy from a female patient who had received male bone marrow transplants, some Purkinje neurons were tetraploid (XXXY) and contained both donor and host chromosomes, consistent with BMDC–Purkinje fusion (Weimann et al. 2003a) (Fig. 8.3).

Genetic lineage-tracing mouse models further confirmed fusion in hepatocytes, cardiomyocytes, and Purkinje neurons. In mice transplanted with GFP^+/Cre^+ BMDCs and carrying a floxed LacZ reporter, LacZ activation required the coexistence of donor and host genomes. $LacZ^+/GFP^+$ cells were observed in multiple tissues, proving low-frequency fusion in vivo (Alvarez-Dolado et al. 2003). In the cerebellum, BMDCs spontaneously fuse with Purkinje cells to form stable binucleated heterokaryons. Donor nuclei initially appear compact but progressively adopt Purkinje-like nuclear morphology and chromatin organization. Reactivation of the Purkinje-specific transgene L7-GFP in donor-derived nuclei provided direct evidence of full neuronal reprogramming in vivo (Weimann et al. 2003b).

BMDC–Purkinje heterokaryons increase with age and injury. In models of chronic brain inflammation (severe dermatitis and autoimmune encephalitis), BMDC–Purkinje fusion events increase 10–100-fold, generating hundreds of heterokaryons. Donor nuclei activate Purkinje genes (Calb1, Pcp2, Kcnc1, and Gsbs) and repress hematopoietic markers (CD45, CD11b, F4/80, and Iba1), indicating full neuronal reprogramming (Johansson et al. 2008). Inflammatory chemokines and cytokines likely promote BMDC recruitment and local engraftment (Priller et al. 2001a; Bajetto et al. 2001).

Alvarez-Buylla and colleagues showed that early postnatal cerebellar irradiation markedly increases BMDC–Purkinje fusion. Abundant $GFP^+/\beta\text{-gal}^+$ binucleated Purkinje neurons were detected when transplantation followed irradiation at postnatal day 4, whereas adult irradiation produced far fewer fusions. Busulfan alone did not trigger fusion, but combined busulfan and irradiation did, emphasizing that local injury, rather than myeloablation, is critical (Espejel et al. 2009). Overall, these studies demonstrate that BMDC–Purkinje fusion forms stable neuronal heterokaryons, whose frequency is strongly enhanced by damage and inflammation.

Beyond Purkinje neurons, BMDCs and HSPCs can fuse with various neural cell types. Donor-derived neurons and glia have been reported after bone marrow transplantation (Mezey et al. 2000; Hess et al. 2004; Eglitis and Mezey 1997; Brazelton et al. 2000; Priller et al. 2001b). In Amyotrophic lateral sclerosis (ALS) models (SOD1-G93A mice), BM transplantation modestly delays disease, improves motor performance, and extends survival. Donor-derived microglia repopulate the Central Nervous System (CNS), and rare binucleated GFP^+ neurons co-express neuronal markers, consistent with fusion. While microglial replacement and trophic support likely drive most benefits, these data confirm that neuronal hybrids can arise in the degenerating CNS (Corti et al. 2004).

BMDC implantation improves outcomes in models of ischemia, trauma, Parkinson's disease, and Huntington's disease, largely via trophic and vascular mechanisms (Lescaudron et al. 2003; Li et al. 2001; Li and Chopp 2009; Momin et al. 2010; Rosser et al. 2007). Still, inflammation can increase fusion rates by an order of magnitude, amplifying the potential contribution of fusion-mediated reprogramming (Alvarez-Dolado et al. 2003; Weimann et al. 2003b; Johansson et al. 2008; Sanges et al. 2013; Nygren et al. 2008).

Stroke is the most intensely studied brain pathology for BMDC therapy (Li and Chopp 2009; Chang et al. 2007; Tang et al. 2007). Transplantation 24 hours after stroke, by various delivery routes, improves function and yields occasional neuron–BMDC hybrids (Chen et al. 2001). BMDCs also contribute endothelial and mural cells to revascularization (Chen et al. 2003; Bailey et al. 2006; Ozerdem and Stallcup

2003; Krueger and Bechmann 2010), with some fusion-derived pericyte-like cells appearing in peri-infarct vasculature (Piquer-Gil et al. 2009).

Other CNS stressors and inherited metabolic diseases also promote neuronal fusion. In intermittent hypobaric hypoxia, increased binuclear pyramidal neurons in the motor cortex suggest neuron–oligodendrocyte fusion, potentially enhancing connectivity and ischemic resilience (Paltsyn et al. 2014). After hemorrhagic stroke, chimeras show binucleated GFP^+ neurons–oligodendrocytes in the cerebral cortex; electron microscopy (EM) confirms true fusion (Paltsyn et al. 2013). In Niemann–Pick disease models, MSCs expressing acid sphingomyelinase fuse with Purkinje neurons, improving motor function and enzyme levels; rescued neurons remain electrically active (Bae et al. 2005; Jin and Schuchman 2003; Bae et al. 2007).

In the retina, transplanted or endogenous HSPCs fuse with Müller glia (MG), generating hybrids that dedifferentiate, reenter the cell cycle, and generate retinal neurons (Sanges et al. 2013; Pesaresi et al. 2018; Sanges et al. 2016). Sanges et al. showed that HSPC–MG hybrids, boosted by Wnt/β-catenin activation (GSK3 inhibition), progress from a transient pluripotent/neural progenitor-like state to photoreceptor precursors and then mature photoreceptors, restoring PDE6B expression and partially rescuing visual function in rd10 mice (Sanges et al. 2016). Bonilla-Pons et al. extended these findings to human retinal explants and organoids, where Müller glia (MG) can spontaneously fuse with adult MSCs or HSPCs. Wnt activation in donor stem cells promoted neuronal differentiation of hybrids, which acquired ganglion-like features and integrated into organoids (Bonilla-Pons et al. 2022). In parallel, CXCL12/SDF-1-mobilized endogenous BMDCs predominantly adopt microglial or endothelial identities and exert trophic, anti-inflammatory effects, indicating that paracrine mechanisms and fusion-mediated reprogramming coexist to support retinal repair (Pesaresi et al. 2018; Enzmann et al. 2017).

8.3.1.2 Liver Regeneration: FAH Deficiency and Beyond

Liver regeneration mediated by cell fusion is the best documented example of fusion-driven repair (Alvarez-Dolado et al. 2003; Vassilopoulos et al. 2003; Willenbring et al. 2004; Camargo et al. 2004; Pedone et al. 2017). In $FAH^{-}/^{-}$ mice, which model fatal tyrosinemia type I, transplantation of BMDCs or enriched Haematopoietic Stem Cell (HSC) fractions (e.g., KTLS: c-kit^{high}, Thy^{low}, Lin^{-}, and Sca-1^{+}) carrying a functional FAH allele restored liver function. Initially attributed to BMDC transdifferentiation, subsequent cytogenetic and Southern blot analyses showed that regenerating hepatocytes carried alleles from both the donor and the host, indicating donor–host fusion (Wang et al. 2003; Vassilopoulos et al. 2003; Lagasse et al. 2000).

Myelomonocytic macrophages are the primary bone marrow subpopulation mediating these fusion events (Willenbring et al. 2004; Camargo et al. 2004). Most fusion-derived hepatocytes are tetraploid and stable, but some undergo ploidy reduction, producing daughter cells with lower ploidy. Marker segregation experiments (β-gal and Y chromosome) suggest that 2–5% of FAH^+ nodules lose one or more markers, consistent with ploidy reduction and chromosome loss (Duncan et al.

2009). This generates genetic diversity that may aid adaptation but also raises oncogenic risk. Interestingly, ploidy reduction is also seen during early development in mouse embryos (Frade et al. 2019).

Human cells can behave similarly in mouse models: Transplanted human HSCs or umbilical cord cells generate hepatocytes via fusion, improving liver function and forming hybrids with both mouse and human genomes (Fujino et al. 2007; Kashofer et al. 2006; Sharma et al. 2005). In some liver injury models, however, hepatocytes arise without detectable fusion, indicating that the balance between fusion and transdifferentiation is context-dependent (Jang et al. 2004; Newsome et al. 2003). Blocking BMDC recruitment or ablating BMDC–hepatocyte hybrids impairs liver regeneration (Pedone et al. 2017), showing that fusion is not merely incidental but can be essential for repair (Fig. 8.4). In humans, postmortem analyses of sex-mismatched transplant recipients revealed hepatocytes with male and female

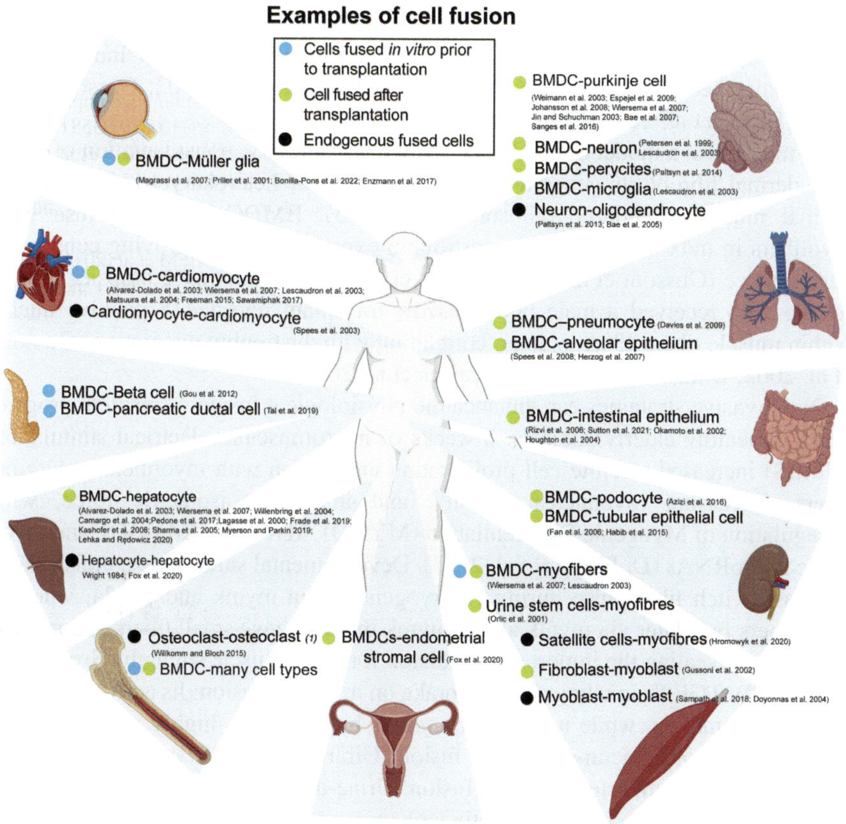

Fig. 8.4 Examples of tissues with reported cell fusion in the field of regeneration. Three main types of cell fusion are included in the examples shown: endogenous cell fusion, transplanted cells that later fuse with endogenous cells, and cells that are fused in vitro before transplant for regenerative purposes

nuclei within the same cytoplasm; donor-marked hepatocytes reached ~10% in some patients with hepatic injury or fibrosis and expressed normal hepatocyte markers. No evidence supported direct hematopoietic-to-hepatocyte transdifferentiation, reinforcing fusion as the main mechanism for donor-derived hepatocyte formation in both mice and humans (Myerson and Parkin 2019).

8.3.1.3 Skeletal Muscle Regeneration

Fusion is fundamental to skeletal muscle biology. During myogenesis, myoblasts fuse to form multinucleated myofibers; satellite cells support postnatal growth and repair (Lehka and Rędowicz 2020; Sampath et al. 2018). Fusion also contributes to normal muscle maintenance. BMDCs fuse with myofibers in injured or stressed muscles, a process largely mediated by myeloid cells and dependent on injury-induced inflammation (Doyonnas et al. 2004; LaBarge and Blau 2002; Camargo et al. 2003)(Fig. 8.4). Transplanted BMDCs were shown to migrate to degenerating muscles and adopt myogenic identity, most likely via fusion, as indicated by sex-mismatched chimeras (Ferrari et al. 1998; Gussoni et al. 2002; Gussoni et al. 1999; Theise et al. 2000; Bittner et al. 1999) (Fig. 8.4).

In mdx mice, a model of Duchenne muscular dystrophy, transplantation of wild-type dermal fibroblasts generated fibroblast–myofiber heterokaryons that restored normal muscle phenotype (Gibson et al. 1995). BMDCs can also fuse with myofibers in mdx mice, restoring dystrophin expression and improving contractile performance (Gussoni et al. 1999; Bittner et al. 1999). In humans, biopsies from a female who received a male bone marrow transplant revealed Y-positive nuclei within muscle fibers 13 years later, correlating with dystrophin restoration (Gussoni et al. 2002; Bittner et al. 1999; Goncalves et al. 2008).

Noninvasive strategies can enhance the physiological fusion of resident progenitors. In healthy elderly subjects, 8 weeks of neuromuscular electrical stimulation (NMES) increased satellite cell proliferation and fusion with myofibers, enlarging fibers, increasing myonuclear density, and improving isometric force, with upregulation of Myogenic Differentiation (MYOD), Myogenin (MYOG), and myogenic microRNAs (Di Filippo et al. 2017). Developmental studies in zebrafish show that fast-twitch fibers fuse during embryogenesis via mymk and jam2a, whereas slow fibers fuse later via mymk alone; mymk mutants have small fibers and muscle weakness, revealing the importance of fusion for adult muscle strength (Hromowyk et al. 2020). TGF-β signaling acts as a brake on myoblast fusion: Its activation limits myotube formation, while its inhibition yields hypertrophic, highly multinucleated fibers by promoting actin-dependent fusion (Girardi et al. 2021). Exogenous stem cells can also aid muscle repair via fusion. Urine-derived stem cells engineered to overexpress IGF1 fuse more efficiently with myocytes and improve muscle recovery in injury models (Yi 2024). Together, these studies highlight both endogenous and transplanted progenitors as fusion-competent contributors to skeletal muscle regeneration.

8.3.1.4 Cardiac Repair After Infarction

BMSCs can generate cardiomyocytes and improve cardiac function after myocardial infarction. Early studies suggested that HSC replaced large fractions of infarcted myocardium by differentiating into smooth muscle cells, endothelial cells, and cardiomyocytes (Orlic et al. 2001; Quaini et al. 2002). Later work, using more stringent methods, reported much lower donor contribution (<1%) (Deb et al. 2003; Laflamme et al. 2002). These donor-derived cardiomyocytes were initially interpreted as transdifferentiation products, but lineage-tracing and karyotypic analyses now support fusion as the principal mechanism (Fig. 8.4). In contrast, smooth muscle and endothelial cells seem to derive mainly from differentiation of BM progenitors (Alvarez-Dolado et al. 2003; Nygren et al. 2004).

Although rare, fused cardiomyocytes can proliferate in vitro and in vivo, suggesting potential for amplifying their reparative contribution (Matsuura et al. 2004). Nonetheless, most functional benefits of BMSC transplantation are explained by paracrine effects: secretion of growth and antiapoptotic factors that protect host cardiomyocytes and BM-derived endothelial precursors that enhance angiogenesis (Leone et al. 2009; White and Chong 2020).

In vitro and *ex vivo* models reveal how fusion with mesenchymal cells can reprogram cardiomyocytes and vice versa. Human multipotent adipose-derived stem cells (hMADS) induce partial dedifferentiation of adult cardiomyocytes via transient cytoplasmic continuity, producing progenitor-like hybrids that upregulate early cardiac genes and proliferate; mitochondrial transfer from hMADS is essential for this reprogramming (Acquistapace et al. 2011). VSV-G-based fusogenic systems and patterned biochips show that direct contact, gap junctions, and occasional fusion between MSCs and cardiomyocytes drive expression of cardiac transcription factors and sarcomere genes and support electrical coupling (Kouris et al. 2012; Ma et al. 2013).

Preconditioning MSCs can further enhance fusion and repair. Interleukin-7 (IL-7) overexpression increases MSC–cardiomyocyte fusion, improves survival, and reduces infarct size in rats, with hybrids expressing cardiac markers in the injured myocardium (Haneef et al. 2018). In vivo imaging revealed that fusion products can disseminate beyond the heart, appearing in the stomach, intestine, and liver, with implications for both efficacy and safety (Freeman et al. 2015).

In zebrafish, fusion-activated reporters reveal transient cardiomyocyte–cardiomyocyte fusion during development and after injury. Fusion-positive cardiomyocytes are highly proliferative and enriched at lesion borders, and jam3b mutants lacking fusion show reduced cardiomyocyte proliferation and impaired cardiac performance (Sawamiphak et al. 2017). Thus, transient or stable cardiomyocyte fusion appears to cooperate with paracrine and angiogenic mechanisms in cardiac growth and repair.

8.3.1.5 Lung and Airway Epithelium

BMDCs can adopt lung epithelial phenotypes via both differentiation and fusion. In vitro, MSCs cocultured with heat-shocked airway epithelium undergo fusion in ~1% of cells, producing heterokaryons that express epithelial markers and show new gene expression profiles (Spees et al. 2003).

In vivo, BMDCs contribute to airway and alveolar epithelium after transtracheal delivery in naphthalene injury models, with ~1% of engrafted cells arising from fusion, confirmed by Fluorescent In Situ Hybridisation (FISH) (Wong et al. 2007). In pulmonary hypertension models, bone marrow progenitors generate donor-derived fibroblasts, Clara cells, endothelial cells, and smooth muscle cells; fusion rates are low (0.52–0.59%), with ~0.3–0.5% showing nuclear fusion during mitosis (Spees et al. 2008). In surfactant protein C (Sp-C)-deficient mice, wt BMDC–pneumocyte fusion produces binucleated cells that restore lung-specific gene expression, including Sp-C (Herzog et al. 2007). Thus, although infrequent, BMDC–lung fusion can contribute to epithelial repair and functional gene restoration (Fig. 8.4).

8.3.1.6 Intestine, Skin, Pancreas, Kidney, and Other Epithelia

In the intestine, BMDC fusion complements stem cell-mediated regeneration in certain injury contexts. Following γ-irradiation, BMDCs fuse with intestinal epithelial cells, forming hybrids that generate goblet, Paneth, and enterocyte lineages. Fusion seems to involve intestinal progenitors, producing stem-like hybrids with multilineage potential. Fusion events are rare at baseline but increase with inflammation or high epithelial turnover; it is also reduced upon anti-inflammatory treatment (Davies et al. 2009; Rizvi et al. 2006; Sutton et al. 2021). However, not all studies find a major contribution of fusion. Using Cre–Lox lineage tracing to detect fusion events and an inducible diphtheria toxin receptor (iDTR) to ablate these hybrids, De Jong et al. failed to detect significant numbers of fusion-derived intestinal cells under homeostasis or injury (γ-irradiation and dextran sulfate sodium colitis), estimating frequencies below 1 in 1000 crypt cells, and ablation of putative hybrids did not impair regeneration (De Jong et al. 2012). These data suggest that fusion-driven gut regeneration is highly model-dependent.

In female recipients of sex-mismatched bone marrow transplants, up to 4.6% of intestinal epithelial cells carried donor Y chromosomes after injury (Silk et al. 2013; Okamoto et al. 2002). Fusion-derived hybrids can persist and clonally expand within crypts and villi, likely originating from stem/progenitor fusion. While functional impact remains to be fully defined, these hybrids may enhance barrier function or immune modulation and, under chronic inflammation, contribute to tumor evolution (Terada et al. 2002; Rizvi et al. 2006; Houghton et al. 2004).

Fusion-derived cells have also been detected in the kidney and skin, although at low frequencies and with uncertain physiological significance (Fang et al. 2005; Fan et al. 2006; Held et al. 2006). In a chronic proximal tubular injury model ($Fah^{-}/^{-}/Hgd^{aku/?t}$ mice), bone marrow transplantation led to extensive repopulation of damaged tubules by donor-derived Fah^+ epithelium; most Fah^+ tubular cells were hybrids containing both donor and host chromosomes, repopulating up to 50% of proximal tubules and normalizing renal function. Some animals instead showed functional correction via loss of heterozygosity in host cells, indicating parallel operation of genetic reversion and fusion-mediated repopulation (Held et al. 2006) (Fig. 8.4).

Preconditioning MSCs with hypoxia or IL-7 enhances their migration and fusion with injured renal epithelia in vitro, via the HIF-1α and CXCR4/SDF-1 pathways (Habib et al. 2015). In Col4a3-deficient Alport syndrome mice, MSCs can undergo podocyte horizontal gene transfer/cell fusion (enhanced by TGF-β1), which is associated with de novo Col4a3 (type IV collagen) expression in the glomerulus and improved renal outcomes (LeBleu et al. 2024).

Similar principles apply to the endocrine pancreas. Optimized PEG/PHA-P (phytohemagglutinin) and electrofusion protocols generate β-MSC hybrids expressing key β-cell transcription factors from both parental genomes, forming islet-like clusters with enhanced and prolonged glucose-stimulated insulin secretion that better corrects hyperglycemia in diabetic rats than unfused controls (Azizi et al. 2016; Yanai et al. 2013). Conversely, fusion of pancreatic ductal cells with bone marrow-derived mesenchymal stem cells (BMSCs) can produce mesenchymal-like tetraploid hybrids with high SIRT1, enhanced proliferation, and invasiveness, highlighting the pro-regenerative yet potentially tumor-promoting nature of such hybrids (Gou et al. 2012). Rare fusion-derived stromal cells have been identified in endometriotic lesions, arising from BMDC–endometrial stromal fusion (Tal et al. 2019). These quiescent mesenchymal-like hybrids may contribute to lesion stability or, under certain conditions, malignancy, underscoring that fusion-generated stromal diversity extends beyond classical regenerative organs (Fig. 8.4).

8.4 Polyploidy

As cell fusion can increase the genomic content of a cell and increased genomic content has been detected without evidence of cell fusion, we consider the broader question of how cells acquire additional chromosome sets, known as polyploidy. Somatic cells can become polyploid either by cell fusion, as discussed in the previous sections, or by undergoing noncanonical cell cycles in which they replicate their DNA but do not divide into two daughter cells (Fig. 8.5). The mechanisms of non-cell fusion-derived polyploidy have been explored extensively elsewhere (Fox et al. 2020; Müller et al. 2021; Choudhury et al. 2024; Darmasaputra et al. 2024). A wide range of terms has been used to describe noncanonical cell cycles leading to non-cell fusion-derived polyploidy. They can be broadly grouped into two categories. The first comprises cycles in which cells alternate between S and G phases without entering mitosis, referred to as endoreplication cycles (Fig. 8.5). The second includes cycles that pass through all canonical cell cycle phases but abort M phase before cytokinesis is initiated or completed, referred to as endomitosis cycles (Fig. 8.5).

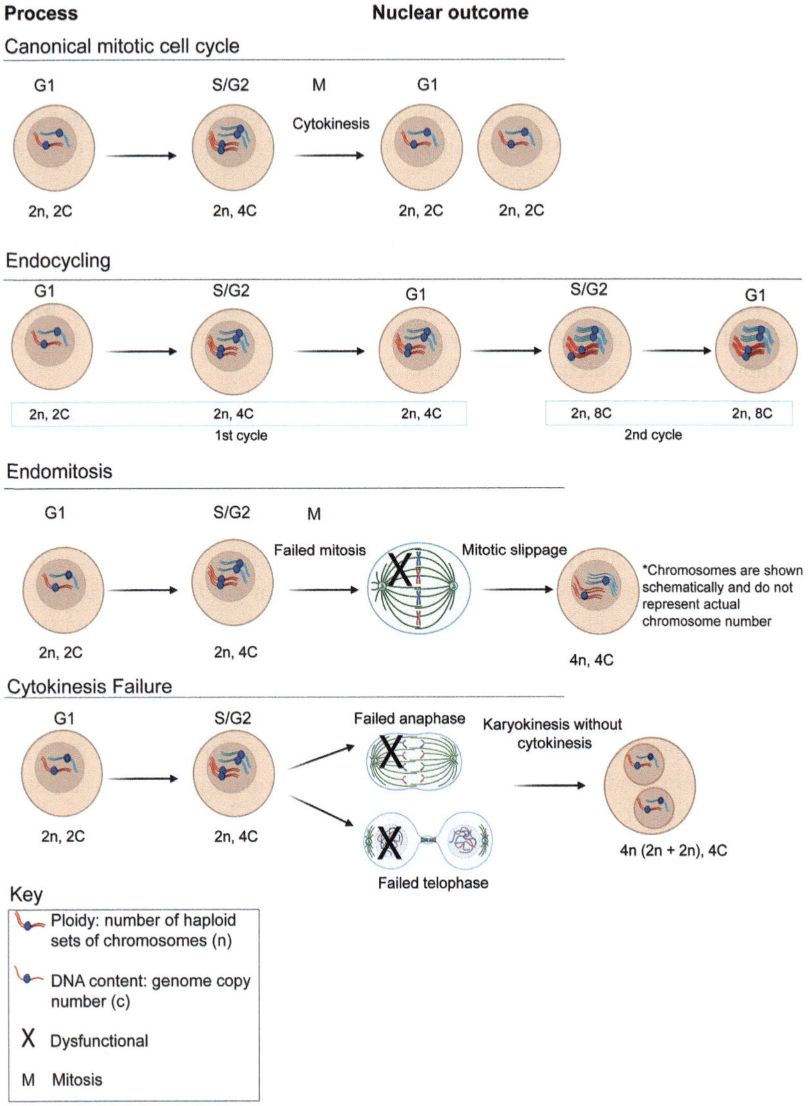

Fig. 8.5 Non-cell fusion-derived mechanisms for the generation of polyploidy. Other than through cell fusion, polyploidy can emerge through several distinct routes, which differ in whether they require the cell to pass through the normal cycle. All remaining pathways involve deviations from the standard eukaryotic cycle, occurring during interphase (G1, S, or G2) or at various stages of mitosis. A major category of these deviations is endoreplication, which includes endocycling and endomitosis. Endocycling consists of recurring rounds of DNA synthesis, interrupted by gaps that never proceed into mitosis. Because mitotic entry is actively blocked—through continuous removal of cyclin B or suppression of M-CDK activity—S-CDK levels oscillate, and the nucleus undergoes progressive increases in ploidy. Endomitosis arises when mitotic errors prevent the completion of normal division. A classic example is prolonged activation of the spindle assembly checkpoint caused by faulty chromosome attachments. During this arrest, cyclin B1 may slowly decline,

Evidence of Polyploidy in Healthy Tissues

Physiological polyploidy is a well-established feature of several normal, non-diseased human tissues, documented largely through postmortem autopsy material, healthy organ donors, and surgical control samples. In the liver, hepatocytes in healthy adult humans commonly exhibit 4n–8n DNA content and frequent binucleation, as shown in donor-derived tissue and cytometric studies (Donne et al. 2020; Duncan et al. 2012). In the heart, classic human autopsy work and modern reviews confirm that adult cardiomyocytes typically contain a single polyploid nucleus, reflecting normal postnatal maturation rather than pathology (Gan et al. 2020). In the nervous system, quantitative cytophotometry of normal human cerebellar autopsy specimens demonstrated that neurons remain diploid while multiple glial populations are polyploid (Mann and Yates 1973). In the bone marrow, normal human megakaryocytes are obligately polyploid, reaching 8 N–64 N in healthy individuals (Kobayashi and Kondo 1999). Together, these studies show that polyploidy is a conserved, physiological cellular state in healthy postmortem human tissues, not restricted to disease contexts.

Endopolyploidy is the property of having more chromatid sets or, less frequently, more chromosome sets than the germ line (Fig. 8.5). Endopolyploidy arises through repeated, parallel replication of chromosomes, such that the centromeric and pericentromeric regions of sister chromatids coalesce into a single, consolidated heterochromatic domain, referred to as the chromocenter. This organization creates a characteristic architectural distinction between polyploid cells and those generated by endopolyploid cycles (Comai 2005) (Fig. 8.5).

Mechanistic evidence already shows that polyploidy is intentionally used in regeneration in several organs. In tissues that have been studied with adequate ploidy tools, polyploidy is clearly part of the regenerative toolkit (Donne et al. 2020; Comai 2005). However, methodological blind spots (no ploidy readouts in scRNA-seq datasets, loss of large cells in dissociation, and ambiguous histology) mean that polyploid steps in many injury models could be mislabeled as generic "hypertrophy" or "senescence" rather than recognized as endoreplication-driven programs. Given the infrequency of studies measuring ploidy directly, it is very plausible that similar polyploid phases are contributing silently in other regenerative contexts and are currently under-recognized rather than absent.

Fig. 8.5 (continued) allowing the cell to abandon mitosis altogether—mitotic slippage—and return to G1 with duplicated genetic content. Failed cytokinesis involves chromosome segregation without cytoplasmic division. Disruptions to any of the structural or regulatory components required for cytokinesis in telophase can yield a binucleate daughter cell. X symbols highlight the points at which these dysfunctions occur, such as lagging chromosomes or defects in cytoskeletal organization. Ploidy number (n) = full haploid set of chromosomes

8.5 Challenges and Future Perspectives

For several years, cell fusion has been a key model to study nuclear reprogramming and the principles of cell plasticity. Early cell fusion studies demostrated that, within only a few hours after fusion, somatic nuclei can undergo extremely rapid changes in gene expression, epigenetics state, and chromatin organization, such as activation of pluripotency genes and X chromosome reactivation. All these findings provided indispensable evidence that the nuclei of differentiated somatic cells are, in fact, reprogrammable.

Despite their historical impact, fusion-based approaches have notable limitations. Fusion is a rare stochastic phenomenon and is commonly inefficient, so generating enough hybrids can be technically challenging. Moreover, another limitation is that the resulting hybrid nuclear environment is inherently artificial because it brings together two genomes and regulatory systems that would not likely coexist physiologically. As a result, the field of nuclear reprogramming has shifted toward newer and more scalable methods such as transcription factor-induced reprogramming, direct lineage conversion, high-throughput clustered regularly interspaced short palindromic repeats (CRISPR) perturbation screens, time-resolved single-cell multi-omics, and live imaging tools. These more recent methods allow investigators to dissect cell fate decisions with a higher degree of resolution and reproducibility and have largely replaced cell fusion as the primary tools for studying cell identity and plasticity. Although these techniques do not exclude one another, further integration of them with cell fusion would be valuable for further elucidating the mechanisms underlying cell reprogramming.

Interestingly, the same biological principles that make cell fusion a powerful model for reprogramming also make it potentially useful for tissue regeneration. Cell fusion between endogenous cell types (such as macrophages, BMSCs, or tissue-resident progenitors) takes place in multiple organs and can contribute to repair by introducing new genetic, epigenetic, or metabolic properties into damaged cells. However, being able to use this natural regenerative process as a controlled therapeutic strategy faces significant conceptual and practical obstacles.

A major challenge is identifying how to induce fusion safely and selectively in vivo. Classical fusogens used in vitro, such as PEG, viral envelopes, and electrofusion, are nonspecific, often toxic, and not very suitable for therapeutic use. Future strategies will require enhancing endogenous fusion programs rather than using artificial fusogens. Although physiological fusion is known to depend on coordinated actions of membrane proteins like syncytins, intracellular regulators, and soluble cues, none of these are sufficient on their own for the efficient induction of fusion.

Cell fusion is a context-dependent process, tightly regulated by tissue state and microenvironmental signals, and this has been observed across multiple organs. Hypoxia, inflammation, and tissue damage significantly increase fusion frequencies, sometimes even 10–100 times more, by recruiting immune stem cells and activating fusogenic pathways. Therefore, any therapeutic approach pursuing the use of cellular fusion will require spatial and temporal control of these cues to ensure that fusion occurs only at intended sites without provoking systemic inflammatory responses.

Selection and delivery of the optimal fusogenic partner cells also remain problematic. Different tissues seem to rely on distinct fusion-competent populations: The liver, for example, uses macrophages and other myeloid cells; the muscle uses myeloid precursors; and the brain and heart use partners that are not well-characterized. Macrophages and BMSCs are particularly promising candidates because they are naturally home to sites of injury and upregulate adhesion molecules and interleukins that also participate in fusion pathways across neuronal, cardiac, and hepatic tissues. Chemokines such as SDF-1, which regulate the homing of HSCs, MSCs, and other $CXCR4^+$ cells, may provide a way to guide fusogenic cells into specific tissues. Combining preselected donor cells, potentially engineered to enhance fusogenic or regenerative properties, with localized modulation of signaling molecules, like chemokines and cytokines, could help control when and where the fusions take place.

Perhaps the most critical obstacle for the use of cell fusion as a regenerative therapy is safety. Cell fusion is not as uncommon or cell type-specific as it was originally believed: Transplanted cells are capable of fusing with a broader range of host cell types than typically detected, partly due to reporter silencing and technical limitations. Therefore, these off-target fusion events could produce unpredictable outcomes. Additionally, fusion-derived hybrids are frequently tetraploid, raising concerns about genomic instability and oncogenic transformation. However, tetraploidy alone is not sufficient to induce malignancy. Tetraploid cells also require additional types of genomic damage, such as p53 loss or severe aneuploidy, to transition to a malignant phenotype. In addition, polyploid cells perform normal physiological roles in tissue homeostasis, like in the liver, bones, and muscles, and fusion-derived cells can contribute to embryonic development without forming tumors. These observations suggest that cell fusion is a risk modifier rather than a tumor initiator. However, in premalignant or highly inflamed environments, fusion may contribute to exacerbating the disease. Hybrids formed between tumor cells and stromal or immune cells can acquire enhanced invasiveness, metastatic capacity, drug resistance, and stem-like traits. Chronic inflammation, which increases fusion rates, could create a permissive environment for clonal evolution of hybrids. New strategies for more controlled induction of cell fusion, such as optically induced cell fusion, which allows targeted fusion of two specific cells, and even precise selection of the fusion sites in each cell, hold promise for improved safety of cell fusion in the future. If these strategies could be expanded to bulk generation, their potential for regenerative therapies would be considerable.

The major limitation in fully understanding the role of cell fusion in regeneration is the intrinsic difficulty of detecting fusion events in vivo. Fusion products can be extremely rare, short-lived, or molecularly unstable, and many hybrids downregulate or silence reporter genes over time. As a result, conventional lineage-tracing approaches could underestimate the true frequency and diversity of fusion events. This means that some regenerative improvements currently attributed to paracrine signaling, transdifferentiation, or niche-driven plasticity may instead reflect transient fusion events that escape detection. Therefore, the regenerative contribution of cell fusion may be significantly greater than what is currently measurable. Recognizing and overcoming detection limitations will be essential, as a more accurate understanding of in vivo fusion could reveal an underexplored and potentially powerful mechanism of tissue repair.

The major regenerative alternative to cell fusion, in vivo reprogramming, also has a similar oncogenic risk, which arises from the delicate balance between achieving reprogramming–regeneration and reprogramming–tumorigenesis. Activation of molecular cascades for cellular reprogramming has the inherent risk of uncontrolled cell proliferation, resulting in neoplasms. To manage the risk of neoplasia, any reprogramming approaches, such as Oct4, Sox2, Klf4, and c-Myc (OSKM) induction, should be carefully controlled. Various articles have shown evidence of both transient and prolonged YAP activation leading to tumor formation, fibrosis, aberrant growth, and muscle hypotrophy and myopathy, and this has been reviewed (Moya and Halder 2019). Modern workflows, including direct lineage conversion strategies that bypass the pluripotent stage, have made great steps to minimize this risk. Key mechanisms allowing safe reprogramming are being studied and adopted from tissues with physiological cell turnover, which evade cancer formation, such as the gastrointestinal tract and the liver. Furthermore, as control mechanisms differentiating constant regeneration from neoplastic uncontrolled cell division are better understood, these can be applied to mammalian cell reprogramming more safely. For example, a study mined RNA-sequencing datasets from two intestinal damage models, and using pharmacological, transcriptomic, and genetic tools, they identified liver X receptor (LXR) pathway activation as a sensor that mediates tissue adaptation to damage that reciprocally regulates intestinal regeneration and tumorigenesis (Das et al. 2025). This study highlights how gene regulation is crucial in the modulation of functional cell renewal and uncontrolled neoplastic cell division and could be targeted in the prevention or treatment of cancer.

Presently, reprogramming approaches are easier to characterize than cell fusion in many regenerative contexts, as the scalability and tracking of effector cells/molecules are feasible. Future developments will continue to revolutionize our understanding of how important cell fusion is in in vivo regeneration through key needed developments, such as live cell in vivo profiling.

8.6 Closing Remarks

This chapter has shed light on the current approaches to cell fusion in reprogramming and regeneration. These studies highlight that even in areas of the human body with well-described endogenous regeneration, there are still many major driving mechanisms being actively characterized in recent literature and many more to be understood and fully integrated to understand the complex dynamic interplay between cells in regenerating tissue and between different tissues and organs in the human body. In the future, improved methodologies will allow the examination of cell fusion in vivo in human tissues. This will greatly aid in clarifying whether observed regeneration in human tissues, which, until now, has only been detected postmortem, is derived from any type of cell fusion. As biotechnology advances, our capacity to dissect the mechanisms underpinning cell fusion will also continue to expand, and new questions and paradigms will be developed.

References

Acquistapace A, Bru T, Lesault PF, Figeac F, Coudert AE, Le Coz O et al (2011 May 1) Human mesenchymal stem cells reprogram adult Cardiomyocytes toward a progenitor-like state through partial cell fusion and mitochondria transfer. Stem Cells 29(5):812–824

Aguilar PS, Baylies MK, Fleissner A, Helming L, Inoue N, Podbilewicz B et al (2013) Genetic basis of cell-cell fusion mechanisms. Trends Genet 29:427–437

Ahkong QF, Howell JI, Lucy JA (1975) Fusion of hen erythrocytes with yeast protoplasts induced by polyethylene glycol. Nature 255:66

Aligianni S, Varga-Weisz P (2006) Chromatin-remodelling factors and the maintenance of transcriptional states through DNA replication. Biochem Soc Symp 73:97

Álvarez-Dolado M, Martínez-Losa M (2011) Cell fusion and tissue regeneration. In: Dittmar T, Zänker KS (eds) Cell fusion in health and disease [Internet]. Springer, Dordrecht, pp 161–175. Placeholder Text [cited 2025 Nov 17] (Advances in Experimental Medicine and Biology; vol. 713). Available from: http://link.springer.com/10.1007/978-94-007-0763-4_10

Alvarez-Dolado M, Pardal R, Garcia-Verdugo JM, Fike JR, Lee HO, Pfeffer K et al (2003) Fusion of bone-marrow-derived cells with Purkinje neurons, cardiomyocytes and hepatocytes. Nature 425:968–973

Andrews PW, Gokhale PJ (2024 Jan 9) A short history of pluripotent stem cells markers. Stem Cell Rep 19(1):1–10

Andrews PW, Goodfellow PN (1980) Antigen expression by somatic cell hybrids of a murine embryonal carcinoma cell with thymocytes and L cells. Somat Cell Genet. 6:271–284

Arnold K, Herrmann A, Pratsch L (1985) The dielectric properties of aqueous solutions of poly (ethylene glycol) and their influence on membrane structure. Biochim Biophys Acta 815:515

Atsumi T, Shirayoshi Y, Takeichi M (1982) Nullipotent teratocarcinoma cells acquire the pluripotency for differentiation by fusion with somatic cells. Differentiation 23:83

Aurade F, Pinset C, Chafey P (1994) Myf5, MyoD, myogenin and MRF4 myogenic derivatives of the embryonic mesenchymal cell line C3H10T1/2 exhibit the same adult muscle phenotype. Differentiation 55:185

Azizi Z, Lange C, Paroni F, Ardestani A, Meyer A, Wu Y et al (2016 Aug 2) β-MSCs: successful fusion of MSCs with β-cells results in a β-cell like phenotype. Oncotarget 7(31):48963–48977

Bae JS, Furuya S, Shinoda Y (2005) Neurodegeneration augments the ability of bone marrow-derived mesenchymal stem cells to fuse with Purkinje neurons in Niemann-pick type C mice. Hum Gene Ther 16:1006

Bae JS, Han HS, Youn DH (2007) Bone marrow-derived mesenchymal stem cells promote neuronal networks with functional synaptic transmission after transplantation into mice with neurodegeneration. Stem Cells Dayt Ohio 25:1307–1316

Bailey AS, Willenbring H, Jiang S (2006) Myeloid lineage progenitors give rise to vascular endothelium. Proc Natl Acad Sci USA 103:13156

Bajetto A, Bonavia R, Barbero S, Florio T, Schettini G (2001) Chemokines and their receptors in the central nervous system. Front Neuroendocrinol 22:147–184

Baron MH, Maniatis T (1986) Rapid reprogramming of globin gene expression in transient heterokaryons. Cell 46:591

Bernstein E, Duncan EM, Masui O (2006) Mouse polycomb proteins bind differentially to methylated histone H3 and RNA and are enriched in facultative heterochromatin. Mol Cell Biol 26:2560

Bhutani N, Brady JJ, Damian M (2010) Reprogramming towards pluripotency requires AID-dependent DNA demethylation. Nature 463:1042

Bittner RE, Schofer C, Weipoltshammer K, Ivanova S, Streubel B, Hauser E et al (1999) Recruitment of bone-marrow-derived cells by skeletal and cardiac muscle inadult dystrophic mdx mice. Anat Embryol 199:391–396

Blau HM (1989) How fixed is the differentiated state? Lessons from heterokaryons. Trends Genet 5: 268–272

Blau HM, Blakely BT (1999) Plasticity of cell fate: insights from heterokaryons. Semin Cell Dev Biol 10:267

Blau HM, Chiu CP, Webster C (1983) Cytoplasmic activation of human nuclear genes in stable heterocaryons. Cell 32:1171

Blau HM, Pavlath GK, Hardeman EC (1985) Plasticity of the differentiated state. Science 230:758

Bonilla-Pons SÀ, Nakagawa S, Bahima EG, Fernández-Blanco Á, Pesaresi M, D'Antin JC et al (2022 Mar) Müller glia fused with adult stem cells undergo neural differentiation in human retinal models. EBioMedicine 77:103914

Bou G, Guo J, Fang Y, Li X, Wei R, Li Y et al (2021 Aug) Interspecies cell fusion between mouse embryonic stem cell and porcine pluripotent cell. Reprod Domest Anim 56(8):1095–1103

Bourc'his D, Le Bourhis D, Patin D (2001) Delayed and incomplete reprogramming of chromosome methylation patterns in bovine cloned embryos. Curr Biol 11:1542–1546

Boyer LA, Lee TI, Cole MF (2005) Core transcriptional regulatory circuitry in human embryonic stem cells. Cell 122:947–956

Boyer LA, Plath K, Zeitlinger J (2006) Polycomb complexes repress developmental regulators in murine embryonic stem cells. Nature 441:349

Brazelton TR, Rossi FM, Keshet GI, Blau HM (2000) From marrow to brain: expression of neuronalphenotypes in adult mice. Science 290:1775–1779

Briggs R, King TJ (1952 May) Transplantation of living nuclei from blastula cells into enucleated frogs' eggs. Proc Natl Acad Sci USA 38(5):455–463

Brown KE, Fisher AG (2021 Oct) Reprogramming lineage identity through cell–cell fusion. Curr Opin Genet Dev 70:15–23

Brown KE, Bagci H, Soza-Ried J, Fisher AG (2013) Atypical heterochromatin organization and replication are rapidly acquired by somatic cells following fusion-mediated reprogramming by mouse ESCs. Cell Cycle 12:3253

Brukman NG, Uygur B, Podbilewicz B, Chernomordik L (2019) How cells fuse. J Cell Biol 218: 1436–1451

Bussmann LH, Schubert A, Vu Manh TP (2009) A robust and highly efficient immune cell reprogramming system. Cell Stem Cell 5:554

Byrne JA, Pedersen DA, Clepper LL (2007) Producing primate embryonic stem cells by somatic cell nuclear transfer. Nature 450:497–502

Camargo FD, Green R, Capetenaki Y, Jackson KA, Goodell MA (2003) Single hematopoietic stem cells generate skeletal muscle through myeloid intermediates. Nat Med 9:1520–1527

Camargo FD, Finegold M, Goodell MA (2004) Hematopoietic myelomonocytic cells are the major source of hepatocyte fusion partners. J Clin Invest 113:1266–1270

Cantone I, Fisher AG (2017) Human X chromosome inactivation and reactivation: implications for cell reprogramming and disease. Philos Trans R Soc B Biol Sci 372(1733):20160358

Cantone I, Bagci H, Dormann D, Dharmalingam G, Nesterova T, Brockdorff N et al (2016) Ordered chromatin changes and human X chromosome reactivation by cell fusion-mediated pluripotent reprogramming. Nat Commun 7:12354

Carlson LL, Page AW, Bestor TH (1992) Properties and localization of DNA methyltransferase in preimplantation mouse embryos: implications for genomic imprinting. Genes Dev 6:2536

Carrette LLG, Blum R, Ma W, Kelleher RJ III, Lee JT (2018) Tsix-Mecp2 female mouse model for Rett syndrome reveals that low-level MECP2 expression extends life and improves neuromotor function. Proc Natl Acad Sci USA 115:8185

Chamberlain SJ, Yee D, Magnuson T (2008) Polycomb repressive complex 2 is dispensable for maintenance of embryonic stem cell pluripotency. Stem Cells 26:1496

Chambers I, Colby D, Robertson M (2003) Functional expression cloning of Nanog, a pluripotency sustaining factor in embryonic stem cells. Cell 113:643

Chang YC, Shyu WC, Lin SZ (2007) Regenerative therapy for stroke. Cell Transpl. 16:171

Chen H, Cheong ML. (2011) Syncytins: molecular aspects. In: Larsson LI (ed) Cell fusions: regulation and control [internet]. Springer, Dordrecht. p 117–137. Available from: https://doi.org/10.1007/978-90-481-9772-9_5

Chen J, Li Y, Wang L (2001) Therapeutic benefit of intravenous administration of bone marrow stromal cells after cerebral ischemia in rats. Stroke 32:1005

Chen J, Zhang ZG, Li Y (2003) Intravenous administration of human bone marrow stromal cells induces angiogenesis in the ischemic boundary zone after stroke in rats. Circ Res 92:692

Chiu DT (2001) A microfluidics platform for cell fusion. Curr Opin Chem Biol 5:609

Chiu CP, Blau HM (1984) Reprogramming cell differentiation in the absence of DNA synthesis. Cell 37:879–887

Chiu CP, Blau HM (1985) 5-Azacytidine permits gene activation in a previously noninducible cell type. Cell 40:417–424

Choi J, Costa ML, Mermelstein CS (1990) MyoD converts primary dermal fibroblasts, chondroblasts, smooth muscle, and retinal pigmented epithelial cells into striated mononucleated myoblasts and multinucleated myotubes. Proc Natl Acad Sci USA 87:7988

Choi HW, Chung MK, Hong YJ, Jang HS, Seo BJ, Jung TH et al (2015) Mitochondrial and metabolic remodelling during reprogramming and differentiation of the reprogrammed cells. Stem Cells Dev 24:1366–1373

Choudhury D, Ghosh D, Mondal M, Singha D, Pothuraju R, Malakar P (2024 Mar 27) Polyploidy and mTOR signaling: a possible molecular link. Cell Commun Signal 22(1):196

Cole MF, Johnstone SE, Newman JJ (2008) Tcf3 is an integral component of the core regulatory circuitry of embryonic stem cells. Genes Dev 22:746

Comai L (2005 Nov) The advantages and disadvantages of being polyploid. Nat Rev Genet 6(11): 836–846

Corti S, Locatelli F, Donadoni C, Guglieri M, Papadimitriou D, Strazzer S et al (2004 Nov) Wild-type bone marrow cells ameliorate the phenotype of SOD1-G93A ALS mice and contribute to CNS, heart and skeletal muscle tissues. Brain 127(11):2518–2532

Cossec J-C, Theurillat I, Chica C, Aguin SB, Gaume X, Andrieux A et al (2018) SUMO safeguards somatic and pluripotent cell identities by enforcing distinct chromatin states. Cell Stem Cell 23: 742

Cowan CA, Atienza J, Melton DA (2005) Nuclear reprogramming of somatic cells after fusion with human embryonic stem cells. Science 309:1369

Darlington GJ, Bernard HP, Ruddle FH (1974) Human serum albumin phenotype activation in mouse hepatoma–human leukocyte cell hybrids. Science 185:859

Darmasaputra GS, Van Rijnberk LM, Galli M (2024 Mar 1) Functional consequences of somatic polyploidy in development. Development 151(5):dev202392

Das S, Parigi SM, Luo X, Fransson J, Kern BC, Okhovat A et al (2025 Jan 30) Liver X receptor unlinks intestinal regeneration and tumorigenesis. Nature 637(8048):1198–1206

Davies PS, Powell AE, Swain JR, Wong MH (2009) Inflammation and proliferation act together to mediate intestinal cell fusion. PLoS One 4:e6530

de Jaime-Soguero A, Abreu de Oliveira WA, Lluis F (2018) The pleiotropic effects of the canonical Wnt pathway in early development and pluripotency. Genes 9(2):93

De Jong JH, Rodermond HM, Zimberlin CD, Lascano V, F DSEM, Richel DJ et al (2012) Fusion of intestinal epithelial cells with bone marrow derived cells is dispensable for tissue homeostasis. Sci Rep 2(1):271

Dean W, Santos F, Stojkovic M (2001) Conservation of methylation reprogramming in mammalian development: aberrant reprogramming in cloned embryos. Proc Natl Acad Sci USA 98:13734

Deb A, Wang S, Skelding KA (2003) Bone marrow-derived cardiomyocytes are present in adult human heart: a study of gender-mismatched bone marrow transplantation patients. Circulation 107:1247

Deisseroth A, Hendrick D (1979) Activation of phenotypic expression of human globin genes from nonerythroid cells by chromosome-dependent transfer to tetraploid mouse erythroleukemia cells. Proc Natl Acad Sci USA 76:2185–2189

Di Filippo ES, Mancinelli R, Marrone M, Doria C, Verratti V, Toniolo L et al (2017) Neuromuscular electrical stimulation improves skeletal muscle regeneration through satellite cell fusion with myofibers in healthy elderly subjects. J Appl Physiol 123(3):501–512

Do JT, Scholer HR (2004) Nuclei of embryonic stem cells reprogram somatic cells. Stem Cells 22: 941

Do JT, Han DW, Gentile L (2007) Erasure of cellular memory by fusion with pluripotent cells. Stem Cells 25:1013

Donne R, Saroul-Aïnama M, Cordier P, Celton-Morizur S, Desdouets C (2020 July) Polyploidy in liver development, homeostasis and disease. Nat Rev Gastroenterol Hepatol 17(7):391–405

Donovan PJ (1994) Growth factor regulation of mouse primordial germ cell development. Curr Top Dev Biol 29:189–225

Dörnen J, Dittmar T (2021) The role of MSCs and cell fusion in tissue regeneration. Int J Mol Sci 22(20):10980

Doyonnas R, LaBarge MA, Sacco A (2004) Hematopoietic contribution to skeletal muscle regeneration by myelomonocytic precursors. Proc Natl Acad Sci USA 101:13507

Duncan AW, Hickey RD, Paulk NK, Culberson AJ, Olson SB, Finegold MJ et al (2009) Ploidy reductions in murine fusion-derived hepatocytes. PLoS Genet 5:e1000385

Duncan AW, Hanlon Newell AE, Smith L, Wilson EM, Olson SB, Thayer MJ et al (2012) Frequent aneuploidy among normal human hepatocytes. Gastroenterology 142:25–28

Eggan K, Baldwin K, Tackett M (2004) Mice cloned from olfactory sensory neurons. Nature 428: 44

Eglitis MA, Mezey E (1997) Hematopoietic cells differentiate into both microglia and macroglia in the brains of adult mice. Proc Natl Acad Sci USA 94:4080–4085

Elling U, Klasen C, Eisenberger T (2006) Murine inner cell mass-derived lineages depend on Sall4 function. Proc Natl Acad Sci USA 103:16319

Enzmann V, Lecaudé S, Kruschinski A, Vater A (2017 Apr) CXCL12/SDF-1-dependent retinal migration of endogenous bone marrow-derived stem cells improves visual function after pharmacologically induced retinal degeneration. Stem Cell Rev Rep 13(2):278–286

Espejel S, Romero R, Alvarez-Buylla A (2009 July) Radiation damage increases Purkinje neuron heterokaryons in neonatal cerebellum. Ann Neurol 66(1):100–109

Evans MJ, Kaufman MH (1981) Establishment in culture of pluripotential cells from mouse embryos. Nature 292:154

Fan Q, Yee CL, Ohyama M, Tock C, Zhang G, Darling TN et al (2006 May) Bone marrow-derived keratinocytes are not detected in normal skin and only rarely detected in wounded skin in two different murine models. Exp Hematol 34(5):672–679

Fang TC, Alison MR, Cook HT (2005) Proliferation of bone marrow-derived cells contributes to regeneration after folic acid-induced acute tubular injury. J Am Soc Nephrol 16:1723

Fang Y, Guo J, Wu S, Li X, Zhao J, Li Y et al (2020 Apr) Cellular reprogramming by single-cell fusion with mouse embryonic stem cells in pig. J Cell Physiol 235(4):3558–3568

Feliciano D, Ott CM, Espinosa-Medina I, Weigel AV, Benedetti L, Milano KM et al (2021 July 23) YAP1 nuclear efflux and transcriptional reprograming follow membrane diminution upon VSV-G-induced cell fusion. Nat Commun 12(1):4502

Ferrari G, Cusella-De Angelis G, Coletta M (1998) Muscle regeneration by bone marrow-derived myogenic progenitors. Science 279:1528–1530

Foshay KM, Looney TJ, Chari S, Mao FF, Lee JH, Zhang L et al (2012 Apr) Embryonic stem cells induce pluripotency in somatic cell fusion through biphasic reprogramming. Mol Cell 46(2): 159–170

Fox DT, Soltis DE, Soltis PS, Ashman TL, Van De Peer Y (2020 Sept) Polyploidy: a biological force from cells to ecosystems. Trends Cell Biol 30(9):688–694

Frade J, Nakagawa S, Cortes P, di Vicino U, Romo N, Lluis F et al (2019) Controlled ploidy reduction of pluripotent 4n cells generates 2n cells during mouse embryo development. Sci Adv 5:eaax4199

Freeman BT, Kouris NA, Ogle BM (2015 June 1) Tracking fusion of human mesenchymal stem cells after transplantation to the heart. Stem Cells Transl Med 4(6):685–694

Frei E, Schuh R, Baumgartner S (1988) Molecular characterization of Spalt, a homeotic gene required for head and tail development in the drosophila embryo. EMBO J 7:197

Fujino H, Hiramatsu H, Tsuchiya A (2007) Human cord blood CD34+ cells develop into hepatocytes in the livers of NOD/SCID/gamma(c)null mice through cell fusion. FASEB J 21:3499

Gan P, Patterson M, Sucov HM (2020 Feb 10) Cardiomyocyte polyploidy and implications for heart regeneration. Annu Rev Physiol 82(1):45–61

Gibson AJ, Karasinski J, Relvas J, Moss J, Sherratt TG, Strong PN et al (1995) Dermal fibroblasts convert to a myogenic lineage in mdx mouse muscle. J Cell Sci 108:207–214

Girardi F, Taleb A, Ebrahimi M, Datye A, Gamage DG, Peccate C et al (2021 Feb 2) TGFβ signaling curbs cell fusion and muscle regeneration. Nat Commun 12(1):750

Goncalves MA, Swildens J, Holkers M (2008) Genetic complementation of human muscle cells via directed stem cell fusion. Mol Ther 16:741–748

Gou S, Liu T, Li X, Cui J, Wan C, Wang C (2012) Pancreatic ductal cells acquire mesenchymal characteristics through cell fusion with bone marrow-derived mesenchymal stem cells and SIRT1 attenuates the apoptosis of hybrid cells. Cells Tissues Organs 196(2):129–136

Gupta SK, Malhotra SS, Malik A, Verma S, Chaudhary P (2015) Cell signaling pathways involved during invasion and Syncytialization of trophoblast cells. Am J Reprod Immunol 75(3):361–371

Gussoni E, Soneoka Y, Strickland CD, Buzney EA, Khan MK, Flint AF et al (1999) Dystrophin expression in the mdx mouse restored by stem cell transplantation. Nature 401:390–394

Gussoni E, Bennett RR, Muskiewicz KR, Meyerrose T, Nolta JA, Gilgoff I et al (2002) Long-term persistence of donor nuclei in a Duchenne muscular dystrophy patient receiving bone marrow transplantation. J Clin Invest 110:807

Habib R, Haneef K, Naeem N, Khan I, Jamall S, Atta-ur-Rahman et al (2015 May) Hypoxic stress and IL-7 gene overexpression enhance the fusion potential of rat bone marrow mesenchymal stem cells with bovine renal epithelial cells. Mol Cell Biochem 403(1–2):125–137

Hajkova P, Erhardt S, Lane N (2002) Epigenetic reprogramming in mouse primordial germ cells. Mech Dev 117:15

Han J, Sidhu KS (2008) Current concepts in reprogramming somatic cells to pluripotent state. Curr Stem Cell Res Ther 3:66

Haneef K, Ali A, Khan I, Naeem N, Jamall S, Salim A (2018 Dec) Role of interleukin-7 in fusion of rat bone marrow mesenchymal stem cells with cardiomyocytes in vitro and improvement of cardiac function in vivo. Cardiovasc Ther 36(6):e12479

Harris H, JF W (1965) Hybrid cells derived from mouse and man: artificial Heterokaryons of mammalian cells from different species. Nature 205(4972):640–646

Hasegawa K, Zhang P, Wei Z, Pomeroy JE, Lu W, Pera MF (2010 Aug 1) Comparison of Reprogramming Efficiency Between Transduction of Reprogramming Factors, Cell–Cell Fusion, and Cytoplast Fusion. Stem Cells 28(8):1338–1348

Held PK, Al-Dhalimy M, Willenbring H, Akkari Y, Jiang S, Torimaru Y et al (2006 Jan) In vivo genetic selection of renal proximal tubules. Mol Ther 13(1):49–58

Herzog EL, Van Arnam J, Hu B (2007) Lung-specific nuclear reprogramming is accompanied by heterokaryon formation and Y chromosome loss following bone marrow transplantation and secondary inflammation. FASEB J 21:2592–2601

Hess DC, Abe T, Hill WD (2004) Hematopoietic origin of microglial and perivascular cells in brain. Exp Neurol 186:134

Hörnblad A, Remeseiro S (2022) Epigenetics, Enhancer function and 3D chromatin Organization in Reprogramming to pluripotency. Cells 11(9):1404

Houghton J, Stoicov C, Nomura S (2004) Gastric cancer originating from bone marrow-derived cells. Science 306:1568

Hromowyk KJ, Talbot JC, Martin BL, Janssen PML, Amacher SL (2020 June) Cell fusion is differentially regulated in zebrafish post-embryonic slow and fast muscle. Dev Biol 462(1): 85–100

Imai H, Kusakabe KT, Kiso Y, Hattori S, Kai C, Ono E et al (2020 Jan) Induction of pluripotency in mammalian fibroblasts by cell fusion with mouse embryonic stem cells. Biochem Biophys Res Commun 521(1):24–30

Jacob F (1978 May 16) The Leeuwenhoek lecture, 1977. Mouse teratocarcinoma and mouse embryo. Proc R Soc Lond B Biol Sci 201(1144):249–270

Jaenisch R, Hochedlinger K, Blelloch R (2004) Nuclear cloning, epigenetic reprogramming, and cellular differentiation. Cold Spring Harb Symp Quant Biol 69:19

Jang YY, Collector MI, Baylin SB, Diehl AM, Sharkis SJ (2004) Hematopoietic stem cells convert into liver cells within days without fusion. Nat Cell Biol 6:532–539

Janiszewski A, Talon I, Chappell J, Collombet S, Song J, De Geest N et al (2019) Dynamic reversal of random X-chromosome inactivation during iPSC reprogramming. Genome Res 29:1659

Jin HK, Schuchman EH (2003) Ex vivo gene therapy using bone marrow-derived cells: combined effects of intracerebral and intravenous transplantation in a mouse model of Niemann-pick disease. Mol Ther 8:876

Johansson CB, Youssef S, Koleckar K, Holbrook C, Doyonnas R, Corbel SY et al (2008) Extensive fusion of haematopoietic cells with Purkinje neurons in response to chronic inflammation. Nat Cell Biol 10:575–583

Kang YK, Koo DB, Park JS (2001) Aberrant methylation of donor genome in cloned bovine embryos. Nat Genet 28:173

Kashofer K, Siapati EK, Bonnet D (2006) In vivo formation of unstable heterokaryons after liver damage and hematopoietic stem cell/progenitor transplantation. Stem Cells Dayt Ohio 24:1104–1112

Kobayashi Y, Kondo M (1999 Oct) Human megakaryocyte ploidy. Histol Histopathol 14(4): 1223–1229

Kouris NA, Schaefer JA, Hatta M, Freeman BT, Kamp TJ, Kawaoka Y et al (2012) Directed fusion of mesenchymal stem cells with Cardiomyocytes via VSV-G facilitates stem cell programming. Stem Cells Int 2012:1–13

Krause DS, Theise ND, Collector MI (2001) Multi-organ, multi-lineage engraftment by a single bone marrow-derived stem cell. Cell 105:369–376

Krueger M, Bechmann I (2010) CNS pericytes: concepts, misconceptions, and a way out. Glia 58:1

LaBarge MA, Blau HM (2002) Biological progression from adult bone marrow to mononucleate muscle stem cell to multinucleate muscle fiber in response to injury. Cell 111:589–601

Laflamme MA, Myerson D, Saffitz JE (2002) Evidence for cardiomyocyte repopulation by extracardiac progenitors in transplanted human hearts. Circ Res 90:634–640

Lagasse E, Connors H, Al-Dhalimy M, Reitsma M, Dohse M, Osborne L et al (2000) Purified hematopoietic stem cells can differentiate into hepatocytes in vivo. Nat Med 6:1229–1234

Larson AG, Elnatan D, Keenen MM, Trnka MJ, Johnston JB, Burlingame AL et al (2017) Liquid droplet formation by HP1a suggests a role for phase separation in heterochromatin. Nature 547: 236

Larsson LI, Bjerregaard B, Talts JF (2008) Cell fusions in mammals. Histochem Cell Biol 129:551

LeBleu VS, Kanasaki K, Lovisa S, Alge JL, Kim J, Chen Y et al (2024 June) Genetic reprogramming with stem cells regenerates glomerular epithelial podocytes in Alport syndrome. Life Sci Alliance 7(6):e202402664

Lee TI, Jenner RG, Boyer LA (2006) Control of developmental regulators by Polycomb in human embryonic stem cells. Cell 125:301
Lehka L, Rędowicz MJ (2020 Aug) Mechanisms regulating myoblast fusion: a multilevel interplay. Semin Cell Dev Biol 104:81–92
Leone AM, Valgimigli M, Giannico MB (2009) From bone marrow to the arterial wall: the ongoing tale of endothelial progenitor cells. Eur Heart J 30:890–899
Lescaudron L, Unni D, Dunbar GL (2003) Autologous adult bone marrow stem cell transplantation in an animal model of Huntington's disease: behavioral and morphological outcomes. Int J Neurosci 113:945
Li Y, Chopp M (2009) Marrow stromal cell transplantation in stroke and traumatic brain injury. Neurosci Lett 456:120
Li Y, Chen J, Wang L (2001) Intracerebral transplantation of bone marrow stromal cells in a 1-methyl-4-phenyl-1,2,3,6-tetrahydropyridine mouse model of Parkinson's disease. Neurosci Lett 316:67
Lluis F, Cosma MP (2009) Somatic cell reprogramming control: signaling pathway modulation versus transcription factor activities. Cell Cycle 8:1138
Lluis F, Pedone E, Pepe S, Cosma MP (2008) Periodic activation of Wnt/beta-catenin signaling enhances somatic cell reprogramming mediated by cell fusion. Cell Stem Cell 3:493–507
Lopes Novo C, Tang C, Ahmed K, Djuric U, Fussner E, Mullin NP et al (2016) The pluripotency factor Nanog regulates pericentromeric heterochromatin organization in mouse embryonic stem cells. Genes Dev 30:1101
Lowe M, Hostager R, Nobuaki K (2016) Preservation of epigenetic memory during DNA replication. J Stem Cell Res Ther Edmond 1:1
Ma DK, Chiang CH, Ponnusamy K (2008) G9a and Jhdm2a regulate embryonic stem cell fusion-induced reprogramming of adult neural stem cells. Stem Cells 26:2131
Ma Z, Yang H, Liu H, Xu M, Runyan RB, Eisenberg CA et al (2013) Mesenchymal stem cell-cardiomyocyte interactions under defined contact modes on laser-patterned biochips. Pesce M (ed). PLoS ONE 8(2):e56554
Magrassi L, Grimaldi P, Ibatici A (2007) Induction and survival of binucleated Purkinje neurons by selective damage and aging. J Neurosci 27:9885
Mai T, Markov G, Brady JJ, Palla A, Zeng H, Sebastiano V et al (2018) NKX3-1 is required for induced pluripotent stem cell reprogramming and can replace OCT4 in mouse and human iPSC induction. Nat Cell Biol 10:900
Mann DM, Yates PO (1973 Feb) Polyploidy in the human nervous system. 1. The DNA content of neurones and glia of the cerebellum. J Neurol Sci 18(2):183–196
Marshak D, Gardner R, Gottlieb D (2001) Stem cell biology. Cold Spring Harbor Laboratory Press, New York
Martens S, HT MM (2008) Mechanisms of membrane fusion: disparate players and common principles. Nat Rev Mol Cell Biol 9(7):543–556
Martin GR (1981) Isolation of a pluripotent cell line from early mouse embryos cultured in mediumconditioned by teratocarcinoma stem cells. Proc Natl Acad Sci USA 78:7634–7638
Martinez-Sarmiento JA, Cosma MP, Lakadamyali M (2024 May 28) Dissecting gene activation and chromatin remodeling dynamics in single human cells undergoing reprogramming. Cell Rep 43(5):114170
Matsuura K, Wada H, Nagai T (2004) Cardiomyocytes fuse with surrounding noncardiomyocytes and reenter the cell cycle. J Cell Biol 167:351
Matveeva NM, Shilov AG, Kaftanovskaya EM (1998) In vitro and in vivo study of pluripotency in intraspecific hybrid cells obtained by fusion of murine embryonic stem cells with splenocytes. Mol Reprod Dev 50:128
Matveeva NM, Kuznetsov SB, Kaftanovskaya EM (2001) Segregation of parental chromosomes in hybrid cells obtained by fusion between embryonic stem cells and differentiated cells of adult animal. Dokl Biol Sci 379:399–401

Matveeva NM, Kizilova EA, Serov OL (2017) Generation of mouse chimeras with high contribution of tetraploid embryonic stem cells and embryonic stem cell-fibroblast hybrid cells. Methods Mol Biol 1313:61

Meir YJJ, Li G (2021) Somatic reprogramming-above and beyond pluripotency. Cells 10(11):2888

Meissner A, Mikkelsen TS, Gu H (2008) Genome-scale DNA methylation maps of pluripotent and differentiated cells. Nature 454:766

Mezey E, Chandross KJ, Harta G (2000) Turning blood into brain: cells bearing neuronal antigens generated in vivo from bone marrow. Science 290:1779–1782

Miller RA, Ruddle FH (1976) Pluripotent teratocarcinoma-thymus somatic cell hybrids. Cell 9:45

Mitsui K, Tokuzawa Y, Itoh H (2003) The homeoprotein Nanog is required for maintenance of pluripotency in mouse epiblast and ES cells. Cell 113:631

Momin EN, Mohyeldin A, Zaidi HA (2010) Mesenchymal stem cells: new approaches for the treatment of neurological diseases. Curr Stem Cell Res Ther 5:326–344

Monk M, Boubelik M, Lehnert S (1987) Temporal and regional changes in DNA methylation in the embryonic, extraembryonic and germ cell lineages during mouse embryo development. Development 99:371

Morgan HD, Santos F, Green K (2005) Epigenetic reprogramming in mammals. Hum Mol Genet 14(1):R47–R58

Moya IM, Halder G (2019 Apr) Hippo–YAP/TAZ signalling in organ regeneration and regenerative medicine. Nat Rev Mol Cell Biol 20(4):211–226

Müller M, May S, Bird TG (2021 Mar 25) Ploidy dynamics increase the risk of liver cancer initiation. Nat Commun 12(1):1896

Munoz JP, Zorzano A (2017) Fission for reprogramming. Cell Cycle 16:159

MVC G (2020) Get out and stay out: new insights into DNA methylation reprogramming in mammals. Front Cell Dev Biol 8:629068

Myerson D, Parkin RK (2019 Mar) Donor-derived hepatocytes in human hematopoietic cell transplant recipients: evidence of fusion. Virchows Arch 474(3):365–374

Naik PP, Birbrair A, Bhutia SK (2019) Mitophagy-driven metabolic switch reprograms cell fate. Cell Mol Life Sci 76(1):27–43

Nakamura T, Inoue K, Ogawa S (2008) Effects of Akt signaling on nuclear reprogramming. Genes Cells 13:1269

Newsome PN, Johannessen I, Boyle S, Dalakas E, McAulay KA, Samuel K et al (2003) Human cord blood-derived cells can differentiate into hepatocytes in the mouse liver with no evidence of cellular fusion. Gastroenterology 124:1891–1900

Nygren JM, Jovinge S, Breitbach M (2004) Bone marrow-derived hematopoietic cells generate cardiomyocytes at a low frequency through cell fusion, but not transdifferentiation. Nat Med 10: 494–501

Nygren JM, Liuba K, Breitbach M, Stott S, Thoren L, Roell W et al (2008) Myeloid and lymphoid contribution to non-haematopoietic lineages through irradiation-induced heterotypic cell fusion. Nat Cell Biol 10:584–592

Okamoto R, Yajima T, Yamazaki M, Kanai T, Mukai M, Okamoto S et al (2002) Damaged epithelia regenerated by bone marrow-derived cells in the human gastrointestinal tract. Nat Med 8:1011–1017

Onishi A, Iwamoto M, Akita T (2000) Pig cloning by microinjection of fetal fibroblast nuclei. Science 289:1188

Orlic D, Kajstura J, Chimenti S, Jakoniuk I, Anderson SM, Li B et al (2001) Bone marrow cells regenerate infarcted myocardium. Nature 410:701–705

Ozerdem U, Stallcup WB (2003) Early contribution of pericytes to angiogenic sprouting and tube formation. Angiogenesis 6:241

Palermo A, Doyonnas R, Bhutani N (2009) Nuclear reprogramming in heterokaryons is rapid, extensive, and bidirectional. FASEB J 23:1431–1440

Paltsyn A, Komissarova S, Dubrovin I, Kubatiev A (2013) Increased cell fusion in cerebral cortex May contribute to Poststroke regeneration. Stroke Res Treat 2013:1–16

Paltsyn AA, Manukhina EB, Goryacheva AV, Downey HF, Dubrovin IP, Komissarova SV et al (2014 May) Intermittent hypoxia stimulates formation of binuclear neurons in brain cortex—a role of cell fusion in neuroprotection? Exp Biol Med 239(5):595–600

Papayannopoulou T, Enver T, Takegawa S (1988) Activation of developmentally mutated human globin genes by cell fusion. Science 242:1056–1058

Pasini D, Bracken AP, Hansen JB (2007) The polycomb group protein Suz12 is required for embryonic stem cell differentiation. Mol Cell Biol 27:3769

Pedone E, Olteanu VA, Marucci L, Munoz-Martin MI, Youssef SA, de Bruin A et al (2017) Modeling dynamics and function of bone marrow cells in mouse liver regeneration. Cell Rep 18: 107–121

Peñalosa-Ruiz G, Bright AR, Mulder KW, Veenstra GJC (2019 Sept) The interplay of chromatin and transcription factors during cell fate transitions in development and reprogramming. Biochim Biophys Acta Gene Regul Mech 1862(9):194407

Pereira CF, Terranova R, Ryan NK (2008) Heterokaryon-based reprogramming of human B lymphocytes for pluripotency requires Oct4 but not Sox2. PLoS Genet 4:e1000170

Pereira CF, Piccolo FM, Tsubouchi T (2010) ESCs require PRC2 to direct the successful reprogramming of differentiated cells toward pluripotency. Cell Stem Cell 6:547

Pesaresi M, Bonilla-Pons SA, Simonte G, Sanges D, Di Vicino U, Cosma MP (2018 Apr) Endogenous mobilization of bone-marrow cells into the murine retina induces fusion-mediated reprogramming of Müller glia cells. EBioMedicine 30:38–51

Petersen BE, Bowen WC, Patrene KD (1999) Bone marrow as a potential source of hepatic oval cells. Science 284:1168–1170

Piccolo F, Bagci H, Brown KE, Landeira D, Soza-Ried J, Feytout A et al (2013) Different roles for Tet1 and Tet2 proteins in reprogramming-mediated erasure of imprints induced by ESC fusion. Mol Cell 49:1023

Piquer-Gil M, Garcia-Verdugo JM, Zipancic I (2009) Cell fusion contributes to pericyte formation after stroke. J Cereb Blood Flow Metab 29:480

Pomerantz JH, Mukherjee S, Palermo AT (2009) Reprogramming to a muscle fate by fusion recapitulates differentiation. J Cell Sci 122:1045–1053

Prieto J, Leon M, Ponsoda X, Sendra R, Bort R, Ferrer-Lorente R et al (2016) Early ERK1/2 activation promotes DRP1-dependent mitochondrial fission necessary for cell reprogramming. Nat Commun 7:11124

Priller J, Flügel A, Wehner T, Boentert M, Haas CA, Prinz M et al (2001a Dec) Targeting gene-modified hematopoietic cells to the central nervous system: use of green fluorescent protein uncovers microglial engraftment. Nat Med 7(12):1356–1361

Priller J, Persons DA, Klett FF (2001b) Neogenesis of cerebellar Purkinje neurons from gene-marked bone marrow cells in vivo. J Cell Biol 155:733

Quaini F, Urbanek K, Beltrami AP (2002) Chimerism of the transplanted heart. N Engl J Med 346:5

Ramos C, Teissie J (2000) Electrofusion: a biophysical modification of cell membrane and a mechanism in exocytosis. Biochimie 82:511

Reubinoff BE, Pera MF, Fong CY (2000) Embryonic stem cell lines from human blastocysts: somatic differentiation in vitro. Nat Biotechnol 18:399

Ringrose L, Paro R (2004) Epigenetic regulation of cellular memory by the Polycomb and Trithorax group proteins. Annu Rev Genet 38:413

Rizvi AZ, Swain JR, Davies PS, Bailey AS, Decker AD, Willenbring H et al (2006) Bone marrow-derived cells fuse with normal and transformed intestinal stem cells. Proc Natl Acad Sci USA 103:6321–6325

Rosser AE, Zietlow R, Dunnett SB (2007) Stem cell transplantation for neurodegenerative diseases. Curr Opin Neurol 20:688

Rousset JP, Dubois P, Lasserre C (1979) Phenotype and surface antigens of mouse teratocarcinoma x fibroblast cell hybrids. Somat Cell Genet 5:739

Rousset JP, Bucchini D, Jami J (1983) Hybrids between F9 nullipotent teratocarcinoma and thymus cells produce multidifferentiated tumors in mice. Dev Biol 96:331

Sampath SC, Sampath SC, Millay DP (2018 Jan 31) Myoblast fusion confusion: the resolution begins. Skelet Muscle 8(1):3

Sanges D, Romo N, Simonte G, Di Vicino U, Tahoces AD, Fernandez E et al (2013) Wnt/beta-catenin signaling triggers neuron reprogramming and regeneration in the mouse retina. Cell Rep 4:271–286

Sanges D, Simonte G, Di Vicino U, Romo N, Pinilla I, Nicolás M et al (2016) Reprogramming Müller glia via in vivo cell fusion regenerates murine photoreceptors. J Clin Invest 126(8): 3104–3116

Sato N, Meijer L, Skaltsounis L (2004) Maintenance of pluripotency in human and mouse embryonic stem cells through activation of Wnt signaling by a pharmacological GSK-3-specific inhibitor. Nat Med 10:55

Sawamiphak S, Kontarakis Z, Filosa A, Reischauer S, Stainier DYR (2017 Nov 15) Transient cardiomyocyte fusion regulates cardiac development in zebrafish. Nat Commun 8(1):1525

Serov OL, Matveeva NM, Khaborova AA (2011) Reprogramming mediated by cell fusion technology. Int Rev Cell Mol Biol 291:155

Sharma AD, Cantz T, Richter R (2005) Human cord blood stem cells generate human cytokeratin 18-negative hepatocyte-like cells in injured mouse liver. Am J Pathol 167:555–564

Silk AD, Gast CE, Davies PS, Fakhari FD, Vanderbeek GE, Mori M et al (2013) Fusion between hematopoietic and epithelial cells in adult human intestine. PLoS One 8:e55572

Silva J, Chambers I, Pollard S (2006) Nanog promotes transfer of pluripotency after cell fusion. Nature 441:997

Silva J, Nichols J, Theunissen TW (2009) Nanog is the gateway to the pluripotent ground state. Cell 138:722

Song JH, Choi J, Hong YJ, La H, Hong TK, Hong K et al (2022 Oct 17) Developmental potency and metabolic traits of extended pluripotency are faithfully transferred to somatic cells via cell fusion-induced reprogramming. Cells 11(20):3266

Sottile F, Aulicino F, Theka I, Cosma MP (2016 Nov 9) Mesenchymal stem cells generate distinct functional hybrids in vitro via cell fusion or entosis. Sci Rep 6(1):36863

Spees JL, Olson SD, Ylostalo J, Lynch PJ, Smith J, Perry A et al (2003) Differentiation, cell fusion, and nuclear fusion during ex vivo repair of epithelium by human adult stem cells from bone marrow stroma. Proc Natl Acad Sci USA 100:2397–2402

Spees JL, Whitney MJ, Sullivan DE (2008) Bone marrow progenitor cells contribute to repair and remodeling of the lung and heart in a rat model of progressive pulmonary hypertension. FASEB J 22:1226

Spurlock BM, Xie Y, Song Y, Ricketts SN, Hua JR, Chi HR et al (2025 Mar 25) Mitochondrial fusion and cristae reorganization facilitate acquisition of cardiomyocyte identity during reprogramming of murine fibroblasts. Cell Rep 44(3):115377

Sumer H, Kim K, Liu J, Ng K, Daley GQ, Verma PJ (2014 June) Functional evaluation of ES–somatic cell hybrids *in vitro* and *in vivo*. Cell Reprogram 16(3):167–174

Sutton TL, Walker BS, Wong MH (2021) Digesting the importance of cell fusion in the intestine. Cell Mol Gastroenterol Hepatol 11(1):299–302

Tada M, Tada T, Lefebvre L (1997) Embryonic germ cells induce epigenetic reprogramming of somatic nucleus in hybrid cells. EMBO J 16:6510

Tada M, Takahama Y, Abe K (2001) Nuclear reprogramming of somatic cells by in vitro hybridization with ES cells. Curr Biol 11:1553

Tada M, Morizane A, Kimura H (2003) Pluripotency of reprogrammed somatic genomes in embryonic stem hybrid cells. Dev Dyn 227:504

Takahashi K, Tanabe K, Ohnuki M, Narita M, Ichisaka T, Tomoda K et al (2007) Induction of pluripotent stem cells from adult humanfibroblasts by defined factors. Cell 131:861–872

Takeda K (2019 Dec 18) Functional consequences of mitochondrial mismatch in reconstituted embryos and offspring. J Reprod Dev 65(6):485–489

Takehashi M, Tada M, Kanatsu-Shinohara M, Morimoto H, Kazuki Y, Oshimura M et al (2012 June 1) Hybridization of testis-derived stem cells with somatic cells and embryonic stem cells in mice. Biol Reprod 86(6):178

Tal A, Tal R, Shaikh S, Gidicsin S, Mamillapalli R, Taylor HS (2019) Characterization of cell fusion in an experimental mouse model of endometriosis†. Biol Reprod 100(2):390–397

Tam WL, Lim CY, Han J (2008) T-cell factor 3 regulates embryonic stem cell pluripotency and self-renewal by the transcriptional control of multiple lineage pathways. Stem Cells 26:2019

Tang Y, Yasuhara T, Hara K (2007) Transplantation of bone marrow-derived stem cells: a promising therapy for stroke. Cell Transpl 16:159

Tapscott SJ, Davis RL, Thayer MJ (1988) MyoD1:a nuclear phosphoprotein requiring a Myc homology region to convert fibroblasts to myoblasts. Science 242:405

Taranger CK, Noer A, Sorensen AL (2005) Induction of dedifferentiation, genomewide transcriptional programming, and epigenetic reprogramming by extracts of carcinoma and embryonic stem cells. Mol Biol Cell 16:5719–5735

Tat PA, Sumer H, Pralong D, Verma PJ (2011 Aug) The efficiency of cell fusion-based reprogramming is affected by the somatic cell type and the *in vitro* age of somatic cells. Cell Reprogram 13(4):331–344

Terada N, Hamazaki T, Oka M, Hoki M, Mastalerz DM, Nakano Y et al (2002) Bone marrow cells adopt the phenotype of other cells by spontaneous cell fusion. Nature 416:542–545

Terranova R, Pereira CF, Du Roure C (2006) Acquisition and extinction of gene expression programs are separable events in heterokaryon reprogramming. J Cell Sci 119:2065–2072

Theise ND, Badve S, Saxena R, Henegariu O, Sell S, Crawford JM et al (2000) Derivation of hepatocytes from bone marrow cells in mice after radiation-induced myeloablation. Hepatology (Baltimore) 31:235–240

Thomson JA, Itskovitz-Eldor J, Shapiro SS (1998) Embryonic stem cell lines derived from human blastocysts. Science 282:1145

Tsubouchi T, Soza-Ried J, Brown K, Piccolo FM, Cantone I, Landeira D et al (2013 Feb) DNA synthesis is required for reprogramming mediated by stem cell fusion. Cell 152(4):873–883

Vassilopoulos G, Russell DW (2003) Cell fusion: an alternative to stem cell plasticity and its therapeut icimplications. Curr Opin Genet Dev 13:480–485

Vassilopoulos G, Wang PR, Russell DW (2003 Apr 24) Transplanted bone marrow regenerates liver by cell fusion. Nature 422(6934):901–904

Vierbuchen T, Ostermeier A, Pang ZP (2010) Direct conversion of fibroblasts to functional neurons by defined factors. Nature 463:1035

Wagers AJ, Weissman IL (2004) Plasticity of adult stem cells. Cell 116:639

Wakayama T, Yanagimachi R (1998) Development of normal mice from oocytes injected with freeze-dried spermatozoa. Nat Biotechnol 16:639

Wakayama T, Perry AC, Zuccotti M (1998) Full-term development of mice from enucleated oocytes injected with cumulus cell nuclei. Nature 394:369

Wang X, Willenbring H, Akkari Y, Torimaru Y, Foster M, Al-Dhalimy M et al (2003) Cell fusion is the principal source of bone-marrow-derived hepatocytes. Nature 422:897–901

Wang Y, Yang C-Y, Rambold A, Mittler G, Grosschedl R (2020) A prion-like domain in transcription factor EBF1 promotes phase separation and enables B cell programming of progenitor chromatin. Immunity 53:1

Watanabe S, Umehara H, Murayama K (2006) Activation of Akt signaling is sufficient to maintain pluripotency in mouse and primate embryonic stem cells. Oncogene 25:2697

Weimann JM, Charlton CA, Brazelton TR, Hackman RC, Blau HM (2003a) Contribution of transplanted bone marrow cells to Purkinje neurons in human adult brains. Proc Natl Acad Sci U S A 100:2088–2093

Weimann JM, Johansson CB, Trejo A (2003b) Stable reprogrammed heterokaryons form spontaneously in Purkinje neurons after bone marrow transplant. Nat Cell Biol 5:959–966

Weintraub H, Tapscott SJ, Davis RL (1989) Activation of muscle-specific genes in pigment, nerve, fat, liver, and fibroblast cell lines by forced expression of MyoD. Proc Natl Acad Sci USA 86: 5434–5438

White SJ, Chong JJH (2020 Oct) Mesenchymal stem cells in cardiac repair: effects on myocytes, vasculature, and fibroblasts. Clin Ther 42(10):1880–1891

Wiersema A, Dijk F, Dontje B, van der Want JJ, de Haan G (2007) Cerebellar heterokaryon formation increases with age and after irradiation. Stem Cell Res 1:150–154

Willenbring H, Bailey AS, Foster M, Akkari Y, Dorrell C, Olson S et al (2004) Myelomonocytic cells are sufficient for therapeutic cell fusion in liver. Nat Med 10:744–748

Willing MC, Nienhuis AW, Anderson WF (1979) Selective activation of human beta-but not gamma-globin gene in human fibroblast × mouse erythroleukaemia cell hybrids. Nature 277: 534–538

Willkomm L, Bloch W (2015) State of the art in cell-cell fusion. Methods Mol Biol 1313:1–19

Wilmut I, Schnieke AE, McWhir J, Kind AJ, Campbell KH (1997) Viable offspring derived from fetal and adult mammalian cells. Nature 385:810–813

Wong AP, Dutly AE, Sacher A (2007) Targeted cell replacement with bone marrow cells for airway epithelial regeneration. Am J Physiol Lung Cell Mol Physiol 293:L740

Wong CC, Gaspar-Maia A, Ramalho-Santos M (2008) High-efficiency stem cell fusion-mediated assay reveals Sall4 as an enhancer of reprogramming. PLoS One 3:e1955

Wong WT, Matrone G, Tian XY, Tomoiaga SA, Au KF, Meng S et al (2017) Discovery of novel determinants of endothelial lineage using chimeric heterokaryons. elife 6:e23588

Wright WE (1984) Control of differentiation in heterokaryons and hybrids involving differentiation-defective myoblast variants. J Cell Biol 98:436

Wu W, Qu Y, Hu N, Zeng Y, Yang J, Xu H et al (2015) A cell electrofusion chip for somatic cells reprogramming. Kues WA (ed). PLOS ONE 10(7):e0131966

Xie H, Ye M, Feng R (2004) Stepwise reprogramming of B cells into macrophages. Cell 117:663

Yamanaka S, Blau HM (2010) Nuclear reprogramming to a pluripotent state by three approaches. Nature 465:704

Yanai G, Hayashi T, Zhi Q, Yang KC, Shirouzu Y, Shimabukuro T et al (2013) Electrofusion of mesenchymal stem cells and islet cells for diabetes therapy: a rat model. Maedler K (ed). PLoS ONE. 8(5):e64499

Yang J, Chai L, Fowles TC (2008) Genome-wide analysis reveals Sall4 to be a major regulator of pluripotency in murine-embryonic stem cells. Proc Natl Acad Sci USA 105:19756

Yang PF, Wang CH, Lee GB (2016 Feb 25) Optically-induced cell fusion on cell pairing micro-structures. Sci Rep 6(1):22036

Yi H (2024) Urine-derived stem cells genetically modified with IGF1 improve muscle regeneration. Am J Clin Exp Urol 12(2):64–87

Yi F, Pereira L, Merrill BJ (2008) Tcf3 functions as a steady-state limiter of transcriptional programs of mouse embryonic stem cell self-renewal. Stem Cells 26:1951

Ying QL, Nichols J, Evans EP, Smith AG (2002) Changing potency by spontaneous fusion. Nature 416:545–548

Yu J, Vodyanik MA, He P (2006) Human embryonic stem cells reprogram myeloid precursors following cell-cell fusion. Stem Cells 24:168

Zaret KS, Mango S (2016) Pioneer transcription factors, chromatin dynamics, and cell fate control. Curr Opin Genet Dev 37:76

Zhang J, Tam WL, Tong GQ (2006) Sall4 modulates embryonic stem cell pluripotency and early embryonic development by the transcriptional regulation of Pou5f1. Nat Cell Biol 8:1114

Chapter 9
How Do Cancer Cells Fuse and Why?

Kayode Komolafe, Oluwatoyin V. Odubanjo, Ariane M. Chitoh, and Felicite K. Noubissi

Abstract Cell fusion is increasingly recognized as a critical event in cancer biology. The process has now been associated with diverse outcomes related to cancer pathophysiology, including tumor heterogeneity, metastasis, immune evasion, therapy resistance, and progression. Evidence from in vitro, in vivo, and clinical studies confirms that both homotypic (cancer–cancer) and heterotypic (cancer–non-cancer) fusion events result in hybrid populations with enhanced fitness. Comparable to what occurs in normal physiological conditions, fusion involves membrane priming, adhesion, merging, and post-fusion resetting. The process of fusion is driven by fusogens such as syncytin-1 and annexins, and it is influenced by external cues, including inflammation, hypoxia, viral infection, and therapeutic stress. Functionally, cell fusion boosts genomic instability and tumor plasticity, increases metastatic potential, allows for stem-like reprogramming, and aids in immune evasion. Clinically, hybrid cells could serve as reliable biomarkers for diagnosis, prognosis, and treatment stratification. Nevertheless, the dual function of fusion in tumor growth and normal physiology necessitates context-specific approaches. Unraveling the fusion machinery or leveraging fusion-induced immunogenicity might yield novel therapeutic opportunities. This chapter discusses the molecular, functional, and clinical aspects of cancer cell fusion, as well as its importance in cancer progression and therapy resistance.

Keywords Cancer cell fusion · Hybrid cells · Tumor heterogeneity · Immune evasion · Fusogens

K. Komolafe · A. M. Chitoh
Department of Biology, Jackson State University, Jackson, MS, USA

Environmental Science PhD Program, Jackson State University, Jackson, MS, USA

O. V. Odubanjo · F. K. Noubissi (✉)
Department of Biology, Jackson State University, Jackson, MS, USA

Research Centers in Minority Institutions (RCMI), Center for Health Disparity Research (RCMI-CHDR), Jackson State University, Jackson, MS, USA
e-mail: felicite.noubissi_kamdem@jsums.edu

© The Author(s), under exclusive license to Springer Nature Switzerland AG 2026

J. Weiler et al. (eds.), *Cell Fusion*, Advances in Experimental Medicine and Biology 1509, https://doi.org/10.1007/978-3-032-22637-2_9

Abbreviations

α2M	Alpha-2-macroglobulin
Akt	Protein kinase B
ALDH1	Aldehyde dehydrogenase 1
AMPK	AMP-activated protein kinase
Arp2/3	Actin-related protein 2/3 complex
ASCT1	Alanine/serine/cysteine/threonine transporter 1 (SLC1A4)
ASCT2	Alanine/serine/cysteine/threonine transporter 2 (SLC1A5)
Bai1/3	Brain-specific angiogenesis inhibitor 1/3
BMDCs	Bone marrow-derived cells
BMT	Bone marrow transplant
CAMLs	Cancer-associated macrophage-like cells
CCL21	C-C motif chemokine ligand 21
CD14	Cluster of differentiation 14
Cre–LoxP	Cyclization recombination–LoxP (site-specific recombination system)
CSC	Cancer stem cell
CS/IC	Cancer stem/initiating cell
CTCs	Circulating tumor cells
CXCR4	C-X-C chemokine receptor type 4
EBV	Epstein–Barr virus
ECM	Extracellular matrix
ELMO	Engulfment and cell motility protein
EMT	Epithelial-to-mesenchymal transition
EpCAM	Epithelial cell adhesion molecule
ERK1/2	Extracellular signal-regulated kinase 1/2
FN1	Fibronectin 1
GFP	Green fluorescent protein
GRP78	Glucose-regulated protein 78
HERV	Human endogenous retrovirus
HLA	Human leukocyte antigen
HSC	Hematopoietic stem cell
HUVECs	Human umbilical vein endothelial cells
ICC	Intrahepatic cholangiocarcinoma
IL	Interleukin
LTR	Long terminal repeat
MAPK	Mitogen-activated protein kinase
MHC	Major histocompatibility complex
MFSD2A	Major facilitator superfamily domain containing 2A
MMP9	*Matrix metallopeptidase 9*
MNGCs	Multinucleated giant cells
MSC(s)	Mesenchymal stem/stromal cell(s)
NF-κB	Nuclear factor kappa-light-chain-enhancer of activated B Cells

NK	Natural killer (cell)
NSCLC	Non-small cell lung cancer
PDAC	Pancreatic ductal adenocarcinoma
PD-1	Programmed cell death protein 1
PD-L1	Programmed cell death ligand 1
PDT	Photodynamic therapy
PFS	Progression-free survival
PHPP	Pre-hybrid preparation program
PHSP	Post-hybrid selection process
PI3K	Phosphoinositide 3-kinase
PGCCs	Polyploid giant cancer cells
RCC	Renal cell carcinoma
RFP	Red fluorescent protein
SIGLEC5	Sialic acid-binding Ig-like lectin 5
SPARC	Secreted protein acidic and rich in cysteine
STR	Short tandem repeat
SU	Surface unit (domain of retroviral Env proteins)
TAMs	Tumor-associated macrophages
THCs	Tumor hybrid cells
TME	Tumor microenvironment
TM	Transmembrane domain
TMEM16E	Transmembrane protein 16E
TNF-α	Tumor necrosis factor-alpha
VCAM-1	Vascular cell adhesion molecule-1
VEGF	Vascular endothelial growth factor
VLA-1	Very late antigen-1

9.1 Introduction

Cancer is a dynamic and heterogeneous disease that is characterized by complex cellular interactions and evolutionary pressures that facilitate adaptation, progression, and survival (Jacquemin et al. 2022). Cell fusion is increasingly getting recognition as one of the mechanisms contributing to the complexity of cancer (Bastida-Ruiz et al. 2016; López-Collazo and Hurtado-Navarro 2025), and existing evidence now identifies fusion events as significant drivers of tumor evolution, heterogeneity, and therapeutic resistance (Lu and Kang 2009a; Shultes et al. 2024). In cancer biology, cell fusion refers to the merging of cancer cells with other cancer or noncancerous cells such as mesenchymal stem/stromal cells (MSCs), macrophages, or endothelial cells (Dittmar and Hass 2023). This process produces hybrid cells possessing novel characteristics and implicated in many hallmarks of cancer (Melzer et al. 2018a; Peng et al. 2021; Sieler et al. 2021; Cozzo et al. 2023; Noubissi et al. 2024). There are lines of evidence showing that cell fusion occurs across in vitro, in vivo, and clinical settings (Hass et al. 2021b; Peng et al. 2021; López-Collazo and Hurtado-Navarro 2025). Laboratory models have shown that cancer cells can fuse with stromal elements and that such fusion events can confer stem-like features, induce epithelial-to-mesenchymal transition (EMT), and boost resistance to radiation and chemotherapy (Hass et al. 2021a, b). In vivo, lineage-tracing and xenograft studies have been used to confirm the presence of hybrid cells (Peng et al. 2021; Sieler et al. 2021; López-Collazo and Hurtado-Navarro 2025). Also in clinical settings, hybrid cells expressing markers from both cancer and noncancerous lineages have been detected in patients' blood and tumor samples (Shabo et al. 2020; Peng et al. 2021).

Similar to what occurs under physiological conditions, cancer cell fusion occurs in four stages, namely priming, membrane apposition, fusion, and post-fusion resetting. This process is controlled by molecular drivers such as fusogenic proteins (e.g., syncytin-1), phosphatidylserine (PS) externalization, and cytoskeletal remodeling (Hass et al. 2021b; Zhang et al. 2022). Furthermore, external factors like inflammation, hypoxia, and therapeutic stress can increase the likelihood of fusion events in the tumor microenvironment (TME). These factors not only allow cells to physically merge but also promote hybrid survival and selection (Zhang et al. 2022; Noubissi et al. 2024).

The physiological consequences of cancer cell fusion are profound, and one of them is increasing genomic diversity by creating aneuploid or polyploid hybrids. Others are accelerating metastatic competence by promoting EMT and motility and enabling cancer cells to evade both therapeutic and immune pressures (Zhou et al. 2015; Søe 2020; Wang et al. 2021). Numerous studies provide evidence that fusion-dependent hybrids are resistant to many chemotherapeutic drugs and acquire features associated with cancer stem cells (CSCs), ultimately contributing to tumor relapse and progression (Kaur et al. 2015; Lindström et al. 2017; Tretyakova et al. 2022). Furthermore, fusion enables tumors to adapt more quickly to environmental pressures than mutation alone, thereby serving as an evolutionary shortcut that promotes malignant plasticity (Goldenberg et al. 2014; Wang et al. 2021; Dittmar and Hass 2023). While most spontaneous fusion events promote the progression of tumors, engineered fusion may hold potential in cancer therapy. For example, purposeful fusion of cancer cells with dendritic cells has been investigated in vaccine development to improve tumor antigen presentation and boost antitumor immune responses (Koido et al. 2009; Boudreau et al. 2011). Such applications take advantage of naturally occurring fusion events, which often result in immune evasion rather than immunogenicity (López-Collazo and Hurtado-Navarro 2025). This chapter seeks to examine the current understanding of how cancer cells fuse, the functional ramifications of cancer cell fusion, and the possibility of harnessing the process to enhance patient outcome potential.

9.1.1 *Evidence Supporting Cancer Cell Fusion*

Cancer cell fusion is gaining increasing acceptance as a critical event in tumor biology. The numerous literature reports on cancer cell fusions under in vitro and in vivo conditions, as well as in clinical settings, and how they contribute to cancer hallmarks such as tumor heterogeneity, progression, metastasis, and therapy resistance are discussed.

9.1.1.1 In Vitro Evidence Supporting Cancer Cell Fusion

Multiple reports about the fusion of cancer cells with both other cancer cells and noncancerous cell types in vitro are found in the literature (Mortensen et al. 2004; Song et al. 2014; Melzer et al. 2019b; Raj et al. 2019; Hass et al. 2021b). A few research studies in this direction reveal that coculturing cancer cells with endothelial cells yields hybrids with proliferative capacity and long-term survival (Song et al. 2014; Peng et al. 2021). Cell-tracing approaches such as fluorescent protein labeling and barcoding have allowed the detection of tumor–tumor fusion events in vitro (Shultes et al. 2024). These hybrids are oftentimes observed to express both parental markers, undergo nuclear fusion, and display drug resistance (Peng et al. 2021). Parallel approaches using more sophisticated markers have confirmed such fusions. In breast cancer, cocultivation of bone- and lung-tropic MDA-MB-231 sublines resistant to distinct antibiotics produced double-resistant hybrids with dual metastatic tropism (Lu and Kang 2009b). The evidence largely shows that fusion events promote tumor cell (TC) persistence through the acquisition of traits that confer increased survival advantage (Hass et al. 2021b; Peng et al. 2021; Noubissi et al. 2024).

In addition, bone marrow-derived cells (BMDCs), including MSCs, hematopoietic stem cells (HSCs), and macrophages, can fuse with cancer cells under in vitro conditions. Such MSC–cancer cell hybrids have been observed across a wide range of models, such as breast, lung, gastric, liver, and glioma (Rappa et al. 2012; Noubissi et al. 2015; Sun et al. 2015; Melzer et al. 2018c; Peng et al. 2021). These hybrids reportedly assume different characteristics, including undergoing EMT, gaining drug resistance, and acquiring stem-like traits. For instance, the fusion of MSCs with breast cancer cells produced hybrids with significantly enhanced migration (Noubissi et al. 2015; Melzer et al. 2018c), just as the fusion of glioma progenitors with MSCs promoted neovascularization (Song et al. 2014). Breast epithelial–cancer cell fusions were further shown to produce hybrids with increased mammosphere formation, high ALDH1 expression, and increased migration toward CCL21, a chemokine linked to lymph node metastasis (Berndt et al. 2013; Hass et al. 2021b).

In other reports, hybrid cells that resulted from the spontaneous fusion of MCF-7 cells with macrophages were found to acquire macrophage traits like CD163 and CD45, which cannot be explained by paracrine signalingalone. These hybrid cells contained genetic material from both parent cells (Lindström et al. 2017; Shabo et al. 2020). The work by Lindstrom and colleagues further demonstrated that such macrophage–cancer hybrids were more radioresistant and that they exhibited greater colony-forming capacity following γ-radiation than parental cancer cells (Lindström et al. 2017). Furthermore, hybrids between breast cancer cells and macrophages were reported to exhibit EMT and tumor stem cell properties modulated by Wnt/β-catenin activation (Chakraborty et al. 2001; Zhang et al. 2019a). Most macrophage–melanoma hybrids showed enhanced invasiveness and upregulated glycosylation-associated enzymes (Chakraborty et al. 2001).

9.1.1.2 In Vivo Evidence Supporting Cancer Cell Fusion

Compared to in vitro studies, in vivo evidence provides stronger support for the occurrence and roles of fusion in tumor behavior, cancer development, and heterogeneity. Even though it is still technically challenging to demonstrate cell–cell fusion in humans at the DNA level, findings from murine models, xenografts, and clinical cases make a compelling case that cell fusion is a recurrent event in malignancy (Lindström et al. 2017; Hass et al. 2021b). Visual confirmation of hybrid cells in xenograft tumors has been achieved using complementary lineage-tracing techniques such as Cre–LoxP recombination (Sprangers et al. 2012; Tal et al. 2019), among others. Evidence of the occurrence of cancer cell fusion in vivo is demonstrated by the co-expression of multiple major histocompatibility complex (MHC) antigens resulting from the spontaneous fusion of sarcoma cells in mice with host cells (Kerbel et al. 1983; Peng et al. 2021). Similar occurrences in melanoma models also resulted in hybrids that acquired lung metastases (Chakraborty et al. 2001). While using genetically tagged systems, Gast et al. (2018) showed that green fluorescent protein (GFP)-labeled macrophages merged with red fluorescent protein (RFP)-tagged colon cancer cells to generate highly motile, invasive, and metastatic hybrids. Furthermore, some in vivo experiments reported the fusion of MSCs with human breast cancer cells after transplantation into mice. The findings showed that there are more hybrids in metastatic lesions than primary tumors (Chitwood et al. 2018). Similarly, breast cancer–MSC hybrids (MDA-MB-231) were demonstrated to exhibit accelerated growth, shortened latency, and occasional dormancy, followed by aggressive relapse (Melzer et al. 2018a; Melzer et al. 2021). In other xenograft investigations, human lymphoma cells were found to successfully fuse with hamster stromal cells to yield hybrids with malignant B cell characteristics (Goldenberg et al. 2013). Similar fusion events were observed between human breast epithelial cells and mouse stroma as well (Jacobsen et al. 2006). Moreover, in vivo data revealed that fusion enables other hallmarks of cancer besides metastasis. Hybrids between glioma stem cells and mouse MSCs, for instance, promoted angiogenesis (Sun et al. 2019), resisted immune clearance, or adapted under selective pressures (López-Collazo and Hurtado-Navarro 2025). In a similar manner, BRAF (V600E) mutations detected in melanoma-associated stromal cells suggest that the resistance traits were transferred via fusion (Peng et al. 2021; Sieler et al.

2021). Taken together, in vivo evidence firmly establishes the occurrence of cancer cell fusion and highlights the process as a mechanism enriching tumor adaptability and malignant progression.

9.1.1.3 Clinical Evidence for Cancer Cell Fusion

Human clinical observations give some of the most direct evidence of cancer cell fusion, and they correlate well with in vitro and in vivo results. The major challenge, however, is that technological constraints still persist in detecting such occurrences under clinical settings. Macrophage–tumor hybrids that co-express macrophage markers (CD14, CD68, and CD163) and melanoma or epithelial markers have been found in patients' blood samples (Clawson et al. 2015; Hass et al. 2021b). These hybrids have been found to produce tumors when implanted into nude mice (Clawson et al. 2015; Hass et al. 2021b). Donor-derived DNA has also been found in recipients' renal, intestinal, and lung malignancies following allogeneic bone marrow or hematopoietic stem cell (HSC) transplantation (Chakraborty et al. 2004; Yilmaz et al. 2005; Pawelek and Chakraborty 2008a; Harkness et al. 2013). Another direct evidence of hybridization was found in sex-mismatched cases where Y-chromosome-positive tumor cells were detected in female patients. Short tandem repeat (STR) analysis revealed donor–recipient alleles inside individual tumor cells in melanoma and brain metastases (Lazova et al. 2013; LaBerge et al. 2017). The observations that patients who developed renal cell carcinoma (RCC) after allogeneic HSC transplantation had donor-specific alleles or Y chromosomes in female recipients are strong indicators of BMDCs fusing with preexisting tumor cells (Chakraborty et al. 2004; Yilmaz et al. 2005; Cogle et al. 2007; López-Collazo and Hurtado-Navarro 2025).

Co-expression of proteins from different biological origins has also been used as proof of fusion. In many instances, tumor cells expressing the macrophage-specific marker CD163 have been identified in breast, colorectal, renal, and bladder malignancies (Maniecki et al. 2012; Aljabery et al. 2018; Shabo et al. 2020). Multiplexed immunohistochemistry has also shown that epithelial tumor cells co-expressed lymphocytic or monocytic proteins (Ruano et al. 2022; Shultes et al. 2024). Further clinical evidence of cancer cell fusion is given by the detection of tumor-associated osteoclasts in myeloma patients. In this scenario, the multinucleated cells contained both malignant and normal nuclei, with the malignant B cell nuclei comprising over 30% of the osteoclast population (Andersen et al. 2007; Andersen et al. 2010).

Multiple studies have provided direct clinical evidence that tumor–immune cell fusion occurs in human cancers and can be reliably detected in patient blood samples across multiple malignancies (Dietz et al. 2021; Patel et al. 2024b). In one study, circulating hybrid cells (CHCs) were found to regularly outnumber conventional circulating tumor cells (CTCs) and could be accurately quantified in peripheral blood, even when CTCs were undetectable (Dietz et al. 2021; Parappilly et al. 2022; Patel et al. 2024b). According to Patel et al. (2024b), the median baseline levels of CHCs identified by the co-expression of pan-cytokeratin (CK) and CD45 in intrahepatic cholangiocarcinoma (ICC) were approximately 13.5 per 50,000 peripheral blood mononuclear cells. These numbers were found to increase with therapeutic resistance or cancer progression and decreased to zero after successful therapy. In pancreatic ductal adenocarcinoma (PDAC), high CHC levels before surgery signaled hidden metastases and rapid recurrence after surgery. Rising levels of CHCs were also found to precede the radiographic detection of metastasis (Patel et al. 2024a). Moreover, CHCs were found in all stage I–III uveal melanoma patients, whereas CTCs were detected in only 58.8% of cases (Parappilly et al. 2022). In these patients, CHC levels were observed to be good predictors of 3-year disease progression-free survival (PFS) and overall survival (Parappilly et al. 2022).

9.1.2 How Do Cancer Cells Fuse?

Cancer cell fusion typically follows fundamental principles akin to those controlling normal fusion events, including fertilization, placentation, and tissue repair (Hass et al. 2021b). The process entails four stages, namely preparation/priming, membrane approach and adhesion, membrane fusion, and post-fusion resetting. During the priming stage, cells become fusion-competent by producing adhesion molecules and modifying lipids (for example, phosphatidylserine externalization). They also respond to extracellular cues that allow them to recognize possible fusion partners. Close contact reduces hydration repulsion across membranes and results in the creation of a fusion stalk. Hemifusion follows at this stage prior to the final pore opening and merging of cytoplasmic contents (Zhang et al. 2022). At times, fusion can fail at the hemifusion stage, making the membranes become partly fused but without cytoplasmic continuity (Zawada et al. 2018). Fusogens play a crucial role in cell fusion events. These proteins are required and sufficient to overcome the energy barrier of fusion, and they operate alone or in synergy (Brukman et al. 2019). Though many of these fusogens have been investigated in model systems, cancer cells appear to use analogous pathways, with several membrane-active components still unknown (Petrany and Millay 2019). Actin cytoskeleton remodeling could also be important, as it was found that inhibiting actin polymerization significantly lowers breast cancer–MSC fusion (Melzer et al. 2019b). Furthermore, host tetraspanin proteins can prevent fusion at the hemifusion-to-pore transition, as demonstrated in HIV-1-induced cell–cell fusion (Symeonides et al. 2014). These typically regulated processes are disrupted in cancer, allowing tumor cells to merge with one another or with stromal and immune cells, resulting in hybrids with increased heterogeneity, motility, and therapeutic resistance (Bastida-Ruiz et al. 2016).

9.1.2.1 Types of Cell–Cell Fusion in Cancer

Cell–cell fusion in cancer can involve identical cancer cell types or different cell types, like cancer cells and stromal cells. These are referred to as homotypic and heterotypic fusions, respectively (Shabo et al. 2020; Hass et al. 2021a). The ability of cancer cells to fuse is assumed to be determined by a pro-fusogenic state, which can be acquired by one or both fusion partners (Hass et al. 2021b; Demin et al., 2022; Dittmar and Hass 2023). Homotypic fusions are known to generate polyploid giant cancer cells with enhanced survival and therapy resistance, just as heterotypic fusion produces hybrids with greater motility and metastatic potential. Together, these fusion events drive tumor progression and plasticity, shaping cancer development and therapy response.

9.1.2.2 Homotypic Fusion

Homotypic cell fusion mostly occurs during physiological conditions such as placentation and myogenesis (Melzer et al. 2018a; Dittmar and Hass 2023). In cancer, this occurs when two tumor cells combine to produce hybrid progeny. Hybrids from homotypic fusions are usually difficult to identify due to their similar genetic makeup with the cells from which they are formed (Shabo et al. 2020). Although such fusion can also result in new phenotypes, it is often difficult to detect because progenies frequently have genomes and exomes that are very much identical to the parental cells. They may undergo ploidy reduction, resulting in "dark matter hybrids" that are largely invisible to standard detection methods (Chakraborty et al. 2004; Yilmaz et al. 2005; Cozzo et al. 2023). Hybrid cells formed by homotypic fusion may exhibit genomic instability occasioned by chromosomal rearrangements, DNA double-strand breaks, and aneuploidy, ultimately resulting in tumor heterogeneity and evolution (Chernomordik and Kozlov 2005; Donaldson et al. 2011; Wang et al. 2021). This is often the case if the hybrid genome is favorable enough as a result of the hybrid progenies overexpressing or inhibiting certain genes. In this situation, they can easily adapt to a variety of tumor microenvironments and promote tumor development. Furthermore, epigenetic modifications caused by fusion events may modulate gene expression, thereby adding another layer of heterogeneity without overt genetic changes (Chernomordik and Kozlov 2005; Donaldson et al. 2011; Wang et al. 2021).

9.1.2.3 Heterotypic Fusion

A typical example of heterotypic fusion in the physiological state is fertilization, in which the sperm and oocyte nuclei combine to form the zygote nucleus. In the context of cancer, it happens when cancer cells interact with other cell types, such as macrophages or MSCs. In this regard, hybrids formed exhibit significantly different phenotypes (Shabo et al. 2020; Dittmar et al. 2021). Heterotypic fusion happens at a greater frequency than homotypic fusion and usually requires two separate fusogenic cell types (Hernández and Podbilewicz 2017; Dittmar et al. 2021). Heterotypic fusion of immune or stromal cells could promote tumorigenesis and metastasis. Indeed, fusion of epithelial tumor cells (TCs) with macrophages can increase migratory and invasive capacities, modulate the immune system, and potentially enable extravasation at inflammatory sites by inducing the expression of gene products that are supposed to be found only in immune cells (Sieler et al. 2021; Cozzo et al. 2023). Heterotypic hybrids are frequently identified through the co-expression of parental markers, the use of fluorescent proteins, or karyotype studies, particularly in cases of sex-mismatched cell fusions (Sieler et al. 2021; Cozzo et al. 2023). Experimental methods, such as Cre–LoxP models and in vitro cocultures, have shown direct transfer of genetic material between tumor and non-tumor cells, confirming fusion over alternative pathways such as vesicle-mediated transfer (Vassilopoulos et al. 2003; Shabo et al. 2020).

9.1.3 Molecular Drivers of Fusion Events in Cancer

Cell fusion is controlled by many intrinsic molecular drivers and is fine-tuned by external environmental factors. The intrinsic factors include fusogenic proteins such as syncytin-1, phosphatidylserine (PS) signaling, annexins, chaperones, and cyto-skeletal remodeling. The extrinsic factors, however, include inflammation, viral infection, hypoxia, and therapeutic stress (Dittmar et al. 2021; Dittmar and Hass 2023; Noubissi et al. 2024). According to Hass et al. (2021a, b), cells are not innately fusogenic, but instead, they go through a "pre-hybrid preparation program" (PHPP). During this time, a pro-fusogenic state is created through the combination of cytoskeletal remodeling, exposure of phosphatidylserine, and fusogen expression. Fusion is strictly regulated under physiological conditions, but cancer cells may exploit dysregulated pathways to acquire fusogenicity and create hybrids that contribute to tumor heterogeneity and metastatic fitness.

9.1.3.1 Fusogens and Their Receptors (Syncytins, Myomakers, and Myomergers)

Fusogens are core players in the fusion of membranes, working both in normal and in diseased conditions. These are endogenous retroviral envelope proteins, gamete recognition proteins, and muscle-specific fusion proteins. Of the well-characterized fusion proteins, syncytin-1 (HERV-W) and syncytin-2 (HERV-FRD) are the most extensively studied. Under physiological conditions, they promote trophoblast union to create the syncytiotrophoblast. This mechanism is essential for optimal placental growth and a healthy pregnancy (Dittmar and Hass 2023; Chen et al. 2025). Syncytin-1, a HERV-W-family envelope (Env) protein, interacts with the amino acid transporters ASCT1 (SLC1A4) and ASCT2 (SLC1A5), whereas syncytin-2, a HERV-FRD-family Env protein, binds to major facilitator superfamily domain containing 2A (MFSD2A) (Chen et al. 2025). Syncytin-1 activity can be inhibited by suppressyn, a HERV-derived protein that binds to ASCT2 (SLC1A5). Disruptions in the equilibrium controlled by the fusion proteins are linked to placental diseases like preeclampsia and miscarriage (Janssen and Medema 2011; Pérez-Vargas et al. 2014; Priščáková et al. 2023). Syncytin-1 is the only fusogen consistently implicated in tumor cell–cell fusion (Bjerregaard et al. 2006; Larsson et al. 2007; Strick et al. 2007; Yan et al. 2017; Dittmar and Hass 2023), and its overexpression has been reported in breast, prostate, and endometrial cancers (Bjerregaard et al. 2006; Strick et al. 2007). The precise mechanism by which syncytin-1 (HERV-W) promotes cancer cell fusion has not yet been fully unraveled. However, it is believed to function in a similar manner as the widely known retroviral envelope (Env) proteins due to their structural similarities. As the viral Env proteins, syncytin-1 surface unit (SU) domain binds to ASCT receptors (ASCT1/ASCT2), and the transmembrane domain (TM) fusion peptide inserts into the target membrane to start fusion (Grandi and Tramontano 2018; Noubissi et al. 2024). In tumor models, inhibiting syncytin-1 or its receptors decreases fusion activity (Bjerregaard et al. 2006; Noubissi et al. 2024; Chen et al. 2025). Syncytin-1 production in malignancies may be due to promoter or long terminal repeat (LTR) alterations, cytokine stimulation, or viral infections such as EBV, HIV, or SARS-CoV-2 (Dittmar et al. 2021; Dittmar and Hass 2023). Syncytin-1 is unique in that it enables cancer cells to fuse not only with each other but also with noncancerous cell types. For example, prostate cancer cells could fuse with myoblasts via syncytin-1 in a way that is not dependent on muscle fusogens like myomaker or myomerger (Uygur et al. 2019).

9.1.3.2 Phosphatidylserine (PS) Externalization

PS is a phospholipid normally sequestered in the inner plasma membrane leaflet but externalized during apoptosis and cell–cell fusion (Whitlock and Chernomordik 2021; Wang et al. 2022; Dittmar and Hass 2023). Emerging evidence suggests that PS is an important driver of cell fusion in cancer. PS exposure in the outer leaflet of the membrane is common to virtually all fusogenic cell types, such as myoblasts, macrophages, trophoblasts, sperm, and cancer cells (Brukman et al. 2019; Whitlock and Chernomordik 2021). Exposure of the phospholipid is regulated by Ca^{2+}-activated scramblases like TMEM16E and TMEM16F, which are very important for trophoblast and myoblast fusion (Nagata et al. 2016; Zhang et al. 2020; Dittmar et al. 2021; Dittmar and Hass 2023). TMEM16F-mediated PS externalization is essential for trophoblast syncytialization under physiological conditions, as mutant animals have been found to show placental failure and perinatal lethality (Zhang et al. 2020). Analogously, TMEM16E deficiency could impair myoblast fusion and muscle growth even though there are strain-dependent variations (Gyobu et al. 2016; Whitlock et al. 2018). In addition, PS exposure on sperm and recognition receptors on oocytes promotes gamete fusion during fertilization, and there is decreased fertility if this is disrupted (Jeong and Conboy 2011; Žigon et al. 2015; Rival et al. 2019; Hass et al. 2021b). Pro-inflammatory cytokines (Interleukin) IL-4 and IL-13 promoted PS externalization and annexin A5 expression in prostate cancer cells. This results in syncytin-1-mediated fusion that can be reversed by silencing annexin A5 (Uygur et al. 2019; Dittmar and Hass 2023). Furthermore, apoptosis appears to operate as a pro-fusogenic signal because MSCs fused more easily with breast cancer cells when apoptosis was activated and produced hybrids with enhanced migratory capacity (Noubissi et al. 2015; Dittmar and Hass 2023). PS exposure is sensed by receptors such as Tim4, Bai1/3, and CD36, which activate ELMO/Dock180/Rac1 signaling and induce actin remodeling needed for fusion (Hochreiter-Hufford et al. 2013; Whitlock and Chernomordik 2021; Noubissi et al. 2024). Thus, PS functions as a conserved "eat-me/fuse-me" signal that tumors use to promote hybridization with stromal partners during conditions of stress.

9.1.3.3 Annexins and Glucose-Regulated Protein 78 (GRP78)

Annexins are PS-binding proteins that are well-known to play vital roles in trophoblast, osteoclast, and myoblast fusion (Whitlock and Chernomordik 2021). In prostate cancer, annexin A5 enhances syncytin-1-mediated fusion, and its silencing reduces hybrid formation (Bjerregaard et al. 2006; Uygur et al. 2019). Annexins might, therefore, act as molecular bridges linking PS exposure to fusogen activation in tumor settings. Glucose-regulated protein 78 (GRP78) is aberrantly expressed in cancer cells and promotes a pro-fusogenic phenotype by activating survival pathways that encourage cell fusion (Misra et al. 2006; Noubissi et al. 2024). The endoplasmic reticulum chaperone also promotes fusion by increasing syncytin levels in trophoblasts (Bastida-Ruiz et al. 2020). In prostate cancer cells, GRP78 binding to α2-macroglobulin activates pro-survival and antiapoptotic pathways that involve ERK1/2, p38 MAPK, PI3K, Akt, and NF-κB (Misra et al. 2006; Ni et al. 2011). According to Bastida-Ruiz et al. (2020), the interaction between α2M and GRP78 in cancer cells may trigger ERK1/2 signaling to induce cell fusion.

9.1.3.4 Actin Cytoskeleton Remodeling

Membrane merger requires not only fusogens and lipids but also structural remodeling of the actin cytoskeleton. The mechanical forces and surface protrusions necessary for cell–cell fusion is provided by actin polymerization (Petrany and Millay 2019). Small GTPases such as Rac1 and Cdc42, together with the Arp2/3–WASP complex, orchestrate actin filament nucleation and protrusion formation, enabling a dynamic cytoskeletal landscape for fusion (Shen et al. 2008; Gruenbaum-Cohen et al. 2012; Zhang et al. 2017).

In myoblasts, podosome-like F-actin structures formed via Arp2/3-mediated polymerization are essential for membrane merger (Abmayr and Pavlath 2012).

Also, during osteoclast fusion, actin filaments organize into thick bundles at contact zones and recruit non-muscle myosin IIA to generate protrusive forces that drive fusion. These actin-rich domains are localized by Rac1 (Takito et al. 2017). Evidence suggests that the mechanisms involved in cancer are similar. Rac1, frequently upregulated in aggressive cancers, couples PS receptor signaling to actin rearrangement at fusion sites, reinforcing the PS–actin axis as a central cancer fusion driver (Noubissi et al. 2024). Furthermore, fusion of breast cancer cells with MSCs requires actin polymerization. In fact, inhibition of actin regulators (i.e., Arp2/3 and formins) using sublethal concentrations of cytochalasin D (blocking filament elongation) and latrunculin B (inhibiting polymerization) markedly reduces cancer cell fusion without affecting the expression of adhesion molecules (Melzer et al. 2019b; Hass 2020).

9.1.3.5 Extracellular Modulators

Extracellular factors function as critical enhancers for fusion in the tumor microenvironment, while the cells are prepared for the fusion event by intrinsic fusogens and signaling proteins. These extracellular modulators involve inflammation, cytokines, hypoxia, viruses, and therapeutic stress.

Inflammation and Cytokines

Inflammation is now being recognized as a potent driver of cancer cell fusion. A typical instance is the induction of fusion events between breast cancer cells MDA-MD-435 and breast epithelial cells M13SV1 by the pro-inflammatory cytokine, TNF-α (Mohr et al. 2015a; Weiler and Dittmar 2019b). The cytokine supports fusion by activating syncytin-1 and ASCT receptors and signaling via the VCAM-1/VLA-1 pathway (Bjerregaard et al. 2006; Song et al. 2012). Increased VCAM-1 expression in the inflammatory tumor microenvironment is known to promote the growth and progression of tumors (Yan et al. 2017; Weiler and Dittmar 2019b). TNF-α also stimulates the NF-κB and Wnt/β-catenin pathways, both of which are known to induce syncytin-1 production and EMT, thereby promoting fusion-driven aggressiveness (reviewed in Noubissi et al. 2024).

Viruses

Viruses are both direct fusogens and indirect inducers of cellular fusion. Certain viruses that infect human cells have been implicated in cancer cell fusion, and they have been found to contribute to both tumor development and progression (Duelli et al. 2005; Duelli and Lazebnik 2007). By promoting fusion, these viruses may induce chromosomal instability and aberrant gene expression. As a result, they may contribute to oncogenesis and facilitate metastasis (Duelli et al. 2005; Duelli and

Lazebnik 2007). Although the exact mechanisms responsible for virus-mediated cell fusion are unclear, two major hypotheses have been presented. These include the activation of viral fusogens during infection and the virus functioning as a physical bridge between cells to stimulate their fusion (reviewed in Noubissi et al. 2024). Furthermore, viral infections trigger endogenous retroviral fusogens; for example, HERV-K is increased in breast, pancreatic, liver, and melanoma malignancies, where its Env protein promotes multinucleation and genomic instability (Zhou et al. 2016; Ko et al. 2021; Dittmar and Hass 2023). Inhibiting HERV-K diminishes tumorigenic traits, pointing to its potential as a therapeutic target (Zhou et al. 2016; Shah et al. 2023; Lin et al. 2025).

Hypoxia and Therapeutic Stress (Chemotherapy, Radiotherapy, etc.)

Hypoxia is a hallmark of rapidly growing solid tumors. The condition creates a stressful microenvironment that can promote the fusion of cancer cells with stromal cells recruited to the tumor niche. Findings from non-cancer models indicate that hypoxia promotes the fusion of several eukaryotic cell types. Notable instances include hypoxia-induced fusions of bone marrow-derived MSCs and cardiomyocytes (Haneef et al. 2014), MSCs and myoblasts (Archacka et al. 2021), and urine-derived stem cells and liver cells (Hu et al. 2021). Likewise, in cancer, hypoxic environments promote fusion events, such as between MSCs and breast cancer cells (T47D and MCF7) (Noubissi et al. 2015; Noubissi and Ogle 2016; Noubissi et al. 2024) and between oral squamous carcinoma cells and human immortalized oral epithelial cells (Huang et al. 2018). When combined with inflammatory signals such as TNF-α, hypoxia significantly promotes fusion in breast cancer cells (MDA-MB-435 and MDA-MB-231) (Mohr et al. 2015a). Hypoxia-driven fusion may involve a number of processes, including the secretion of growth factors such as VEGF, upregulation of chemokine receptors such as CXCR4, activation of apoptotic signaling pathways, and induction of EMT (Noubissi et al. 2015; Huang et al. 2018; Archacka et al. 2021; Hu et al. 2021).

Chemotherapy and radiotherapy are stressors that can induce cancer cell fusion, leading to the formation of polyploid giant cancer cells (PGCCs). PGCCs are large, atypical cells with multiple DNA copies that contribute to tumor heterogeneity (Zhou et al. 2022; Noubissi et al. 2024). Even though PGCCs may arise from incomplete mitosis, studies in ovarian and colon cancer cells have revealed that cell fusion is a significant mechanism driving their formation (Zhang et al. 2014a; Fei et al. 2019). Treatments such as paclitaxel, capecitabine, oxaliplatin, and irradiation increase PGCC prevalence, particularly in metastases, and this is associated with chemoresistance, recurrence, and aggressive mesenchymal phenotypes (Song et al. 2021). These findings suggest that certain conventional cancer therapies aimed at tumor cells may paradoxically promote fusion-mediated heterogeneity and tumor advancement.

9.2 Why Do Cancer Cells Fuse?

The scientific explanation for cancer cell fusion can be understood by considering the selection advantages it offers. Cancer cell fusion promotes tumor evolution by increasing genetic diversity, improving invasion and treatment resistance, providing stem-like and immune-evasive features, and allowing for adaptation to adverse microenvironments. These possibilities reveal why cancer cells fuse: The process can afford a tremendous evolutionary shortcut for tumors to acquire several cancer-related characteristics in a single biological event.

Fusion-mediated genetic recombination can yield malignant transformation. Immortalized fibroblasts expressing various oncogenic drivers generated hybrids with numerous oncogenes, aneuploidy, and tumorigenicity (Hass et al. 2021b; Peng et al. 2021). There are records of sarcoma-derived hybrids with unique metastatic features and pleomorphic histology that are not found in the parental cells (Delespaul et al. 2020; Lartigue et al. 2020). It was established in prostate cancer cocultures that stromal–cancer cell hybrids exhibited genomic hybridization and divergent fate changes (Wang et al. 2020).

9.2.1 *Enhancing Genetic Diversity and Tumor Plasticity*

Cancer cell fusion usually brings about tumor heterogeneity by promoting genetic instability and phenotypic variation. Fusion frequently produces tetraploid intermediates, which may progress into aneuploid or polyploid neoplastic cells with aberrant chromosome numbers. This is consistent with models proposing tetraploid precursors in carcinogenesis (Margolis 2005; Kroemer 2010; Melzer et al. 2020). Such chromosomal instability serves as a substrate for clonal convergence and diversification, resulting in surviving hybrids that quickly outgrow parental cancer cells (Dittmar and Hass 2023).

According to the heterokaryon-to-synkaryon transition hypothesis, fused cells begin as multinucleated heterokaryons and then undergo chromosomal rearrangements to create genetically distinct synkaryons (Bjerkvig et al. 2005; Mohr et al. 2015b). Such rearrangements could produce instability as a result of DNA double-strand breaks and translocations (Delespaul et al. 2019; Wang et al. 2021). The resulting progenies may overexpress or lose specific genes to enhance their tumor adaptability to various microenvironments (Goldenberg et al. 2014). Cell fusion not only promotes genetic variety through chromosomal rearrangements and aneuploidy but can also increase tumor plasticity via epigenetic reprogramming and the acquisition of stem cell-like features (Wang et al. 2021). Some of the potential consequences are tumor progression, resistance to therapy, and poor patient outcomes (Duncan et al. 2012; Skinner et al. 2012; Zhou et al. 2015; Delespaul et al. 2019; Delespaul et al. 2020; Søe 2020).

It is, however, important to note that tumors formed by fused cells can only advance more rapidly if the resulting genomic architecture gives survival benefits. In certain scenarios, MSCs regularly fuse with breast cancer cells to form tumor hybrid cells (THCs) with enhanced migration, invasion, and tumorigenicity and infrequent reductions in proliferation (Ding et al. 2012; Melzer et al. 2020; Tretyakova et al. 2022). Intratumoral diversity or heterogeneity from cell fusion may lead to the failure of therapy and poor disease prognosis. This is because some hybrid subclones adapt better to chemotherapy or immune pressures than their parental cells. A critical step in this event is the post-hybrid selection process (PHSP), during which time many hybrid cells undergo apoptosis, necroptosis, or dormancy. It is during this time that survivors acquire malignant traits such as cancer stem/initiating cell (CS/IC) properties, altered therapeutic responsiveness, and enhanced metastatic behavior (Passerini et al. 2016; Chunduri and Storchová 2019; Melzer et al. 2020; Hass et al. 2021b).

9.2.2 *Facilitating Metastasis*

The multistep process of metastasis enables tumor cells to spread from their original location and begin growing in other organs. A key event in metastasis is the epithelial-to-mesenchymal transition, during which time tumor cells lose epithelial adhesion molecules, adopt an interstitial phenotype, and gain motility (Kalluri and Neilson 2003; Pawelek and Chakraborty 2008b). Cell–cell fusion has now been recognized as a mechanism that accelerates these transitions and provides cancer cells with additional metastatic fitness. Fusion between tumor cells and stromal partners such as MSCs, fibroblasts, and BMDCs generates hybrids with increased heterogeneity, adaptability, and metastatic ability (Tretyakova et al. 2022). In multiple studies, spontaneous fusion of MSCs with MDA-MB-231 breast cancer cells resulted in aneuploid populations displaying more tumorigenicity in vivo and enriched in mesenchymal markers like FN1, SNAI2, and MMP9 (Melzer et al. 2018a; Melzer et al. 2021). MSC fusion with MCF-7 and T47D breast cancer cells produced similar results, with hybrids exhibiting increased migration (George et al. 2017; Chitwood et al. 2018). The observation in murine models that lung metastases contained elevated levels of MSC–tumor hybrids compared to primary tumors is an indication that fusion plays a direct role in the colonization of secondary sites (Chitwood et al. 2018).

BMDCs, particularly macrophages, could also serve as essential fusion partners with cancer cells. Spontaneous fusion of macrophages with tumor cells was shown to occur in both mice (Chakraborty et al. 2000) and humans (Yilmaz et al. 2005). These hybrids exhibited genes related to invasion and metastasis, including SPARC, MCR1, and MET (Pawelek and Chakraborty 2008b). They also retained macrophage characteristics like phagocytosis (Seyfried and Huysentruyt 2013). These two properties could explain why aggressive cancer frequently emulates BMDC activity. Indeed, macrophage–tumor fusion cells can be detected in patients' blood, and their abundance correlates with stage and prognosis (Gast et al. 2018). The "BMDC fusion hypothesis" elucidates organotropism, indicating that metastatic colonization frequently occurs in the liver, lungs, and bone regions abundant in BMDCs that create a favorable milieu for hybrids (Wang et al. 2021). In another study, fibroblast fusion with precancerous cells produced unstable clones exhibiting increased metastatic characteristics, whereas sarcoma cell hybrids formed undifferentiated pleomorphic sarcoma-like tumors in vivo (Lartigue et al. 2020). Endothelial cells and tumor-associated fibroblasts merged with cancer cells to promote invasion and proliferation (Kalluri 2016; Choi and Moon 2018). In prostate cancer cocultures, mesenchymal–tumor hybrids sustained growth and migration capacity (Wang et al. 2012). Furthermore, fusion promotes EMT and stemness, which are two traits central to metastasis (Celià-Terrassa and Jolly 2020). Macrophage fusion with MCF-7 cells was shown to produce hybrids with upregulated EMT markers (Snail1, Snail2, Vimentin) and downregulated E-cadherin. This is in addition to the expansion of stem-like CD44+/CD24− populations and higher tumorigenicity (Ding et al. 2012). Similar observations have been reported in lung, gastric, and prostate cancers as well (Xu et al. 2014; Tretyakova et al. 2022). In breast cancer, the expression of the macrophage marker CD163 was correlated with poor patient survival (Shabo et al. 2015). Conclusively, these findings show that hybrid cells provide a viable opportunity for metastasis and organ-specific colonization by combining tumor aggressiveness with stromal cell migratory traits (Pawelek 2000; Gast et al. 2018; Sieler et al. 2021).

9.2.3 Promotion of Drug Resistance

It is established that fusion enables genetic and phenotypic reprogramming that might support the production of heterogeneous THCs with multidrug resistance traits (Tretyakova et al. 2022). In metastatic colon carcinoma models for instance, fused cells were resistant to both 5-fluorouracil (5-FU) and oxaliplatin, which are standard-of-care drugs for cancer (Carloni et al. 2013). This is closely related to other studies showing that spontaneous fusion between the 168FAR and 44FTO subpopulations produced hybrids resistant to melphalan and methotrexate (Miller et al. 1989; Tretyakova et al. 2022). Cell fusion allows for intercellular gene exchange, which can increase cancer cells' rapid adaptability within the tumor microenvironment. Resistant subpopulations can prevail by fusion because multidrug resistance is conferred even on originally susceptible cells (Vasan et al. 2019). Nonresistant tumor cells, for example, can develop resistance by fusing with BMDCs (Nagler et al. 2011) or with local stromal elements such as skeletal or smooth muscle cells. This was demonstrated in prostate cancer, where hybrids exhibited doxorubicin and cisplatin resistance as well as stem cell-like properties (Uygur et al. 2019). Analogously, hybridization of oral cancer cells (SCC9) and endothelial cells (HUVECs) increased chemoresistance while retaining parental phenotypes (Song et al. 2014). In addition, melanoma hybrids formed with fibroblasts or macrophages showed functional gene exchange that reinforced resistance (Searles et al. 2018).Drug resistance in cancer hybrids is not confined to chemotherapy. In certain studies, radiation-induced homotypic fusion in glioblastoma cells resulted in multinucleated giant cells (MNGCs) with enhanced survival and DNA repair capabilities (Kaur et al. 2015; Lindström et al. 2017). Hybrids resulting from the fusion of M2 macrophages and MCF-7 breast cancer cells were also more radioresistant than parental lines, as they tend to grow colonies more efficiently and with less DNA damage (Lindström et al. 2017). It has also been demonstrated that therapeutic interventions can cause resistance-promoting fusions, as evidenced by doxorubicin-resistant hybrids resulting from doxorubicin-induced fusion in MCF-7 cells (Yang et al. 2010).

Similarly, senescent multinucleated giant cells formed by radiation-induced fusions exhibited pro-survival signals and resistance markers (Kaur et al. 2015). Importantly, cell fusion not only introduces medication resistance but also promotes tumor plasticity. Surviving hybrids frequently undergo a PHSP, where most fused cells die, but selected clones acquire cancer stem/initiating cell (CS/IC) traits that heighten chemoresistance (Yang et al. 2010; Nagler et al. 2011; Searles et al. 2018; Hass et al. 2021a; Liu et al. 2021). Hybrids of squamous cell carcinoma with MSCs, for example, were markedly resistant to paclitaxel (Liu et al. 2021), just as breast cancer–MSC hybrids displayed enhanced proliferation despite reduced metastatic potential (Melzer et al. 2019a). This variability reflects the heterogeneous nature of THCs; that is, while some exhibit reduced aggressiveness, others display heightened drug tolerance and tumor-initiating capacity (Fahlbusch et al. 2020).

Metabolic reprogramming also plays a role in cell fusion-dependent drug resistance. Fusion-driven cytoplasmic mixing can worsen mitochondrial dysfunction—already a common feature of malignancies—by shifting cells toward glycolysis under hypoxic conditions. This Warburg-like phenotype is associated with increased medication resistance that is reversible by inhibiting glycolysis (Xu et al. 2005; Seyfried et al. 2020). These findings demonstrate that cancer cell fusion is a central mechanism promoting multidrug resistance across cancer types. This is made possible by the chromosomal rearrangements, acquisition of CS/IC traits, enhanced DNA repair, and metabolic rewiring that accompany cell fusion.

9.2.4 Acquisition of Stem Cell-Like Properties

The acquisition of stem cell-like properties is also a crucial mechanism by which cancer cell fusion contributes to tumor progression, therapeutic resistance, and recurrence. Aichel's century-old hypothesis that fusion between leukocytes and cancer cells drives metastasis (Warner 1975) has gained renewed support through modern evidence showing that fusion with macrophages, MSCs, or other stromal cells can generate THCs with cancer stem cell (CSC)-like traits. Earlier studies showed that leukocyte–cancer cell hybrids retaining donor–recipient DNA were detected in bone marrow transplant (BMT) patients (Chakraborty et al. 2004;

Yilmaz et al. 2005; LaBerge et al. 2017). Similarly, fusion of M2 macrophages with breast cancer cells was demonstrated to produce CD44+/CD24− stem-like hybrids having EMT markers like Snail1, Snail2, and vimentin and decreased E-cadherin levels (Ding et al. 2012; Celià-Terrassa and Jolly 2020). The stem-like THCs were found to have strong migratory and invasion capacities but a lower ability for proliferation. This indicates the role of fusion in promoting intratumoral heterogeneity and enriching cancer stem cell subpopulations (Tretyakova et al. 2022).

There have been many reports of CSC-like properties in fusion-derived cells across cancer types. In lung cancer, spontaneous fusion with MSCs resulted in progenies with decreased proliferation but increased stemness (Wei et al. 2014). Hybrids of breast epithelial cells and cancer cells exhibited common stem cell-like characteristics such as self-renewal and resistance to apoptosis (Dittmar et al. 2011; Kreso and Dick 2014; Ye et al. 2015). Similar hybrid populations were produced when MSCs were fused with lung or ovarian cancer cells. They resulted in hybrid cells with altered tumorigenic and metastatic properties (Xu et al. 2014; Gauck et al. 2017; Melzer et al. 2020). Molecular characterization of these hybrids strongly supported their CSC phenotype. Additionally, fusion-derived lung cancer/MSC hybrids have been demonstrated to display increased CD44 and CD133, as well as pluripotency markers Oct4, Nanog, Sox2, Bmi1, and Kif4 (Zhang et al. 2019b). Hybrids of non-small cell lung cancer cells and MSCs were found to show an EMT phenotype with downregulated E-cadherin and upregulated vimentin, α-smooth muscle actin, and fibronectin. This could be taken as an indication of a partial EMT state associated with stemness (Xu et al. 2014; Grosse-Wilde et al. 2015; Ye et al. 2015; Hass et al. 2021a). Similarly, breast epithelial–breast cancer hybrids were shown to display EMT transcription factors Snail and Zeb1, which stabilize combined epithelial/mesenchymal phenotypes (Jolly et al. 2015; Hass et al. 2021a).

The stem-like behavior of hybrids also provides functional advantages in terms of therapy resistance and relapse. CSC-like hybrids frequently enter dormancy or "hibernation"-like states in hypoxic and nutrient-poor environments, endowing them with long-term survival and contributing to recurrence (Bartosh et al. 2016). For example, when breast cancer cells fused with MSCs, they entered a metabolically suppressed state similar to quiescence, which caused them to avoid chemotherapeutic harm (Bartosh et al. 2016). Similarly, prostate cancer–muscle cell hybrids were found to exhibit enhanced stemness, as shown by upregulated CD133 expression (Uygur et al. 2019). Hybrid cell lines like MDA-hyb5 are known to exhibit continuous proliferation, self-renewal, and apoptotic resistance in a manner characteristic of CSCs (Wei et al. 2014; Melzer et al. 2018b). In a nutshell, these findings suggest that cancer cell fusion could be vital for the acquisition of CSC features in many malignancies (Beck and Blanpain 2013; Ye et al. 2015; Gauck et al. 2017).

9.2.5 Evasion of Immune Surveillance

Evasion of immune surveillance is one of the key features required for cancer progression. Cancer cells must learn to survive in the circulation, where they are continuously exposed to immune surveillance, before they can disseminate and successfully colonize distant sites (López-Collazo and Hurtado-Navarro 2025). Fusion events have been found to confer immune-modulating features that suppress cytotoxic responses and promote metastatic spread of tumor hybrids. Lung cancer cell–monocyte hybrids could suppress the cytotoxicity of natural killer cells and perforin synthesis by upregulating inhibitory HLA class I molecules such as HLA-B and HLA-E (Aguirre et al. 2020). These hybrids have also been demonstrated to be capable of impairing CD4+ and CD8+ T cell proliferation and inducing the expression of immune checkpoints PD-1, CTLA-4, and SIGLEC5 (Aguirre et al. 2020; Montalbán-Hernández et al. 2022). The fact that inhibiting SIGLEC5 can partially restore CD4+ proliferation suggests that hybrids could be involved in immunosuppressive processes (Montalbán-Hernández et al. 2022). In support, glioblastoma and macrophage hybrids were shown to inhibit CD8+ T cell responses by increasing YKL-40 and B7-2 (CD86) expression (Shao et al. 2022). In addition, in vitro-derived hybrids were demonstrated to express high PD-L1 and limit the activity of CD8+ via PD-L1/PD-1 signaling (Aguirre et al. 2020). Furthermore, the fusogenic proteins (syncytin-1 and syncytin-2) involved in cell fusion were found to possess intrinsic immunosuppressive properties that support hybrid-mediated immune escape (Dervan et al. 2021; Gao et al. 2021). Multicellular dynamics can also be used to evade immune responses. CTCs are known to frequently move in clusters, sometimes alongside tumor-associated macrophages (TAMs) or cancer-associated macrophage-like cells (CAMLs). These clusters have increased metastatic seeding efficiency and decreased immune recognition, which are thought to be aided by the downregulation of MHC class II antigen presentation in hybrids (Cheung et al. 2016; Nasr and Lynch 2023; López-Collazo and Hurtado-Navarro 2025). This cooperative action confers anoikis resistance and increased survival in circulation, thus transforming hybrids into metastatic "seeds." Interestingly, not all fusion events eventually result in immunological suppression. Experimental fusion of cancer cells with dendritic cells could produce hybrid vaccines that are capable of activating CD4+ and CD8+ T cells and inducing robust anticancer responses (Koido et al. 2014a; Koido et al. 2014b; Zhang et al. 2014b). This notwithstanding, immunosuppression remains a common outcome of cell fusion in the tumor microenvironment. Here, hybrids exploit checkpoint signaling, altered antigen presentation, and immune-evasive clustering to avoid destruction and accelerate metastatic development.

9.2.6 Tumor Microenvironment Adaptation

The dynamic tumor microenvironment (TME) contributes both structurally and biochemically to regulate many aspects of cancer growth and progression (Weber and Kuo 2012; Bolouri 2015). It comprises stromal fibroblasts, immunological cells, vasculature, extracellular matrix, cytokines, and soluble mediators. Adverse factors in this environment, such as hypoxia, food deprivation, low pH, and chronic inflammation, encourage abnormal fusion of tumor cells with stromal or immune cells (Davies et al. 2009; Böhme and Bosserhoff 2016; Anastasiou 2017; Jiang et al. 2019). The TME resembles chronic inflammatory tissue, with persistent cytokine signaling, immune cell infiltration, tissue remodeling, angiogenesis, and oxidative stress (Dittmar and Hass 2023). These ongoing inflammatory processes promote genomic instability, immunological suppression, and tumor progression; hence, chronic inflammation is frequently cited as a primary driver of cancer development (Jiang et al. 2019; Singh et al. 2019). The resulting tumor milieu can promote cell–cell fusion by attracting macrophages and MSC (Davies et al. 2009; Dittmar and Hass 2023). The hybrids so formed can also show genetic instability and heterogeneity that may foster rapid evolutionary selection (Jiang et al. 2019). They might gain motility through macrophage-associated genes, become stem-like through MSC fusion, and develop drug- or radioresistance by acquiring resistant phenotypes.

It is plausible that fusion confers cancer cells with adaptability traits that extend beyond what gradual mutations alone can achieve (Manjunath et al. 2020; López-Collazo and Hurtado-Navarro 2025). Since fusion also supports immune evasion and metabolic reprogramming, particularly glycolytic shifts that favor survival under hypoxia, it allows malignant cells to remodel their phenotype and thrive within the hostile TME. This event further promotes tumor progression, therapeutic resistance, and metastasis (Lu and Kang 2009a).

9.3 Clinical Implications of Cell Fusion

The presence of hybrid cells in the blood could be used as a potential biomarker for tumor staging and outcome prediction, just as their frequency can also be correlated with disease aggressiveness and resistance. It is also plausible to harness the process of cell fusion as a complementary measure to improve treatment on the therapeutic front. This might involve modulating fusion machinery, addressing its role in radiation resistance, and leveraging fusion-induced immunogenicity. It is imperative that efforts are channeled toward precise and context-specific therapeutics since the process of fusion also drives malignant progression. A better molecular understanding of the process and the players involved is crucial to translating these findings into successful clinical applications.

9.3.1 Diagnostic and Prognostic Relevance

9.3.1.1 Detection in Patient Samples

The detection of tumor–normal (TN) hybrid cells in cancer patients could offer valuable insights into disease stage and progression (Anderson et al. 2023; Tanjak et al. 2024).

Studies by Gast et al. (2018) identified macrophage–tumor cell hybrids in the peripheral blood of mice and demonstrated that their frequency correlated with tumor stage and overall survival. Such putative hybrids have been isolated and tracked in a few studies using fluorescent proteins, drug resistance markers, and sex-mismatched transplantation (Suetsugu et al. 2016; Weiler and Dittmar 2019a). However, detecting these hybrid cells in humans is more complex, and researchers are making use of markers such as CD45, CD14, CD163 (hematopoietic/macrophage), and tumor-specific markers like EpCAM or cytokeratins to carry out this (Clawson et al. 2012; Ramakrishnan et al. 2013; Shabo et al. 2015; Weiler and Dittmar 2019a). Recent clinical studies in PDAC, ICC, uveal melanoma, and other epithelial and non-epithelial cancers suggest that circulating hybrid cells (CHCs) have the potential to serve as extremely sensitive, minimally invasive biomarkers for the detection and assessment of a wide range of tumor types (Dietz et al. 2021; Parappilly et al. 2022; Patel et al. 2024a; Patel et al. 2024b). In PDAC, CHCs were consistently found to be more abundant than conventional CTCs, and they could be reliably detected even when CTCs were absent (Patel et al. 2024b). Because CHCs exhibit tumor and immune cell markers, tumor-derived genetic features, and stem-like properties, they can mirror the molecular and phenotypic heterogeneity of parental tumors and can be useful as liquid surrogates for tissue biopsies (Dietz et al. 2021). Another advantage is that CHCs can be assessed with simple blood draws, thus enabling real-time surveillance and early tumor diagnosis. CHCs could also provide the benefit of detecting occult disease or metastatic risk long before conventional imaging or biomarkers could indicate progression (Dietz et al. 2021; Parappilly et al. 2022; Patel et al. 2024a).

Sex-mismatched bone marrow transplant (BMT) studies provide strong evidence of cancer cell fusion because they allow the detection of the Y chromosome in female recipients (Chakraborty et al. 2004; Yilmaz et al. 2005; Gast et al. 2018). However, the greatest evidence for the occurrence of cell fusion events in human cancers is provided by STR analysis (Weiler and Dittmar 2019a). This method makes it possible to unequivocally identify TN-hybrid cells through parallel determination of parental alleles located on different chromosomes (Yilmaz et al. 2005; Lazova et al. 2013).

Even though considerable progress has been made in the detection of cancer fusion hybrids, there are still some limitations. For example, fetal cell microchimerism can mimic fusion events. The scarcity of BMT cases, loss of markers over time, and tumor genetic instability complicate detection using STR analysis, even though it remains the most reliable method (Weiler and Dittmar

2019a). Moreover, some hybrids, termed "dark matter hybrids," may be undetectable due to the loss of specific markers or genetic changes. This further makes the identification of these potentially crucial cells in tumor progression very challenging (Scarpa and Mafficini 2018; Weiler and Dittmar 2019a).

9.3.1.2 Prognostic Significance

The prognostic value of cancer cell hybrids, particularly CHCs and hybrid epithelial/mesenchymal (E/M) cells, holds for multiple cancers (Godin et al. 2020; Dietz et al. 2021; Parappilly et al. 2022; Patel et al. 2024a; Patel et al. 2024b). The frequency of hybrid cells was shown to be closely associated with tumor grade and prognosis; hence, monitoring their presence could help stratify patients based on disease severity (Gast et al. 2018). For instance, CD163-positive tumor cells, arising from radiation-induced macrophage–tumor cell fusion, have been detected in breast cancer patients undergoing radiotherapy and are frequently associated with poor prognosis (Garvin et al. 2018). In breast cancer cohorts also, 48% of patients had CD163-positive cancer cells, and such prevalence was associated with lower recurrence-free survival and progression of disease (Shabo et al. 2008). CD163 expression, in addition, was associated with earlier local recurrence, shorter survival, and decreased apoptosis in rectal cancer. This effect was exacerbated in individuals undergoing preoperative radiation (Aljabery et al. 2018; Shabo et al. 2020).

In ICC and PDAC, CHC counts were shown to be associated with therapeutic response and disease progression. In these malignancies, rising levels of CHCs often precede radiographic evidence of recurrence or metastasis, while reductions are associated with treatment response (Patel et al. 2024a; Patel et al. 2024b). In uveal melanoma, the quantity of CTCs accurately predicted 3-year progression-free survival (PFS) and overall survival, while traditional CTCs failed to show a similar predictive value (Parappilly et al. 2022). In addition, tissue-based hybrid E/M cells quantified by using sequential immunohistochemistry in urothelial carcinoma were found to be strongly associated with aggressive clinicopathological features, disease severity, and poor outcomes (Godin et al. 2020).

The consistency of these findings with the identification of donor-derived chromosomal markers in tumor cells from transplant patients reveals that fusion is indeed a clinically significant aspect of tumor progression (López-Collazo and Hurtado-Navarro 2025). Beyond prognosis, fusion events have an impact on tumor biology. It was reported that macrophage–tumor hybrids in the breast could actually reduce estrogen receptor (ER) expression and limit response to ER-targeted therapy (Ding et al. 2012). Fusion-mediated plasticity allows cancers to acquire new qualities such as increased survival, invasiveness, and stem-like properties while losing lineage-specific markers via dedifferentiation (Shabo et al. 2015; Brukman et al. 2021). Based on the numerous clinical findings on cancer cell fusion events and their outcomes, it is expedient to probe further into the mechanisms of cancer cell fusion with a view to developing new types of drugs providing better treatment options and preventing the formation of aggressive hybrid clones (Rajah et al. 2022; López-Collazo and Hurtado-Navarro 2025).

9.3.2 *Therapeutic Implications*

9.3.2.1 Targeting Fusogens and Fusion Machinery

Many have proposed that blocking fusion could present an attractive therapeutic approach because of the role of cell fusion in enhancing cancer cell heterogeneity, stemness, drug resistance, and EMT. Fusogens are important components of the fusion machinery, and targeting them could provide novel avenues for therapy. However, these fusion mediators are structurally and functionally complex. Their loss of function is linked to varied conditions such as infertility, muscular dystrophies, and neurocutaneous syndromes (Segev et al. 2018). Despite these challenges, clarification of fusogen biology is crucial for the design of reliable and selective fusion inhibitors (Fédry et al. 2017). The report by Li et al. (2013) reveals that the progression of rhabdomyoblastoma in vivo could be successfully blocked by inhibiting IL-4 receptor-mediated myoblast fusion. Inhibition of cell fusion in colon cancer models has also shown favorable results (Zhou et al. 2015). Nonetheless, not all fusion events are pathological, and in some cases, hybrids may acquire more benign phenotypes (Platt and Cascalho 2019). To avoid interfering with physiological fusion processes, therapeutic efforts must be highly targeted. Nikanjam et al. (2020) found that greater response rates and longer progression-free survival (PFS) are achieved if oncogenic fusion genes are targeted in cancer treatment. In a study of 6189 patients, those with fusion-positive tumors had a 68% response rate compared with 50% for non-fusion tumors. In fact, patients with fusion targets recorded longer PFS (11.6 months) than those without them (4.9 months). However, fusions with additional tumor alterations were not found to respond to treatment (Nikanjam et al. 2020). This implies that fusion events may constitute significant therapeutic targets. Nonetheless, further studies to refine treatment strategies, particularly in the presence of additional alterations, may be needed.

9.3.2.2 Radiotherapy Considerations

Radiotherapy remains a cornerstone of cancer treatment, but radiation itself is a potent inducer of cell fusion. Rizvi et al. (2006) showed that γ-ray exposure prompted the fusion of small intestinal stem cells with BMDCs at rates directly proportional to the radiation dose (Davies et al. 2009). The detection of CD163-positive tumor cells in breast cancer patients receiving breast-conserving surgery and radiotherapy are indications of macrophage–tumor cell fusion. The hybrids showed strong radioresistance and poorer clinical outcomes (Garvin et al. 2018; Wang et al. 2021). The efficacy of treatment is further complicated by experimental findings that macrophage–tumor hybrids (e.g., MCF-7 cells fused with macrophages) exhibit enhanced DNA repair and survival under radiation stress (Lindström et al. 2017). In addition, Yeh et al. (2016) reported that radiation-induced fusion between macrophages and intestinal stromal cells contributed to chronic fibrosis. This suggests the potential long-term effects of radiotherapy. The findings expose the dual role of radiation as both a therapeutic tool and a driver of tumor-promoting fusion events. It is thus important to integrate fusion biology into radiotherapy planning and resistance management.

9.3.2.3 Immunotherapeutic Considerations

The immunomodulatory consequences of cell fusion offer opportunities for therapeutic exploitation. Cell fusion vaccines have been developed by fusing bone marrow-derived dendritic cells with tumor cells. These vaccines can expose tumor-associated antigens, activate both CD8+ and CD4+ T cells, and induce powerful anticancer immunity (Koido 2016; Wang et al. 2021). They also have the advantage of being susceptible to necrosis and capable of releasing proteins that may act as additional tumor antigens presented by dendritic cells. This concept has been expanded by advances in biomaterial technology to bring about a viable future in fusion-based immunotherapy. For example, Liu et al. (2019) developed immunotherapeutic nanoplatforms from cancer–dendritic cell hybrid membranes to enhance the precision and efficacy of photodynamic therapy (PDT). These hybrid-derived nanoplatforms use tumor tropism to enrich therapeutic molecules at tumor locations.

9.4 Conclusion and Future Directions

Cancer cell fusion is a biologically significant and clinically actionable process that promotes many cancer hallmarks. The events proceed under the influence of both intrinsic and extrinsic environmental factors to give rise to hybrid cells with enhanced survival and adaptability. If well-harnessed, fusion might present unique opportunities for therapeutic targeting and immunomodulation. As evidence continues to accumulate, a precise understanding of the mechanisms driving cancer cell fusion is required to effectively target its vulnerabilities and allow the translation of mechanistic insights into clinical practice. This would significantly enhance cancer treatment efficacy and the reduction of cancer mortality and morbidity in our populations.

Acknowledgments This work was supported by funding from NIH/NIMHD U54MD015929, NIH 1R16GM153544-01A1, and NSF award IDs 2142465 and 2417643 to FNK.

References

Abmayr SM, Pavlath GK (2012) Myoblast fusion: lessons from flies and mice. Development 139(4):641–656. https://doi.org/10.1242/dev.068353

Aguirre LA, Montalbán-Hernández K, Avendaño-Ortiz J, Marín E, Lozano R, Toledano V et al (2020) Tumor stem cells fuse with monocytes to form highly invasive tumor-hybrid cells. Onco Targets Ther 9(1):1773204. https://doi.org/10.1080/2162402x.2020.1773204

Aljabery F, Olsson H, Gimm O, Jahnson S, Shabo I (2018) M2-macrophage infiltration and macrophage traits of tumor cells in urinary bladder cancer. Urol Oncol 36(4):159.e119–159.e126. https://doi.org/10.1016/j.urolonc.2017.11.020

Anastasiou D (2017) Tumour microenvironment factors shaping the cancer metabolism landscape. Br J Cancer 116(3):277–286. https://doi.org/10.1038/bjc.2016.412

Andersen TL, Boissy P, Sondergaard TE, Kupisiewicz K, Plesner T, Rasmussen T et al (2007) Osteoclast nuclei of myeloma patients show chromosome translocations specific for the myeloma cell clone: a new type of cancer-host partnership? J Pathol 211(1):10–17. https://doi.org/10.1002/path.2078

Andersen TL, Søe K, Sondergaard TE, Plesner T, Delaisse JM (2010) Myeloma cell-induced disruption of bone remodelling compartments leads to osteolytic lesions and generation of osteoclast-myeloma hybrid cells. Br J Haematol 148(4):551–561. https://doi.org/10.1111/j.1365-2141.2009.07980.x

Anderson AN, Conley P, Klocke CD, Sengupta SK, Pang A, Farley HC et al (2023) Detection of neoplastic-immune hybrid cells with metastatic properties in uveal melanoma. Res Sq. https://doi.org/10.21203/rs.3.rs-3694879/v1

Archacka K, Grabowska I, Mierzejewski B, Graffstein J, Górzyńska A, Krawczyk M et al (2021) Hypoxia preconditioned bone marrow-derived mesenchymal stromal/stem cells enhance myoblast fusion and skeletal muscle regeneration. Stem Cell Res Ther 12(1):448. https://doi.org/10.1186/s13287-021-02530-3

Bartosh TJ, Ullah M, Zeitouni S, Beaver J, Prockop DJ (2016) Cancer cells enter dormancy after cannibalizing mesenchymal stem/stromal cells (MSCs). Proc Natl Acad Sci USA 113(42):E6447–E6456. https://doi.org/10.1073/pnas.1612290113

Bastida-Ruiz D, Van Hoesen K, Cohen M (2016) The dark side of cell fusion. Int J Mol Sci 17(5). https://doi.org/10.3390/ijms17050638

Bastida-Ruiz D, Wuillemin C, Pederencino A, Yaron M, Martinez de Tejada B, Pizzo SV, Cohen M (2020) Activated α(2)-macroglobulin binding to cell surface GRP78 induces trophoblastic cell fusion. Sci Rep 10(1):9666. https://doi.org/10.1038/s41598-020-66554-0

Beck B, Blanpain C (2013) Unravelling cancer stem cell potential. Nat Rev Cancer 13(10):727–738. https://doi.org/10.1038/nrc3597

Berndt B, Haverkampf S, Reith G, Keil S, Niggemann B, Zänker KS, Dittmar T (2013) Fusion of CCL21 non-migratory active breast epithelial and breast cancer cells give rise to CCL21 migratory active tumor hybrid cell lines. PLoS One 8(5):e63711. https://doi.org/10.1371/journal.pone.0063711

Bjerkvig R, Tysnes BB, Aboody KS, Najbauer J, Terzis AJ (2005) Opinion: the origin of the cancer stem cell: current controversies and new insights. Nat Rev Cancer 5(11):899–904. https://doi.org/10.1038/nrc1740

Bjerregaard B, Holck S, Christensen IJ, Larsson LI (2006) Syncytin is involved in breast cancer-endothelial cell fusions. Cell Mol Life Sci 63(16):1906–1911. https://doi.org/10.1007/s00018-006-6201-9

Böhme I, Bosserhoff AK (2016) Acidic tumor microenvironment in human melanoma. Pigment Cell Melanoma Res 29(5):508–523. https://doi.org/10.1111/pcmr.12495

Bolouri H (2015) Network dynamics in the tumor microenvironment. Semin Cancer Biol 30:52–59. https://doi.org/10.1016/j.semcancer.2014.02.007

Boudreau JE, Bonehill A, Thielemans K, Wan Y (2011) Engineering dendritic cells to enhance cancer immunotherapy. Mol Ther 19(5):841–853. https://doi.org/10.1038/mt.2011.57

Brukman NG, Uygur B, Podbilewicz B, Chernomordik LV (2019) How cells fuse. J Cell Biol 218(5):1436–1451. https://doi.org/10.1083/jcb.201901017

Brukman NG, Li X, Podbilewicz B (2021) Fusexins, HAP2/GCS1 and evolution of gamete fusion. Front Cell Dev Biol 9:824024. https://doi.org/10.3389/fcell.2021.824024

Carloni V, Mazzocca A, Mello T, Galli A, Capaccioli S (2013) Cell fusion promotes chemoresistance in metastatic colon carcinoma. Oncogene 32(21):2649–2660. https://doi.org/10.1038/onc.2012.268

Celià-Terrassa T, Jolly MK (2020) Cancer stem cells and epithelial-to-mesenchymal transition in cancer metastasis. Cold Spring Harb Perspect Med 10(7). https://doi.org/10.1101/cshperspect.a036905

Chakraborty AK, Sodi S, Rachkovsky M, Kolesnikova N, Platt JT, Bolognia JL, Pawelek JM (2000) A spontaneous murine melanoma lung metastasis comprised of host x tumor hybrids. Cancer Res 60(9):2512–2519

Chakraborty AK, Pawelek J, Ikeda Y, Miyoshi E, Kolesnikova N, Funasaka Y et al (2001) Fusion hybrids with macrophage and melanoma cells up-regulate N-acetylglucosaminyltransferase V, beta1-6 branching, and metastasis. Cell Growth Differ 12(12):623–630

Chakraborty A, Lazova R, Davies S, Bäckvall H, Ponten F, Brash D, Pawelek J (2004) Donor DNA in a renal cell carcinoma metastasis from a bone marrow transplant recipient. Bone Marrow Transplant 34(2):183–186. https://doi.org/10.1038/sj.bmt.1704547

Chen Y, Xu Y, Zhang Y, Yang D, Sun Y (2025) Functions of the fusogenic and non-fusogenic activities of syncytin-1 in human physiological and pathological processes. Biochem Biophys Res Commun 761:151746. https://doi.org/10.1016/j.bbrc.2025.151746

Chernomordik LV, Kozlov MM (2005) Membrane hemifusion: crossing a chasm in two leaps. Cell 123(3):375–382. https://doi.org/10.1016/j.cell.2005.10.015

Cheung KJ, Padmanaban V, Silvestri V, Schipper K, Cohen JD, Fairchild AN et al (2016) Polyclonal breast cancer metastases arise from collective dissemination of keratin 14-expressing tumor cell clusters. Proc Natl Acad Sci USA 113(7):E854–E863. https://doi.org/10.1073/pnas.1508541113

Chitwood CA, Dietzsch C, Jacobs G, McArdle T, Freeman BT, Banga A et al (2018) Breast tumor cell hybrids form spontaneously in vivo and contribute to breast tumor metastases. APL Bioeng 2(3):031907. https://doi.org/10.1063/1.5024744

Choi H, Moon A (2018) Crosstalk between cancer cells and endothelial cells: implications for tumor progression and intervention. Arch Pharm Res 41(7):711–724. https://doi.org/10.1007/s12272-018-1051-1

Chunduri NK, Storchová Z (2019) The diverse consequences of aneuploidy. Nat Cell Biol 21(1): 54–62. https://doi.org/10.1038/s41556-018-0243-8

Clawson GA, Kimchi E, Patrick SD, Xin P, Harouaka R, Zheng S et al (2012) Circulating tumor cells in melanoma patients. PLoS One 7(7):e41052. https://doi.org/10.1371/journal.pone.0041052

Clawson GA, Matters GL, Xin P, Imamura-Kawasawa Y, Du Z, Thiboutot DM et al (2015) Macrophage-tumor cell fusions from peripheral blood of melanoma patients. PLoS One 10(8): e0134320. https://doi.org/10.1371/journal.pone.0134320

Cogle CR, Theise ND, Fu D, Ucar D, Lee S, Guthrie SM et al (2007) Bone marrow contributes to epithelial cancers in mice and humans as developmental mimicry. Stem Cells 25(8):1881–1887. https://doi.org/10.1634/stemcells.2007-0163

Cozzo AJ, Coleman MF, Hursting SD (2023) You complete me: tumor cell-myeloid cell nuclear fusion as a facilitator of organ-specific metastasis. Front Oncol 13:1191332. https://doi.org/10.3389/fonc.2023.1191332

Davies PS, Powell AE, Swain JR, Wong MH (2009) Inflammation and proliferation act together to mediate intestinal cell fusion. PLoS One 4(8):e6530. https://doi.org/10.1371/journal.pone.0006530

Delespaul L, Merle C, Lesluyes T, Lagarde P, Le Guellec S, Pérot G et al (2019) Fusion-mediated chromosomal instability promotes aneuploidy patterns that resemble human tumors. Oncogene 38(33):6083–6094. https://doi.org/10.1038/s41388-019-0859-6

Delespaul L, Gélabert C, Lesluyes T, Le Guellec S, Pérot G, Leroy L et al (2020) Cell-cell fusion of mesenchymal cells with distinct differentiations triggers genomic and transcriptomic remodelling toward tumour aggressiveness. Sci Rep 10(1):21634. https://doi.org/10.1038/s41598-020-78502-z

Dervan E, Bhattacharyya DD, McAuliffe JD, Khan FH, Glynn SA (2021) Ancient adversary – HERV-K (HML-2) in cancer. Front Oncol 11:658489. https://doi.org/10.3389/fonc.2021.658489

Dietz MS, Sutton TL, Walker BS, Gast CE, Zarour L, Sengupta SK et al (2021) Relevance of circulating hybrid cells as a non-invasive biomarker for myriad solid tumors. Sci Rep 11(1): 13630. https://doi.org/10.1038/s41598-021-93053-7

Ding J, Jin W, Chen C, Shao Z, Wu J (2012) Tumor associated macrophage × cancer cell hybrids may acquire cancer stem cell properties in breast cancer. PLoS One 7(7):e41942. https://doi.org/10.1371/journal.pone.0041942

Dittmar T, Hass R (2023) Intrinsic signalling factors associated with cancer cell-cell fusion. Cell Commun Signaling 21(1):68. https://doi.org/10.1186/s12964-023-01085-5

Dittmar T, Schwitalla S, Seidel J, Haverkampf S, Reith G, Meyer-Staeckling S et al (2011) Characterization of hybrid cells derived from spontaneous fusion events between breast epithelial cells exhibiting stem-like characteristics and breast cancer cells. Clin Exp Metastasis 28(1): 75–90. https://doi.org/10.1007/s10585-010-9359-3

Dittmar T, Weiler J, Luo T, Hass R (2021) Cell-cell fusion mediated by viruses and HERV-derived Fusogens in cancer initiation and progression. Cancers (Basel) 13(21). https://doi.org/10.3390/cancers13215363

Donaldson SH, Lee CT, Chmelka BF, Israelachvili JN (2011) General hydrophobic interaction potential for surfactant/lipid bilayers from direct force measurements between light-modulated bilayers. Proc Natl Acad Sci USA 108(38):15699–15704. https://doi.org/10.1073/pnas.1112411108

Duelli D, Lazebnik Y (2007) Cell-to-cell fusion as a link between viruses and cancer. Nat Rev Cancer 7(12):968–976. https://doi.org/10.1038/nrc2272

Duelli DM, Hearn S, Myers MP, Lazebnik Y (2005) A primate virus generates transformed human cells by fusion. J Cell Biol 171(3):493–503. https://doi.org/10.1083/jcb.200507069

Duncan AW, Hanlon Newell AE, Bi W, Finegold MJ, Olson SB, Beaudet AL, Grompe M (2012) Aneuploidy as a mechanism for stress-induced liver adaptation. J Clin Invest 122(9): 3307–3315. https://doi.org/10.1172/jci64026

Fahlbusch SS, Keil S, Epplen JT, Zänker KS, Dittmar T (2020) Comparison of hybrid clones derived from human breast epithelial cells and three different cancer cell lines regarding in vitro cancer stem/ initiating cell properties. BMC Cancer 20(1):446. https://doi.org/10.1186/s12885-020-06952-9

Fédry J, Liu Y, Péhau-Arnaudet G, Pei J, Li W, Tortorici MA et al (2017) The ancient gamete Fusogen HAP2 is a eukaryotic class II fusion protein. Cell 168(5):904–915.e910. https://doi.org/10.1016/j.cell.2017.01.024

Fei F, Zhang M, Li B, Zhao L, Wang H, Liu L et al (2019) Formation of polyploid giant cancer cells involves in the prognostic value of neoadjuvant chemoradiation in locally advanced rectal cancer. J Oncol 2019:2316436. https://doi.org/10.1155/2019/2316436

Gao Y, Yu XF, Chen T (2021) Human endogenous retroviruses in cancer: expression, regulation and function. Oncol Lett 21(2):121. https://doi.org/10.3892/ol.2020.12382

Garvin S, Oda H, Arnesson LG, Lindström A, Shabo I (2018) Tumor cell expression of CD163 is associated to postoperative radiotherapy and poor prognosis in patients with breast cancer treated with breast-conserving surgery. J Cancer Res Clin Oncol 144(7):1253–1263. https://doi.org/10.1007/s00432-018-2646-0

Gast, C. E., Silk, A. D., Zarour, L., Riegler, L., Burkhart, J. G., Gustafson, K. T., . . . Wong, M. H. (2018). Cell fusion potentiates tumor heterogeneity and reveals circulating hybrid cells that correlate with stage and survival. Sci Adv, 4(9), eaat7828. doi:https://doi.org/10.1126/sciadv.aat7828

Gauck D, Keil S, Niggemann B, Zänker KS, Dittmar T (2017) Hybrid clone cells derived from human breast epithelial cells and human breast cancer cells exhibit properties of cancer stem/initiating cells. BMC Cancer 17(1):515. https://doi.org/10.1186/s12885-017-3509-9

George JT, Jolly MK, Xu S, Somarelli JA, Levine H (2017) Survival outcomes in cancer patients predicted by a partial EMT gene expression scoring metric. Cancer Res 77(22):6415–6428. https://doi.org/10.1158/0008-5472.can-16-3521

Godin L, Balsat C, Van Eycke YR, Allard J, Royer C, Remmelink M et al (2020) A novel approach for quantifying cancer cells showing hybrid epithelial/mesenchymal states in large series of tissue samples: towards a new prognostic marker. Cancers (Basel) 12(4). https://doi.org/10.3390/cancers12040906

Goldenberg DM, Gold DV, Loo M, Liu D, Chang CH, Jaffe ES (2013) Horizontal transmission of malignancy: in-vivo fusion of human lymphomas with hamster stroma produces tumors retaining human genes and lymphoid pathology. PLoS One 8(2):e55324. https://doi.org/10.1371/journal.pone.0055324

Goldenberg DM, Rooney RJ, Loo M, Liu D, Chang CH (2014) In-vivo fusion of human cancer and hamster stromal cells permanently transduces and transcribes human DNA. PLoS One 9(9):e107927. https://doi.org/10.1371/journal.pone.0107927

Grandi N, Tramontano E (2018) HERV envelope proteins: physiological role and pathogenic potential in cancer and autoimmunity. Front Microbiol 9:462. https://doi.org/10.3389/fmicb.2018.00462

Grosse-Wilde A, Fouquier d'Hérouël A, McIntosh E, Ertaylan G, Skupin A, Kuestner RE et al (2015) Stemness of the hybrid epithelial/mesenchymal state in breast cancer and its association with poor survival. PLoS One 10(5):e0126522. https://doi.org/10.1371/journal.pone.0126522

Gruenbaum-Cohen Y, Harel I, Umansky K-B, Tzahor E, Snapper SB, Shilo B-Z, Schejter ED (2012) The actin regulator N-WASp is required for muscle-cell fusion in mice. Proc Natl Acad Sci 109(28):11211–11216. https://doi.org/10.1073/pnas.1116065109

Gyobu S, Miyata H, Ikawa M, Yamazaki D, Takeshima H, Suzuki J, Nagata S (2016) A role of TMEM16E carrying a scrambling domain in sperm motility. Mol Cell Biol 36(4):645–659. https://doi.org/10.1128/mcb.00919-15

Haneef K, Naeem N, Khan I, Iqbal H, Kabir N, Jamall S et al (2014) Conditioned medium enhances the fusion capability of rat bone marrow mesenchymal stem cells and cardiomyocytes. Mol Biol Rep 41(5):3099–3112. https://doi.org/10.1007/s11033-014-3170-1

Harkness T, Weaver BA, Alexander CM, Ogle BM (2013) Cell fusion in tumor development: accelerated genetic evolution. Crit Rev Oncog 18(1–2):19–42. https://doi.org/10.1615/critrevoncog.v18.i1-2.30

Hass R (2020) Role of MSC in the tumor microenvironment. Cancers (Basel) 12(8). https://doi.org/10.3390/cancers12082107

Hass R, von der Ohe J, Dittmar T (2021a) Cancer cell fusion and post-hybrid selection process (PHSP). Cancers (Basel) 13(18). https://doi.org/10.3390/cancers13184636

Hass R, von der Ohe J, Dittmar T (2021b) Hybrid formation and fusion of cancer cells in vitro and in vivo. Cancers (Basel) 13(17). https://doi.org/10.3390/cancers13174496

Hernández JM, Podbilewicz B (2017) The hallmarks of cell-cell fusion. Development 144(24):4481–4495. https://doi.org/10.1242/dev.155523

Hochreiter-Hufford AE, Lee CS, Kinchen JM, Sokolowski JD, Arandjelovic S, Call JA et al (2013) Phosphatidylserine receptor BAI1 and apoptotic cells as new promoters of myoblast fusion. Nature 497(7448):263–267. https://doi.org/10.1038/nature12135

Hu C, He Y, Liu D, Zhao L, Fang S, Tan B et al (2021) Hypoxia preconditioning promotes the proliferation and migration of human urine-derived stem cells in chronically injured liver of mice by upregulating CXCR4. Stem Cells Dev 30(10):526–536. https://doi.org/10.1089/scd.2021.0008

Huang CM, Yan TL, Xu Z, Wang M, Zhou XC, Jiang EH et al (2018) Hypoxia enhances fusion of oral squamous carcinoma cells and epithelial cells partly via the epithelial-mesenchymal transition of epithelial cells. Biomed Res Int 2018:5015203. https://doi.org/10.1155/2018/5015203

Jacobsen BM, Harrell JC, Jedlicka P, Borges VF, Varella-Garcia M, Horwitz KB (2006) Spontaneous fusion with, and transformation of mouse stroma by, malignant human breast cancer epithelium. Cancer Res 66(16):8274–8279. https://doi.org/10.1158/0008-5472.can-06-1456

Jacquemin V, Antoine M, Dom G, Detours V, Maenhaut C, Dumont JE (2022) Dynamic cancer cell heterogeneity: diagnostic and therapeutic implications. Cancers (Basel) 14(2). https://doi.org/10.3390/cancers14020280

Janssen A, Medema RH (2011) Entosis: aneuploidy by invasion. Nat Cell Biol 13(3):199–201. https://doi.org/10.1038/ncb0311-199

Jeong J, Conboy IM (2011) Phosphatidylserine directly and positively regulates fusion of myoblasts into myotubes. Biochem Biophys Res Commun 414(1):9–13. https://doi.org/10.1016/j.bbrc.2011.08.128

Jiang E, Yan T, Xu Z, Shang Z (2019) Tumor microenvironment and cell fusion. Biomed Res Int 2019:5013592. https://doi.org/10.1155/2019/5013592

Jolly MK, Boareto M, Huang B, Jia D, Lu M, Ben-Jacob E et al (2015) Implications of the hybrid epithelial/mesenchymal phenotype in metastasis. Front Oncol 5:155. https://doi.org/10.3389/fonc.2015.00155

Kalluri R (2016) The biology and function of fibroblasts in cancer. Nat Rev Cancer 16(9):582–598. https://doi.org/10.1038/nrc.2016.73

Kalluri R, Neilson EG (2003) Epithelial-mesenchymal transition and its implications for fibrosis. J Clin Invest 112(12):1776–1784. https://doi.org/10.1172/jci20530

Kaur E, Rajendra J, Jadhav S, Shridhar E, Goda JS, Moiyadi A, Dutt S (2015) Radiation-induced homotypic cell fusions of innately resistant glioblastoma cells mediate their sustained survival and recurrence. Carcinogenesis 36(6):685–695. https://doi.org/10.1093/carcin/bgv050

Kerbel RS, Lagarde AE, Dennis JW, Donaghue TP (1983) Spontaneous fusion in vivo between normal host and tumor cells: possible contribution to tumor progression and metastasis studied with a lectin-resistant mutant tumor. Mol Cell Biol 3(4):523–538. https://doi.org/10.1128/mcb.3.4.523-538.1983

Ko EJ, Song KS, Ock MS, Choi YH, Kim S, Kim HS, Cha HJ (2021) Expression profiles of human endogenous retrovirus (HERV)-K and HERV-R Env proteins in various cancers. BMB Rep 54(7):368–373. https://doi.org/10.5483/BMBRep.2021.54.7.246

Koido S (2016) Dendritic-tumor fusion cell-based cancer vaccines. Int J Mol Sci 17(6). https://doi.org/10.3390/ijms17060828

Koido S, Hara E, Homma S, Namiki Y, Ohkusa T, Gong J, Tajiri H (2009) Cancer vaccine by fusions of dendritic and cancer cells. Clin Dev Immunol 2009:657369. https://doi.org/10.1155/2009/657369

Koido S, Enomoto Y, Apostolopoulos V, Gong J (2014a) Tumor regression by CD4 T-cells primed with dendritic/tumor fusion cell vaccines. Anticancer Res 34(8):3917–3924

Koido S, Homma S, Kan S, Takakura K, Namiki Y, Kobayashi H et al (2014b) Induction of antigen-specific cytotoxic T lymphocytes by fusion cells generated from allogeneic plasmacytoid dendritic and tumor cells. Int J Oncol 45(1):470–478. https://doi.org/10.3892/ijo.2014.2433

Kreso A, Dick JE (2014) Evolution of the cancer stem cell model. Cell Stem Cell 14(3):275–291. https://doi.org/10.1016/j.stem.2014.02.006

Kroemer G (2010) Tetraploid cancer cell precursors. Nat Rev Mol Cell Biol 11(8):539–539. https://doi.org/10.1038/nrm2924

LaBerge GS, Duvall E, Grasmick Z, Haedicke K, Pawelek J (2017) A melanoma lymph node metastasis with a donor-patient hybrid genome following bone marrow transplantation: a second case of leucocyte-tumor cell hybridization in cancer metastasis. PLoS One 12(2):e0168581. https://doi.org/10.1371/journal.pone.0168581

Larsson LI, Holck S, Christensen IJ (2007) Prognostic role of syncytin expression in breast cancer. Hum Pathol 38(5):726–731. https://doi.org/10.1016/j.humpath.2006.10.018

Lartigue L, Merle C, Lagarde P, Delespaul L, Lesluyes T, Le Guellec S et al (2020) Genome remodeling upon mesenchymal tumor cell fusion contributes to tumor progression and metastatic spread. Oncogene 39(21):4198–4211. https://doi.org/10.1038/s41388-020-1276-6

Lazova R, Laberge GS, Duvall E, Spoelstra N, Klump V, Sznol M et al (2013) A melanoma brain metastasis with a donor-patient hybrid genome following bone marrow transplantation: first evidence for fusion in human cancer. PLoS One 8(6):e66731. https://doi.org/10.1371/journal.pone.0066731

Li G, Kikuchi K, Radka M, Abraham J, Rubin BP, Keller C (2013) IL-4 receptor blockade abrogates satellite cell: rhabdomyosarcoma fusion and prevents tumor establishment. Stem Cells 31(11):2304–2312. https://doi.org/10.1002/stem.1491

Lin Y, Satta RR, Simula ER, Tang S, Molicotti P, Cossu A et al (2025) The role of human endogenous retroviruses in the initiation and progression of melanoma. Biomedicine 13-(7) Retrieved from. https://doi.org/10.3390/biomedicines13071662

Lindström A, Midtbö K, Arnesson LG, Garvin S, Shabo I (2017) Fusion between M2-macrophages and cancer cells results in a subpopulation of radioresistant cells with enhanced DNA-repair capacity. Oncotarget 8(31):51370–51386. https://doi.org/10.18632/oncotarget.17986

Liu WL, Zou MZ, Liu T, Zeng JY, Li X, Yu WY et al (2019) Expandable immunotherapeutic nanoplatforms engineered from cytomembranes of hybrid cells derived from cancer and dendritic cells. Adv Mater 31(18):e1900499. https://doi.org/10.1002/adma.201900499

Liu C, Billet S, Choudhury D, Cheng R, Haldar S, Fernandez A et al (2021) Bone marrow mesenchymal stem cells interact with head and neck squamous cell carcinoma cells to promote cancer progression and drug resistance. Neoplasia 23(1):118–128. https://doi.org/10.1016/j.neo.2020.11.012

López-Collazo E, Hurtado-Navarro L (2025) Cell fusion as a driver of metastasis: re-evaluating an old hypothesis in the age of cancer heterogeneity. Front Immunol 16. https://doi.org/10.3389/fimmu.2025.1524781

Lu X, Kang Y (2009a) Cell fusion as a hidden force in tumor progression. Cancer Res 69(22): 8536–8539. https://doi.org/10.1158/0008-5472.can-09-2159

Lu X, Kang Y (2009b) Efficient acquisition of dual metastasis organotropism to bone and lung through stable spontaneous fusion between MDA-MB-231 variants. Proc Natl Acad Sci USA 106(23):9385–9390. https://doi.org/10.1073/pnas.0900108106

Maniecki MB, Etzerodt A, Ulhøi BP, Steiniche T, Borre M, Dyrskjøt L et al (2012) Tumor-promoting macrophages induce the expression of the macrophage-specific receptor CD163 in malignant cells. Int J Cancer 131(10):2320–2331. https://doi.org/10.1002/ijc.27506

Manjunath Y, Porciani D, Mitchem JB, Suvilesh KN, Avella DM, Kimchi ET et al (2020) Tumor-cell-macrophage fusion cells as liquid biomarkers and tumor enhancers in cancer. Int J Mol Sci 21(5). https://doi.org/10.3390/ijms21051872

Margolis RL (2005) Tetraploidy and tumor development. Cancer Cell 8(5):353–354. https://doi.org/10.1016/j.ccr.2005.10.017

Melzer C, von der Ohe J, Hass R (2018a) Enhanced metastatic capacity of breast cancer cells after interaction and hybrid formation with mesenchymal stroma/stem cells (MSC). Cell Commun Signal 16(1):2. https://doi.org/10.1186/s12964-018-0215-4

Melzer C, von der Ohe J, Hass R (2018b) In vitro fusion of normal and neoplastic breast epithelial cells with human mesenchymal stroma/stem cells partially involves tumor necrosis factor receptor signaling. Stem Cells 36(7):977–989. https://doi.org/10.1002/stem.2819

Melzer C, von der Ohe J, Hass R (2018c) MSC stimulate ovarian tumor growth during intercellular communication but reduce tumorigenicity after fusion with ovarian cancer cells. Cell Communication and Signaling 16(1):67. https://doi.org/10.1186/s12964-018-0279-1

Melzer C, von der Ohe J, Hass R (2019a) In vivo cell fusion between mesenchymal stroma/stem-like cells and breast cancer cells. Cancers (Basel) 11(2). https://doi.org/10.3390/cancers11020185

Melzer C, von der Ohe J, Hass R (2019b) Involvement of actin cytoskeletal components in breast cancer cell fusion with human mesenchymal stroma/stem-like cells. Int J Mol Sci 20(4). https://doi.org/10.3390/ijms20040876

Melzer C, Ohe JV, Hass R (2020) Altered tumor plasticity after different cancer cell fusions with MSC. Int J Mol Sci 21(21). https://doi.org/10.3390/ijms21218347

Melzer C, Ohe JV, Luo T, Hass R (2021) Spontaneous fusion of MSC with breast cancer cells can generate tumor dormancy. Int J Mol Sci 22(11). https://doi.org/10.3390/ijms22115930

Miller FR, Mohamed AN, McEachern D (1989) Production of a more aggressive tumor cell variant by spontaneous fusion of two mouse tumor subpopulations. Cancer Res 49(15):4316–4321

Misra UK, Deedwania R, Pizzo SV (2006) Activation and cross-talk between Akt, NF-kappaB, and unfolded protein response signaling in 1-LN prostate cancer cells consequent to ligation of cell surface-associated GRP78. J Biol Chem 281(19):13694–13707. https://doi.org/10.1074/jbc.M511694200

Mohr M, Tosun S, Arnold WH, Edenhofer F, Zänker KS, Dittmar T (2015a) Quantification of cell fusion events human breast cancer cells and breast epithelial cells using a Cre-LoxP-based double fluorescence reporter system. Cell Mol Life Sci 72(19):3769–3782. https://doi.org/10.1007/s00018-015-1910-6

Mohr M, Zaenker KS, Dittmar T (2015b) Fusion in cancer: an explanatory model for aneuploidy, metastasis formation, and drug resistance. Methods Mol Biol 1313:21–40. https://doi.org/10.1007/978-1-4939-2703-6_2

Montalbán-Hernández K, Cantero-Cid R, Casalvilla-Dueñas JC, Avendaño-Ortiz J, Marín E, Lozano-Rodríguez R et al (2022) Colorectal cancer stem cells fuse with monocytes to form tumour hybrid cells with the ability to migrate and evade the immune system. Cancers (Basel) 14(14). https://doi.org/10.3390/cancers14143445

Mortensen K, Lichtenberg J, Thomsen PD, Larsson LI (2004) Spontaneous fusion between cancer cells and endothelial cells. Cell Mol Life Sci 61(16):2125–2131. https://doi.org/10.1007/s00018-004-4200-2

Nagata S, Suzuki J, Segawa K, Fujii T (2016) Exposure of phosphatidylserine on the cell surface. Cell Death Differ 23(6):952–961. https://doi.org/10.1038/cdd.2016.7

Nagler C, Hardt C, Zänker KS, Dittmar T (2011) Co-cultivation of murine BMDCs with 67NR mouse mammary carcinoma cells give rise to highly drug resistant cells. Cancer Cell Int 11(1): 21. https://doi.org/10.1186/1475-2867-11-21

Nasr MM, Lynch CC (2023) How circulating tumor cluster biology contributes to the metastatic cascade: from invasion to dissemination and dormancy. Cancer Metastasis Rev 42(4): 1133–1146. https://doi.org/10.1007/s10555-023-10124-z

Ni M, Zhang Y, Lee AS (2011) Beyond the endoplasmic reticulum: atypical GRP78 in cell viability, signalling and therapeutic targeting. Biochem J 434(2):181–188. https://doi.org/10.1042/bj20101569

Nikanjam M, Okamura R, Barkauskas DA, Kurzrock R (2020) Targeting fusions for improved outcomes in oncology treatment. Cancer 126(6):1315–1321. https://doi.org/10.1002/cncr.32649

Noubissi FK, Ogle BM (2016) Cancer cell fusion: mechanisms slowly unravel. Int J Mol Sci 17(9). https://doi.org/10.3390/ijms17091587

Noubissi FK, Harkness T, Alexander CM, Ogle BM (2015) Apoptosis-induced cancer cell fusion: a mechanism of breast cancer metastasis. FASEB J 29(9):4036–4045. https://doi.org/10.1096/fj.15-271098

Noubissi FK, Odubanjo OV, Ogle BM, Tchounwou PB (2024) Mechanisms of cell fusion in cancer. Results Probl Cell Differ 71:407–432. https://doi.org/10.1007/978-3-031-37936-9_19

Parappilly MS, Chin Y, Whalen RM, Anderson AN, Robinson TS, Strgar L et al (2022) Circulating neoplastic-immune hybrid cells predict metastatic progression in uveal melanoma. Cancers (Basel) 14(19). https://doi.org/10.3390/cancers14194617

Passerini V, Ozeri-Galai E, de Pagter MS, Donnelly N, Schmalbrock S, Kloosterman WP et al (2016) The presence of extra chromosomes leads to genomic instability. Nat Commun 7(1): 10754. https://doi.org/10.1038/ncomms10754

Patel RK, Parappilly M, Farley HC, Latour EJ, Wang LG, Nair AM et al (2024a) Circulating neoplastic-immune hybrid cells are biomarkers of occult metastasis and treatment response in pancreatic cancer. Cancers (Basel) 16(21). https://doi.org/10.3390/cancers16213650

Patel RK, Parappilly MS, Walker BS, Heussner RT, Fung A, Chang YH et al (2024b) Exploratory analyses of circulating neoplastic-immune hybrid cells as prognostic biomarkers in advanced intrahepatic cholangiocarcinoma. Int J Mol Sci 25(17). https://doi.org/10.3390/ijms25179198

Pawelek JM (2000) Tumour cell hybridization and metastasis revisited. Melanoma Res 10(6): 507–514. https://doi.org/10.1097/00008390-200012000-00001

Pawelek JM, Chakraborty AK (2008a) Chapter 10 the cancer cell–leukocyte fusion theory of metastasis. In: Advances in cancer research, vol 101. Academic Press, pp 397–444

Pawelek JM, Chakraborty AK (2008b) Fusion of tumour cells with bone marrow-derived cells: a unifying explanation for metastasis. Nat Rev Cancer 8(5):377–386. https://doi.org/10.1038/nrc2371

Peng XC, Zhang M, Meng YY, Liang YF, Wang YY, Liu XQ et al (2021) Cell-cell fusion as an important mechanism of tumor metastasis (review). Oncol Rep 46(1). https://doi.org/10.3892/or.2021.8096

Pérez-Vargas J, Krey T, Valansi C, Avinoam O, Haouz A, Jamin M et al (2014) Structural basis of eukaryotic cell-cell fusion. Cell 157(2):407–419. https://doi.org/10.1016/j.cell.2014.02.020

Petrany MJ, Millay DP (2019) Cell fusion: merging membranes and making muscle. Trends Cell Biol 29(12):964–973. https://doi.org/10.1016/j.tcb.2019.09.002

Platt JL, Cascalho M (2019) Cell fusion in malignancy: a cause or consequence? A provocateur or cure? Cells 8(6). https://doi.org/10.3390/cells8060587

Priščáková P, Svoboda M, Feketová Z, Hutník J, Repiská V, Gbelcová H, Gergely L (2023) Syncytin-1, syncytin-2 and suppressyn in human health and disease. J Mol Med (Berl) 101(12):1527–1542. https://doi.org/10.1007/s00109-023-02385-6

Raj AT, Kheur S, Patil VR, Gupta AA (2019) Assessing the role of cell fusion in cancer metastasis. Oral Oncol 90:124–125. https://doi.org/10.1016/j.oraloncology.2019.01.006

Rajah MM, Bernier A, Buchrieser J, Schwartz O (2022) The mechanism and consequences of SARS-CoV-2 spike-mediated fusion and syncytia formation. J Mol Biol 434(6):167280. https://doi.org/10.1016/j.jmb.2021.167280

Ramakrishnan M, Mathur SR, Mukhopadhyay A (2013) Fusion-derived epithelial cancer cells express hematopoietic markers and contribute to stem cell and migratory phenotype in ovarian carcinoma. Cancer Res 73(17):5360–5370. https://doi.org/10.1158/0008-5472.can-13-0896

Rappa G, Mercapide J, Lorico A (2012) Spontaneous formation of tumorigenic hybrids between breast cancer and multipotent stromal cells is a source of tumor heterogeneity. Am J Pathol 180(6):2504–2515. https://doi.org/10.1016/j.ajpath.2012.02.020

Rival CM, Xu W, Shankman LS, Morioka S, Arandjelovic S, Lee CS et al (2019) Phosphatidylserine on viable sperm and phagocytic machinery in oocytes regulate mammalian fertilization. Nat Commun 10(1):4456. https://doi.org/10.1038/s41467-019-12406-z

Rizvi AZ, Swain JR, Davies PS, Bailey AS, Decker AD, Willenbring H et al (2006) Bone marrow-derived cells fuse with normal and transformed intestinal stem cells. Proc Natl Acad Sci USA 103(16):6321–6325. https://doi.org/10.1073/pnas.0508593103

Ruano APC, Gadelha Guimarães AP, Braun AC, Flores B, Tariki MS, Abdallah EA et al (2022) Fusion cell markers in circulating tumor cells from patients with high-grade ovarian serous carcinoma. Int J Mol Sci 23(23). https://doi.org/10.3390/ijms232314687

Scarpa A, Mafficini A (2018) Non-coding regulatory variations: the dark matter of pancreatic cancer genomics. Gut 67(3):399–400. https://doi.org/10.1136/gutjnl-2017-314310

Searles SC, Santosa EK, Bui JD (2018) Cell-cell fusion as a mechanism of DNA exchange in cancer. Oncotarget 9(5):6156–6173. https://doi.org/10.18632/oncotarget.23715

Segev N, Avinoam O, Podbilewicz B (2018) Fusogens. Curr Biol 28(8):R378–R380. https://doi.org/10.1016/j.cub.2018.01.024
Seyfried TN, Huysentruyt LC (2013) On the origin of cancer metastasis. Crit Rev Oncog 18(1–2): 43–73. https://doi.org/10.1615/critrevoncog.v18.i1-2.40
Seyfried TN, Arismendi-Morillo G, Mukherjee P, Chinopoulos C (2020) On the origin of ATP synthesis in cancer. iScience 23(11):101761. https://doi.org/10.1016/j.isci.2020.101761
Shabo I, Stål O, Olsson H, Doré S, Svanvik J (2008) Breast cancer expression of CD163, a macrophage scavenger receptor, is related to early distant recurrence and reduced patient survival. Int J Cancer 123(4):780–786. https://doi.org/10.1002/ijc.23527
Shabo I, Midtbö K, Andersson H, Åkerlund E, Olsson H, Wegman P et al (2015) Macrophage traits in cancer cells are induced by macrophage-cancer cell fusion and cannot be explained by cellular interaction. BMC Cancer 15:922. https://doi.org/10.1186/s12885-015-1935-0
Shabo I, Svanvik J, Lindström A, Lechertier T, Trabulo S, Hulit J et al (2020) Roles of cell fusion, hybridization and polyploid cell formation in cancer metastasis. World J Clin Oncol 11(3): 121–135. https://doi.org/10.5306/wjco.v11.i3.121
Shah AH, Rivas SR, Doucet-O'Hare TT, Govindarajan V, DeMarino C, Wang T et al (2023) Human endogenous retrovirus K contributes to a stem cell niche in glioblastoma. J Clin Invest 133(13). https://doi.org/10.1172/JCI167929
Shao R, Han X, Zhu B, Luo J, Zhang X, Chen H et al (2022) Fusion of invasive tumor cells with infiltrating macrophages fuels epithelial-mesenchymal transition and adaptive immune evasion. Res. Sq. https://doi.org/10.21203/rs.3.rs-2251764/v1
Shen Q, Rahn JJ, Zhang J, Gunasekera N, Sun X, Shaw AR et al (2008) MUC1 initiates Src-CrkL-Rac1/Cdc42-mediated actin cytoskeletal protrusive motility after ligating intercellular adhesion molecule-1. Mol Cancer Res 6(4):555–567. https://doi.org/10.1158/1541-7786.mcr-07-2033
Shultes PV, Weaver DT, Tadele DS, Barker-Clarke RJ, Scott JG (2024) Cell-cell fusion in cancer: the next cancer hallmark? Int J Biochem Cell Biol 175:106649. https://doi.org/10.1016/j.biocel.2024.106649
Sieler M, Weiler J, Dittmar T (2021) Cell-cell fusion and the roads to novel properties of tumor hybrid cells. Cells 10(6). https://doi.org/10.3390/cells10061465
Singh N, Baby D, Rajguru JP, Patil PB, Thakkannavar SS, Pujari VB (2019) Inflammation and cancer. Ann Afr Med 18(3):121–126. https://doi.org/10.4103/aam.aam_56_18
Skinner AM, Grompe M, Kurre P (2012) Intra-hematopoietic cell fusion as a source of somatic variation in the hematopoietic system. J Cell Sci 125(Pt 12):2837–2843. https://doi.org/10.1242/jcs.100123
Søe K (2020) Osteoclast fusion: physiological regulation of multinucleation through heterogeneity-potential implications for drug sensitivity. Int J Mol Sci 21(20). https://doi.org/10.3390/ijms21207717
Song K, Zhu F, Zhang HZ, Shang ZJ (2012) Tumor necrosis factor-α enhanced fusions between oral squamous cell carcinoma cells and endothelial cells via VCAM-1/VLA-4 pathway. Exp Cell Res 318(14):1707–1715. https://doi.org/10.1016/j.yexcr.2012.05.022
Song K, Song Y, Zhao XP, Shen H, Wang M, Yan TL et al (2014) Oral cancer/endothelial cell fusion experiences nuclear fusion and acquisition of enhanced survival potential. Exp Cell Res 328(1):156–163. https://doi.org/10.1016/j.yexcr.2014.07.006
Sprangers AJ, Freeman BT, Kouris NA, Ogle BM (2012) A Cre-Lox P recombination approach for the detection of cell fusion in vivo. J Vis Exp (59):e3581. https://doi.org/10.3791/3581
Strick R, Ackermann S, Langbein M, Swiatek J, Schubert SW, Hashemolhosseini S et al (2007) Proliferation and cell-cell fusion of endometrial carcinoma are induced by the human endogenous retroviral Syncytin-1 and regulated by TGF-beta. J Mol Med (Berl) 85(1):23–38. https://doi.org/10.1007/s00109-006-0104-y
Suetsugu A, Matsumoto T, Hasegawa K, Nakamura M, Kunisada T, Shimizu M et al (2016) Color-coded live imaging of heterokaryon formation and nuclear fusion of hybridizing cancer cells. Anticancer Res 36(8):3827

Sun C, Zhao D, Dai X, Chen J, Rong X, Wang H et al (2015) Fusion of cancer stem cells and mesenchymal stem cells contributes to glioma neovascularization. Oncol Rep 34(4):2022–2030. https://doi.org/10.3892/or.2015.4135

Sun C, Dai X, Zhao D, Wang H, Rong X, Huang Q, Lan Q (2019) Mesenchymal stem cells promote glioma neovascularization in vivo by fusing with cancer stem cells. BMC Cancer 19(1):1240. https://doi.org/10.1186/s12885-019-6460-0

Symeonides M, Lambelé M, Roy NH, Thali M (2014) Evidence showing that tetraspanins inhibit HIV-1-induced cell-cell fusion at a post-hemifusion stage. Viruses 6(3):1078–1090. https://doi.org/10.3390/v6031078

Takito J, Otsuka H, Inoue S, Kawashima T, Nakamura M (2017) Symmetrical retrograde actin flow in the actin fusion structure is involved in osteoclast fusion. Biol Open 6(7):1104–1114. https://doi.org/10.1242/bio.025460

Tal A, Tal R, Shaikh S, Gidicsin S, Mamillapalli R, Taylor HS (2019) Characterization of cell fusion in an experimental mouse model of endometriosis†. Biol Reprod 100(2):390–397. https://doi.org/10.1093/biolre/ioy221

Tanjak P, Chaiboonchoe A, Suwatthanarak T, Thanormjit K, Acharayothin O, Chanthercrob J et al (2024) Tumor-immune hybrid cells evade the immune response and potentiate colorectal cancer metastasis through CTLA4. Clin Exp Med 25(1):2. https://doi.org/10.1007/s10238-024-01515-9

Tretyakova MS, Subbalakshmi AR, Menyailo ME, Jolly MK, Denisov EV (2022) Tumor hybrid cells: nature and biological significance. Front Cell Dev Biol 10:814714. https://doi.org/10.3389/fcell.2022.814714

Uygur B, Leikina E, Melikov K, Villasmil R, Verma SK, Vary CPH, Chernomordik LV (2019) Interactions with muscle cells boost fusion, stemness, and drug resistance of prostate cancer cells. Mol Cancer Res 17(3):806–820. https://doi.org/10.1158/1541-7786.mcr-18-0500

Vasan N, Baselga J, Hyman DM (2019) A view on drug resistance in cancer. Nature 575(7782): 299–309. https://doi.org/10.1038/s41586-019-1730-1

Vassilopoulos G, Wang PR, Russell DW (2003) Transplanted bone marrow regenerates liver by cell fusion. Nature 422(6934):901–904. https://doi.org/10.1038/nature01539

Wang R, Sun X, Wang CY, Hu P, Chu CY, Liu S et al (2012) Spontaneous cancer-stromal cell fusion as a mechanism of prostate cancer androgen-independent progression. PLoS One 7(8): e42653. https://doi.org/10.1371/journal.pone.0042653

Wang R, Lewis MS, Lyu J, Zhau HE, Pandol SJ, Chung LWK (2020) Cancer-stromal cell fusion as revealed by fluorescence protein tracking. Prostate 80(3):274–283. https://doi.org/10.1002/pros.23941

Wang HF, Xiang W, Xue BZ, Wang YH, Yi DY, Jiang XB et al (2021) Cell fusion in cancer hallmarks: current research status and future indications. Oncol Lett 22(1):530. https://doi.org/10.3892/ol.2021.12791

Wang J, Yu C, Zhuang J, Qi W, Jiang J, Liu X et al (2022) The role of phosphatidylserine on the membrane in immunity and blood coagulation. Biomark Res 10(1):4. https://doi.org/10.1186/s40364-021-00346-0

Warner TFCS (1975) Cell hybridization: an explanation for the phenotypic diversity of certain tumours. Med Hypotheses 1(1):51–57. https://doi.org/10.1016/0306-9877(75)90042-0

Weber CE, Kuo PC (2012) The tumor microenvironment. Surg Oncol 21(3):172–177. https://doi.org/10.1016/j.suronc.2011.09.001

Wei HJ, Nickoloff JA, Chen WH, Liu HY, Lo WC, Chang YT et al (2014) FOXF1 mediates mesenchymal stem cell fusion-induced reprogramming of lung cancer cells. Oncotarget 5(19): 9514–9529. https://doi.org/10.18632/oncotarget.2413

Weiler J, Dittmar T (2019a) Cell fusion in human cancer: the dark matter hypothesis. Cells 8(2). https://doi.org/10.3390/cells8020132

Weiler J, Dittmar T (2019b) Minocycline impairs TNF-α-induced cell fusion of M13SV1-Cre cells with MDA-MB-435-pFDR1 cells by suppressing NF-κB transcriptional activity and its induction of target-gene expression of fusion-relevant factors. Cell Commun Signal 17(1):71. https://doi.org/10.1186/s12964-019-0384-9

Whitlock JM, Chernomordik LV (2021) Flagging fusion: phosphatidylserine signaling in cell-cell fusion. J Biol Chem 296:100411. https://doi.org/10.1016/j.jbc.2021.100411

Whitlock JM, Yu K, Cui YY, Hartzell HC (2018) Anoctamin 5/TMEM16E facilitates muscle precursor cell fusion. J Gen Physiol 150(11):1498–1509. https://doi.org/10.1085/jgp.201812097

Xu RH, Pelicano H, Zhou Y, Carew JS, Feng L, Bhalla KN et al (2005) Inhibition of glycolysis in cancer cells: a novel strategy to overcome drug resistance associated with mitochondrial respiratory defect and hypoxia. Cancer Res 65(2):613–621

Xu MH, Gao X, Luo D, Zhou XD, Xiong W, Liu GX (2014) EMT and acquisition of stem cell-like properties are involved in spontaneous formation of tumorigenic hybrids between lung cancer and bone marrow-derived mesenchymal stem cells. PLoS One 9(2):e87893. https://doi.org/10.1371/journal.pone.0087893

Yan T-L, Wang M, Xu Z, Huang C-M, Zhou X-C, Jiang E-H et al (2017) Up-regulation of syncytin-1 contributes to TNF-α-enhanced fusion between OSCC and HUVECs partly via Wnt/β-catenin-dependent pathway. Sci Rep 7(1):40983. https://doi.org/10.1038/srep40983

Yang JY, Ha SA, Yang YS, Kim JW (2010) P-glycoprotein ABCB5 and YB-1 expression plays a role in increased heterogeneity of breast cancer cells: correlations with cell fusion and doxorubicin resistance. BMC Cancer 10:388. https://doi.org/10.1186/1471-2407-10-388

Ye X, Tam WL, Shibue T, Kaygusuz Y, Reinhardt F, Ng Eaton E, Weinberg RA (2015) Distinct EMT programs control normal mammary stem cells and tumour-initiating cells. Nature 525(7568):256–260. https://doi.org/10.1038/nature14897

Yeh MH, Chang YH, Tsai YC, Chen SL, Huang TS, Chiu JF, Ch'ang HJ (2016) Bone marrow derived macrophages fuse with intestine stromal cells and contribute to chronic fibrosis after radiation. Radiother Oncol 119(2):250–258. https://doi.org/10.1016/j.radonc.2016.01.025

Yilmaz Y, Lazova R, Qumsiyeh M, Cooper D, Pawelek J (2005) Donor Y chromosome in renal carcinoma cells of a female BMT recipient: visualization of putative BMT-tumor hybrids by FISH. Bone Marrow Transplant 35(10):1021–1024. https://doi.org/10.1038/sj.bmt.1704939

Zawada KE, Okamoto K, Kasson PM (2018) Influenza hemifusion phenotype depends on membrane context: differences in cell-cell and virus-cell fusion. J Mol Biol 430(5):594–601. https://doi.org/10.1016/j.jmb.2018.01.006

Zhang S, Mercado-Uribe I, Xing Z, Sun B, Kuang J, Liu J (2014a) Generation of cancer stem-like cells through the formation of polyploid giant cancer cells. Oncogene 33(1):116–128. https://doi.org/10.1038/onc.2013.96

Zhang Y, Luo W, Wang Y, Liu Y, Zheng L (2014b) Purified dendritic cell-tumor fusion hybrids supplemented with non-adherent dendritic cells fraction are superior activators of antitumor immunity. PLoS One 9(1):e86772. https://doi.org/10.1371/journal.pone.0086772

Zhang Y, Yang Y, Zhu Z, Ou G (2017) WASP-Arp2/3-dependent actin polymerization influences fusogen localization during cell-cell fusion in Caenorhabditiselegans embryos. Biol Open 6(9): 1324–1328. https://doi.org/10.1242/bio.026807

Zhang LN, Huang YH, Zhao L (2019a) Fusion of macrophages promotes breast cancer cell proliferation, migration and invasion through activating epithelial-mesenchymal transition and Wnt/β-catenin signaling pathway. Arch Biochem Biophys 676:108137. https://doi.org/10.1016/j.abb.2019.108137

Zhang LN, Kong CF, Zhao D, Cong XL, Wang SS, Ma L, Huang YH (2019b) Fusion with mesenchymal stem cells differentially affects tumorigenic and metastatic abilities of lung cancer cells. J Cell Physiol 234(4):3570–3582. https://doi.org/10.1002/jcp.27011

Zhang Y, Le T, Grabau R, Mohseni Z, Kim H, Natale DR et al (2020) TMEM16F phospholipid scramblase mediates trophoblast fusion and placental development. Sci Adv 6(19):eaba0310. https://doi.org/10.1126/sciadv.aba0310

Zhang H, Ma H, Yang X, Fan L, Tian S, Niu R et al (2022) Cell fusion-related proteins and signaling pathways, and their roles in the development and progression of cancer. Front Cell Dev Biol 9:809668. https://doi.org/10.3389/fcell.2021.809668

Zhou X, Merchak K, Lee W, Grande JP, Cascalho M, Platt JL (2015) Cell fusion connects oncogenesis with tumor evolution. Am J Pathol 185(7):2049–2060. https://doi.org/10.1016/j.ajpath.2015.03.014

Zhou F, Li M, Wei Y, Lin K, Lu Y, Shen J et al (2016) Activation of HERV-K Env protein is essential for tumorigenesis and metastasis of breast cancer cells. Oncotarget 7(51): 84093–84117. https://doi.org/10.18632/oncotarget.11455

Zhou X, Zhou M, Zheng M, Tian S, Yang X, Ning Y et al (2022) Polyploid giant cancer cells and cancer progression. Front Cell Dev Biol 10:1017588. https://doi.org/10.3389/fcell.2022.1017588

Žigon P, Perdan Pirkmajer K, Tomšič M, Kveder T, Božič B, Sodin Šemrl S et al (2015) Anti-phosphatidylserine/prothrombin antibodies are associated with adverse pregnancy outcomes. J Immunol Res 2015:975704. https://doi.org/10.1155/2015/975704

Chapter 10
Polyploidy and Senescence of Cancer Cells: Impact of Cell Fusion

Soham Bhattacharyya, Debopriya Choudhury, Brahmachari Vedeshachaitanya, and Pushkar Malakar

Abstract This chapter explores both the normal physiological roles of cell fusion and its pathological implications in cancer. We focus on the emerging concept of fusion-induced polyploidy and senescence as a "double-edged sword"—a process that can suppress malignancy through growth arrest yet also promote tumor progression when polyploid cells escape senescence and acquire stem-like properties. In addition, we examine the interconnection between polyploidy and senescence, highlighting their similarities, differences, and dynamic interplay.

Special attention is given to the molecular signaling pathways governing these processes, with particular emphasis on the central role of mTOR and the regulatory functions of long noncoding ribonucleic acids (lncRNAs). We also discuss how these insights inform potential therapeutic strategies, including the use of senolytic agents, the modulation of the senescence-associated secretory phenotype (SASP), and inhibitors targeting mTOR or mitotic regulators. Finally, we outline key unanswered questions in the field, emphasizing the need for reliable biomarkers to identify fusion-derived hybrids and for a deeper mechanistic understanding of cell fusion-derived polyploidy and senescence.

A more comprehensive grasp of these mechanisms may pave the way for novel therapeutic approaches aimed at limiting tumor plasticity, overcoming therapy resistance, and ultimately improving patient outcomes.

Keywords Cell fusion · Polyploidy · Senescence · Signaling pathways · mTORC1 · mTORC2 · Long noncoding RNAs · Cancer

---

S. Bhattacharyya · D. Choudhury · B. Vedeshachaitanya · P. Malakar (✉)
Department of Biomedical Science and Technology, School of Biological Sciences, Ramakrishna Mission Vivekananda Educational Research Institute (RKMVERI), Kolkata, India
e-mail: pushkarbt@gm.rkmvu.ac.in

© The Author(s), under exclusive license to Springer Nature Switzerland AG 2026

J. Weiler et al. (eds.), *Cell Fusion*, Advances in Experimental Medicine and Biology 1509, https://doi.org/10.1007/978-3-032-22637-2_10

10.1 Introduction

Cell fusion, the process in which two or more cells merge their plasma membranes and combine their cytoplasmic and nuclear contents, is a fundamental biological mechanism with diverse roles in development, regeneration, and immunity (Zhang et al. 2022). In recent years, it has also emerged as a key contributor to cancer progression (Zhang et al. 2022).

Within the tumor microenvironment, stressors such as hypoxia, chronic inflammation, and therapy-induced damage can provoke aberrant fusion events between tumor cells and stromal or immune cells (Landskron et al. 2014; Jiang et al. 2019). The resulting hybrid cells often become polyploid, acquiring characteristics of cellular senescence, including growth arrest, DNA damage, and the secretion of pro-inflammatory mediators (Hayashi et al. 2024). This senescent phenotype can serve as a tumor-suppressive barrier, but it may also create a reservoir of tumor-promoting factors via the senescence-associated secretory phenotype (SASP) (Dong et al. 2024). Crucially, some polyploid cells bypass senescence, reenter the cell cycle, and generate genomically unstable, stem-like progeny that fuel therapy resistance, metastasis, and tumor recurrence (Song et al. 2023).

Thus, cell fusion is a double-edged sword in cancer biology—it can suppress tumor growth through senescence or promote malignancy through polyploidy and genomic instability (Saini et al. 2022). Understanding the factors that tilt this balance is central to devising effective therapies.

Polyploidy and senescence represent distinct yet interconnected stress responses in cancer. While senescence primarily acts as a tumor-suppressive checkpoint, polyploidy enhances genetic diversity and drives tumor evolution. Both processes, however, present exploitable vulnerabilities for therapeutic intervention, particularly in the context of drug resistance (Mosieniak and Sikora 2010; Conway et al. 2024).

At the molecular level, cancer cell fusion is regulated by several signaling cascades, including the phosphoinositide 3-kinase (PI3K)/ protein kinase B (Akt), mitogen-activated protein kinase (MAPK), Wnt/β-catenin, nuclear factor kappa-light chain-enhancer of activated B cells (NF-κB), transforming growth factor beta (TGF-β), and Rho GTPase pathways. These pathways influence fusogenic protein expression, actin cytoskeleton remodeling, and epithelial-to-mesenchymal transition (EMT), thereby facilitating cell fusion and hybrid formation (Wu et al. 2021; Ning et al. 2024). Targeting these pathways offers opportunities to reduce polyploidy, metastasis, and resistance to conventional therapies.

Beyond signaling proteins, long noncoding RNAs (lncRNAs)—a class of nonprotein-coding RNAs known to regulate nearly every step of gene expression—have also been implicated in the control of cancer cell fusion. Their regulatory roles are only beginning to be uncovered, but they hold promise as novel therapeutic and diagnostic targets (Sahu et al. 2015).

Progress has been made in characterizing the markers of polyploidy and senescence, such as DNA content, γ-H2AX, p21, p16, and SA-β-gal (Valdiglesias et al. 2013; Huang et al. 2022). However, clinically validated biomarkers for detecting fusion-derived hybrid cells remain elusive. Filling this gap is essential for translating the basic insights into clinical applications.

In summary, cell fusion represents a paradoxical force in cancer, capable of both constraining and driving tumor progression. Deciphering the interplay between polyploidy and senescence—and the signaling and noncoding RNA networks that regulate them—will be key to exploiting this process for future therapeutic strategies.

10.2 Cell Fusion in Development and Tissue Repair

Cell fusion is essential for various biological processes during development, tissue regeneration, and immune responses (Fig. 10.1) (Zhang et al. 2022; Shultes et al. 2024).

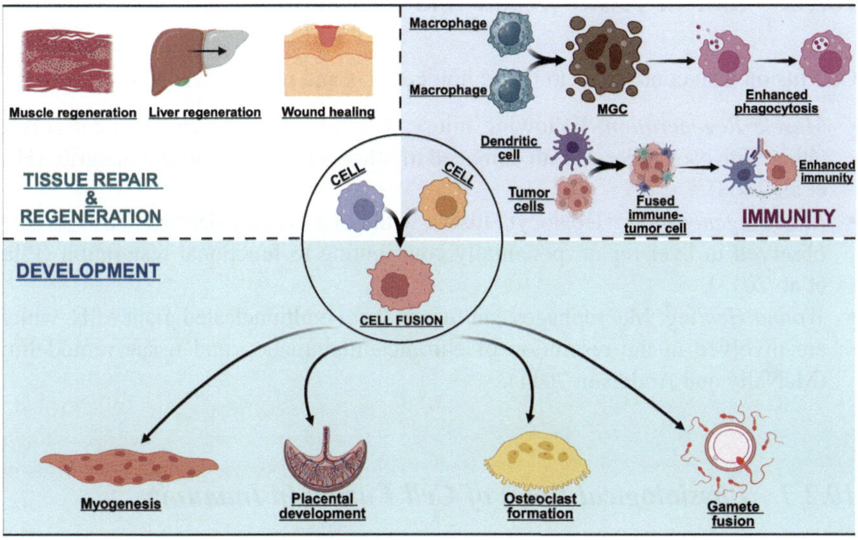

Fig. 10.1 Role of cell fusion in normal physiological processes. Cell fusion contributes to tissue repair and regeneration through processes such as muscle regeneration, liver regeneration, and wound healing. During development, it plays key roles in myogenesis, placental development, osteoclast formation, and gamete fusion. In immunity, fusion of activated macrophages generates multinucleated giant cells (MGCs), which aid in the clearance of pathogens or particles too large for phagocytosis by a single macrophage. Additionally, the fusion of dendritic cells with tumor cells produces hybrid immune-tumor cells capable of presenting tumor antigens to immune cells. (Figure created with BioRender)

10.2.1 Role in Development

During embryogenesis and organ formation, cell fusion contributes to tissue morphogenesis and functional maturation:

- *Myogenesis*: Myoblasts (muscle precursor cells) fuse to form multinucleated myotubes, which are essential for skeletal muscle development (Taylor 2000).
- *Placental Development*: In the human placenta, cytotrophoblasts fuse to form the syncytiotrophoblast layer—a multinucleated structure crucial for nutrient exchange between mother and fetus (Renaud and Jeyarajah 2022).
- *Osteoclast Formation*: Fusion of monocyte/macrophage lineage cells gives rise to multinucleated osteoclasts, which are involved in bone remodeling (Sabe et al. 2024).
- *Gamete Fusion*: Fertilization itself is a specialized form of cell fusion where a sperm and an egg unite to form a zygote (Deneke and Pauli 2021).

10.2.2 Role in Tissue Repair and Regeneration

Cell fusion also contributes to tissue homeostasis and repair after injury:

- *Muscle Regeneration*: Following injury, satellite cells (muscle stem cells) fuse either with each other or with damaged myofibers to restore muscle integrity (Hill et al. 2003).
- *Liver Regeneration*: Hepatocyte fusion with bone marrow-derived cells has been observed in liver repair, potentially contributing to functional restoration (Pilat et al. 2013).
- *Wound Healing*: Macrophages can fuse to form multinucleated giant cells, which are involved in the resolution of chronic inflammation and tissue remodeling (McNally and Anderson 2011).

10.2.3 Physiological Roles of Cell Fusion in Immunity

In the immune system, fusion events are observed under physiological conditions such as granuloma formation, as well as in pathological states including viral infections, chronic inflammation, and cancer progression (Chen et al. 2023).

Multinucleated Giant Cells (MGCs)
MGCs form through the fusion of activated macrophages, particularly during chronic inflammatory responses (Milde et al. 2015). These cells are a hallmark of granulomatous diseases such as tuberculosis, sarcoidosis, and foreign body reactions (Okamoto et al. 2003). MGCs assist in sequestering and degrading pathogens or particles that are too large for phagocytosis by single macrophages.

Dendritic Cell-Tumor Cell Hybrids
In experimental settings, fusion of dendritic cells (DCs) with tumor cells creates immunogenic hybrid cells capable of presenting tumor antigens to T cells. These fused cells are being explored in cancer vaccine strategies for enhancing antitumor immune responses (Koido et al. 2009).

10.3 Mechanisms and Regulation of Normal Cell Fusion

Normal cell fusion is a highly coordinated and regulated biological process. For the successful fusion between two or more cells, several key steps and molecular components must act in concert.

10.3.1 Membrane Proximity and Alignment

The initial step involves bringing the plasma membranes of the two fusing cells into close proximity. This is achieved through chemotactic signals or cell adhesion molecules that facilitate tight apposition of membranes, overcoming the repulsive forces between lipid bilayers (Mayer 2001).

10.3.2 Cytoskeletal Rearrangements

Actin cytoskeleton remodeling plays a pivotal role in membrane juxtaposition and fusion pore formation (Lee and Dominguez 2010). Dynamic actin polymerization generates protrusions (e.g., filopodia) that promote contact between fusion partners, while localized disassembly creates membrane flexibility at the fusion site (Rodríguez-Pérez et al. 2021).

10.3.3 Expression of Fusogens

Fusogens are specialized membrane-bound proteins that directly mediate the merging of lipid bilayers. Examples include *syncytin* proteins in placental fusion and *myomaker* and *myomerger* in myoblast fusion. These proteins facilitate hemifusion and the subsequent full fusion of membranes, enabling cytoplasmic continuity (Segev et al. 2018).

10.3.4 Signaling Inputs (Cytokines, Growth Factors, and Mechanical Stress)

A diverse array of extrinsic signals—including cytokines, growth factors (e.g., TGF-β, IL-4), and biomechanical cues—modulate the fusion process by activating downstream signaling cascades (e.g., MAPK, PI3K-Akt) that regulate cytoskeletal dynamics, gene expression, and fusogen availability (Garcia et al. 2016).

Cell fusion is tightly controlled both spatially and temporally to ensure that it occurs only in specific physiological contexts. This regulation involves the precise expression of fusion machinery, receptor-ligand interactions, and feedback from intracellular signaling networks (Zhang et al. 2022).

However, when fusion is misregulated or occurs in inappropriate contexts—such as during inflammation, tissue damage, or within tumors—it can lead to pathological consequences. These include:

- *Aneuploidy:* The fusion between genetically distinct cells can produce hybrids with abnormal chromosomal numbers, leading to genomic instability (Dörnen et al. 2020).
- *Cellular Senescence:* Abnormal fusion can trigger DNA damage responses and a senescence-associated secretory phenotype (SASP), impairing tissue function (Gal and Krizhanovsky 2014).
- *Tumorigenesis:* In cancer, the aberrant fusion between tumor cells and stromal or immune cells may promote metastasis, drug resistance, and tumor heterogeneity (Lindström et al. 2017).

10.4 Mechanisms and Consequences of Cell Fusion in Cancer

Cell fusion, a process whereby two or more cells merge their plasma membranes and cytoplasm to form a hybrid cell, is a physiological event during development and tissue regeneration (Dörnen et al. 2020; Zhang et al. 2022; Shultes et al. 2024). However, in cancer, this process can be aberrantly activated, contributing to genomic instability, therapy resistance, and tumor progression (Mosieniak and Sikora 2010; Saini et al. 2022). Cancer cell fusion is driven by a range of intrinsic and extrinsic factors and is mediated by specific molecular machinery and signaling pathways (Noubissi and Ogle 2016).

10.4.1 *Triggers of Cell Fusion in Tumors*

Cancer cells exhibit enhanced fusogenic potential in response to several stimuli originating from the tumor microenvironment or as a consequence of therapy (Fig. 10.2). Major fusion-inducing triggers include:

10.4.1.1 Microenvironmental Stressors

The tumor microenvironment is often characterized by hypoxia, inflammation, and oxidative stress—all of which enhance the fusogenic capacity of tumor cells (Chen et al. 2022; Nishida and Andoh 2025). Hypoxia and oxidative stress destabilize cell membranes and increase the expression of fusogenic proteins. Pro-inflammatory cytokines such as TNF-α, IL-6, and TGF-β have been shown to promote membrane fusion between cancer cells or between cancer and stromal cells (Landskron et al. 2014).

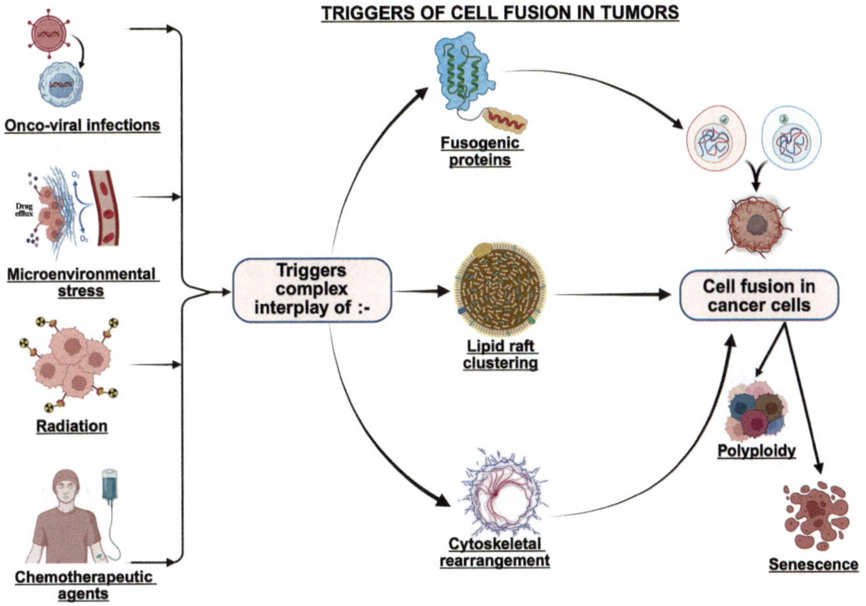

Fig. 10.2 Factors and mechanisms of cell fusion in tumors. Oncoviral infections, microenvironmental stress, radiation, and chemotherapeutic agents can trigger cell fusion via a complex interplay of fusogenic proteins, lipid raft clustering, and cytoskeletal rearrangements. In cancer cells, such fusion events lead to polyploidy and senescence. (Figure created with BioRender)

10.4.1.2 Radiation and Chemotherapeutic Agents

Genotoxic stress caused by DNA-damaging treatments—such as ionizing radiation, cisplatin, or bortezomib—can paradoxically promote cell fusion (Swift and Golsteyn 2014). This may represent a survival strategy under stress, whereby fusion generates polyploid or hybrid cells with increased resilience and adaptability (Velegzhaninov et al. 2018). Stress-induced senescence or mitotic failure may further predispose cells to fusogenic reprogramming (Dikovskaya et al. 2015).

10.4.1.3 Oncogenic Viral Infections

Certain viruses known to be associated with cancer—such as human papillomavirus (HPV), Epstein-Barr Virus (EBV), HIV, and SARS-CoV-2—can induce cell fusion. This is often mediated by viral fusogenic proteins (e.g., HIV-1 gp41 or spike protein), which hijack the host membrane fusion machinery (Gao and Zheng 2010; Negi et al. 2022). Endogenous retroviral proteins such as syncytins, co-opted from ancient viral genes, may be re-expressed in cancer and contribute to fusion (Kitsou et al. 2023).

10.4.1.4 Macrophage-Tumor Cell Fusion

Tumor cells can fuse with tumor-associated macrophages (TAMs), creating hybrid cells with dual phenotypes. These hybrids often display enhanced immune evasion, migratory capacity, and epithelial-to-mesenchymal transition (EMT) characteristics (Shabo et al. 2020). Such fusion-derived hybrids may contribute to the circulating tumor cell (CTC) pool and facilitate metastasis (Shabo et al. 2020).

10.4.2 Molecular Machinery of Cell Fusion

The fusion of cancer cells is orchestrated by a complex interplay of membrane-associated proteins, cytoskeletal regulators, and signaling molecules (Fig. 10.2).

10.4.2.1 Fusogenic Proteins

Syncytin-1 and Syncytin-2: Originally viral envelope proteins, they mediate membrane fusion in the placenta and certain virus-associated tumors (Durnaoglu et al. 2022). Annexins (A1, A5): Involved in calcium-dependent membrane curvature and vesicle fusion, supporting membrane remodeling (Lizarbe et al. 2013). E-cadherin and N-cadherin: Cell adhesion molecules that facilitate membrane alignment and stabilization prior to fusion (Perez and Nelson 2004; Radice 2013).

10.4.2.2 Lipid Raft Clustering

Fusion preferentially occurs in cholesterol- and sphingolipid-rich microdomains of the membrane, known as lipid rafts (Shimanouchi et al. 2013; Grassi et al. 2020). These domains concentrate fusogenic molecules and adhesion complexes, lowering the energy barrier for fusion.

10.4.2.3 Cytoskeletal Rearrangements

Actin remodeling and recruitment of Rho GTPases (e.g., Rac1, Cdc42) support fusion site formation (Tapon and Hall 1997; Abreu-Blanco et al. 2014). Integrins coordinate cell-cell interactions and cytoskeletal remodeling during membrane merging (Delon and Brown 2007).

10.5 Consequences of Abnormal Cell Fusion in Cancer: Polyploidy and Senescence

Fusion between cancer cells, or between cancer and stromal cells, often results in hybrid cells with multiple nuclei (multinucleation), or enlarged genomes (polyploidy), or senescent cells (senescence) (Fig. 10.2) (Hass et al. 2021a, 2021b; Tretyakova et al. 2022).

10.5.1 Fusion-Induced Polyploidy

After fusion, hybrid cells may fail to undergo proper mitosis or nuclear division, leading to the formation of polyploid or aneuploid progeny. Key mechanisms for fusion-induced polyploidy include:

- *Nuclear Reprogramming and Epigenetic Instability*
  Chromatin remodeling and transcriptional reprogramming in fused cells can disrupt the cell cycle control and promote genome duplication without division (Martinez-Sarmiento et al. 2024).
- *Defective Cytokinesis (Mitotic Slippage)*
  Polyploidization often arises from a failed cytokinesis. In these cells, mitosis is exited without completion of division, resulting in binucleated or tetraploid cells (Was et al. 2022). Furthermore, repeated rounds of endoreduplication without

cytokinesis further expand the genomic content of fused cells (Øvrebø and Edgar 2018).

- *Centrosome Amplification and Aneuploidy*

  Extra centrosomes in polyploid cells disrupt spindle bipolarity, increasing the risk of chromosomal missegregation and aneuploidy. This contributes to genomic instability, a hallmark of aggressive tumors (Duensing and Duensing 2010).

Polyploid cells formed due to abnormal cell fusion may follow divergent fates:

- Senescence and Growth Arrest

  Many polyploid cells enter a senescent state, marked by cell cycle arrest, SA-β-gal expression, and the secretion of pro-inflammatory cytokines (Dong et al. 2024). This can exert tumor-suppressive effects in some contexts but also promote a pro-tumorigenic microenvironment (Song et al. 2023).
- *Escape from Senescence and Clonal Expansion*

  A subset of fused polyploid cells may evade senescence and reenter the cell cycle, generating aneuploid daughter cells with diverse karyotypes. These cells often acquire cancer stem-like properties, enhanced drug resistance, and metastatic potential (Conway et al. 2024).

10.5.1.1 Molecular Markers of Polyploidy

Polyploid cells exhibit distinct molecular markers that are typically associated with this state (Fig. 10.3a).

(a) *DNA Content and Chromosomal Markers*

Increased DNA Content (≥4 N): Measured by flow cytometry using propidium iodide (PI) or DAPI staining (Chang et al. 2024). *Aneuploid Karyotypes:* Polyploid cells often show chromosomal instability (CIN), detected using spectral karyotyping (SKY) or comparative genomic hybridization (CGH) (Nacheva et al. 1998; Varella-Garcia et al. 2007). *γ-H2AX (Phosphorylated H2AX):* Marker of DNA damage and chromosomal instability, often elevated in polyploid cells (Valdiglesias et al. 2013).

(b) *Cell Cycle and Checkpoint Dysregulation Markers*

Cyclin D1 (CCND1) and Cyclin E1 (CCNE1): Promote G1/S transition, often upregulated in polyploid tumors (Eliades et al. 2010; Crockford et al. 2016). *Cyclin B1 (CCNB1):* Essential for the G2/M transition, frequently overexpressed in polyploid cells. *p53 (TP53):* Mutations or loss of p53 allow polyploid cells to bypass mitotic checkpoints (Chen 2016). *p21 (CDKN1A):* Inhibitor of cyclin-dependent kinases (CDKs); its overexpression can lead to G1 arrest and inhibition of polyploidization (Cazzalini et al. 2010).

(c) *Mitotic Slippage and Centrosome Amplification Markers*

Aurora Kinase A/B (AURKA, AURKB): Regulate mitotic spindle assembly; their overexpression is linked to polyploidy (Den Hollander et al. 2010). *PLK1 (Polo-like kinase 1):* Regulates mitotic progression; upregulated in polyploid

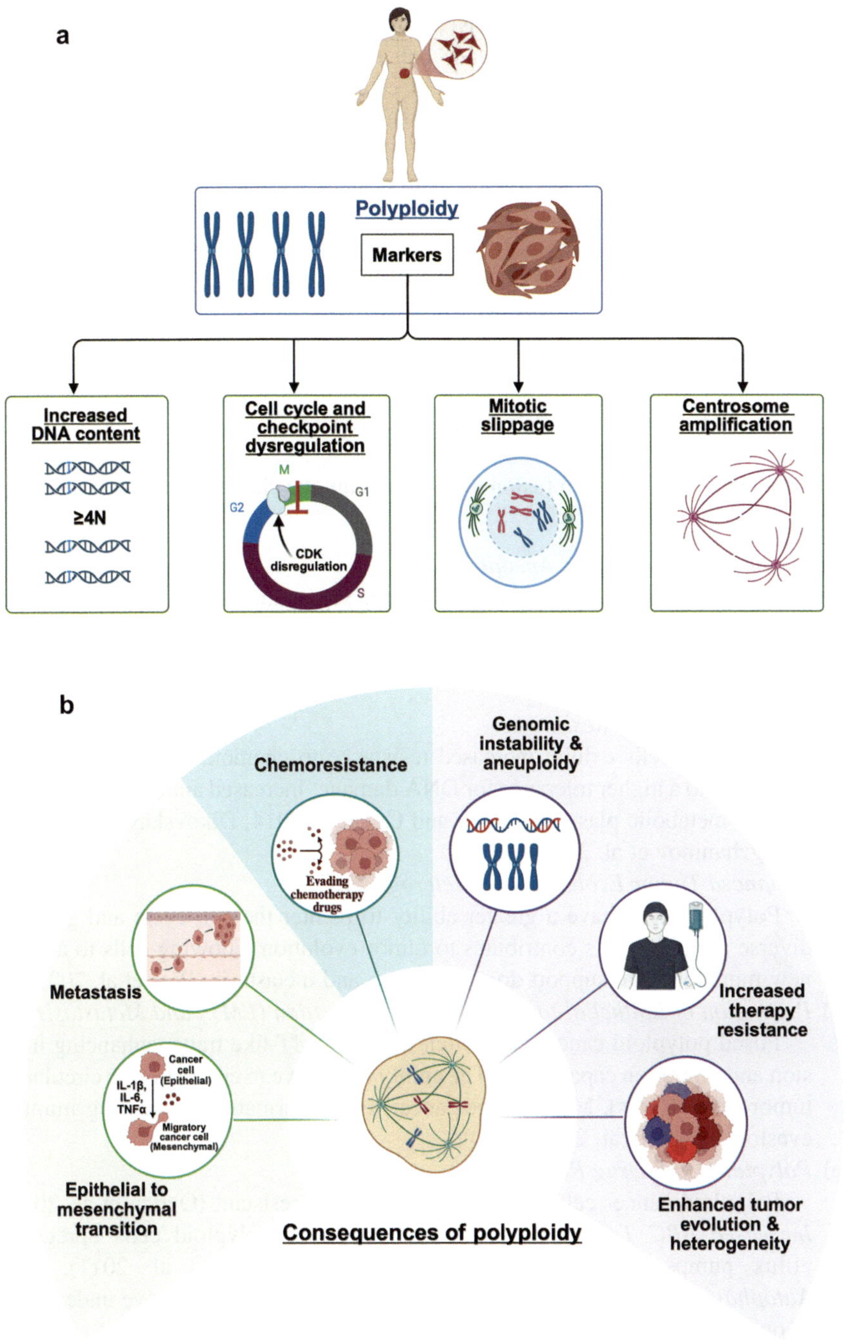

Fig. 10.3 Markers and consequences of polyploidy. (**a**) Markers of polyploidy include increased DNA content, cell cycle and checkpoint dysregulation, mitotic slippage, and centromere amplification. (**b**) Consequences of polyploidy in cancer include epithelial-mesenchymal transition (EMT), metastasis, chemoresistance, genomic instability, and aneuploidy, increased therapy resistance, and enhanced tumor evolution and heterogeneity. (Figure created with BioRender)

tumors (Mok et al. 2012). *Survivin (BIRC5):* Inhibitor of apoptosis protein (IAP), helps polyploid cells evade cell death (Antonio Cheung et al. 2013). *γ-Tubulin and Pericentrin:* Markers of centrosome amplification, often seen in polyploid cancer cells (Denu et al. 2016).

(d) *Polyploidy-Associated Transcriptional Changes*

E2F1/E2F3: Regulate S-phase entry; their dysregulation promotes polyploidy (Duronio et al. 1995). *YAP/TAZ (Hippo Pathway Effectors):* Overactivation enhances polyploidy and cancer stemness (Zanconato et al. 2016). *c-MYC:* Drives cell cycle progression and polyploidy in various cancers (Li and Dang 1999; Vazquez-Martin et al. 2016).

10.5.1.2 Consequences of Polyploidy in Cancer

Polyploid cells derived from fusion events have unique characteristics that influence tumor behavior (Fig. 10.3b):

(a) *Genomic Instability and Aneuploidy*

Polyploidy disrupts normal chromosome segregation during mitosis, leading to chromosomal instability (CIN). This instability results in aneuploidy, in which cells have an abnormal number of chromosomes, a hallmark of aggressive cancers (Dörnen et al. 2020).

(b) *Increased Therapy Resistance*

Polyploid cells exhibit increased resistance to chemotherapy and radiation, likely due to a higher tolerance for DNA damage. Increased antioxidant capacity. Greater metabolic plasticity (Swift and Golsteyn 2014; Dikovskaya et al. 2015; Velegzhaninov et al. 2018).

(c) *Enhanced Tumor Evolution and Heterogeneity*

Polyploid cells have a greater ability to reenter the cell cycle and generate diverse progeny. This contributes to tumor evolution, allowing cells to acquire new mutations that support drug resistance and metastasis (Was et al. 2022).

(d) *Promotion of Epithelial-to-Mesenchymal Transition (EMT) and Metastasis*

Fused polyploid cancer cells often display EMT-like traits, enhancing invasion and migration capabilities. The ability to survive in circulation as circulating tumor cells (CTCs). Macrophage-cancer hybrid formation, increasing immune evasion (Shabo et al. 2015, 2020).

(e) *Polyploidy and Drug Resistance*

Polyploid cancer cells are often more chemo-resistant (Ogden et al. 2015). *Increased ABC Transporters (ABCB1, ABCC1):* Polyploid cells upregulate efflux pumps, reducing drug accumulation (Tamaki et al. 2011). *High Autophagy Markers (LC3, Beclin-1):* Helps polyploid cells survive under stress (Kong and Yao 2022).

10.5.1.3 Experimental Evidence of Cell Fusion-Induced Polyploidy

Researchers use various methods to study how cell fusion leads to polyploidy in cancer:

(a) *In Vitro Studies*
Labeling Techniques (GFP/RFP Fusion Assays): Track the fusion of differentially labeled cancer cells (Crivat and Taraska 2012; Chen et al. 2016). *Fluorescence In Situ Hybridization (FISH)*: Detect polyploidy by analyzing chromosome numbers (Shakoori 2017). *Time-lapse microscopy*: Observe multinucleation and mitotic abnormalities (Zaninovic et al. 2017).
(b) *In Vivo Evidence*
Xenograft Models: Tumor cells engineered to express distinct markers fuse in vivo, forming polyploid hybrids (Zhang et al. 2015). *Lineage Tracing*: Used in mouse models to track fused tumor cells (Nathans et al. 2024; Deng et al. 2025).

10.5.1.4 Therapeutic Implications: Targeting Polyploid Cancer Cells

Since polyploid cancer cells are associated with therapy resistance and metastasis, potential strategies to target them include:

(a) *Targeting Mitotic Vulnerabilities*
KIF11 inhibitors (Eg5 inhibitors) to induce mitotic catastrophe (Wang et al. 2017). *Aurora kinase inhibitors* to prevent the abnormal division of polyploid cells.
(b) *Exploiting Metabolic Stress*
Polyploid cells often have high metabolic demands; targeting glucose or oxidative phosphorylation pathways may selectively kill them (Cunha et al. 2023).
(c) *DNA Damage Response (DDR) Inhibitors*
Polyploid cells rely on robust DNA repair mechanisms. Poly (ADP-ribose) Polymerase (PARP) inhibitors may exploit their genomic instability (Fu et al. 2024).
(d) *Senescence Induction and Immune Clearance*
Senolytics or immune checkpoint inhibitors may be used to eliminate senescent polyploid cells (Czajkowski et al. 2025).

10.5.2 Cell Fusion Leading to Senescence in Cancer Cells

Cell fusion is an important biological process that can significantly influence cancer progression by generating cells with increased genomic content and altered phenotypes. In tumors, fusion events between cancer cells or between cancer and stromal cells often lead to the formation of binucleated or polyploid cells, which are prone to genomic instability. As a protective response, such aberrant cells frequently undergo cellular senescence, a stable state of cell-cycle arrest mediated by pathways such as p53/p21 and p16/Rb. While this fusion-induced senescence acts as a barrier to uncontrolled proliferation, the accumulation of senescent cells and their associated secretory phenotype may paradoxically contribute to tumor progression and remodeling of the tumor microenvironment.

10.5.2.1 Mechanisms of Cell Fusion-Induced Senescence

When cancer cells undergo fusion, they experience several stressors that can drive senescence:

(a) *Genomic Instability and DNA Damage Response (DDR) Activation*
Fusion results in multinucleation, chromosome missegregation, and replication stress. This activates the DDR pathways, primarily involving ATM/ATR kinases, leading to p53 stabilization and cell cycle arrest. γ-H2AX foci, markers of DNA double-strand breaks, accumulate in fused cells (Valdiglesias et al. 2013).

(b) *Mitotic Dysfunction and Polyploidy*
Fused cells often fail cytokinesis, becoming polyploid (Gentric and Desdouets 2014). Polyploid cells activate the p53/p21 and p16/Rb pathways, enforcing senescence (Giménez-Bastida et al. 2019). If p53 is nonfunctional (as in many cancers), fused cells may escape senescence, leading to tumor progression instead (Chen 2016).

(c) *Oxidative and Metabolic Stress*
Mitochondrial dysfunction in fused cells leads to ROS (reactive oxygen species) accumulation, which can induce senescence (Wang et al. 2022). Oxidative stress activates p38 MAPK and NF-κB, reinforcing the senescence-associated secretory phenotype (SASP) (Yasuda et al. 2022).

(d) *Epigenetic and Transcriptional Changes*
Fusion alters chromatin states, leading to the global heterochromatin formation (H3K9me3 accumulation) (Grewal 2023). Activation of senescence-associated genes (CDKN1A/p21, CDKN2A/p16, IL6, CXCL8) (Leontieva and Blagosklonny 2013).

10.5.2.2 Features of Fusion-Induced Senescence

Cell fusion-induced senescent cells display characteristic features (Fig. 10.4a), as outlined below:

(a) *Cell Cycle Arrest*
G1-phase arrest mediated by p16INK4a/Rb and p53/p21 pathways (Giménez-Bastida et al. 2019). Polyploidy-linked G2/M arrest due to mitotic stress (Sazonova et al. 2021).

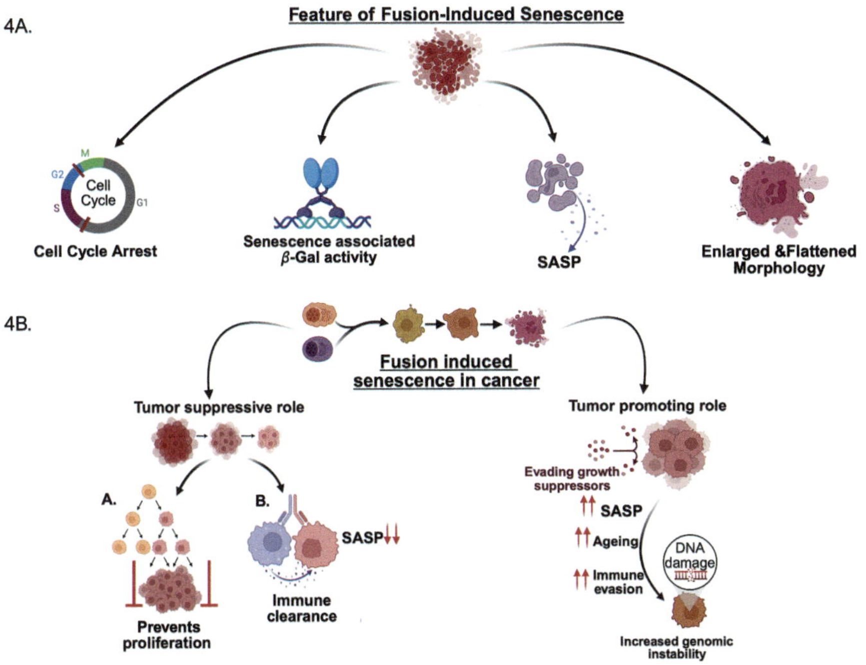

Fig. 10.4 (**a**) Key features include cell cycle arrest, senescence-associated β-galactosidase activity, senescence-associated secretory phenotype (SASP), and enlarged, flattened morphology.(**b**) Fusion-induced senescence has dual roles in cancer. Its tumor-suppressive effects include reduced proliferation and decreased SASP formation, leading to immune clearance. Conversely, its tumor-promoting effects include SASP induction, aging, immune evasion, and increased genomic instability. (Figure created with BioRender)

(b) *Senescence-Associated β-Galactosidase (SA-β-Gal) Activity*

Fused cells express SA-β-Gal, a common marker of senescence (Valieva et al. 2022).

(c) *Senescence-Associated Secretory Phenotype (SASP)*

Fused senescent cells secrete pro-inflammatory factors, including: Cytokines: IL-6, IL-8 (promote inflammation and immune recruitment) (Orange et al. 2023). Growth factors: VEGF, IGF (can paradoxically promote tumor growth). Matrix metalloproteinases (MMPs): Enhance tissue remodeling and metastasis (Page-McCaw et al. 2007).

(d) *Enlarged and Flattened Morphology*

Fused senescent cells exhibit hypertrophy, vacuolation, and irregular nuclei (Kloc et al. 2022).

10.5.2.3 Dual Role of Fusion-Induced Senescence in Cancer

Cell fusion-induced senescence exhibits both tumor-suppressive and tumor-promoting roles (Fig. 10.4b).

(a) *Tumor-Suppressive Role*
Senescence prevents fused cancer cells from proliferating. Senescent cells can trigger immune clearance via SASP (recruitment of NK cells, macrophages) (Giannoula et al. 2023).
(b) *Tumor-Promoting Role*
SASP factors can create a pro-tumorigenic environment by promoting angiogenesis and immune evasion (Takasugi et al. 2022). Senescent cells that escape arrest often reenter the cell cycle with increased genomic instability, leading to aggressive tumor subclones (Shultes et al. 2024).

10.5.2.4 Experimental Evidence of Cell Fusion-Induced Senescence

In Vitro Fusion Models Cancer cells engineered with differentially fluorescent tags (GFP/RFP) show fusion, multinucleation, and senescence markers (Crivat and Taraska 2012; Chen et al. 2016).

Senescence Markers
SA-β-Gal staining (Piechota et al. 2016). p16INK4a and p21 expression (Cazzalini et al. 2010; Giménez-Bastida et al. 2019). γ-H2AX foci (DNA damage response activation) (Valdiglesias et al. 2013).

Xenograft Studies Fused cells injected into mice show increased SASP secretion and immune infiltration (Zhang et al. 2015).

10.5.2.5 Therapeutic Implications: Exploiting Fusion-Induced Senescence

(a) *Senolytic Therapy (Eliminating Senescent Cancer Cells)*
Navitoclax (BCL-2 inhibitor): Shown to target apoptosis-resistant senescent cells (Zhu et al. 2016). Dasatinib + Quercetin: This combination promotes the selective clearance of senescent cells (Hickson et al. 2019).
(b) *Immunotherapy Targeting SASP Factors*
IL-6/IL-8 Inhibitors: Reduce the tumor-promoting effects of the SASP by suppressing pro-inflammatory cytokines (Dal Bo et al. 2025). Checkpoint inhibitors (PD-1/PD-L1, CTLA-4): Enhance immune recognition and clearance of senescent tumor cells (Tang et al. 2022).

(c) *Inducing Terminal Senescence in Polyploid Cells*

CDK4/6 inhibitors (Palbociclib): Strengthen p16-mediated senescence in fused cancer cells (Kim and Scott 2017). DNA damage enhancers (PARP inhibitors): Force senescence into permanent growth arrest (Chen 2011).

10.6 Polyploidy and Senescence: A Linked Process

Polyploidy and senescence are interconnected processes in cancer cells, influencing tumor progression, therapy resistance, and cellular fate. Here is how they are linked:

I. *Polyploidy and Senescence as Stress Responses*

Cancer cells often undergo polyploidization (whole-genome duplication) as an adaptation to stress, such as chemotherapy or radiation (Dörnen et al. 2020). Similarly, senescence is a response to stress (DNA damage, oxidative stress, telomere shortening) which cells enter a stable growth-arrest state (Kong et al. 2011).

II. *Polyploidy Can Induce Senescence*

Polyploid cells often experience mitotic stress, leading to genomic instability and activation of the DNA damage response (DDR) (Bretscher and Fox 2016). DDR can trigger senescence-associated pathways (e.g., p53-p21, p16-Rb) to prevent uncontrolled proliferation (Giménez-Bastida et al. 2019). Telomere dysfunction in polyploid cells may also reinforce senescence (Rossiello et al. 2022).

III. *Escape from Senescence and Polyploidy-Driven Tumor Evolution*

Senescent cells are not always permanently arrested; they can reenter the cell cycle and generate new, more aggressive cancer clones (Was et al. 2022). Some polyploid senescent cells undergo depolyploidization, leading to the emergence of aneuploid cells with increased tumorigenic potential (Vitale et al. 2010).

IV. *Therapeutic Implications*

Many anticancer drugs induce polyploidy or senescence as a mechanism to suppress tumor growth (Fitsiou et al. 2022). However, if polyploid senescent cells evade arrest, they can lead to therapy resistance (Song et al. 2023). Targeting the senescence-associated secretory phenotype (SASP) or exploiting vulnerabilities in polyploid cells (e.g., mitotic catastrophe) is a potential therapeutic approach (Dal Bo et al. 2025).

10.6.1 Similarities Between Polyploidy and Senescence

Polyploidy and senescence in cancer cells share several similarities, particularly in how they act as stress responses and influence tumor progression (Fig. 10.5). Here are the key similarities:

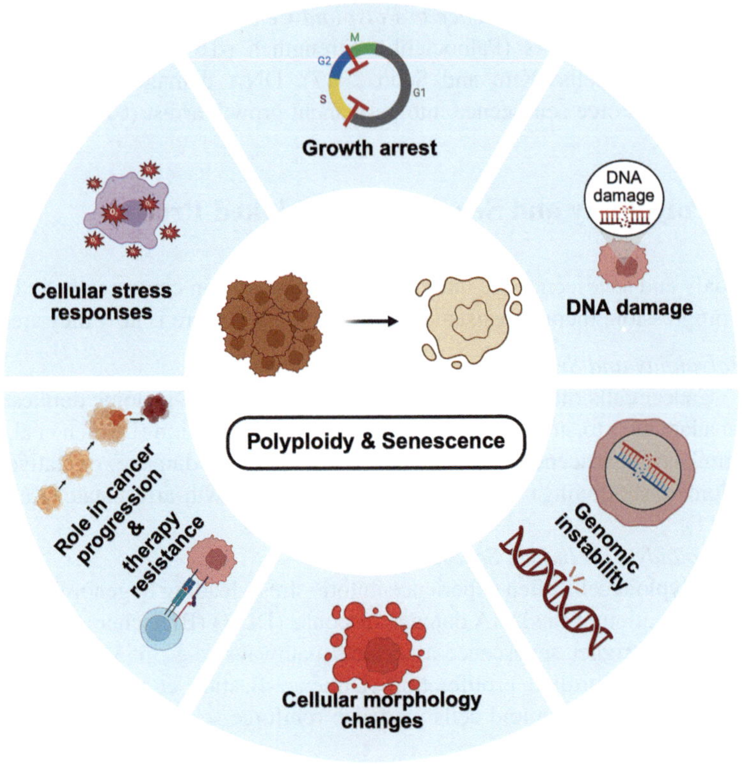

Fig. 10.5 Interconnection between polyploidy and senescence. Polyploidy and senescence are interconnected processes that contribute to growth arrest, DNA damage, cellular stress responses, and morphological changes. Both promote genomic instability and are implicated in cancer progression and therapy resistance. (Figure created with BioRender)

I. *Cellular Stress Responses*

Both polyploidy and senescence can be triggered by genotoxic stress, such as chemotherapy, radiation, or oxidative stress (Swift and Golsteyn 2014; Dikovskaya et al. 2015). They act as protective mechanisms against uncontrolled proliferation but can also contribute to therapy resistance (Song et al. 2023).

II. *Growth Arrest and Cell Cycle Checkpoints*

Senescent cells undergo a stable G1/S arrest, mediated by the p53/p21 or p16/Rb pathways (Giménez-Bastida et al. 2019). Polyploid cells often experience mitotic failure and become arrested at the G2/M checkpoint due to the activation of the DNA damage response (DDR) (Sazonova et al. 2021).

III. *DNA Damage and Genomic Instability*

Both conditions are associated with DNA damage accumulation and activation of DDR (ATM/ATR, p53 pathways) (Chen 2016). Genomic instability in polyploid cells can lead to chromosomal aberrations, while senescence is often a response to such instability (Lee and Ong 2021).

IV. *Cellular Morphology Changes*
Polyploid cells become enlarged due to an increase in DNA content (Fang et al. 2023). Senescent cells also exhibit enlarged, flattened morphology, often with increased β-galactosidase activity (Cho et al. 2004).

V. *Role in Cancer Progression and Therapy Resistance*
Both can act as barriers to cancer progression by preventing rapid proliferation. However, if cancer cells escape senescence or polyploidy-induced arrest, they can evolve into more aggressive and therapy-resistant clones (Saleh et al. 2022).

VI. *Involvement in the Tumor Microenvironment*
Senescent cells secrete senescence-associated secretory phenotype (SASP) factors, which promote inflammation and tumor progression (Takasugi et al. 2022). Polyploid cells can contribute to a similar pro-tumorigenic environment by releasing DNA and inflammatory cytokines (Shabo et al. 2020; Orange et al. 2023).

Polyploidy and senescence in cancer cells share some similarities but have distinct characteristics in terms of their mechanisms, outcomes, and roles in cancer progression. Here are the key differences:

10.6.2 Polyploidy and Senescence: Differences

I. *Definition and Nature*
Polyploidy: A state where a cell has more than the normal diploid set of chromosomes due to errors in mitosis or cytokinesis failure (Ghosh et al. 2024). Senescence: A state of stable, irreversible cell cycle arrest triggered by stress or damage (Varela-Eirín and Demaria 2022).

II. *Cell Cycle Status*
Polyploidy: Polyploid cells may remain in the G2/M phase or undergo aberrant mitosis. Some may eventually reenter the cell cycle (Bretscher and Fox 2016). Senescence: Cells are permanently arrested in the G1 phase and do not divide (Giménez-Bastida et al. 2019).

III. *Reversibility and Tumorigenic Potential*
Polyploidy: Can be reversible; polyploid cells may undergo depolyploidization, leading to aneuploid progeny with increased tumorigenic potential (Vitale et al. 2010). Senescence: Typically irreversible, but some cancer cells can bypass senescence and regain proliferative capacity (Reimann et al. 2024).

IV. *Genetic and Molecular Pathways*
Polyploidy: Often associated with mitotic checkpoint defects, spindle assembly checkpoint failure, Aurora kinases, and p53 inactivation (Den

Table 10.1 Morphological and functional differences between polyploidy and senescence

Feature	Polyploidy	Senescence
Cell size	Enlarged due to increased DNA content (Fang et al. 2023)	Enlarged, flattened shape (Cho et al. 2004)
Nuclear features	Multiple or giant nuclei (Was et al. 2022)	Enlarged nucleus with heterochromatin foci (Aird and Zhang 2013)
Metabolism	High metabolic flexibility (Anatskaya and Vinogradov 2022)	Increased β-galactosidase activity, altered metabolism (Valieva et al. 2022)
Secretory activity	Less secretion of inflammatory factors (Hayashi et al. 2024)	High SASP secretion (IL-6, IL-8, MMPs) (Yasuda et al. 2022)

Hollander et al. 2010; Chen 2016). Senescence: Primarily regulated by p53/p21 and p16/Rb pathways, which enforce growth arrest (Beauséjour et al. 2003).

V. *Impact on Tumor Progression*

Polyploidy: Can promote genomic instability, leading to increased mutation rates and tumor evolution (Shultes et al. 2024). Senescence: Acts as a tumor-suppressive mechanism, preventing damaged cells from proliferating, but can contribute to cancer progression through SASP (Senescence-Associated Secretory Phenotype) (Takasugi et al. 2022).

VI. *Morphological and Functional Features*

Polyploidy and senescence exhibit differences in their morphological and functional features, as highlighted in Table 10.1.

VII. *Response to Therapy*

Polyploidy: May contribute to therapy resistance by providing genetic diversity for cancer evolution. Senescence: Initially enhances therapy response, but senescent cells can promote tumor relapse via SASP-driven inflammation and microenvironment remodeling (Takasugi et al. 2022).

VIII. *Fate of Cells*

Polyploid Cells: Can undergo mitotic catastrophe, apoptosis, or generate new cancer clones through depolyploidization (Vitale et al. 2010). Senescent Cells: Typically remain in an arrested state but may later undergo apoptosis or immune clearance (Reimann et al. 2024).

10.7 Signaling Pathways Responsible for Cell Fusion in Cancer

Cell fusion in cancer is regulated by multiple signaling pathways that control membrane dynamics, cytoskeletal rearrangements, and transcriptional programs. These pathways contribute to tumor progression by promoting tumor evolution, therapy resistance, and metastasis (Zhang et al. 2022).

I. *Key Signaling Pathways Involved in Cancer Cell Fusion*

A. *Phosphoinositide 3-Kinase (PI3K)/Akt Pathway*

Function: Regulates cell survival, migration, and membrane dynamics in cancer cells (Fresno Vara et al. 2004).

- *Mechanism:*
  - PI3K activation → Akt phosphorylation → mTOR signaling (Glaviano et al. 2023).
  - Enhances fusogenic protein (Syncytin-1) expression (Wang et al. 2024).
  - Promotes actin remodeling, facilitating fusion (Eitzen 2003).
- *Evidence:*
  - PI3K/Akt activation increases cell fusion rates in glioblastoma and breast cancer (Zheng et al. 2009; Behrooz et al. 2022).
  - Inhibitors (e.g., LY294002, Wortmannin) block fusion events (Fukuchi et al. 2000).

B. *Mitogen-Activated Protein Kinase (MAPK) Pathway*

Function: Controls cell proliferation, stress responses, and fusion dynamics.

- *Mechanism:*
  - ERK1/2 activation: Increases cadherin expression (E-cadherin, N-cadherin), promoting fusion (Yun et al. 2025).
  - p38 MAPK activation: Triggers senescence or stress-induced fusion (Shimada et al. 2009).
- *Evidence:*
  - ERK signaling enhances fusion of melanoma cells and macrophages (Clawson et al. 2015).
  - MAPK inhibitors (U0126, SB203580) reduce cell fusion (Henklova et al. 2008).

C. *Wnt/β-Catenin Signaling*

Function: Promotes cell adhesion, EMT, and cancer stem-like properties after fusion (Basu et al. 2018).

- Mechanism:
  - Wnt ligand binding stabilizes β-catenin, increasing cadherin expression (Basu et al. 2018).

- Enhances fusion-driven EMT, leading to increased motility (Xue et al. 2024).

- Evidence:
  - Wnt3a stimulation increases fusion in colorectal cancer (Qi et al. 2014).
  - β-catenin inhibitors (ICG-001) reduce fusion events (Danieau et al. 2021).

D. *NF-κB Signaling Pathway*

Function: Regulates inflammatory cytokines that promote fusion (Lawrence 2009).

- Mechanism:
  - TNF-α, IL-6, IL-1β activate NF-κB, leading to upregulation of fusogenic proteins (Brasier 2010).
  - Increases Syncytin-1, annexins, and EMT markers (Yan et al. 2017).
- Evidence:
  - NF-κB inhibition (BAY 11-7082) reduces tumor cell fusion (Nasrollahzadeh et al. 2022).
  - Macrophage-tumor fusion is NF-κB-dependent (Mancino and Lawrence 2010).

E. *TGF-β/SMAD Signaling*

Function: Induces fusion in epithelial-to-mesenchymal transition (EMT) (Miyazono 2009).

- Mechanism:
  - TGF-β binds TGF-βR → SMAD2/3 phosphorylation → EMT activation (Cai et al. 2019).
  - Suppresses E-cadherin, promotes N-cadherin → enhances fusion (Liao et al. 2024).
- Evidence:
  - TGF-β increases fusion in breast and lung cancers (Fong et al. 2009).
  - TGF-β inhibitors (SB431542) block EMT-related fusion (Halder et al. 2005).

F. *Rho GTPase (Ras-related C3 botulinum toxin substrate 1 (Rac1), Cell division cycle 42 (Cdc42), Ras homolog family member A (RhoA)) Signaling*

Function: Regulates cytoskeletal changes required for membrane merging (Haga and Ridley 2016).

- Mechanism:
  - Rac1 and Cdc42 promote actin polymerization, leading to membrane fusion (Srinivasan et al. 2003).
  - RhoA enhances cell contraction, bringing membranes closer (Cavanaugh et al. 2020).
- Evidence:
  - Rac1 activation enhances fusion in glioblastoma (Xu et al. 2020).
  - RhoA inhibitors (Y-27632) reduce fusion events (Imamura et al. 2000).

II. *Cross Talk Between Pathways*

- PI3K/Akt and NF-κB → Both pathways regulate Syncytin-1, enhancing fusion (Yan et al. 2017; Wang et al. 2024).
- MAPK and Wnt/β-Catenin → Work together in EMT-driven fusion (Xue et al. 2024; Yun et al. 2025).
- TGF-β and NF-κB → Cooperate to induce macrophage-cancer fusion (Mancino and Lawrence 2010).

III. *Therapeutic Strategies to Target Fusion Pathways*

Inhibitors of cell signaling pathways have been explored as therapeutic strategies to target cell fusion, as presented in Table 10.2.

Table 10.2 Effect of cell signaling inhibitors on cell fusion

Pathway	Targeted therapy	Effect on fusion
PI3K/Akt	LY294002, Wortmannin	Blocks fusion by downregulating p21 (Kaliman et al. 1996)
MAPK (ERK)	SB203580	Inhibits cadherin-mediated fusion (Henklova et al. 2008)
NF-κB	BAY 11-7082	Suppresses TNF-α-driven fusion (Nasrollahzadeh et al. 2022)
TGF-β	SB431542	Blocks EMT-associated fusion (Halder et al. 2005)
Wnt/β-catenin	ICG-001	Prevents β-catenin-driven fusion (Danieau et al. 2021)
Rho GTPase	Y-27632 (RhoA inhibitor)	Disrupts cytoskeletal rearrangement (Imamura et al. 2000)

10.7.1 mTOR Signaling in Polyploidy and Senescence in Cancer Cells

The mechanistic target of rapamycin (mTOR) is a central regulator of cell growth, metabolism, and survival (Mehta et al. 2024). It plays a crucial role in the development of polyploidy and senescence in cancer cells, affecting tumor progression and drug resistance (Choudhury et al. 2024).

I. *mTOR Pathway Overview*

mTOR functions through two main complexes:

- *mTORC1 (Raptor-dependent)* → Controls cell growth, protein synthesis, and metabolism (Takahara et al. 2020).
- *mTORC2 (Rictor-dependent)* → Regulates cytoskeletal dynamics and cell survival via Akt signaling (Li et al. 2015).

mTOR is activated by growth factors (e.g., IGF-1, insulin), amino acids, and cellular stress, and it integrates inputs from PI3K- Akt, AMPK, and p53 signaling (Huang and Fingar 2014).

II. *mTOR in Polyploidy*

Polyploid cancer cells arise due to mitotic slippage, failed cytokinesis, or endoreduplication (Was et al. 2022). mTOR contributes to polyploidization through:

A. *mTORC1 and Cell Cycle Dysregulation*

- Promotes G1-S and G2-M Transition
- mTORC1 upregulates Cyclin D1 (CCND1) and Cyclin E1 (CCNE1), driving DNA replication and allowing polyploidization (Lamm et al. 2019). mTORC1 suppresses p27 (CDKN1B) and p21 (CDKN1A), enabling cell cycle progression even under stress conditions (Patra et al. 2024).
- Mitotic Slippage and Cytokinesis Failure
- Hyperactivation of mTORC1 disrupts the spindle assembly checkpoints (SAC), leading to defective mitosis and polyploidization (Yamada et al. 2022). Increased Aurora Kinase A/B (AURKA/B) and PLK1 expression promote mitotic bypass (Vats et al. 2025).

B. *mTORC2 and Cell Growth in Polyploid Cells*

mTORC2 activates Akt and SGK1, supporting polyploid cell survival under stress (Ragupathi et al. 2024). Polyploid cells show increased metabolic demand, and mTORC1/mTORC2 balance helps maintain their viability (Linke et al. 2017).

C. *Autophagy and Polyploidy*

mTOR inhibition (e.g., rapamycin) induces autophagy, which can either support polyploid cell survival or trigger cell death (Kim and Guan 2015). Polyploid cells often exhibit high autophagic flux, aiding in drug resistance (Patra et al. 2024).

III. *mTOR in Senescence of Cancer Cells*

Cellular senescence is a permanent cell cycle arrest triggered by stress, oncogene activation, or chemotherapy (Zhang and Yang 2011). The mTOR pathway influences senescence in two ways:

A. *mTORC1-Driven Senescence*

- *SASP (Senescence-Associated Secretory Phenotype)*
  mTORC1 regulates the NF-κB and IL-6/IL-8 signaling, promoting a pro-inflammatory environment (Hu et al. 2016). SASP factors (IL-1β, CXCLs) create a tumor-supportive niche and may lead to therapy resistance (Luo et al. 2025).
- *mTORC1 and p53/p21 Axis*
  Hyperactive mTORC1 can reinforce senescence by upregulating p53 and p21, preventing further proliferation (Jung et al. 2019). Rapamycin-mediated mTOR inhibition delays senescence by maintaining quiescence instead of full senescence (Blagosklonny 2018).

B. *mTORC2 and Escape from Senescence*

mTORC2- Akt activation allows some senescent cells to reenter the cell cycle, leading to tumor recurrence (Son et al. 2024). Polyploid cells can use mTORC2 to evade senescence and survive under stress (Choudhury et al. 2024).

IV. *Implications for Cancer Therapy*

Since chemotherapy-resistant cells often display polyploidy and senescence-like states, targeting mTOR could enhance therapeutic outcomes:

- *mTOR Inhibitors (e.g., Rapamycin, Everolimus)* → May reduce polyploid cell survival (Ciołczyk-Wierzbicka et al. 2019).
- *Dual mTORC1/mTORC2 Inhibitors (e.g., Torin-1, AZD8055)* → Prevent escape from senescence and re-proliferation (Leontieva et al. 2015).
- *Autophagy Modulation* → Combining bortezomib with autophagy inhibitors (e.g., chloroquine) could disrupt polyploid cell survival (Di Lernia et al. 2020).

10.7.2 mTOR Signaling in Cell Fusion

The mechanistic target of rapamycin (mTOR) pathway plays a crucial role in regulating cell growth, metabolism, and survival, all of which are essential for cell fusion (Murugan 2019; Mehta et al. 2024). In cancer, cell fusion contributes to tumor heterogeneity, metastasis, and drug resistance (López-Collazo and Hurtado-Navarro 2025). mTOR signaling influences this process by controlling membrane dynamics, cytoskeletal reorganization, and metabolic adaptation—key factors in successful cell fusion.

I. *mTORC1 and Cell Fusion*

mTORC1 is the nutrient-sensing arm of the pathway that integrates growth signals and metabolic cues (Murugan 2019). It contributes to cell fusion by:

A. *Regulating Membrane Fusion and Cytoskeletal Dynamics*

- *mTORC1 controls actin remodeling*, which is critical for membrane merging that occurs during fusion (Eitzen 2003).
- *Upregulates Rho GTPases (RhoA, Rac1, Cdc42)*, which drive membrane protrusions and fusion (Chircop 2014).
- *Promotes lipid biosynthesis via SREBP1 activation,* providing membrane components for fusion events (Porstmann et al. 2008).

B. *Enhancing Metabolic Adaptation for Fusion*

- *Increases adenosine triphosphate (ATP) and lipid metabolism,* ensuring the energy supply needed for fusion (Liu and Sabatini 2020).
- *Activates HIF-1α under hypoxia,* which supports tumor cell fusion and survival (Dodd et al. 2015).

C. *Regulating Fusogenic Protein Expression*

mTORC1 affects the expression of:

- *Syncytin-1 and Syncytin-2* → fusogenic proteins involved in cell fusion (Furuta et al. 2022).
- *E-cadherin and N-cadherin* → regulate adhesion and membrane fusion potential (Barnes et al. 2010).

II. *mTORC2 and Cell Fusion*

mTORC2 is involved in cytoskeletal remodeling and cell survival, making it critical for fusion-driven tumor progression.

A. *PI3K/mTORC2- Akt Activation Enhances Fusion Potential*

- Phosphorylates Akt (Ser473), leading to increased fusion events (Li et al. 2022).
- Activates integrin signaling, enhancing adhesion and fusion between cells (Torti et al. 2020).

B. *FAK (Focal Adhesion Kinase) and Cell Fusion*

- mTORC2 stabilizes FAK, which promotes fusion in mesenchymal-like cancer cells (Farhan et al. 2015).
- FAK/mTORC2 signaling enhances EMT (epithelial-to-mesenchymal transition), increasing fusion potential (Gonzalez and Medici 2014).

III. *mTOR, Cell Fusion, and Chemoresistance*

Cancer cell fusion generates hybrid cells that are more resistant to chemotherapy (Hass et al. 2021a, 2021b). mTORC1/MYC-driven hybrids show increased survival under drug stress (Pourdehnad et al. 2013). mTORC2/Akt activation enhances fusion-mediated resistance by increasing antiapoptotic signaling (Dormond et al. 2007). Targeting mTOR (e.g., using rapamycin or dual mTORC1/mTORC2 inhibitors) could reduce cell fusion and enhance drug sensitivity in resistant cancer cells.

10.8 Role of lncRNAs in Cell Fusion

Long noncoding RNAs (lncRNAs) are emerging as key regulators of cell fusion in various biological processes, including cancer progression (Uppaluri et al. 2023). They influence cell fusion by modulating gene expression, signaling pathways, and epigenetic modifications. Here is how lncRNAs contribute to cell fusion:

A. *Regulation of Fusogenic Gene Expression*

lncRNAs can regulate the expression of genes involved in membrane fusion, such as:

- *Syncytins (Syncytin-1, Syncytin-2):* These are fusogenic proteins initially identified in placental development but are also implicated in cancer cell fusion (Talotta 2023).
- *E-cadherin and N-cadherin:* EMT-associated cadherins influence cell fusion events, often controlled by lncRNAs like metastasis -associated lung adenocarcinoma transcript 1 (MALAT1) and HOX transcript antisense Intergenic RNA (HOTAIR) (Ahmad et al. 2023).

Example:

- lncRNA HOTAIR promotes epithelial-to-mesenchymal transition (EMT), which increases cancer cell fusion potential (Jarroux et al. 2021).

B. *EMT and Cell Fusion*

The epithelial-to-mesenchymal transition (EMT) enhances the plasticity of cancer cells, making them more prone to fusion. Several lncRNAs regulate EMT and may indirectly drive fusion events.

- *LncRNA MALAT1* promotes EMT by regulating ZEB1/ZEB2, facilitating cell fusion (Gugnoni and Ciarrocchi 2019).
- *LncRNA ZFAS1* enhances EMT in breast cancer, which could contribute to fusion-mediated metastasis (Sharma et al. 2021).

C. *Tumor-Stromal and Tumor-Immune Cell Fusion*

Cancer cells can fuse with macrophages, mesenchymal stem cells (MSCs), or fibroblasts, leading to hybrid cells with enhanced metastasis and immune evasion (Shabo et al. 2020).

- *LncRNA NEAT1* regulates macrophage polarization and fusion with tumor cells (Gao et al. 2020).
- *LncRNA LINK-A* modulates interactions between cancer cells and stromal components, potentially affecting fusion (Agostini et al. 2022).

10.9 Unanswered Questions in the Field of Polyploidy

Polyploidy is a fascinating and complex phenomenon with several unanswered questions in cancer biology and beyond. Some of the key unknowns in the field include:

I. *Polyploidy and Tumor Evolution*

- *How does polyploidy contribute to long-term tumor evolution?*
  - While polyploidy is linked to genomic instability, it is unclear how frequently polyploid tumors give rise to more aggressive subpopulations through depolyploidization or aneuploidy.
- *What factors determine whether a polyploid cell will undergo mitotic catastrophe, senescence, or tumorigenic evolution?*
  - Some polyploid cells die, while others fuel cancer progression. The molecular switches governing these outcomes remain poorly understood.

II. *Polyploidy and Drug Resistance*

- *What are the specific mechanisms through which polyploid cells resist chemotherapy?*
  - Some evidence suggests that polyploidy enhances drug efflux, DNA repair efficiency, and metabolic adaptation (Tamaki et al. 2011; Cadart et al. 2023), but the exact molecular pathways are still being explored.
- *Can polyploid cells serve as a reservoir for tumor relapse after therapy?*
  - It is hypothesized that polyploid giant cancer cells (PGCCs) can survive treatment and later produce resistant progeny, but their exact role in recurrence remains to be fully established (Zhou et al. 2022).

III. *Polyploidy and Cellular Plasticity*

- *How do polyploid cells regulate cell fate decisions?*
  - Polyploid cells exhibit stem-like properties (Choudhury et al. 2024; Ghosh et al. 2024), but it is unclear whether this is an inherent feature or is induced by external stressors like hypoxia and therapy.

- *Can polyploidy induce epithelial-to-mesenchymal transition (EMT) and metastasis?*
  - Some studies suggest that polyploid cells have increased migratory potential, but direct evidence linking polyploidization to EMT remains incomplete (Shabo et al. 2015, 2020).

IV. *Role of Polyploidy in Normal Physiology vs. Cancer*

- *Why do some normal tissues (e.g., liver, heart, placenta) tolerate polyploidy while cancer cells use it for survival?*
  - Polyploidy is beneficial in some physiological contexts, yet it is often detrimental in cancer. The mechanistic differences between physiological and pathological polyploidy remain unclear.
- *Are there unknown tumor-suppressor mechanisms that selectively eliminate polyploid cells?*
  - Some polyploid cells are cleared by immune surveillance, but how the immune system differentiates harmful from normal polyploid cells is still an open question.

V. *Therapeutic Targeting of Polyploidy*

- *Can we selectively target polyploid cells without harming normal diploid cells?*
  - Drugs like AURKA inhibitors and mitotic checkpoint modulators show promise, but a selective strategy to eliminate polyploid cancer cells remains elusive.
- *What are the vulnerabilities of polyploid cancer cells that can be exploited for therapy?*
  - Polyploid cells may have specific metabolic dependencies or stress responses that could be targeted, but these remain poorly characterized.

VI. *Polyploidy, Aging, and Senescence*

- *What is the relationship between polyploidy and cellular senescence?*
  - Some polyploid cells enter senescence, while others escape and proliferate (Mosieniak and Sikora 2010). The key regulators of this decision are still unknown.
- *Does polyploidy contribute to age-related diseases beyond cancer?*
  - While polyploid cells are found in aging tissues, their impact on organ dysfunction, fibrosis, and degenerative diseases is still unclear.

10.10 Unanswered Questions in the Field of Senescence

Senescence in cancer cells is a rapidly evolving field with many unresolved questions. Despite its well-established role as a tumor-suppressive mechanism, senescence also contributes to cancer progression, therapy resistance, and immune evasion. Here are some key unknowns:

I. *Mechanisms Regulating Senescence Entry and Escape*

- *What determines whether a cancer cell undergoes senescence or apoptosis in response to stress?*
  - Many cancer therapies induce senescence, but some cells instead undergo apoptosis. The precise molecular cues that dictate this decision remain unclear.
- *How do cancer cells escape senescence, and what triggers their reentry into the cell cycle?*
  - Some senescent cancer cells regain proliferative capacity, potentially leading to more aggressive tumor cells (Feng et al. 2025). The molecular drivers of senescence reversal ("senescence escape") are not fully understood.
- *Is therapy-induced senescence truly irreversible, or can it always be reversed under certain conditions?*
  - While senescence is considered permanent, senescence-associated stemness (SAS) suggests that cells may regain proliferative potential under specific circumstances (Milanovic et al. 2018).

II. *Senescence-Associated Secretory Phenotype (SASP) and Tumor Progression*

- *What determines whether SASP promotes tumor suppression or tumor progression?*
  - The inflammatory secretome (IL-6, IL-8, MMPs) can have opposing effects on tumors, but the mechanisms governing this duality are not well understood (López de Andrés et al. 2020).
- *How does SASP influence the tumor microenvironment in different cancers?*
  - SASP can attract immune cells, promote fibrosis, or enhance angiogenesis, but the context-dependent effects of SASP on different tumor types remain unclear (Colucci et al. 2025).

- *Can SASP be selectively inhibited without disrupting beneficial senescence effects?*
  - Targeting SASP to prevent tumor progression is a promising approach, but selectively modulating SASP components without affecting senescence remains a challenge.

III. *Interactions Between Senescence and the Immune System*

- *How do senescent cancer cells evade immune clearance?*
  - The immune system can clear senescent cells, yet many persist in tumors. Understanding senescence-associated immune evasion is crucial for developing senolytic therapies.
- *What are the key immune cells responsible for eliminating senescent cancer cells?*
  - While natural killer (NK) cells, macrophages, and T cells are implicated in senescence clearance, their precise roles in different cancers remain an open question.
- *Can immunotherapy be combined with senescence-inducing drugs to improve cancer treatment?*
  - Some studies suggest that senescence makes tumors more immunogenic, but the best strategies for integrating senescence-based therapy with immunotherapy remain unclear.

IV. *Senescence in Therapy Resistance and Relapse*

- *Is therapy-induced senescence a mechanism of resistance in cancer?*
  - While senescence initially prevents tumor growth, therapy-induced senescent cells (TISCs) can later drive recurrence (Song et al. 2023). The exact role of senescence in residual disease is still unclear.
- *How do senescent cells contribute to drug resistance in heterogeneous tumors?*
  - Some senescent cells remain viable and resistant to therapy, but the mechanisms that allow this survival are poorly understood.
- *Can senolytic drugs effectively eliminate senescent cancer cells without harming normal tissue?*
  - Senolytics (drugs targeting senescent cells, like navitoclax or fisetin) are promising, but their selectivity and effectiveness in cancer therapy need further research (Zhu et al. 2016; Hickson et al. 2019).

V. *Senescence and Polyploidy Interconnection*

- *Does polyploidy promote or suppress senescence in cancer cells?*
  - Some polyploid cells enter senescence, while others escape and drive tumor progression. The molecular factors determining these outcomes remain unclear.
- *Can senescent polyploid cells contribute to therapy resistance and relapse?*
  - Polyploid senescent cells may act as a reservoir for resistant clones, but their exact role in tumor recurrence is unknown.

VI. *Senescence in the Tumor Microenvironment (TME)*

- *How do senescent stromal cells influence cancer progression?*
  - Senescent fibroblasts and endothelial cells can secrete factors that either suppress or support tumor growth, but their precise role is unclear.
- *Can we develop therapies that selectively target senescent cells in the TME without affecting tumor-suppressive senescence?*
  - Senescence in cancer cells and the surrounding stroma may have different consequences, making selective targeting challenging.

VII. *Senescence and Metabolism*

- *How does metabolic reprogramming support senescent cancer cells?*
  - Senescent cells often undergo metabolic shifts (e.g., increased oxidative stress or mitochondrial dysfunction), but the metabolic vulnerabilities of these cells remain unclear.
- *Can metabolic interventions enhance senolytic therapies?*
  - Targeting metabolism could make senescent cancer cells more susceptible to clearance, but the best strategies are still unknown.

VIII. *Senescence, Aging, and Cancer Progression*

- *How does age-related accumulation of senescent cells impact cancer development?*
  - Older individuals have more senescent cells, but whether they promote or prevent tumorigenesis is still debated.
- *Does senescence in noncancerous cells contribute to cancer aggressiveness?*
  - Senescent immune cells, fibroblasts, or endothelial cells could enhance tumor progression via SASP, but their exact role is unclear.

10.11 Future Perspectives

Although significant progress has been made in understanding the role of cell fusion in cancer, many critical questions remain unresolved. First, it is unclear how tumors regulate the balance between fusion-induced senescence as a protective barrier and the escape of polyploid cells that drive malignant progression. Identifying the molecular "switches" that determine whether fused cells remain arrested or reenter the cell cycle will be essential.

Second, the lack of reliable biomarkers to specifically identify fusion-derived hybrids in patient samples limits both mechanistic studies and clinical translation. Advances in single-cell sequencing, lineage tracing, and imaging technologies may provide novel tools to address this gap.

Third, while the tumor microenvironment—including immune cells, stromal signals, and inflammatory factors—clearly influences cell fusion and the fate of polyploid cells, its precise contribution is not yet fully understood. Deciphering these interactions could reveal therapeutic vulnerabilities.

Finally, clinical strategies need to be developed to exploit these insights. Senolytic drugs, mTOR and Aurora kinase inhibitors, or SASP modulators may hold promise, but their safety and efficacy in selectively targeting fusion-derived senescent or polyploid cancer cells must be rigorously tested. Future therapies may combine approaches that enforce permanent senescence with those that eliminate harmful polyploid escapees, offering a novel way to limit tumor plasticity, therapy resistance, and recurrence.

10.12 Conclusion

Cell fusion represents a paradoxical process in cancer biology—capable of enforcing growth arrest through polyploidy and senescence, yet equally capable of driving tumor evolution when fused cells escape this arrest. The emergence of polyploid giant cells following fusion highlights a critical adaptive mechanism by which tumors survive stress, resist therapy, and acquire stem-like plasticity. Senescence induced by fusion can act as a natural barrier to tumor growth; however, the persistence of senescent cells and their pro-inflammatory secretome (SASP) may inadvertently support a tumor-promoting microenvironment.

The mTOR pathway sits at the nexus of these events, coordinating signals that determine whether fusion leads to permanent arrest or malignant progression. Understanding the molecular switches that govern this balance is essential for therapeutic innovation. Strategies that selectively eliminate harmful senescent or polyploid cells, or that stabilize their tumor-suppressive functions, hold promise for improving cancer treatment outcomes.

Ultimately, fusion-induced polyploidy and senescence must be viewed as a "double-edged sword"—a process that can both suppress and fuel cancer. Future work should focus on developing reliable biomarkers of fusion-derived hybrids, dissecting the role of the tumor microenvironment in their fate, and designing therapies that exploit this vulnerability to halt cancer progression.

Acknowledgments Pushkar Malakar would like to thank Science and Engineering Research Board (SERB), India, for the Ramanujan Fellowship (RJF/2021/000123), DST, Government of India.

References

Abreu-Blanco MT, Verboon JM, Parkhurst SM (2014) Coordination of rho family GTPase activities to orchestrate cytoskeleton responses during cell wound repair. Curr Biol 24:144. https://doi.org/10.1016/J.CUB.2013.11.048

Agostini M, Mancini M, Candi E (2022) Long non-coding RNAs affecting cell metabolism in cancer. Biol direct 17(1):1–12. https://doi.org/10.1186/S13062-022-00341-X

Ahmad M, Weiswald LB, Poulain L, Denoyelle C, Meryet-Figuiere M (2023) Involvement of lncRNAs in cancer cells migration, invasion and metastasis: cytoskeleton and ECM crosstalk. J Exp Clin Cancer Res *42*(*1*):1–14. https://doi.org/10.1186/S13046-023-02741-X

Aird KM, Zhang R (2013) Detection of senescence-associated heterochromatin foci (SAHF). Methods Mol Biol 965:185–196. https://doi.org/10.1007/978-1-62703-239-1_12

Anatskaya OV, Vinogradov AE (2022) Polyploidy and Myc proto-oncogenes promote stress adaptation via epigenetic plasticity and gene regulatory network rewiring. Int J Mol Sci 23. https://doi.org/10.3390/IJMS23179691

Antonio Cheung CH, Huang CC, Tsai FY, Lee JYC, Cheng SM, Chang YC et al (2013) Survivin – biology and potential as a therapeutic target in oncology. Onco Targets Ther 6:1453. https://doi.org/10.2147/OTT.S33374

Barnes EA, Kenerson HL, Jiang X, Yeung RS (2010) Tuberin regulates E-cadherin localization: implications in epithelial-mesenchymal transition. Am J Pathol 177:1765. https://doi.org/10.2353/AJPATH.2010.090233

Basu S, Cheriyamundath S, Ben-Ze'ev A (2018) Cell–cell adhesion: linking wnt/β-catenin signaling with partial emt and stemness traits in tumorigenesis. F1000Res 7. https://doi.org/10.12688/F1000RESEARCH.15782.1/DOI

Beauséjour CM, Krtolica A, Galimi F, Narita M, Lowe SW, Yaswen P et al (2003) Reversal of human cellular senescence: roles of the p53 and p16 pathways. EMBO J 22:4212–4222. https://doi.org/10.1093/EMBOJ/CDG417

Behrooz AB, Talaie Z, Jusheghani F, Łos MJ, Klonisch T, Ghavami S (2022) Wnt and PI3K/Akt/mTOR survival pathways as therapeutic targets in glioblastoma. Int J Mol Sci 23. https://doi.org/10.3390/IJMS23031353

Blagosklonny MV (2018) Rapamycin, proliferation and geroconversion to senescence. Cell Cycle 17:2655–2665. https://doi.org/10.1080/15384101.2018.1554781

Brasier AR (2010) The nuclear factor-B-interleukin-6 signalling pathway mediating vascular inflammation. Cardiovasc Res 86:211–218. https://doi.org/10.1093/CVR/CVQ076

Bretscher HS, Fox DT (2016) Proliferation of double Strand break resistant polyploid cells requires drosophila FANCD2. Dev Cell 37:444. https://doi.org/10.1016/J.DEVCEL.2016.05.004

Cadart C, Bartz J, Oaks G, Liu MZ, Heald R (2023) Polyploidy in Xenopus lowers metabolic rate by decreasing total cell surface area. Curr Biol 33:1744–1752.e7. https://doi.org/10.1016/j.cub.2023.03.071

Cai J, Xia L, Li J, Ni S, Song H, Wu X (2019) Tumor-associated macrophages derived TGF-β-induced epithelial to mesenchymal transition in colorectal cancer cells through

Smad2,3–4/Snail signaling pathway. Cancer Res Treat 51:252–256. https://doi.org/10.4143/CRT.2017.613

Cavanaugh KE, Staddon MF, Munro E, Banerjee S, Gardel ML (2020) RhoA mediates epithelial cell shape changes via mechanosensitive endocytosis. Dev Cell 52:152–166.e5. https://doi.org/10.1016/j.devcel.2019.12.002

Cazzalini O, Scovassi AI, Savio M, Stivala LA, Prosperi E (2010) Multiple roles of the cell cycle inhibitor p21CDKN1A in the DNA damage response. Mutat Res Rev Mutat Res 704:12–20. https://doi.org/10.1016/j.mrrev.2010.01.009

Chang YC, Davis MJ, Kwon-Chung KJ (2024) Determination of ploidy levels and nuclear DNA content in *Cryptococcus neoformans* by flow cytometry: drawbacks with variability. J Fungi 10: 296. https://doi.org/10.3390/JOF10040296

Chen A (2011) PARP inhibitors: its role in treatment of cancer. Chin J Cancer 30:463. https://doi.org/10.5732/CJC.011.10111

Chen J (2016) The cell-cycle arrest and apoptotic functions of p53 in tumor initiation and progression. Cold Spring Harb Perspect Med 6. https://doi.org/10.1101/CSHPERSPECT.A026104

Chen Q, Wang X, Wu H, Wang H, Zhu M, Wang R et al (2016) Establishment of a dual-color fluorescence tracing orthotopic transplantation model of hepatocellular carcinoma. Mol Med Rep 13:762–768. https://doi.org/10.3892/MMR.2015.4624

Chen G, Wu K, Li H, Xia D, He T (2022) Role of hypoxia in the tumor microenvironment and targeted therapy. Front Oncol 12:961637. https://doi.org/10.3389/FONC.2022.961637

Chen S, Xing Z, Geng M, Zhao R, Yang X, Zhu X et al (2023) Macrophage fusion event as one prerequisite for inorganic nanoparticle-induced antitumor response. Sci Adv 9. https://doi.org/10.1126/SCIADV.ADD9871/SUPPL_FILE/SCIADV.ADD9871_SM.PDF

Chircop M (2014) Rho GTPases as regulators of mitosis and cytokinesis in mammalian cells. Small GTPases 5. https://doi.org/10.4161/SGTP.29770

Cho KA, Sung JR, Yoon SO, Ji HP, Jung WL, Kim HP et al (2004) Morphological adjustment of senescent cells by modulating caveolin-1 status. J Biol Chem 279:42270–42278. https://doi.org/10.1074/jbc.M402352200

Choudhury D, Ghosh D, Mondal M, Singha D, Pothuraju R, Malakar P (2024) Polyploidy and mTOR signaling: a possible molecular link. Cell Commun Signal 22. https://doi.org/10.1186/S12964-024-01526-9

Ciołczyk-Wierzbicka D, Zarzycka M, Gil D, Laidler P (2019) mTOR inhibitor everolimus-induced apoptosis in melanoma cells. J Cell Commun Signal 13:357. https://doi.org/10.1007/S12079-019-00510-0

Clawson GA, Matters GL, Xin P, Imamura-Kawasawa Y, Du Z, Thiboutot DM et al (2015) Macrophage-tumor cell fusions from peripheral blood of melanoma patients. PLoS One 10: e0134320. https://doi.org/10.1371/JOURNAL.PONE.0134320

Colucci M, Sarill M, Maddalena M, Valdata A, Troiani M, Massarotti M et al (2025) Senescence in cancer. Cancer Cell 43:1204–1226. https://doi.org/10.1016/j.ccell.2025.05.015

Conway PJ, Dao J, Kovalskyy D, Mahadevan D, Dray E (2024) Polyploidy in cancer: causal mechanisms, cancer-specific consequences, and emerging treatments. Mol Cancer Ther 23:638–647. https://doi.org/10.1158/1535-7163.MCT-23-0578

Crivat G, Taraska JW (2012) Imaging proteins inside cells with fluorescent tags. Trends Biotechnol 30:8–16. https://doi.org/10.1016/j.tibtech.2011.08.002

Crockford A, Zalmas LP, Grönroos E, Dewhurst SM, McGranahan N, Cuomo ME et al (2016) Cyclin D mediates tolerance of genome-doubling in cancers with functional p53. Ann Oncol 28: 149. https://doi.org/10.1093/ANNONC/MDW612

Cunha A, Silva PMA, Sarmento B, Queirós O (2023) Targeting glucose metabolism in cancer cells as an approach to overcoming drug resistance. Pharmaceutics 15. https://doi.org/10.3390/PHARMACEUTICS15112610

Czajkowski K, Herbet M, Murias M, Piątkowska-Chmiel I (2025) Senolytics: charting a new course or enhancing existing anti-tumor therapies? Cell Oncol 48:351–371. https://doi.org/10.1007/S13402-024-01018-5

Dal Bo M, Gambirasi M, Vruzhaj I, Cecchin E, Pishdadian A, Toffoli G et al (2025) Targeting aging hallmarks with monoclonal antibodies: a new era in cancer immunotherapy and geriatric medicine. Int J Mol Sci *26:4982*. https://doi.org/10.3390/IJMS26114982

Danieau G, Morice S, Renault S, Brion R, Biteau K, Amiaud J et al (2021) Icg-001, an inhibitor of the β-catenin and camp response element-binding protein dependent gene transcription, decreases proliferation but enhances migration of osteosarcoma cells. Pharmaceuticals 14. https://doi.org/10.3390/PH14050421

Delon I, Brown NH (2007) Integrins and the actin cytoskeleton. Curr Opin Cell Biol 19:43–50. https://doi.org/10.1016/j.ceb.2006.12.013

Den Hollander J, Rimpi S, Doherty JR, Rudelius M, Buck A, Hoellein A et al (2010) Aurora kinases A and B are up-regulated by Myc and are essential for maintenance of the malignant state. Blood 116:1498–1505. https://doi.org/10.1182/BLOOD-2009-11-251074

Deneke VE, Pauli A (2021) The fertilization enigma: how sperm and egg Fuse. Annu Rev Cell Dev Biol 37:391–414. https://doi.org/10.1146/ANNUREV-CELLBIO-120219-021751/CITE/REFWORKS

Deng LH, Li MZ, Huang XJ, Zhao XY (2025) Single-cell lineage tracing techniques in hematology: unraveling the cellular narrative. J Transl Med 23:1–10. https://doi.org/10.1186/S12967-025-06318-4/FIGURES/1

Denu RA, Zasadil LM, Kanugh C, Laffin J, Weaver BA, Burkard ME (2016) Centrosome amplification induces high grade features and is prognostic of worse outcomes in breast cancer. BMC Cancer 16:1–13. https://doi.org/10.1186/S12885-016-2083-X/FIGURES/5

Di Lernia G, Leone P, Solimando AG, Buonavoglia A, Saltarella I, Ria R et al (2020) Bortezomib treatment modulates autophagy in multiple myeloma. J Clin Med 9:552. https://doi.org/10.3390/JCM9020552

Dikovskaya D, Cole JJ, Mason SM, Nixon C, Karim SA, McGarry L et al (2015) Mitotic stress is an integral part of the oncogene-induced senescence program that promotes multinucleation and cell cycle arrest. Cell Rep 12:1483–1496. https://doi.org/10.1016/j.celrep.2015.07.055

Dodd KM, Yang J, Shen MH, Sampson JR, Tee AR (2015) mTORC1 drives HIF-1α and VEGF-A signalling via multiple mechanisms involving 4E-BP1, S6K1 and STAT3. Oncogene 34:2239–2250. https://doi.org/10.1038/ONC.2014.164

Dong Z, Luo Y, Yuan Z, Tian Y, Jin T, Xu F (2024) Cellular senescence and SASP in tumor progression and therapeutic opportunities. Mol Cancer 23. https://doi.org/10.1186/S12943-024-02096-7

Dormond O, Madsen JC, Briscoe DM (2007) The effects of mTOR-Akt interactions on anti-apoptotic signaling in vascular endothelial cells. J Biol Chem 282:23679–23686. https://doi.org/10.1074/jbc.M700563200

Dörnen J, Sieler M, Weiler J, Keil S, Dittmar T (2020) Cell fusion-mediated tissue regeneration as an inducer of polyploidy and aneuploidy. Int J Mol Sci 21. https://doi.org/10.3390/IJMS21051811

Duensing A, Duensing S (2010) Centrosomes, polyploidy and cancer. Adv Exp Med Biol 675:93–103. https://doi.org/10.1007/978-1-4419-6199-0_6

Durnaoglu S, Lee SK, Ahnn J (2022) Syncytin, envelope protein of human endogenous retrovirus (HERV): no longer 'fossil' in human genome. Anim Cells Syst (Seoul) 25:358. https://doi.org/10.1080/19768354.2021.2019109

Duronio RJ, O'Farrell PH, Xie JE, Brook A, Dyson N (1995) The transcription factor E2F is required for S phase during drosophila embryogenesis. Genes Dev 9:1445–1455. https://doi.org/10.1101/GAD.9.12.1445

Eitzen G (2003) Actin remodeling to facilitate membrane fusion. Biochim Biophys Acta 1641:175–181. https://doi.org/10.1016/S0167-4889(03)00087-9

Eliades A, Papadantonakis N, Ravid K (2010) New roles for cyclin E in megakaryocytic Polyploidization. J Biol Chem 285:18909. https://doi.org/10.1074/JBC.M110.102145

Fang J, de Bruin A, Villunger A, Schiffelers R, Lei Z, Sluijter JPG (2023) Cellular polyploidy in organ homeostasis and regeneration. Protein Cell 14:560–578. https://doi.org/10.1093/PROCEL/PWAC064

Farhan MA, Carmine-Simmen K, Lewis JD, Moore RB, Murray AG (2015) Endothelial cell mTOR complex-2 regulates sprouting angiogenesis. PLoS One 10:e0135245. https://doi.org/10.1371/JOURNAL.PONE.0135245

Feng T, Xie F, Lee LMY, Lin Z, Tu Y, Lyu Y et al (2025) Cellular senescence in cancer: from mechanism paradoxes to precision therapeutics. Mol Cancer 24. https://doi.org/10.1186/S12943-025-02419-2

Fitsiou E, Soto-Gamez A, Demaria M (2022) Biological functions of therapy-induced senescence in cancer. Semin Cancer Biol 81:5–13. https://doi.org/10.1016/j.semcancer.2021.03.021

Fong YC, Hsu SF, Wu CL, Li TM, Te Kao S, Tsai FJ et al (2009) Transforming growth factor-β1 increases cell migration and β1 integrin up-regulation in human lung cancer cells. Lung Cancer 64:13–21. https://doi.org/10.1016/j.lungcan.2008.07.010

Fresno Vara JÁ, Casado E, de Castro J, Cejas P, Belda-Iniesta C, González-Barón M (2004) P13K/Akt signalling pathway and cancer. Cancer Treat Rev 30:193–204. https://doi.org/10.1016/j.ctrv.2003.07.007

Fu X, Li P, Zhou Q, He R, Wang G, Zhu S et al (2024) Mechanism of PARP inhibitor resistance and potential overcoming strategies. Genes Dis 11:306–320. https://doi.org/10.1016/j.gendis.2023.02.014

Fukuchi K, Watanabe H, Tomoyasu S, Ichimura S, Tatsumi K, Gomi K (2000) Phosphatidylinositol 3-kinase inhibitors, Wortmannin or LY294002, inhibited accumulation of p21 protein after γ-irradiation by stabilization of the protein. Biochim Biophys Acta, Mol Cell Res 1496:207–220. https://doi.org/10.1016/S0167-4889(00)00018-5

Furuta A, Shima T, Kawaguchi M, Yamaki-Ushijima A, Yasuda I, Tsuda S et al (2022) The autophagy-lysosomal machinery enhances cytotrophoblast–syncytiotrophoblast fusion process. Reprod Med 3:112–126. https://doi.org/10.3390/REPRODMED3020010/S1

Gal H, Krizhanovsky V (2014) Cell fusion induced senescence. Aging 6:353–354. https://doi.org/10.18632/AGING.100670

Gao P, Zheng J (2010) High-risk HPV E5-induced cell fusion: A critical initiating event in the early stage of HPV-associated cervical cancer. Virol J 7:1–3. https://doi.org/10.1186/1743-422X-7-238/METRICS

Gao Y, Fang P, Li WJ, Zhang J, Wang GP, Jiang DF et al (2020) LncRNA NEAT1 sponges miR-214 to regulate M2 macrophage polarization by regulation of B7-H3 in multiple myeloma. Mol Immunol 117:20–28. https://doi.org/10.1016/j.molimm.2019.10.026

Garcia E, Aguilar-Cevallos J, Silva-Garcia R, Ibarra A (2016) Cytokine and growth factor activation in vivo and in vitro after spinal cord injury. Mediat Inflamm 2016. https://doi.org/10.1155/2016/9476020

Gentric G, Desdouets C (2014) Polyploidization in liver tissue. Am J Pathol 184:322–331. https://doi.org/10.1016/j.ajpath.2013.06.035

Ghosh S, Choudhury D, Ghosh D, Mondal M, Singha D, Malakar P (2024) Characterization of polyploidy in cancer: current status and future perspectives. Int J Biol Macromol 268:131706. https://doi.org/10.1016/J.IJBIOMAC.2024.131706

Giannoula Y, Kroemer G, Pietrocola F (2023) Cellular senescence and the host immune system in aging and age-related disorders. Biom J 46:100581. https://doi.org/10.1016/J.BJ.2023.02.001

Giménez-Bastida JA, Ávila-Gálvez MÁ, Espín JC, González-Sarrías A (2019) Conjugated physiological resveratrol metabolites induce senescence in breast cancer cells: role of p53/p21 and p16/Rb pathways, and ABC transporters. Mol Nutr Food Res 63:1900629. https://doi.org/10.1002/MNFR.201900629;WGROUP:STRING:PUBLICATION

Glaviano A, Foo ASC, Lam HY, Yap KCH, Jacot W, Jones RH et al (2023) PI3K/AKT/mTOR signaling transduction pathway and targeted therapies in cancer. Mol Cancer 22:1–37. https://doi.org/10.1186/S12943-023-01827-6/TABLES/1

Gonzalez DM, Medici D (2014) Signaling mechanisms of the epithelial-mesenchymal transition. Sci Signal 7:re8. https://doi.org/10.1126/SCISIGNAL.2005189

Grassi S, Giussani P, Mauri L, Prioni S, Sonnino S, Prinetti A (2020) Lipid rafts and neurodegeneration: structural and functional roles in physiologic aging and neurodegenerative diseases. J Lipid Res 61:636–654. https://doi.org/10.1194/jlr.TR119000427

Grewal SIS (2023) The molecular basis of heterochromatin assembly and epigenetic inheritance. Mol Cell 83:1767–1785. https://doi.org/10.1016/j.molcel.2023.04.020

Gugnoni M, Ciarrocchi A (2019) Long noncoding RNA and epithelial mesenchymal transition in cancer. Int J Mol Sci 20:1924. https://doi.org/10.3390/IJMS20081924

Haga RB, Ridley AJ (2016) Rho GTPases: regulation and roles in cancer cell biology. Small GTPases 7:207–221. https://doi.org/10.1080/21541248.2016.1232583

Halder SK, Beauchamp RD, Datta PK (2005) A specific inhibitor of TGF-β receptor kinase, SB-431542, as a potent antitumor agent for human cancers. Neoplasia 7:509–521. https://doi.org/10.1593/neo.04640

Hass R, von der Ohe J, Dittmar T (2021a) Cancer cell fusion and post-hybrid selection process (PHSP). Cancers (Basel) 13. https://doi.org/10.3390/CANCERS13184636

Hass R, von der Ohe J, Dittmar T (2021b) Hybrid formation and fusion of cancer cells in vitro and in vivo. Cancers (Basel) 13. https://doi.org/10.3390/CANCERS13174496

Hayashi K, Horisaka K, Harada Y, Ogawa Y, Yamashita T, Kitano T et al (2024) Polyploidy mitigates the impact of DNA damage while simultaneously bearing its burden. Cell Death Discov 10:1–13. https://doi.org/10.1038/S41420-024-02206-W. SUBJMETA=208,509,631,68,80;KWRD=CANCER+GENETICS,SENESCENCE

Henklova P, Vrzal R, Papouskova B, Bednar P, Jancova P, Anzenbacherova E et al (2008) SB203580, a pharmacological inhibitor of p38 MAP kinase transduction pathway activates ERK and JNK MAP kinases in primary cultures of human hepatocytes. Eur J Pharmacol 593: 16–23. https://doi.org/10.1016/J.EJPHAR.2008.07.007

Hickson LTJ, Langhi Prata LGP, Bobart SA, Evans TK, Giorgadze N, Hashmi SK et al (2019) Senolytics decrease senescent cells in humans: preliminary report from a clinical trial of Dasatinib plus quercetin in individuals with diabetic kidney disease. EBioMedicine 47:446–456. https://doi.org/10.1016/j.ebiom.2019.08.069

Hill M, Wernig A, Goldspink G (2003) Muscle satellite (stem) cell activation during local tissue injury and repair. J Anat 203:89–99. https://doi.org/10.1046/J.1469-7580.2003.00195.X; PAGE:STRING:ARTICLE/CHAPTER

Hu Y, Lou J, Mao YY, Lai TW, Liu LY, Zhu C et al (2016) Activation of MTOR in pulmonary epithelium promotes LPS-induced acute lung injury. Autophagy 12:2286–2299. https://doi.org/10.1080/15548627.2016.1230584

Huang K, Fingar DC (2014) Growing knowledge of the mTOR signaling network. Semin Cell Dev Biol 36:79–90. https://doi.org/10.1016/j.semcdb.2014.09.011

Huang W, Hickson LTJ, Eirin A, Kirkland JL, Lerman LO (2022) Cellular senescence: the good, the bad and the unknown. Nat Rev Nephrol 18:611–627. https://doi.org/10.1038/S41581-022-00601-Z

Imamura F, Mukai M, Ayaki M, Akedo H (2000) Y-27632, an inhibitor of rho-associated protein kinase, suppresses tumor cell invasion via regulation of focal adhesion and focal adhesion kinase. Jpn J Cancer Res 91:811–816. https://doi.org/10.1111/J.1349-7006.2000.TB01018.X

Jarroux J, Foretek D, Bertrand C, Gabriel M, Szachnowski U, Saci Z et al (2021) HOTAIR lncRNA promotes epithelial–mesenchymal transition by redistributing LSD1 at regulatory chromatin regions. EMBO Rep 22. https://doi.org/10.15252/EMBR.202050193/SUPPL_FILE/EMBR202050193-SUP-0011-SDATAEV.ZIP

Jiang E, Yan T, Xu Z, Shang Z (2019) Tumor microenvironment and cell fusion. Biomed Res Int 2019. https://doi.org/10.1155/2019/5013592

Jung SH, Hwang HJ, Kang D, Park HA, Lee HC, Jeong D et al (2019) mTOR kinase leads to PTEN-loss-induced cellular senescence by phosphorylating p53. Oncogene 38:1639–1650. https://doi.org/10.1038/S41388-018-0521-8

Kaliman P, Viñals F, Testar X, Palacín M, Zorzano A (1996) Phosphatidylinositol 3-kinase inhibitors block differentiation of skeletal muscle cells. J Biol Chem 271:19146–19151. https://doi.org/10.1074/jbc.271.32.19146

Kim YC, Guan KL (2015) MTOR: a pharmacologic target for autophagy regulation. J Clin Invest 125:25–32. https://doi.org/10.1172/JCI73939

Kim ES, Scott LJ (2017) Palbociclib: a review in HR-positive, HER2-negative, advanced or metastatic breast cancer. Target Oncol 12:373–383. https://doi.org/10.1007/S11523-017-0492-7

Kitsou K, Lagiou P, Magiorkinis G (2023) Human endogenous retroviruses in cancer: oncogenesis mechanisms and clinical implications. J Med Virol 95. https://doi.org/10.1002/JMV.28350

Kloc M, Uosef A, Subuddhi A, Kubiak JZ, Piprek RP, Ghobrial RM (2022) Giant multinucleated cells in aging and senescence—an abridgement. Biology (Basel) 11. https://doi.org/10.3390/BIOLOGY11081121

Koido S, Hara E, Homma S, Namiki Y, Ohkusa T, Gong J et al (2009) Cancer vaccine by fusions of dendritic and cancer cells. Clin Dev Immunol 2009. https://doi.org/10.1155/2009/657369

Kong Z, Yao T (2022) Role for autophagy-related markers Beclin-1 and LC3 in endometriosis. BMC Womens Health 22:1–7. https://doi.org/10.1186/S12905-022-01850-7/TABLES/6

Kong Y, Cui H, Ramkumar C, Zhang H (2011) Regulation of senescence in cancer and aging. J Aging Res 2011:963172. https://doi.org/10.4061/2011/963172

Lamm N, Rogers S, Cesare AJ (2019) The mTOR pathway: implications for DNA replication. Prog Biophys Mol Biol 147:17–25. https://doi.org/10.1016/J.PBIOMOLBIO.2019.04.002

Landskron G, De La Fuente M, Thuwajit P, Thuwajit C, Hermoso MA (2014) Chronic inflammation and cytokines in the tumor microenvironment. J Immunol Res 2014. https://doi.org/10.1155/2014/149185

Lawrence T (2009) The nuclear factor NF-kappaB pathway in inflammation. Cold Spring Harb Perspect Biol 1. https://doi.org/10.1101/CSHPERSPECT.A001651

Lee SH, Dominguez R (2010) Regulation of actin cytoskeleton dynamics in cells. Mol Cells 29:311. https://doi.org/10.1007/S10059-010-0053-8

Lee JW, Ong EBB (2021) Genomic instability and cellular senescence: lessons from the budding yeast. Front Cell Dev Biol 8:619126. https://doi.org/10.3389/FCELL.2020.619126/XML

Leontieva OV, Blagosklonny MV (2013) CDK4/6-inhibiting drug substitutes for p21 and p16 in senescence: duration of cell cycle arrest and MTOR activity determine geroconversion. Cell Cycle 12:3063–3069. https://doi.org/10.4161/CC.26130

Leontieva OV, Demidenko ZN, Blagosklonny MV (2015) Dual mTORC1/C2 inhibitors suppress cellular geroconversion (a senescence program). Oncotarget 6:23238–23248. https://doi.org/10.18632/ONCOTARGET.4836

Li Q, Dang CV (1999) C-Myc overexpression uncouples DNA replication from mitosis. Mol Cell Biol 19:5339–5351. https://doi.org/10.1128/MCB.19.8.5339

Li J, Ren J, Liu X, Jiang L, He W, Yuan W et al (2015) Rictor/mTORC2 signaling mediates TGFβ1-induced fibroblast activation and kidney fibrosis. Kidney Int 88:515–527. https://doi.org/10.1038/ki.2015.119

Li ML, Ragupathi A, Patel N, Hernandez T, Magsino J, Werlen G et al (2022) The RNA-binding protein AUF1 facilitates Akt phosphorylation at the membrane. J Biol Chem 298. https://doi.org/10.1016/j.jbc.2022.102437

Liao J, Chen R, Lin B, Deng R, Liang Y, Zeng J et al (2024) Cross-talk between the TGF-β and cell adhesion signaling pathways in cancer. Int J Med Sci 21:1307–1320. https://doi.org/10.7150/IJMS.96274

Lindström A, Midtbö K, Arnesson LG, Garvin S, Shabo I (2017) Fusion between M2-macrophages and cancer cells results in a subpopulation of radioresistant cells with enhanced DNA-repair capacity. Oncotarget 8:51370–51386. https://doi.org/10.18632/ONCOTARGET.17986

Linke M, Fritsch SD, Sukhbaatar N, Hengstschläger M, Weichhart T (2017) mTORC1 and mTORC2 as regulators of cell metabolism in immunity. FEBS Lett 591:3089. https://doi.org/10.1002/1873-3468.12711

Liu GY, Sabatini DM (2020) mTOR at the nexus of nutrition, growth, ageing and disease. Nat Rev Mol Cell Biol 21:183–203. https://doi.org/10.1038/S41580-019-0199-Y

Lizarbe MA, Barrasa JI, Olmo N, Gavilanes F, Turnay J (2013) Annexin-phospholipid interactions. Functional implications. Int J Mol Sci 14:2652–2683. https://doi.org/10.3390/IJMS14022652

López de Andrés J, Griñán-Lisón C, Jiménez G, Marchal JA (2020) Cancer stem cell secretome in the tumor microenvironment: a key point for an effective personalized cancer treatment. *Journal of Hematology & Oncology 13(1)*:1–22. https://doi.org/10.1186/S13045-020-00966-3

López-Collazo E, Hurtado-Navarro L (2025) Cell fusion as a driver of metastasis: re-evaluating an old hypothesis in the age of cancer heterogeneity. Front Immunol 16:1524781. https://doi.org/10.3389/FIMMU.2025.1524781/XML

Luo J, Sun T, Liu Z, Liu Y, Liu J, Wang S et al (2025) Persistent accumulation of therapy-induced senescent cells: an obstacle to long-term cancer treatment efficacy. Int J Oral Sci 17. https://doi.org/10.1038/S41368-025-00380-W

Mancino A, Lawrence T (2010) Nuclear factor-κB and tumor-associated macrophages. Clin Cancer Res 16:784–789. https://doi.org/10.1158/1078-0432.CCR-09-1015

Martinez-Sarmiento JA, Cosma MP, Lakadamyali M (2024) Dissecting gene activation and chromatin remodeling dynamics in single human cells undergoing reprogramming. Cell Rep 43. https://doi.org/10.1016/j.celrep.2024.114170

Mayer A (2001) What drives membrane fusion in eukaryotes? Trends Biochem Sci 26:717–723. https://doi.org/10.1016/S0968-0004(01)01984-3

McNally AK, Anderson JM (2011) Macrophage fusion and multinucleated giant cells of inflammation. Adv Exp Med Biol 713:97–111. https://doi.org/10.1007/978-94-007-0763-4_7

Mehta D, Rajput K, Jain D, Bajaj A, Dasgupta U (2024) Unveiling the role of mechanistic target of rapamycin kinase (MTOR) signaling in cancer progression and the emergence of MTOR inhibitors as therapeutic strategies. ACS Pharmacol Transl Sci 7:3758–3779. https://doi.org/10.1021/ACSPTSCI.4C00530

Milanovic M, Yu Y, Schmitt CA (2018) The senescence–stemness alliance – a cancer-hijacked regeneration principle. Trends Cell Biol 28:1049–1061. https://doi.org/10.1016/J.TCB.2018.09.001

Milde R, Ritter J, Tennent GA, Loesch A, Martinez FO, Gordon S et al (2015) Multinucleated Giant cells are specialized for complement-mediated phagocytosis and large target destruction. Cell Rep 13:1937–1948. https://doi.org/10.1016/j.celrep.2015.10.065

Miyazono K (2009) Transforming growth factor-β signaling in epithelial-mesenchymal transition and progression of cancer. Proc Jpn Acad Ser B Phys Biol Sci 85:314–323. https://doi.org/10.2183/PJAB.85.314

Mok WC, Wasser S, Tan T, Lim SG (2012) Polo-like kinase 1, a new therapeutic target in hepatocellular carcinoma. World J Gastroenterol: WJG 18:3527. https://doi.org/10.3748/WJG.V18.I27.3527

Mosieniak G, Sikora E (2010) Polyploidy: the link between senescence and cancer. Curr Pharm Des 16:734–740. https://doi.org/10.2174/138161210790883714

Murugan AK (2019) mTOR: role in cancer, metastasis and drug resistance. Semin Cancer Biol 59: 92–111. https://doi.org/10.1016/J.SEMCANCER.2019.07.003

Nacheva EP, Grace CD, Bittner M, Ledbetter DH, Jenkins RB, Green AR (1998) Comparative genomic hybridization: A comparison with molecular and cytogenetic analysis. Cancer Genet Cytogenet 100:93–105. https://doi.org/10.1016/S0165-4608(97)00021-6

Nasrollahzadeh A, Momeny M, Bashash D, Yousefi H, Mousavi SA, Ghaffari SH (2022) Blockade of nuclear factor-kb (NF-kb) pathway using bay 11-7082 enhances arsenic trioxide-induced antiproliferative activity in U87 glioblastoma cells. Rep Biochem Mol Biol 10:602–613. https://doi.org/10.52547/RBMB.10.4.602

Nathans JF, Ayers JL, Shendure J, Simpson CL (2024) Genetic tools for cell lineage tracing and profiling developmental trajectories in the skin. J Invest Dermatol 144:936–949. https://doi.org/10.1016/J.JID.2024.02.006

Negi G, Sharma A, Dey M, Dhanawat G, Parveen N (2022) Membrane attachment and fusion of HIV-1, influenza A, and SARS-CoV-2: resolving the mechanisms with biophysical methods. Biophys Rev 14:1109–1140. https://doi.org/10.1007/S12551-022-00999-7

Ning Y, Zheng M, Zhang Y, Jiao Y, Wang J, Zhang S (2024) RhoA-ROCK2 signaling possesses complex pathophysiological functions in cancer progression and shows promising therapeutic potential. Cancer Cell Int 24. https://doi.org/10.1186/S12935-024-03519-7

Nishida A, Andoh A (2025) The role of inflammation in cancer: mechanisms of tumor initiation, progression, and metastasis. Cells 14. https://doi.org/10.3390/CELLS14070488

Noubissi FK, Ogle BM (2016) Cancer cell fusion: mechanisms slowly unravel. Int J Mol Sci 17. https://doi.org/10.3390/IJMS17091587

Ogden A, Rida PCG, Knudsen BS, Kucuk O, Aneja R (2015) Docetaxel-induced polyploidization may underlie chemoresistance and disease relapse. Cancer Lett 367:89. https://doi.org/10.1016/J.CANLET.2015.06.025

Okamoto H, Mizuno K, Horio T (2003) Monocyte-derived multinucleated giant cells and sarcoidosis. J Dermatol Sci 31:119–128. https://doi.org/10.1016/S0923-1811(02)00148-2

Orange ST, Leslie J, Ross M, Mann DA, Wackerhage H (2023) The exercise IL-6 enigma in cancer. Trends Endocrinol Metab 34:749–763. https://doi.org/10.1016/j.tem.2023.08.001

Øvrebø JI, Edgar BA (2018) Polyploidy in tissue homeostasis and regeneration. Development 145: dev156034. https://doi.org/10.1242/DEV.156034

Page-McCaw A, Ewald AJ, Werb Z (2007) Matrix metalloproteinases and the regulation of tissue remodelling. Nat Rev Mol Cell Biol 8:221. https://doi.org/10.1038/NRM2125

Patra S, Naik PP, Mahapatra KK, Alotaibi MR, Patil S, Patro BS et al (2024) Recent advancement of autophagy in polyploid giant cancer cells and its interconnection with senescence and stemness for therapeutic opportunities. Cancer Lett 590. https://doi.org/10.1016/j.canlet.2024.216843

Perez TD, Nelson WJ (2004) Cadherin adhesion: mechanisms and molecular interactions. Handb Exp Pharmacol:3–21. https://doi.org/10.1007/978-3-540-68170-0_1

Piechota M, Sunderland P, Wysocka A, Nalberczak M, Sliwinska MA, Radwanska K et al (2016) Is senescence-associated β-galactosidase a marker of neuronal senescence? Oncotarget 7:81099–81109. https://doi.org/10.18632/ONCOTARGET.12752

Pilat N, Unger L, Berlakovich GA (2013) Implication for bone marrow derived stem cells in hepatocyte regeneration after orthotopic liver transplantation. Int J Hepatol 2013:310612. https://doi.org/10.1155/2013/310612

Porstmann T, Santos CR, Griffiths B, Cully M, Wu M, Leevers S et al (2008) SREBP activity is regulated by mTORC1 and contributes to Akt-dependent cell growth. Cell Metab 8:224–236. https://doi.org/10.1016/j.cmet.2008.07.007

Pourdehnad M, Truitt ML, Siddiqi IN, Ducker GS, Shokat KM, Ruggero D (2013) Myc and mTOR converge on a common node in protein synthesis control that confers synthetic lethality in Myc-driven cancers. Proc Natl Acad Sci USA 110:11988–11993. https://doi.org/10.1073/PNAS.1310230110

Qi L, Sun B, Liu Z, Cheng R, Li Y, Zhao X (2014) Wnt3a expression is associated with epithelial-mesenchymal transition and promotes colon cancer progression. J Exp Clin Cancer Res 33. https://doi.org/10.1186/S13046-014-0107-4

Radice GL (2013) N-cadherin-mediated adhesion and signaling from development to disease: lessons from mice. Prog Mol Biol Transl Sci 116:263. https://doi.org/10.1016/B978-0-12-394311-8.00012-1

Ragupathi A, Kim C, Jacinto E (2024) The mTORC2 signaling network: targets and cross-talks. Biochem J 481:45–91. https://doi.org/10.1042/BCJ20220325

Reimann M, Lee S, Schmitt CA (2024) Cellular senescence: neither irreversible nor reversible. J Exp Med 221. https://doi.org/10.1084/JEM.20232136

Renaud SJ, Jeyarajah MJ (2022) How trophoblasts fuse: an in-depth look into placental syncytiotrophoblast formation. Cell Mol Life Sci 79:433. https://doi.org/10.1007/S00018-022-04475-Z

Rodríguez-Pérez F, Manford AG, Pogson A, Ingersoll AJ, Martínez-González B, Rape M (2021) Ubiquitin-dependent remodeling of the actin cytoskeleton drives cell fusion. Dev Cell 56:588–601.e9. https://doi.org/10.1016/j.devcel.2021.01.016

Rossiello F, Jurk D, Passos JF, d'Adda di Fagagna F (2022) Telomere dysfunction in ageing and age-related diseases. Nat Cell Biol 24:135–147. https://doi.org/10.1038/S41556-022-00842-X

Sabe H, Yahara Y, Ishii M (2024) Cell fusion dynamics: mechanisms of multinucleation in osteoclasts and macrophages. Inflamm Regen 44. https://doi.org/10.1186/S41232-024-00360-3

Sahu A, Singhal U, Chinnaiyan AM (2015) Long noncoding RNAs in cancer: from function to translation. Trends Cancer 1:93–109. https://doi.org/10.1016/J.TRECAN.2015.08.010

Saini G, Joshi S, Garlapati C, Li H, Kong J, Krishnamurthy J et al (2022) Polyploid giant cancer cell characterization: new frontiers in predicting response to chemotherapy in breast cancer. Semin Cancer Biol 81:220–231. https://doi.org/10.1016/j.semcancer.2021.03.017

Saleh T, Carpenter VJ, Bloukh S, Gewirtz DA (2022) Targeting tumor cell senescence and polyploidy as potential therapeutic strategies. Semin Cancer Biol 81:37–47. https://doi.org/10.1016/j.semcancer.2020.12.010

Sazonova EV, Petrichuk SV, Kopeina GS, Zhivotovsky B (2021) A link between mitotic defects and mitotic catastrophe: detection and cell fate. Biol Direct 16. https://doi.org/10.1186/S13062-021-00313-7

Segev N, Avinoam O, Podbilewicz B (2018) Fusogens. Curr Biol 28:R378–R380. https://doi.org/10.1016/j.cub.2018.01.024

Shabo I, Midtbö K, Andersson H, Åkerlund E, Olsson H, Wegman P et al (2015) Macrophage traits in cancer cells are induced by macrophage-cancer cell fusion and cannot be explained by cellular interaction. BMC Cancer 15:1–11. https://doi.org/10.1186/S12885-015-1935-0/FIGURES/6

Shabo I, Svanvik J, Lindström A, Lechertier T, Trabulo S, Hulit J et al (2020) Roles of cell fusion, hybridization and polyploid cell formation in cancer metastasis. World J Clin Oncol 11:121–135. https://doi.org/10.5306/WJCO.V11.I3.121

Shakoori AR (2017) Fluorescence in situ hybridization (FISH) and its applications. In: Chromosome Structure and Aberrations, p 343. https://doi.org/10.1007/978-81-322-3673-3_16

Sharma U, Barwal TS, Khandelwal A, Malhotra A, Rana MK, Singh Rana AP et al (2021) LncRNA ZFAS1 inhibits triple-negative breast cancer by targeting STAT3. Biochimie 182:99–107. https://doi.org/10.1016/j.biochi.2020.12.026

Shimada N, Rios I, Moran H, Sayers B, Hubbard K (2009) p38 MAP kinase-dependent regulation of the expression level and subcellular distribution of heterogeneous nuclear ribonucleoprotein A1 and its involvement in cellular senescence in normal human fibroblasts. RNA Biol 6:293–304. https://doi.org/10.4161/RNA.6.3.8497

Shimanouchi T, Kawasaki H, Fuse M, Umakoshi H, Kuboi R (2013) Membrane fusion mediated by phospholipase C under endosomal pH conditions. Colloids Surf B Biointerfaces 103:75–83. https://doi.org/10.1016/j.colsurfb.2012.10.011

Shultes PV, Weaver DT, Tadele DS, Barker-Clarke RJ, Scott JG (2024) Cell-cell fusion in cancer: the next cancer hallmark? Int J Biochem Cell Biol 175:106649. https://doi.org/10.1016/J.BIOCEL.2024.106649

Son B, Lee W, Kim H, Shin H, Park HH (2024) Targeted therapy of cancer stem cells: inhibition of mTOR in pre-clinical and clinical research. Cell Death Dis 15. https://doi.org/10.1038/S41419-024-07077-8

Song KX, Wang JX, Huang D (2023) Therapy-induced senescent tumor cells in cancer relapse. J Natl Cancer Cent 3:273–278. https://doi.org/10.1016/j.jncc.2023.09.001

Srinivasan S, Wang F, Glavas S, Ott A, Hofmann F, Aktories K et al (2003) Rac and Cdc42 play distinct roles in regulating PI(3,4,5)P3 and polarity during neutrophil chemotaxis. J Cell Biol 160:375–385. https://doi.org/10.1083/JCB.200208179

Swift LH, Golsteyn RM (2014) Genotoxic anti-cancer agents and their relationship to DNA damage, mitosis, and checkpoint adaptation in proliferating cancer cells. Int J Mol Sci 15: 3403. https://doi.org/10.3390/IJMS15033403

Takahara T, Amemiya Y, Sugiyama R, Maki M, Shibata H (2020) Amino acid-dependent control of mTORC1 signaling: a variety of regulatory modes. J Biomed Sci 27. https://doi.org/10.1186/S12929-020-00679-2

Takasugi M, Yoshida Y, Ohtani N (2022) Cellular senescence and the tumour microenvironment. Mol Oncol 16:3333–3351. https://doi.org/10.1002/1878-0261.13268

Talotta R (2023) Interaction between long noncoding RNAs and Syncytin-1/Syncytin-2 genes and transcripts: how noncoding RNAs may affect pregnancy in patients with systemic lupus erythematosus. Int J Mol Sci 24. https://doi.org/10.3390/IJMS24032259

Tamaki A, Ierano C, Szakacs G, Robey RW, Bates SE (2011) The controversial role of ABC transporters in clinical oncology. Essays Biochem 50:209–232. https://doi.org/10.1042/BSE0500209

Tang S, Qin C, Hu H, Liu T, He Y, Guo H et al (2022) Immune checkpoint inhibitors in non-small cell lung cancer: Progress, challenges, and prospects. Cells 11. https://doi.org/10.3390/CELLS11030320

Tapon N, Hall A (1997) Rho, Rac and Cdc42 GTPases regulate the organization of the actin cytoskeleton. Curr Opin Cell Biol 9:86–92. https://doi.org/10.1016/S0955-0674(97)80156-1

Taylor MV (2000) Muscle development: molecules of myoblast fusion. Curr Biol 10:R646–R648. https://doi.org/10.1016/S0960-9822(00)00664-3

Torti M, Manganaro D, Visconte C, Zarà M, Canino J, Vismara M et al (2020) Stimulation of mTORC2 by integrin αIIbβ3 is required for PI3Kβ-dependent activation of Akt but is dispensable for platelet spreading on fibrinogen. Platelets 31:521–529. https://doi.org/10.1080/09537104.2019.1663806

Tretyakova MS, Subbalakshmi AR, Menyailo ME, Jolly MK, Denisov EV (2022) Tumor hybrid cells: nature and biological significance. Front Cell Dev Biol 10. https://doi.org/10.3389/FCELL.2022.814714

Uppaluri KR, Challa HJ, Gaur A, Jain R, Krishna Vardhani K, Geddam A et al (2023) Unlocking the potential of non-coding RNAs in cancer research and therapy. Transl Oncol 35. https://doi.org/10.1016/j.tranon.2023.101730

Valdiglesias V, Giunta S, Fenech M, Neri M, Bonassi S (2013) γH2AX as a marker of DNA double strand breaks and genomic instability in human population studies. Mutat Res Rev Mutat Res 753:24–40. https://doi.org/10.1016/j.mrrev.2013.02.001

Valieva Y, Ivanova E, Fayzullin A, Kurkov A, Igrunkova A (2022) Senescence-associated β-galactosidase detection in pathology. Diagnostics 12. https://doi.org/10.3390/DIAGNOSTICS12102309

Varela-Eirín M, Demaria M (2022) Cellular senescence. Curr Biol 32:R448–R452. https://doi.org/10.1016/j.cub.2022.04.003

Varella-Garcia M, Chen L, Powell RL, Hirsch FR, Kennedy TC, Keith R et al (2007) Spectral karyotyping detects chromosome damage in bronchial cells of smokers and patients with cancer. Am J Respir Crit Care Med 176:505. https://doi.org/10.1164/RCCM.200609-1329OC

Vats P, Saini C, Baweja B, Srivastava SK, Kumar A, Kushwah AS et al (2025) Aurora kinases signaling in cancer: from molecular perception to targeted therapies. Mol Cancer 24(1):1–44. https://doi.org/10.1186/S12943-025-02353-3

Vazquez-Martin A, Anatskaya OV, Giuliani A, Erenpreisa J, Huang S, Salmina K et al (2016) Somatic polyploidy is associated with the upregulation of c-MYC interacting genes and EMT-like signature. Oncotarget 7:75235–75260. https://doi.org/10.18632/ONCOTARGET.12118

Velegzhaninov IO, Ievlev VA, Pylina YI, Shadrin DM, Vakhrusheva OM (2018) Programming of cell resistance to genotoxic and oxidative stress. Biomedicine 6. https://doi.org/10.3390/BIOMEDICINES6010005

Vitale I, Galluzzi L, Senovilla L, Criollo A, Jemaá M, Castedo M, et al (2010) Illicit survival of cancer cells during polyploidization and depolyploidization. Cell Death Differ 2011 18:9, 1403–1413. doi:https://doi.org/10.1038/cdd.2010.145

Wang Y, Wu X, Du M, Chen X, Ning X, Chen H et al (2017) Eg5 inhibitor YL001 induces mitotic arrest and inhibits tumor proliferation. Oncotarget 8:42510. https://doi.org/10.18632/ONCOTARGET.17207

Wang DK, Zheng HL, Zhou WS, Duan ZW, Jiang SD, Li B et al (2022) Mitochondrial dysfunction in oxidative stress-mediated intervertebral disc degeneration. Orthop Surg 14:1569–1582. https://doi.org/10.1111/OS.13302

Wang YN, Chen XL, Yang J, Gong XX, Zhang HF, Zhang YM et al (2024) Reduced syncytin-1 regulates trophoblast invasion and apoptosis in preeclampsia. Placenta 155:32–41. https://doi.org/10.1016/j.placenta.2024.07.311

Was H, Borkowska A, Olszewska A, Klemba A, Marciniak M, Synowiec A et al (2022) Polyploidy formation in cancer cells: how a Trojan horse is born. Semin Cancer Biol 81:24–36. https://doi.org/10.1016/J.SEMCANCER.2021.03.003

Wu F, Yang J, Liu J, Wang Y, Mu J, Zeng Q et al (2021) Signaling pathways in cancer-associated fibroblasts and targeted therapy for cancer. Signal Transduct Target Ther 6. https://doi.org/10.1038/S41392-021-00641-0

Xu J, Galvanetto N, Nie J, Yang Y, Torre V (2020) Rac1 promotes cell motility by controlling cell mechanics in human glioblastoma. Cancers (Basel) 12:1667. https://doi.org/10.3390/CANCERS12061667

Xue W, Yang L, Chen C, Ashrafizadeh M, Tian Y, Sun R (2024) Wnt/β-catenin-driven EMT regulation in human cancers. Cell Mol Life Sci 81. https://doi.org/10.1007/S00018-023-05099-7

Yamada C, Morooka A, Miyazaki S, Nagai M, Mase S, Iemura K et al (2022) TORC1 inactivation promotes APC/C-dependent mitotic slippage in yeast and human cells. *iScience* 25. https://doi.org/10.1016/J.ISCI.2021.103675

Yan TL, Wang M, Xu Z, Huang CM, Zhou XC, Jiang EH et al (2017) Up-regulation of syncytin-1 contributes to TNF-α-enhanced fusion between OSCC and HUVECs partly via Wnt/β-catenin-dependent pathway. Sci Rep 7(1):1–12. https://doi.org/10.1038/srep40983

Yasuda S, Horinaka M, Iizumi Y, Goi W, Sukeno M, Sakai T (2022) Oridonin inhibits SASP by blocking p38 and NF-κB pathways in senescent cells. Biochem Biophys Res Commun 590:55–62. https://doi.org/10.1016/J.BBRC.2021.12.098

Yun H, Han GH, Wee DJ, Chay DB, Chung JY, Kim JH et al (2025) Loss of E-cadherin activates EGFR-MEK/ERK signaling, promoting cervical cancer progression. Cancer Genomics Proteomics 22:271–284. https://doi.org/10.21873/CGP.20501

Zanconato F, Cordenonsi M, Piccolo S (2016) YAP/TAZ at the roots of cancer. Cancer Cell 29: 783–803. https://doi.org/10.1016/j.ccell.2016.05.005

Zaninovic N, Irani M, Meseguer M (2017) Assessment of embryo morphology and developmental dynamics by time-lapse microscopy: is there a relation to implantation and ploidy? Fertil Steril 108:722–729. https://doi.org/10.1016/J.FERTNSTERT.2017.10.002

Zhang Y, Yang JM (2011) The impact of cellular senescence in cancer therapy: is it true or not? Acta Pharmacol Sin 32:1199–1207. https://doi.org/10.1038/APS.2011.108

Zhang B, Xuan C, Ji Y, Zhang W, Wang D (2015) Zebrafish xenotransplantation as a tool for in vivo cancer study. Fam Cancer 14:487–493. https://doi.org/10.1007/S10689-015-9802-3

Zhang H, Ma H, Yang X, Fan L, Tian S, Niu R et al (2022) Cell fusion-related proteins and signaling pathways, and their roles in the development and progression of cancer. Front Cell Dev Biol 9:809668. https://doi.org/10.3389/FCELL.2021.809668/XML

Zheng X, Jiang F, Katakowski M, Zheng GZ, Lu QE, Chopp M (2009) ADAM17 promotes breast cancer cell malignant phenotype through EGFR-PI3K-AKT activation. Cancer Biol Ther 8: 1045–1054. https://doi.org/10.4161/CBT.8.11.8539

Zhou X, Zhou M, Zheng M, Tian S, Yang X, Ning Y et al (2022) Polyploid giant cancer cells and cancer progression. Front Cell Dev Biol 10:1017588. https://doi.org/10.3389/FCELL.2022.1017588/XML

Zhu Y, Tchkonia T, Fuhrmann-Stroissnigg H, Dai HM, Ling YY, Stout MB et al (2016) Identification of a novel senolytic agent, navitoclax, targeting the Bcl-2 family of anti-apoptotic factors. Aging Cell 15:428–435. https://doi.org/10.1111/ACEL.12445

Chapter 11
Cell Fusion-Mediated Induction of Polyploidy, Aneuploidy, and Genomic Instability in Cancer Cells

Tomonori Matsumoto

Abstract Cell fusion, which occurs in both tissue injury and cancer, inevitably results in polyploidization. While polyploid cells often fail to proliferate, polyploidization does not invariably lead to growth arrest, and polyploid cells can resume proliferation under certain conditions. Notably, the proliferation of polyploid cells is frequently accompanied by genomic instability, including replication stress and chromosomal instability. Furthermore, some polyploid cells have the capacity to reduce their ploidy and revert to a diploid state. Growing evidence indicates that such genomic instability and ploidy fluctuations act as driving forces of clonal evolution. This chapter discusses ploidy changes and genomic instability arising from cell fusion and examines how these alterations influence cellular behavior.

Keywords Polyploidization · Ploidy reduction · Chromosomal instability · Genomic instability · Cell fusion · Polyploid giant cancer cell · Multipolar mitosis · Budding

11.1 Introduction

Cancer is a genomic disease driven by the stepwise or sometimes abrupt accumulation of genomic abnormalities that fuel clonal evolution. These alterations vary in scale, from point mutations and small insertions/deletions to large structural changes such as copy number variations, aneuploidy at the chromosome or chromosome-arm level, and even whole-genome doubling (WGD)/polyploidy (Consortium 2020). While different tumor types exhibit distinct patterns of genomic alterations, pan-cancer analyses have uncovered shared features across diverse cancers, including WGD and chromosomal instability (CIN) (Bielski et al. 2018; Quinton et al. 2021). Analyses using multiregion and longitudinal sampling further highlight the heterogeneity and evolutionary history of cancer genomes (Gerlinger et al. 2012;

T. Matsumoto (✉)
Laboratory of Ploidy Pathology, Graduate School of Frontier Biosciences, The University of Osaka, Suita, Osaka, Japan
e-mail: matsumoto.tomonori.fbs@osaka-u.ac.jp

© The Author(s), under exclusive license to Springer Nature Switzerland AG 2026

J. Weiler et al. (eds.), *Cell Fusion*, Advances in Experimental Medicine and Biology 1509, https://doi.org/10.1007/978-3-032-22637-2_11

McPherson et al. 2025; Turajlic et al. 2018). Together, these findings suggest that cancer evolves through repeated rounds of clonal selection, driven by gradual genomic changes or punctuated events involving multiple alterations.

Ploidy abnormalities represent one of the earliest recognized hallmarks of cancer and continue to be a fundamental driver of cancer genome evolution. Over a century ago, Theodor Boveri hypothesized that aberrant chromosomal segregation caused by defects such as multipolar spindle formation could lead to extensive chromosomal aberrations, driving malignant transformation (Boveri 1914). This concept has since been validated by large-scale cancer genome studies showing that approximately 90% of cancers harbor some form of chromosomal abnormality, whether numerical (such as whole-chromosome or arm-level aneuploidy) or structural (Sdeor et al. 2024). It is now widely accepted that these chromosomal abnormalities play a critical role in tumor development and progression. In particular, cellular alterations such as cell fusion, WGD, and centrosome amplification are known to promote chromosomal instability, thereby fostering a genomic environment conducive to oncogenesis (Ben-David and Amon 2020; Lens and Medema 2019; Vasudevan et al. 2021; Vittoria et al. 2023). These processes not only precipitate large-scale karyotypic changes but also enhance intratumoral genetic heterogeneity, accelerating clonal evolution and promoting phenotypic plasticity that supports therapy resistance and disease progression (Al-Rawi and Bakhoum 2022; Bakhoum and Cantley 2018). In this review, we discuss how cell fusion, through its associated ploidy alterations, contributes to genomic instability and shapes the evolutionary trajectory of cancer.

11.2 Definition of Polyploidy, Aneuploidy, and CIN

Ploidy refers to the number of complete chromosome sets within a cell. Mammalian cells are typically diploid, inheriting one chromosome set from each parent. However, cells exhibiting increased ploidy can arise both under normal physiological conditions and in the context of various diseases. These cells, containing more than two complete chromosome sets, such as triploid (3n), tetraploid (4n), or octoploid (8n) cells, are termed polyploid. A subset of physiologically relevant polyploid cells originates through cell fusion in vivo. Examples include skeletal muscle cells, formed by the fusion of myoblasts, and syncytiotrophoblasts, arising from trophoblast cell fusion during placental development. These specialized cell types exploit polyploidy to meet the heightened metabolic and functional demands associated with their roles in tissue development and homeostasis (Darmasaputra et al. 2024; Øvrebø and Edgar 2018).

Cells typically maintain a strict chromosomal complement. Diploid cells contain two complete sets of chromosomes, while polyploid cells harbor an integer multiple beyond the diploid state. In contrast, aneuploidy describes a deviation from these integer multiples, involving the gain or loss of specific chromosomes, resulting in an imbalanced karyotype (Ben-David and Amon 2020; Schukken and Foijer 2018).

Aneuploidy primarily stems from errors during cell division, such as chromosome missegregation during mitosis. These defects perturb the normal distribution of genetic material, potentially leading to significant cellular consequences, including altered gene expression and disrupted cellular function.

Aneuploidy is closely related to, yet distinct from, CIN. CIN describes a cellular state characterized by the inability to faithfully maintain chromosome number and structure across successive cell divisions (Schukken and Foijer 2018; Targa and Rancati 2018). This instability results in continuous chromosomal missegregation events, generating a high rate of chromosomal abnormalities over time. Importantly, while CIN frequently leads to aneuploidy, the converse is not always true. Some aneuploid cells maintain a stable karyotype despite their numerical aberrations. For example, in down syndrome, trisomy 21 (an extra copy of chromosome 21) represents an altered yet stable chromosomal state across cell generations. This distinction highlights that aneuploidy does not inherently indicate ongoing CIN.

In cancer, polyploidy, aneuploidy, and CIN are intricately linked, yet distinct phenomena. Polyploidization is often observed as a precursor to CIN and aneuploidy. While many cancers exhibit both aneuploidy and CIN, their interplay varies across tumor types and stages of progression. CIN often fuels tumor evolution by generating genetic diversity, enabling adaptation to selective pressures such as immune surveillance and therapy. In contrast, stable aneuploidy can also contribute to tumorigenesis by altering gene dosage and disrupting cellular homeostasis. For instance, certain aneuploidies can activate oncogenes or silence tumor suppressor genes. While the relationships and implications of these chromosomal abnormalities are explored further below, understanding their interplay is crucial for identifying and exploiting tumor vulnerabilities.

11.3 Polyploidization Caused by Cell Fusion

When considering ploidy alterations induced by cell fusion, the most fundamental event is the generation of tetraploid cells through the fusion of two diploid cells. Indeed, cell fusion represents one of the key pathways to polyploidization in both physiological and pathological contexts. Notably, polyploid cells themselves can also participate in cell fusion. Hepatocytes in the liver frequently become polyploid under physiological conditions and in response to tissue damage (Matsumoto 2022). Polyploid cells can also arise in damaged tissues, as seen in inflammatory bowel disease and acute kidney injury (De Chiara et al. 2022; Suzuki et al. 1990). In these settings, the presence of polyploid epithelial cells raises the possibility that fusion events may occur between polyploid and diploid cells, further contributing to ploidy diversification within the tissue. Whether fusion occurs between two diploid cells or between a diploid and a polyploid cell, the resulting progeny retains a polyploid state with a complete set of chromosomes, provided that both parental cells possess intact chromosome complements. Such polyploidization occurs in various physiological contexts, including skeletal muscle cells, osteoclasts, and syncytiotrophoblasts

(Dörnen et al. 2020). Therefore, polyploidization resulting from cell fusion is not inherently abnormal.

11.4 Cell Cycle Progression After Polyploidization

Fusion-derived cells are multinucleated polyploid cells. Polyploidization is widely regarded as being closely linked to cell cycle arrest, as physiological polyploidization typically occurs alongside terminal differentiation and cessation of cell division (Aylon and Oren 2011; Darmasaputra et al. 2024). In contrast, tissue stem cells, such as hematopoietic and intestinal stem cells, which continuously proliferate, remain diploid. The activation of p53, leading to p21 upregulation, plays a key role in restricting cell proliferation after polyploidization (Andreassen et al. 2001; Lens and Medema 2019; Vittoria et al. 2023). One mechanism underlying this p53 activation after polyploidization is DNA damage, which is induced by polyploidization, as discussed later. Polyploidization also leads to an increase in centrosomes, which serve as poles during cell division. This centrosome amplification activates the PIDDosome and the Hippo pathway, both of which contribute to the further activation of p53 and induce cell cycle arrest (Fava et al. 2017; Ganem et al. 2014).

As long as polyploid cells remain dormant and do not proliferate, cell fusion does not contribute to tumorigenesis and may, as discussed later, even exert a buffering effect against genomic damage. However, polyploidization does not necessarily equate to irreversible cell cycle arrest. In fact, while not undergoing cell division, physiologically polyploid cardiomyocytes can continue the cell cycle and reach even higher ploidy states in response to cardiac injury (Darmasaputra et al. 2024; Øvrebø and Edgar 2018). Megakaryocytes also undergo repeated rounds of DNA replication without division during their normal development, ultimately achieving ploidy levels as high as 64 N (Lee et al. 2009). Moreover, in cases of severe liver injury, cell fusion between hepatocytes and bone marrow-derived monocytes has been reported to contribute to liver regeneration, and polyploid hepatocytes arising from such fusion events can proliferate to support tissue repair (Vassilopoulos et al. 2003; Wang et al. 2003). Furthermore, physiological polyploid hepatocytes can also undergo proliferation during aging and liver regeneration (Matsumoto et al. 2020, 2021a). These examples clearly indicate that the cell cycle following polyploidization is not necessarily arrested, and cells derived from fusion events can continue to proliferate when provided with appropriate proliferative stimuli.

11.5 Buffering Effect of Cell Fusion Against Genomic Abnormalities

In cancerous tissues or precancerous environments, various genomic abnormalities frequently arise in cells due to persistent stressors such as chronic inflammation, replication stress, telomere shortening, or chromosomal missegregation. Importantly, by harboring redundant genomic content, cell fusion may exert a buffering effect against such genomic insults (Fig. 11.1). For example, when a cell carrying genomic mutations fuses with a normal cell, the increased copy number of the normal allele resulting from fusion can weaken the functional impact of those mutations in the fused cell. Similarly, the effects of copy number alterations can be attenuated, as their relative proportion is reduced within the fused cell.

The buffering effect of genomic abnormalities through cell fusion may mitigate genomic alterations that lead to carcinogenesis. In particular, for tumor suppressor genes, the addition of normal copies may delay the complete loss of function. On the other hand, this buffering effect may also function as a survival strategy for cancer cells. Actively proliferating cancer cells are constantly exposed to genomic stress, which increases the likelihood of acquiring new genetic aberrations. While these genomic alterations play a central role in driving cancer clonal evolution and tumor progression, they often have immediate detrimental effects on cellular fitness. Newly acquired genomic abnormalities, particularly large-scale chromosomal aberrations, can abruptly disrupt the expression of key regulators of intracellular signaling and metabolism, thereby impairing cellular homeostasis and reducing proliferation and survival (Ben-David and Amon 2020). In fact, these alterations can activate cellular defense mechanisms such as cell cycle arrest, apoptosis, or senescence. Even well-known oncogenic mutations, such as activating mutations in Kirsten rat sarcoma viral oncogene homolog (KRAS) or loss-of-function mutations in phosphatase and tensin homolog (PTEN), can induce cellular senescence, suppressing tumorigenesis in the early stages (Schmitt et al. 2022). In this context, the increased ploidy resulting from cell fusion, which can mitigate the harmful impact of genomic abnormalities, may help precancerous or cancer cells tolerate genomic insults that would otherwise be lethal.

Importantly, while cell fusion can buffer the impact of genomic abnormalities, it does not eliminate them. As a result, fusion may allow genomically unstable cells to survive, providing a window for the gradual accumulation of additional oncogenic alterations. This process not only enhances the adaptability of cancer cells under stress but also accelerates tumor evolution, underscoring the dual nature of cell fusion as both a safeguard against and a driver of malignant progression.

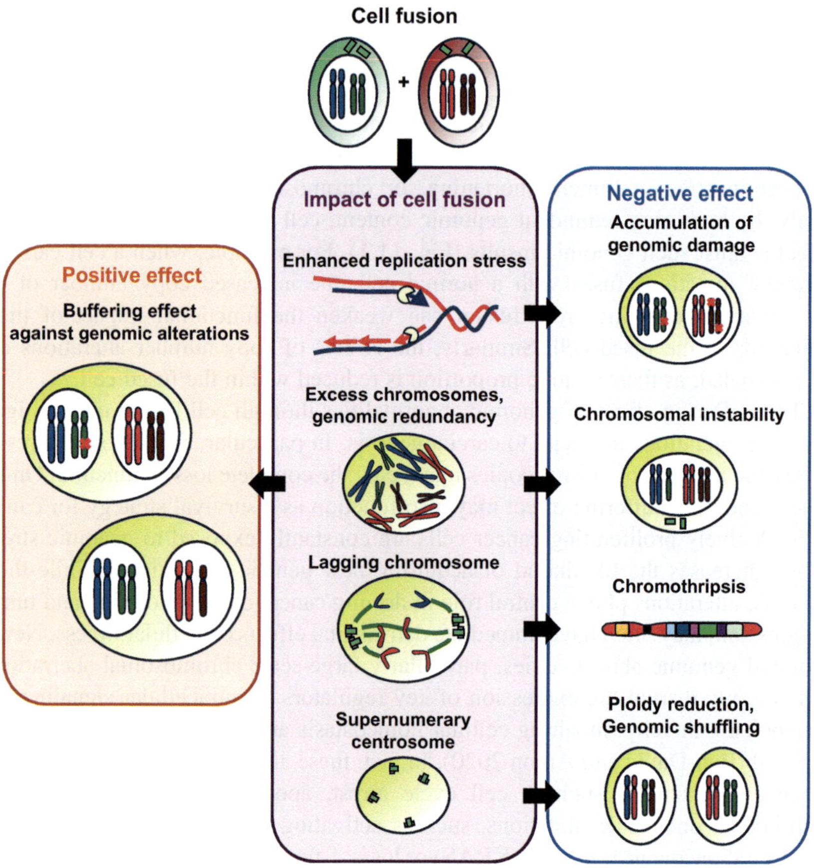

Fig. 11.1 Impact of cell fusion on genomic stability. Cell fusion leads to an increase in genomic content, enhanced replication stress, and supernumerary centrosomes, each of which can influence the maintenance of genomic stability in fused cells in both positive and negative ways. While these changes can predispose cells to numerous genomic abnormalities, including mutations, aneuploidy, and chromothripsis, the redundant genome content may also mitigate such defects. This duality suggests that post-fusion cells carry a substantial risk of serving as the origin of tumorigenesis due to their elevated burden of genomic alterations. Furthermore, fused cells can undergo ploidy reduction, for instance through multipolar mitosis, a process that may induce extensive genomic rearrangements

11.6 Enhancement of Genomic Instability During the Cell Cycle After Polyploidization

Polyploid cells generated through cell fusion are often in a dormant state with cell cycle arrest. However, once these polyploid cells begin to proliferate continuously, they can become genomically unstable and potentially contribute to cancer development.

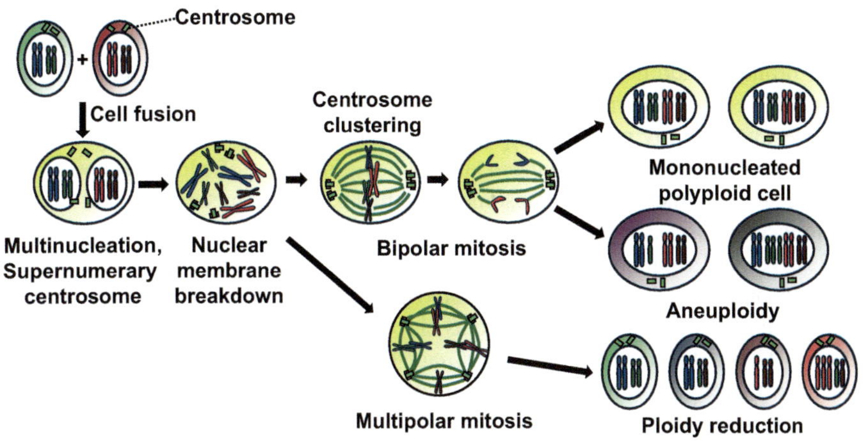

Fig. 11.2 Cell cycle progression after cell fusion. Fused cells are inevitably multinucleated and exhibit centrosome amplification. When the post-fusion cell cycle progresses and enters mitosis, the nuclear envelopes break down, allowing the genomic contents from the separate nuclei to intermingle. All chromosomes then align at the metaphase plate, and bipolar division produces mononucleated polyploid daughter cells. However, the presence of excessive chromosomes often leads to segregation errors, resulting in aneuploidy. Although cells attempt to undergo bipolar mitosis through centrosome clustering, failure of this process results in multipolar mitosis, which in turn leads to ploidy reduction

Recent studies have shown that polyploid cells are more prone to double-strand DNA breaks than would be expected based solely on their increased ploidy (Gemble et al. 2022; Hayashi et al. 2024). One underlying reason for this vulnerability is that polyploid cells often enter subsequent cell cycles without fully establishing the necessary mechanisms to ensure accurate genome replication during the S phase. The increased DNA content places additional demands on the replication machinery, which may not be sufficiently upregulated to accommodate the higher ploidy state (Gemble et al. 2022). As a result, incomplete DNA synthesis and the accumulation of replication-associated DNA damage can occur, further exacerbating genomic instability throughout the cell cycle in polyploid cells. In the context of cell fusion, these risks may be enhanced, particularly when the two fusing cells are at different cell cycle phases (Johnson and Rao 1970). Asynchronous fusion events can lead to conflicts in cell cycle regulation, forcing cells into abnormal replication dynamics that heighten replication errors and genomic instability. Thus, while cell fusion-induced ploidy increase can buffer pre-existing genomic abnormalities, it may simultaneously drive new genomic instability through replication stress and defective genome maintenance (Zhou et al. 2015).

Moreover, one of the most significant consequences of ploidy increase due to cell fusion is its strong association with the induction of CIN (Fig. 11.2). After cell fusion, the resulting cells contain multiple nuclei derived from the parent cells. During subsequent cell proliferation, the nuclear membranes break down in the mitotic phase, and the genomes contained within each nucleus lose their partitioned walls. All chromosomes must then align correctly at the metaphase plate, but the large chromosome number in polyploid cells makes this process inherently more error prone. In addition, cell fusion commonly leads to centrosome amplification, characterized by supernumerary centrosomes that can generate defective mitotic spindles, particularly multipolar spindles, resulting in severely unbalanced chromosome segregation and pronounced aneuploidy. Indeed, polyploid cells typically require more time for chromosome alignment during metaphase and exhibit prolonged mitosis compared with diploid cells (Hayashi et al. 2024; Meyer et al. 2023), yet still frequently display errors such as multipolar divisions and lagging chromosomes. In addition, lagging chromosomes may experience massive, clustered rearrangements, a process referred to as chromothripsis (Simovic-Lorenz and Ernst 2025). These damaged chromosomes can occasionally be re-integrated into the nucleus, thereby promoting genomic instability in polyploid cells (Papathanasiou et al. 2023).

To safeguard chromosome segregation, cells employ the spindle assembly checkpoint (SAC) during metaphase, which prevents premature progression to anaphase by inhibiting the activation of the anaphase-promoting complex/cyclosome (APC/C) until all chromosomes are correctly attached to spindle microtubules under proper tension. However, in cancer cells, SAC function is often impaired by mechanisms such as genetic alterations in CDC27, a component of the APC/C, or in BUB1B, a key SAC regulator (Kazemi-Sefat et al. 2021; Malumbres and Villarroya-Beltri 2024). As a result, polyploid cancer cells, including those arising from cell fusion, are more susceptible to abnormal segregation events. Cells also possess a centrosome-clustering mechanism that can gather extra centrosomes into two functional spindle poles, a process frequently observed at the onset of mitosis in fused cells (Goupil et al. 2020). Nevertheless, the uneven spatial distribution of chromosomes in large polyploid cells can hinder complete bipolarization, allowing multipolar divisions to occur despite initial clustering (Goupil et al. 2020). Collectively, these defects, including centrosome amplification, impaired SAC function, and incomplete centrosome clustering, make post-fusion polyploid cells highly prone to CIN and the generation of progeny with novel chromosomal abnormalities, which can contribute to tumorigenesis.

11.7 Diploid Cell Formation via Multipolar Mitosis in Somatic Polyploid Cells

Polyploid cells arising from cell fusion often experience CIN and replication stress during subsequent proliferation, leaving them in a genomically unstable state (Fig. 11.2). While the surplus genomic content in these cells can buffer the impact of genetic alterations and thereby support the survival of genomically unstable populations, polyploidy also imposes a proliferative burden: each division requires replication of a duplicated or larger genome and accurate alignment of a substantially greater number of chromosomes at the metaphase plate. As a result, when cells with doubled or higher ploidy continue to proliferate under conditions of CIN, those with chromosome losses are more likely to be selected than those with gains. Consistent with this, progressive chromosome loss is frequently observed in cancer cells following polyploidization, and mathematical modeling studies predict that polyploid cancer cells tend to converge toward a near-triploid state (Laughney et al. 2015).

In addition to the stepwise loss of chromosomes following polyploidization, interestingly, polyploid cells can sometimes reduce their ploidy in a single event and generate diploid cells. Although the mechanisms underlying such ploidy reduction are still poorly understood, this phenomenon has been reported in fused cells under certain conditions. For example, during severe liver injury, hematopoietic cells, particularly monocytes, have been shown to fuse with hepatocytes, contributing to the generation of functional hepatocytes and liver regeneration (Vassilopoulos et al. 2003; Wang et al. 2003). Notably, polyploid hepatocytes that have undergone cell fusion can reduce their ploidy during subsequent proliferation for regeneration and give rise to diploid hepatocytes (Duncan et al. 2009). This ploidy reduction in hepatocytes has been observed in several mouse models, and one proposed mechanism for this process is multipolar cell division (Duncan et al. 2010; Matsumoto et al. 2020). Similar instances of ploidy reduction have also been observed in mouse ES cells (Frade et al. 2019). Fusion-derived cells, generated by fusing neural stem cells with ES cells, can give rise to chimeric mice when co-injected with normal ES cells into pseudopregnant mice. During their development, daughter cells originating from the fused cells contribute to the formation of various organs, including the brain, heart, liver, and lungs. Notably, some of these fusion-derived cells are diploid. Furthermore, in vitro time-lapse imaging has demonstrated that multipolar division occurs in ES cells with amplified centrosomes following cell fusion, leading to ploidy reduction (Frade et al. 2019).

This mode of ploidy reduction via multipolar mitosis occurs in polyploid somatic cells and represents a cell behavior entirely distinct from meiosis, in which diploid germ cells give rise to haploid gametes. Centrosome amplification caused by cell fusion likely contributes to the increased occurrence of multipolar mitosis. Although the frequency of such multipolar mitosis remains largely unknown, various lines of evidence indicate that somatic cell ploidy can not only increase to become polyploid but also decrease to the diploid state.

11.8 Regulation of Chromosome Segregation During Ploidy Reduction

One of the major challenges during ploidy reduction in somatic cells through multipolar mitosis is the accurate segregation of chromosomes into daughter cells. In contrast to bipolar mitosis, where chromosomes are evenly divided between two poles, multipolar mitosis requires each chromosome to be assigned to two of several poles. If this process happens randomly, it becomes statistically difficult for any resulting daughter cell to inherit a full, balanced set of chromosomes. Even if daughter cells are initially formed through this process, those lacking a proper chromosome set may undergo cell death. In fact, time-lapse imaging studies conducted in vitro have shown that although daughter cells are generated through multipolar division from polyploid cancer cells, many of them fail to proliferate and eventually die (Ganem et al. 2009).

However, interestingly, some cells that undergo ploidy reduction via multipolar mitosis in vivo are able to survive and contribute to development, tissue regeneration, or tumorigenesis (Duncan et al. 2009, 2010; Frade et al. 2019; Matsumoto et al. 2020b). Proper chromosome segregation during cell division has been shown to depend on cell polarity maintained in a three-dimensional environment (Knouse et al. 2018). Therefore, it is possible that multipolar mitosis in vivo occurs under a more tightly regulated setting than in two-dimensional culture, allowing for viable daughter cells to be generated through more accurate chromosome segregation.

Then, how exactly are chromosomes distributed during ploidy reduction? Several studies have tried to address this question. In *Fah* knockout mice, a model of hereditary tyrosinemia, the absence of the therapeutic agent nitisinone (NTBC) results in liver failure due to hepatotoxic tyrosine metabolites. However, when these mice receive bone marrow transplants from healthy donors carrying a functional Fah gene, donor-derived cells, mainly monocytes, fuse with Fah-deficient hepatocytes (Vassilopoulos et al. 2003; Wang et al. 2003). The resulting fusion cells express the functional *Fah* gene from the donor genome, enabling them to metabolize the toxic intermediates and avoid cell injury. These fusion-derived hepatocytes then serve as a source of liver regeneration (Vassilopoulos et al. 2003; Wang et al. 2003). Notably, during this regenerative process, fusion-derived hepatocytes often undergo ploidy reduction, likely through multipolar mitosis (Duncan et al. 2009). A study examining the chromosome numbers of ploidy-reduced daughter cells derived from polyploid hepatocytes found that, although many cells approach a diploid karyotype, they often exhibit numerical abnormalities (Duncan et al. 2010). This indicates that ploidy reduction through multipolar mitosis does not necessarily produce perfectly diploid cells. Moreover, analyses using marker genes to trace whether the resulting pseudo-diploid cells inherited chromosomes from the donor or host hepatocyte revealed that inheritance patterns vary by chromosome (Duncan et al. 2009). This suggests that daughter cells do not consistently inherit a complete chromosome set from either parent genome.

Additionally, a study using mouse SNP arrays to analyze chromosomal abnormalities showed that regenerative nodules derived from polyploid hepatocytes often carry chromosomal number aberrations, which are likely the result of CIN (Matsumoto et al. 2021b). Importantly, nodules that had undergone ploidy reduction exhibited an even higher frequency of such abnormalities (Matsumoto et al. 2021b). This finding indicates that viable daughter cells can be generated through ploidy reduction, while the process increases the risk of chromosome missegregation to some extent.

On the other hand, a study focusing on mouse development from fusion cells generated by combining neural stem cells and ES cells showed an intriguing aspect of ploidy reduction during multipolar mitosis (Matsumoto et al. 2021b). A combination of EdU pulse-labeling of parental DNA and SNP analyses detected both random and nonrandom segregation of parental chromosomes during ploidy reduction via multipolar mitosis. While some daughter cells inherited a mixed genome from both parent cells, others contained the genome from only one parent, indicating that an entire genome set from one parent could be selectively retained during ploidy reduction (Frade et al. 2019).

Taken together, accumulating evidence suggests that ploidy reduction through multipolar mitosis does not always result in completely nonrandom chromosome segregation, but rather that the allocation of chromosome sets is at least partially regulated. One possible factor enabling more accurate segregation may be the presence of binucleation, as seen in fusion cells. Ploidy reduction has been shown to occur more frequently in binucleated polyploid cells shortly after polyploidization, whereas it becomes rare in mononucleated cells that have continuously proliferated in a polyploid state (Matsumoto et al. 2020). In addition, although nuclear envelope breakdown during M phase typically allows the mixing of chromosomes from each nucleus, some binucleated polyploid cells have been observed to proceed through anaphase while maintaining spatial separation of the two nuclear contents (Duncan et al. 2010). This results in a phenomenon known as double mitosis, and in such cases, the chromosome sets from each nucleus can be separately inherited by daughter cells.

11.9 Various Modes of Ploidy Reduction

There have also been reports of ploidy reduction independent of multipolar mitosis (Fig. 11.3). Polyploid giant cancer cells (PGCCs), defined as mono- or multinucleated cancer cells with nuclei markedly larger than those of normal diploid cells, represent an important cell type capable of undergoing ploidy reduction through amitotic divisions such as budding or bursting (Chen et al. 2019; Saini et al. 2022). PGCCs have been shown to arise through a variety of unusual cellular processes, including entosis—a process in which neighboring cells are engulfed (Li et al. 2023). Following entosis, the internalized cell may undergo decellularization, resulting in fusion between the inner and outer cells (Li et al. 2023). Intriguingly, these PGCCs have been observed to produce small daughter cells with drastically reduced ploidy through striking, yet still poorly understood, mechanisms such as budding, nuclear fragmentation, and membrane bursting. Such highly distinctive cellular dynamics involved in ploidy reduction can only be observed through time-lapse imaging, and their true nature remains largely unknown. However, given the reported similarities between the unusual behavior of PGCCs and the cellular events observed during embryonic development from fertilized eggs (Li et al. 2023; Liu et al. 2022), it is plausible that the mechanisms

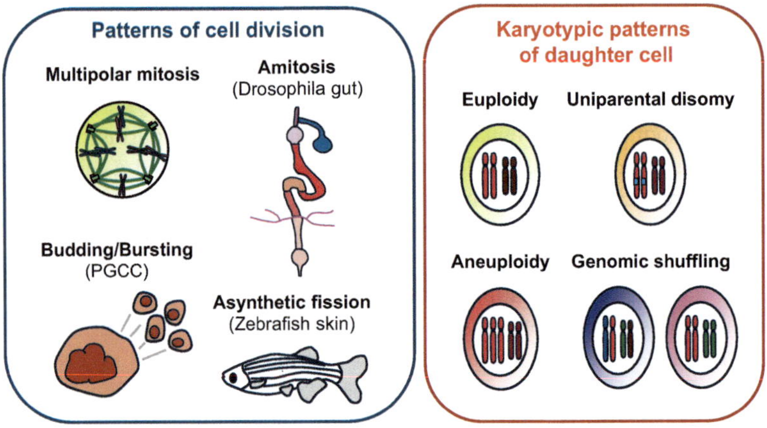

Fig. 11.3 Modes of ploidy reduction. In addition to multipolar mitosis, ploidy reduction can also occur through mechanisms such as budding and bursting in PGCCs, which generate small diploid daughter cells. In the Drosophila gut, an amitotic process that proceeds without spindle formation has been reported, whereas in zebrafish skin cells, asynthetic fission that occurs without DNA replication has been observed. The genomic composition of daughter cells produced by ploidy reduction can vary widely depending on the combination of chromosomes distributed. Possible outcomes include diploid cells with a normal chromosome set, uniparental disomy, in which one parental chromosome set is duplicated without changes in copy number, as well as aneuploidy and more complex genomic rearrangements

underlying ploidy reduction in PGCCs reflect evolutionarily conserved programs with shared features across species and pathological contexts.

In the Drosophila gut, polyploid intestinal epithelial cells produce daughter cells with reduced ploidy via amitosis during recovery from starvation (Lucchetta and Ohlstein 2017). In amitosis, spindle formation does not occur as it does in mitosis. Instead, nuclear division (karyokinesis) proceeds through genome partitioning by invagination of the intact nuclear lamina. This process is followed by a binucleate phase, after which the ingression of the cell membrane produces two daughter cells that contribute to the regeneration of new intestinal stem cells. Moreover, in the superficial epithelial cells of zebrafish skin, daughter cells with reduced ploidy are generated through a process known as "asynthetic fission," which does not involve DNA replication (Chan et al. 2022). This process is particularly active when growing fish experience increased skin tension. Without synthesizing DNA, cells expand clonally and produce up to four daughter cells from a single cell. Because asynthetic fission deviates from conventional mitosis, it has been reported to involve chromosome segregation abnormalities, such as lagging chromosomes and multipolar mitoses, and to produce daughter cells whose DNA ploidy can fall below the diploid level. Whether such unusual forms of ploidy reduction occur in mammalian cells following cell fusion remains unknown. However, they may share molecular mechanisms with ploidy reduction in mammalian cells and are worthy of attention.

11.10 Cellular Evolution Through Ploidy Reduction

Ploidy reduction plays a critically important role in generating cells with novel characteristics. During the process of ploidy reduction, daughter cells do not necessarily receive an exact, complete set of chromosomes identical to that of the parent cell. Instead, they may inherit a genomic composition somewhat different from that of the parent, leading to the acquisition of new features.

One likely genomic change in daughter cells is a gain or loss of chromosomes. While the induction of aneuploidy may have detrimental effects on the survival or proliferation of cells, it may enhance adaptability to environmental pressures and thereby drive clonal evolution. Beyond aneuploidy, ploidy reduction can also promote cellular adaptation and evolution by inducing uniparental disomy (UPD), a state in which both copies of a given chromosome originate from the same parent. UPD does not change the total chromosome number and may enable daughter cells to remain viable. However, if the parental cell already carries a mutation in a tumor suppressor gene, UPD can result in the complete loss of tumor suppressor function. Similarly, UPD can lead to the enrichment of an oncogenic mutant allele, thereby increasing tumor aggressiveness. Indeed, UPD involving cancer-related genes is frequently observed in various cancer types, including ovarian and colorectal cancers (Makishima and Maciejewski 2011).

A compelling example of clonal evolution driven by ploidy reduction is liver regeneration from polyploid hepatocytes in chronically injured livers. As described earlier, when functional *Fah*-bearing donor bone marrow transplant into *Fah*-deficient mice, recipient hepatocytes fuse with donor-derived monocytes, generating metabolically competent hepatocytes. During subsequent liver regeneration via ploidy reduction, hepatocytes inheriting the *Fah* gene, which is essential for cell survival, are preferentially expanded (Duncan et al. 2009). Although other donor- or recipient-derived chromosomes may be variably retained, donor-derived sex chromosomes are relatively prone to loss (Duncan et al. 2009). This model illustrates how ploidy reduction after cell fusion can selectively favor chromosomal combinations that promote survival and proliferation, driving clonal evolution.

On the other hand, cell evolution through ploidy reduction can also have detrimental consequences, including the promotion of tumorigenesis. Indeed, analysis using mouse models that trace the proliferation of polyploid cells showed that ploidy reduction facilitates the induction of allele imbalance between maternally and paternally derived chromosomes and enhances liver cancer development (Matsumoto et al. 2021b). Another study has also shown that polyploid preneoplastic lesions give rise to diploid hepatocellular carcinomas in a separate mouse model (Lin et al. 2021). During tumorigenesis, ploidy reduction may enhance cancer development by increasing the dosage of oncogenic mutations or causing the loss of tumor suppressor genes.

Furthermore, ploidy reduction in PGCCs has been identified as a key driver of clonal evolution in cancer. PGCCs, often induced by chemotherapy or radiation therapy, survive as therapy-resistant cancer cells owing to their high stress tolerance, partly supported by the genomic redundancy of polyploidy (Liu 2018; Saini et al. 2022). One critical role of ploidy reduction in PGCCs is facilitating the transition from cell cycle arrest to proliferation. Evidence indicates that both prior polyploidization and subsequent ploidy reduction are required for PGCCs to re-enter the cell cycle, highlighting the importance of ploidy alterations in switching between proliferative and non-proliferative states (Liu et al. 2022; Puig et al. 2008). In addition, smaller daughter cells with reduced ploidy, produced via budding from PGCCs, often express high levels of CD133 and CD44, display stem-like properties, and exhibit aggressive phenotypes (Zhang et al. 2014). Budding has also been associated with the acquisition of mesenchymal traits (Zhang et al. 2014). These findings suggest that ploidy reduction can also drive phenotypic changes, such as the emergence of stem cell-like features and epithelial–mesenchymal transition, although it remains unclear whether these new properties arise from genetic alterations induced by ploidy reduction or from other mechanisms.

11.11 Conclusion

Cell fusion inevitably leads to polyploidization and centrosome amplification. While polyploidy can buffer genomic damage and enhance cellular stress tolerance, subsequent proliferation often results in genomic instability, with dramatic cellular transformations reported particularly during ploidy reduction. However, the regulation and disruption of the cell cycle following cell fusion remain insufficiently characterized, and the mechanisms underlying ploidy reduction, in particular, remain poorly defined. Post-fusion cells are typically multinucleated, with each genome set spatially segregated into separate nuclei, which may critically influence the outcome of ploidy reduction. Moreover, the in vivo frequency of ploidy reduction events and the proportion that generate daughter cells capable of sustained proliferation and function remain entirely unknown. Future studies will be essential to elucidate how cell fusion and subsequent ploidy alterations shape cellular fate and contribute to both physiological and pathological processes.

Conflict of Interest The authors declare that they have no conflict of interest.

References

Al-Rawi DH, Bakhoum SF (2022) Chromosomal instability as a source of genomic plasticity. Curr Opin Genet Dev 74:101913. https://doi.org/10.1016/j.gde.2022.101913

Andreassen PR, Lohez OD, Lacroix FB, Margolis RL (2001) Tetraploid state induces p53-dependent arrest of nontransformed mammalian cells in G1. Mol Biol Cell 12(5): 1315–1328. https://doi.org/10.1091/mbc.12.5.1315

Aylon Y, Oren M (2011) p53: guardian of ploidy. Mol Oncol 5(4):315–323. https://doi.org/10.1016/j.molonc.2011.07.007

Bakhoum SF, Cantley LC (2018) The multifaceted role of chromosomal instability in cancer and its microenvironment. Cell 174(6):1347–1360. https://doi.org/10.1016/j.cell.2018.08.027

Ben-David U, Amon A (2020) Context is everything: aneuploidy in cancer. Nat Rev Genet 21(1): 44–62. https://doi.org/10.1038/s41576-019-0171-x

Bielski CM, Zehir A, Penson AV, Donoghue MTA, Chatila W, Armenia J, Taylor BS (2018) Genome doubling shapes the evolution and prognosis of advanced cancers. Nat Genet 50(8): 1189–1195. https://doi.org/10.1038/s41588-018-0165-1

Boveri T (1914) Zur Frage der Entstehung maligner Tumoren. Gustav Fischer Verlag

Chan KY, Yan CS, Roan HY, Hsu SC, Tseng TL, Hsiao CD, Chen CH (2022) Skin cells undergo asynthetic fission to expand body surfaces in zebrafish. Nature 605(7908):119–125. https://doi.org/10.1038/s41586-022-04641-0

Chen J, Niu N, Zhang J, Qi L, Shen W, Donkena KV, Liu J (2019) Polyploid Giant cancer cells (PGCCs): the evil roots of cancer. Curr Cancer Drug Targets 19(5):360–367. https://doi.org/10.2174/1568009618666180703154233

Consortium, I. T. P.-C. A. o. W. G (2020) Pan-cancer analysis of whole genomes. Nature 578(7793):82–93. https://doi.org/10.1038/s41586-020-1969-6

Darmasaputra GS, van Rijnberk LM, Galli M (2024) Functional consequences of somatic polyploidy in development. Development 151(5). https://doi.org/10.1242/dev.202392

De Chiara L, Conte C, Semeraro R, Diaz-Bulnes P, Angelotti ML, Mazzinghi B, Romagnani P (2022) Tubular cell polyploidy protects from lethal acute kidney injury but promotes consequent chronic kidney disease. Nat Commun 13(1):5805. https://doi.org/10.1038/s41467-022-33110-5

Duncan AW, Hickey RD, Paulk NK, Culberson AJ, Olson SB, Finegold MJ, Grompe M (2009) Ploidy reductions in murine fusion-derived hepatocytes. PLoS Genet 5(2):e1000385. https://doi.org/10.1371/journal.pgen.1000385

Duncan AW, Taylor MH, Hickey RD, Hanlon Newell AE, Lenzi ML, Olson SB, Grompe M (2010) The ploidy conveyor of mature hepatocytes as a source of genetic variation. Nature 467(7316): 707–710. https://doi.org/10.1038/nature09414

Dörnen J, Sieler M, Weiler J, Keil S, Dittmar T (2020) Cell fusion-mediated tissue regeneration as an inducer of polyploidy and aneuploidy. Int J Mol Sci 21(5). https://doi.org/10.3390/ijms21051811

Fava LL, Schuler F, Sladky V, Haschka MD, Soratroi C, Eiterer L, Villunger A (2017) The PIDDosome activates p53 in response to supernumerary centrosomes. Genes Dev 31(1): 34–45. https://doi.org/10.1101/gad.289728.116

Frade J, Nakagawa S, Cortes P, di Vicino U, Romo N, Lluis F, Cosma MP (2019) Controlled ploidy reduction of pluripotent 4n cells generates 2n cells during mouse embryo development. Sci Adv 5(10):eaax4199. https://doi.org/10.1126/sciadv.aax4199

Ganem NJ, Cornils H, Chiu SY, O'Rourke KP, Arnaud J, Yimlamai D, Pellman D (2014) Cytokinesis failure triggers hippo tumor suppressor pathway activation. Cell 158(4):833–848. https://doi.org/10.1016/j.cell.2014.06.029

Ganem NJ, Godinho SA, Pellman D (2009) A mechanism linking extra centrosomes to chromosomal instability. Nature 460(7252):278–282. https://doi.org/10.1038/nature08136

Gemble S, Wardenaar R, Keuper K, Srivastava N, Nano M, Macé AS, Basto R (2022) Genetic instability from a single S phase after whole-genome duplication. Nature 604(7904):146–151. https://doi.org/10.1038/s41586-022-04578-4

Gerlinger M, Rowan AJ, Horswell S, Math M, Larkin J, Endesfelder D, Swanton C (2012) Intratumor heterogeneity and branched evolution revealed by multiregion sequencing. N Engl J Med 366(10):883–892. https://doi.org/10.1056/NEJMoa1113205

Goupil A, Nano M, Letort G, Gemble S, Edwards F, Goundiam O, Basto R (2020) Chromosomes function as a barrier to mitotic spindle bipolarity in polyploid cells. J Cell Biol 219(4). https://doi.org/10.1083/jcb.201908006

Hayashi K, Horisaka K, Harada Y, Ogawa Y, Yamashita T, Kitano T, Matsumoto T (2024) Polyploidy mitigates the impact of DNA damage while simultaneously bearing its burden. Cell Death Discov 10(1):436. https://doi.org/10.1038/s41420-024-02206-w

Johnson RT, Rao PN (1970) Mammalian cell fusion: induction of premature chromosome condensation in interphase nuclei. Nature 226(5247):717–722. https://doi.org/10.1038/226717a0

Kazemi-Sefat GE, Keramatipour M, Talebi S, Kavousi K, Sajed R, Kazemi-Sefat NA, Mousavizadeh K (2021) The importance of CDC27 in cancer: molecular pathology and clinical aspects. Cancer Cell Int 21(1):160. https://doi.org/10.1186/s12935-021-01860-9

Knouse KA, Lopez KE, Bachofner M, Amon A (2018) Chromosome segregation Fidelity in epithelia requires tissue architecture. Cell 175(1):200–211.e213. https://doi.org/10.1016/j.cell.2018.07.042

Laughney AM, Elizalde S, Genovese G, Bakhoum SF (2015) Dynamics of tumor heterogeneity derived from clonal Karyotypic evolution. Cell Rep 12(5):809–820. https://doi.org/10.1016/j.celrep.2015.06.065

Lee HO, Davidson JM, Duronio RJ (2009) Endoreplication: polyploidy with purpose. Genes Dev 23(21):2461–2477. https://doi.org/10.1101/gad.1829209

Lens SMA, Medema RH (2019) Cytokinesis defects and cancer. Nat Rev Cancer 19(1):32–45. https://doi.org/10.1038/s41568-018-0084-6

Li X, Zhong Y, Zhang X, Sood AK, Liu J (2023) Spatiotemporal view of malignant histogenesis and macroevolution via formation of polyploid giant cancer cells. Oncogene 42(9):665–678. https://doi.org/10.1038/s41388-022-02588-0

Lin H, Huang YS, Fustin JM, Doi M, Chen H, Lai HH, Chao HW (2021) Hyperpolyploidization of hepatocyte initiates preneoplastic lesion formation in the liver. Nat Commun 12(1):645. https://doi.org/10.1038/s41467-020-20572-8

Liu J (2018) The dualistic origin of human tumors. Semin Cancer Biol 53:1–16. https://doi.org/10.1016/j.semcancer.2018.07.004

Liu J, Niu N, Li X, Zhang X, Sood AK (2022) The life cycle of polyploid giant cancer cells and dormancy in cancer: opportunities for novel therapeutic interventions. Semin Cancer Biol 81:132–144. https://doi.org/10.1016/j.semcancer.2021.10.005

Lucchetta EM, Ohlstein B (2017) Amitosis of Polyploid cells regenerates functional stem cells in the drosophila intestine. Cell Stem Cell 20(5):609–620.e606. https://doi.org/10.1016/j.stem.2017.02.012

Makishima H, Maciejewski JP (2011) Pathogenesis and consequences of uniparental disomy in cancer. Clin Cancer Res 17(12):3913–3923. https://doi.org/10.1158/1078-0432.CCR-10-2900

Malumbres M, Villarroya-Beltri C (2024) Mosaic variegated aneuploidy in development, ageing and cancer. Nat Rev Genet 25(12):864–878. https://doi.org/10.1038/s41576-024-00762-6

Matsumoto T (2022) Implications of polyploidy and ploidy alterations in hepatocytes in liver injuries and cancers. Int J Mol Sci 23(16). https://doi.org/10.3390/ijms23169409

Matsumoto T, Wakefield L, Grompe M (2021a) The significance of Polyploid hepatocytes during aging process. Cell Mol Gastroenterol Hepatol 11(5):1347–1349. https://doi.org/10.1016/j.jcmgh.2020.12.011

Matsumoto T, Wakefield L, Peters A, Peto M, Spellman P, Grompe M (2021b) Proliferative polyploid cells give rise to tumors via ploidy reduction. Nat Commun 12(1):646. https://doi.org/10.1038/s41467-021-20916-y

Matsumoto T, Wakefield L, Tarlow BD, Grompe M (2020) In vivo lineage tracing of Polyploid hepatocytes reveals extensive proliferation during liver regeneration. Cell Stem Cell 26(1):34–47.e33. https://doi.org/10.1016/j.stem.2019.11.014

McPherson A, Vázquez-García I, Myers MA, Al-Rawi DH, Zatzman M, Weiner AC, Shah SP (2025) Ongoing genome doubling shapes evolvability and immunity in ovarian cancer. Nature. https://doi.org/10.1038/s41586-025-09240-3

Meyer RE, Sartin A, Gish M, Harsha J, Wilkie E, Haworth D, Dawson DS (2023) Polyploid yeast are dependent on elevated levels of Mps1 for successful chromosome segregation. bioRxiv. https://doi.org/10.1101/2023.01.09.523325

Øvrebø JI, Edgar BA (2018) Polyploidy in tissue homeostasis and regeneration. Development 145(14). https://doi.org/10.1242/dev.156034

Papathanasiou S, Mynhier NA, Liu S, Brunette G, Stokasimov E, Jacob E, Pellman D (2023) Heritable transcriptional defects from aberrations of nuclear architecture. Nature 619(7968): 184–192. https://doi.org/10.1038/s41586-023-06157-7

Puig PE, Guilly MN, Bouchot A, Droin N, Cathelin D, Bouyer F, Chauffert B (2008) Tumor cells can escape DNA-damaging cisplatin through DNA endoreduplication and reversible polyploidy. Cell Biol Int 32(9):1031–1043. https://doi.org/10.1016/j.cellbi.2008.04.021

Quinton RJ, DiDomizio A, Vittoria MA, Kotýnková K, Ticas CJ, Patel S, Ganem NJ (2021) Whole-genome doubling confers unique genetic vulnerabilities on tumour cells. Nature 590(7846): 492–497. https://doi.org/10.1038/s41586-020-03133-3

Saini G, Joshi S, Garlapati C, Li H, Kong J, Krishnamurthy J, Aneja R (2022) Polyploid giant cancer cell characterization: new frontiers in predicting response to chemotherapy in breast cancer. Semin Cancer Biol 81:220–231. https://doi.org/10.1016/j.semcancer.2021.03.017

Schmitt CA, Wang B, Demaria M (2022) Senescence and cancer—role and therapeutic opportunities. Nat Rev Clin Oncol 19(10):619–636. https://doi.org/10.1038/s41571-022-00668-4

Schukken KM, Foijer F (2018) CIN and aneuploidy: different concepts, different consequences. BioEssays 40(1). https://doi.org/10.1002/bies.201700147

Sdeor E, Okada H, Saad R, Ben-Yishay T, Ben-David U (2024) Aneuploidy as a driver of human cancer. Nat Genet 56(10):2014–2026. https://doi.org/10.1038/s41588-024-01916-2

Simovic-Lorenz M, Ernst A (2025) Chromothripsis in cancer. Nat Rev Cancer 25(2):79–92. https://doi.org/10.1038/s41568-024-00769-5

Suzuki K, Muto T, Masaki T, Morioka Y (1990) Microspectrophotometric DNA analysis in ulcerative colitis with special reference to its application in diagnosis of carcinoma and dysplasia. Gut 31(11):1266–1270. https://doi.org/10.1136/gut.31.11.1266

Targa A, Rancati G (2018) Cancer: a CINful evolution. Curr Opin Cell Biol 52:136–144. https://doi.org/10.1016/j.ceb.2018.03.007

Turajlic S, Xu H, Litchfield K, Rowan A, Horswell S, Chambers T, Consortium, T. R (2018) Deterministic evolutionary trajectories influence primary tumor growth: TRACERx renal. Cell 173(3):595–610.e511. https://doi.org/10.1016/j.cell.2018.03.043

Vassilopoulos G, Wang PR, Russell DW (2003) Transplanted bone marrow regenerates liver by cell fusion. Nature 422(6934):901–904. https://doi.org/10.1038/nature01539

Vasudevan A, Schukken KM, Sausville EL, Girish V, Adebambo OA, Sheltzer JM (2021) Aneuploidy as a promoter and suppressor of malignant growth. Nat Rev Cancer 21(2): 89–103. https://doi.org/10.1038/s41568-020-00321-1

Vittoria MA, Quinton RJ, Ganem NJ (2023) Whole-genome doubling in tissues and tumors. Trends Genet 39(12):954–967. https://doi.org/10.1016/j.tig.2023.08.004

Wang X, Willenbring H, Akkari Y, Torimaru Y, Foster M, Al-Dhalimy M, Grompe M (2003) Cell fusion is the principal source of bone-marrow-derived hepatocytes. Nature 422(6934):897–901. https://doi.org/10.1038/nature01531

Zhang S, Mercado-Uribe I, Xing Z, Sun B, Kuang J, Liu J (2014) Generation of cancer stem-like cells through the formation of polyploid giant cancer cells. Oncogene 33(1):116–128. https://doi.org/10.1038/onc.2013.96

Zhou X, Merchak K, Lee W, Grande JP, Cascalho M, Platt JL (2015) Cell fusion connects Oncogenesis with tumor evolution. Am J Pathol 185(7):2049–2060. https://doi.org/10.1016/j.ajpath.2015.03.014

Chapter 12
Genomic Reshuffling After Cell Fusion: Catalyst for Cancer Evolution and Progression

Julian Weiler and Thomas Dittmar

Abstract The fundamental relevance of cell–cell fusion processes is well studied and involves the precise spatiotemporal coordination of fusion-competent cells and the orchestration of typical fusion stages, ultimately leading to the genesis of a functional new cellular entity. While most physiological cell fusion events culminate in the formation of a multinucleated non-dividing cell, whereby nuclei remain spatially distinct, non-physiological fusion events are frequently coupled with the merging and mixing of the parental genomes. Upon cell division, fusion-induced whole-genome doubling and centrosome amplification drive chromosomal instability linked with multipolar mitoses, chromosome mis-segregation, non-disjunction and structural-associated alterations. In this context, aberrant cell fusion events can accelerate genomic heterogeneity that can potentiate the risk of cancer development and further propel cancer evolution by producing genetically diverse progeny via sub-clonal expansion. In this chapter, we provide insights into how unscheduled cell fusion events rapidly drive cancer evolution by promoting genetic and epigenetic instability, thereby creating a tumour landscape in which intratumoural heterogeneity and phenotypic plasticity increase the acquisition of traits linked with therapy resistance, stemness and immune evasion.

Keywords Aberrant cell fusion · Genomic heterogeneity · Chromosomal instability · Genetic/epigenetic instability · Cancer evolution

12.1 Introduction

Cell fusion is a tightly regulated spatiotemporal mechanism essential for normal development, tissue homeostasis, and physiological processes such as fertilisation (Kaji und Kudo 2004), muscle fibre formation (Hindi et al. 2013) or bone remodelling (Sabe et al. 2024). In order to guarantee a successful cell–cell fusion

J. Weiler (✉) · T. Dittmar
Immunology and Tumour Biology, Center for Biomedical Education and Research (ZBAF), Witten/Herdecke University, Witten, Germany
e-mail: Julian.Weiler@uni-wh.de

© The Author(s), under exclusive license to Springer Nature Switzerland AG 2026

J. Weiler et al. (eds.), *Cell Fusion*, Advances in Experimental Medicine and Biology 1509, https://doi.org/10.1007/978-3-032-22637-2_12

operation, the fusion partners involved have to coordinate molecular and cellular events in the correct order and upon precise timing indicated by the distinct cell fusion stages—cell recognition, adhesion, membrane juxtaposition and fusion pore formation (Zhang et al. 2021a; Brukman et al. 2019). Further demands for the precise orchestration of the fusion process include the spatial correct location and matching fusion partners, which otherwise can result in unintended merger events with pathological consequences for the organism. Under normal physiological circumstances, cell fusion culminates in the formation of a multinucleated syncytium prerequisite, for cell mass expansion, barrier protection and biosynthesis production (Fox und Duronio 2013; Shu et al. 2018). Similar to the pre-fusion state, in which cells undergo transcriptional reprogramming to become fusion-competent and reach a certain degree of differentiation, the post-fusion phase in numerous physiological processes is characterised by a switch towards a non-proliferation state, which hinders the merging and mixing of the genomes and the subsequent acquisition of detrimental structural genomic changes that otherwise could be conducive to a malignant transformation into a cancer cell (Hindi et al. 2013; Goldman-Wohl und Yagel 2014). The post-fusion cell fate varies by cell type and environmental cues and depends on extrinsic and intrinsic signals that ultimately determine whether if the newly formed cellular entity falls into a terminal stage of differentiation and specialisation or undergoes a transformative process towards a cancerous phenotype. Nevertheless, a recent study has shown that the fusion process itself can be sufficient to convert the multinucleated cell into a cell-cycle arrest, whereby a transient change in energy metabolism occurs upon membrane remodelling that ultimately leads to transcriptional reprogramming, proliferative arrest and supports a differentiation-like cell state (Feliciano et al. 2021). As addressed above, somatic cell fusion processes in mammals are defined by the occurrence of syncytial structures without the accompaniment of nuclear fusion (Goldman-Wohl und Yagel 2014). However, under certain conditions, cell fusion also gives rise to nuclear fusion, wherein both parental genomes are reorganised into one single nucleus during cell division. This has been shown, for example, in stem cell-induced tissue regeneration or in response to liver injury. Even though cell fusion of pluripotent stem cells with somatic cells has been documented in several studies, mononucleation after the fusion process in vivo seems to be a rather rare event and has been observed primarily in ex vivo or parabiosis experiments (Alvarez-Dolado et al. 2003; Spees et al. 2003; Nern et al. 2009; Skinner et al. 2012). One exception might be the liver, where cell fusion of hepatocytes with stem cells or with themselves (homotypic cell fusion) gives rise to the formation of mononucleated hybrid cells with proliferation potential (Duncan et al. 2009; Quintana-Bustamante et al. 2012). Hepatocytes can extend their chromosomal content (ploidy level) through various mechanisms, such as cell fusion, endocycling and endomitosis or in the absence of cytokinesis (Darmasaputra et al. 2024; Duncan et al. 2009; Miyaoka et al. 2012). It has been reported that hepatocytes can undergo polyploidisation and de-polyploidisation in a context-dependent manner. For instance, hepatocytes frequently undergo polyploidisation under stress situations triggered by oxidative stress, alcohol intoxication, a response to metabolic stress (fatty liver disease) or viral infection (Gentric et al. 2015; Machida et al. 2009;

Ahodantin et al. 2019; Nawroth et al. 2021). Beyond that, hepatocytes are prone to age-dependent polyploidisation, where the overall ploidy level increases throughout the lifespan, ranging from 5% in newborns to 20% in middle-aged adults, and 40–50% in elderly individuals (Matsumoto 2022; Wang et al. 2017; Duncan et al. 2012). Otherwise, polyploid hepatocytes are also capable of undergoing ploidy reversal through multipolar mitosis or through a synchronised double mitotic event. This phenomenon, known as "ploidy conveyor", is a dynamic ploidy mechanism and is suggested to play a role in liver tissue homeostasis (Duncan et al. 2010; Wilson und Duncan 2023). Thus, genome expansion by ploidy induction can support the metabolic turnover under stress or injury, act as a buffer against deleterious mutations and compensate for the loss of genes during polyploid cell division. On the other hand, ploidy reduction to a diploid state accelerate cell division and promotes the repopulation of the cells after injury, leading to a better regenerative potential and maintenance of tissue integrity. However, the dynamic change in ploidy is a double-edged sword, since alterations in ploidy are linked with both tumour promotion and tumour suppression (Zhang et al. 2018; Matsumoto et al. 2021; Müller et al. 2021). Therefore, polyploid hepatocytes with more than two sets of each chromosome have an advantage against the frequent loss of essential genes or deleterious mutations, and so inactivation/loss of tumour suppressor genes can be avoided, which is necessary to counteract the so-called Muller's ratchet effect (López et al. 2020). In addition, controlled ploidy reduction can produce daughter cells with fitness alterations, wherein the fittest without or less deleterious mutations or chromosomal aberrations proliferate faster than hepatocytes with elevated DNA damage and rapidly restore the liver by outcompeting the other (Frade et al. 2019; Wilkinson et al. 2019). Conversely, mitosis of polyploid cells is linked with aneuploidy and chromosomal instability, and so polyploid hepatocytes are more prone to undergo mitosis in a rather error-prone manner, which raises the incidence of hepatocellular carcinoma (HCC) (Müller et al. 2021; Matsumoto et al. 2021). Moreover, genetic alterations in driver genes may also foster the genesis of cells with tumour-initiating capabilities, suspected to acquire survival advantages that promote clonal expansion and propel cancer evolution (Shlush und Hershkovitz 2015). In the following sections, we will discuss how unscheduled deviating cell fusion events can trigger tumourigenesis and serve as precursors to a catastrophic mechanism that causes extensive and multiple genomic changes at a rapid pace. This instability can then propel genomic reshuffling and drive evolution towards a malignant phenotype.

12.2 Cell Fusion and the Balancing Act of Genome Integrity

A crucial challenge following cell fusion is maintaining genome integrity to avert cell death and prevent the acquisition of chromosomal alterations that may promote an oncogenic transformation. In physiological fusion processes, several mechanisms have evolved in parallel to counteract a potential loss of genome integrity, such as regulation by the checkpoint control mechanism (Warmerdam und Kanaar 2010), DNA repair mechanisms (DNA damage response system) (Ciccia und Elledge 2010), or through epitranscriptomic regulation (Janic et al. 2025). The most obvious one to avoid genome instability is the merging of both nuclei upon entering the post-fusion stage to forestall genome reshuffling linked with numerical and structural changes that may occur during the first mitotic division. Examples indicative of this are given by myoblast fusion, syncytiotrophoblast formation or multinucleated giant cell formation during osteoclastogenesis (Zhou und Platt 2011; Sabe et al. 2024). In this context, cell fusion result to the generation of bi- or multinucleated cells that are committed to a non-dividing, terminally differentiated state. For instance, myoblast fusion is preceded by the upregulation of transcription factors like MyoD, myogenin and MEF2, inducing a transcriptional program that ultimately results in cell-cycle arrest, terminal differentiation and the transition into a fusion-competent cell state (Hindi et al. 2013), particularly through the upregulation of the muscle-specific fusogenes myomaker (TMEM8C) and myomerger (MINION/TMEM92) (Chen et al. 2020). The acquisition of a multinucleated cell state enables the cell to meet the metabolic turnover attributed to copy number gain and elevated transcriptional RNA processing while concurrently maintaining genome stability by the absence of mitosis (Mukund und Subramaniam 2020; Miyamoto 2006). However, in some cases, cell fusion is also coupled to nuclear fusion, as for instance during fertilisation or in liver regenerative processes (Kaji und Kudo 2004; Faggioli et al. 2008). However, a direct comparison of both processes is only possible to a limited extent, as fertilisation includes the merging of haploid gametes (karyogamy), while cell fusion of hepatocytes includes the merging of somatic cells that already possess a diploid genotype. Notwithstanding, both fusion processes have to sustain genome stability, which otherwise can lead to (genetic) diseases, disorders or cancer development. (Coticchio et al. 2023; Wang et al. 2016; Middelkamp et al. 2020) Especially oocyte fertilisation followed by embryogenesis underlies a highly coordinated series of molecular and cellular mechanisms that prevent DNA damage, ensure accurate chromosome segregation and maintain genome stability (Coticchio et al. 2023). This coordinated cell fusion progression is elementary, since failures upon fertilisation or right afterwards would have serious consequences for the organism, as all following cells derive from this singular event. Therefore, genetic fidelity upon fertilisation is a prerequisite for the development of the beginning live and mechanisms, such as hardening of the outer oocyte layer, the *zona pellucida* after sperm penetration, termed “polyspermy block”, or pronucleus maturation before karyogamy, prevent the zygote from polyploidy-associated segregation defects and genetic imbalance (Trebichalská und Holubcová 2020; Coticchio et al. 2023). On the other hand, cell fusion in the liver occurs under normal and damaged conditions, but to a far superior degree in damaged liver tissue than vice versa. The study by Faggioli et al. revealed that homotypic cell fusion among hepatocytes might be more frequent than is usually suggested (Faggioli et al. 2008). By creating chimeric mice with dual reporter genes, they assessed fusion incidence within the liver, observing co-expression of both markers in 13–26% of cases among hepatocytes. In a separate experiment within the same study, fluorescent labelled male XY embryonic cells were implanted into a female blastocyst containing BAC-labelled XX sex chromosomes. The outcome was that 11–16% of binucleated hepatocytes bore an XX/XY constitution, which confirmed that those cells were truly raised by cell fusion instead of failed cytokinesis or endoreplication. In addition, in both setups, they were able to detect mononucleated hepatocytes, either carrying both reporter genes or bearing an unequal distribution of X/Y sex chromosomes and concluded that both spatially separated parent nuclei had fused, in a process declared as hetero-to-synkaryon transition. Even though cell fusion studies in humans are rare and polyploidy induction via cell fusion is not exclusively the sole mechanism, diverse xenograft and bone marrow transplant studies suggested that cell fusion occurs, at least partially, in the human liver (Myerson und Parkin 2019; Zhou et al. 2009; Kashofer et al. 2006). However, irrespective of the mechanisms that drive polyploidisation in hepatocytes, the propensity for error-prone mitotic progression should be somehow comparable, since a higher ploidy level is linked with a higher incidence of mitotic aberrations, such as multipolar cell division, kinetochore-microtubule attachment errors or chromosomal non-disjunction resulting in aneuploid and chromosomal instable progeny (Duncan et al. 2012). But how do hepatocytes sustain the integrity of the genome regarding the apparently high frequency of erroneous mitoses upon polyploidy division? Although aneuploidy is commonly connected with cancer, whereby cancer cells become abundantly aggressive through loss of tumour suppressor genes or oncogene amplifications, aneuploidy in hepatocytes does not primarily contribute to malignancy. According to Duncan et al., polyploidy and aneuploidy in the liver can positively contribute to genetic diversity, thereby enhancing adaptability to stress or injury within the liver (Duncan et al. 2012). Furthermore, hepatocytes are capable of undergoing de-polyploidisation to generate near-diploid offspring that exhibit enhanced mitotic fidelity (Matsumoto et al. 2020). Conversely, other studies demonstrated that the dynamic nature of ploidy in liver cells is itself a facilitator of genomic instability by contributing to error-prone mitoses and the accumulation of malignant traits, which in turn negate the argument for an elevated mitotic fidelity upon ploidy reduction (Müller et al. 2021; Matsumoto et al. 2021). Because polyploid hepatocytes are less proliferative than their diploid counterparts, polyploidisation upon liver injury may serve as a mechanism to curb uncontrolled hepatic cell proliferation (Wilkinson et al. 2019). Research concerning this contradictory phenomenon has shown that the extent of cellular ploidy in the liver is limited by the activation of the multi-protein complex PIDDosome (Sladky et al. 2020a, b; Fava et al. 2017). The PIDDosome complex is activated in response to supernumerary centrosomes, consequently inducing a caspase-2-mediated p53/p21 cell-cycle arrest to prevent chromosomally mis-segregation driven by the excessive number of centrosomes and thereby safeguard the cell against genomic instability (Sladky et al. 2020a; Fava et al. 2017). This protective mechanism might also explain why the majority of polyploid liver cells remain binucleated and only a minor subpopulation achieves a larger ploidy level (Heinke et al. 2022; Kreutz et al. 2017). Even though the polyploid-dependent restriction mechanism by PIDDosome engagement has hitherto been examined in hepatocytes undergoing cytokine failures, it is plausible that a comparable mechanism operates within binucleated hepatocytes originating from cell–cell fusion events, whereby centrosome overduplication is also prevalent. Another study, however, noted that under genotoxic stress, PIDDosome activation can paradoxically promote hepatocarcinogenesis by facilitating the loss of heterozygosity (LOH) of certain tumour suppressors such as p53, leading to its inactivation and consequently triggering genomic instability and cancer progression (Sladky et al. 2020b). Using PIDDosome-deficiency mice, they could demonstrate that non-tumour tissues predominantly consist of polyploid hepatocytes, whereas liver tumours exhibit a significant prevalence of low-ploidy cells. The authors concluded that polyploidy, in the presence of genotoxic stress, mitigates the risk of tumour suppressor loss-of-function and thus prevents HCC. The question of how hepatocytes protect genome stability during cell division when possessing extra genomic content and centrosomes is not readily elucidated. Cell fusion can be an underestimated contributor to both hepatic liver health and the pathogenesis of liver cancer in humans. Whether polyploidisation and aneuploidisation in the liver are harmful or harmless depends on several factors, conditions, and their extent. Most studies addressing this query have revealed that a dynamic balance between diploid and polyploid hepatocytes is crucial for maintaining liver health (Wilkinson and Duncan 2021; Donne et al. 2021; Duncan et al. 2010). Overall, a disruption of the regulatory mechanisms can tip the homeostatic balance from a healthy post-fusion condition to a precancerous state, which, if not re-established in time, can lead to complete pathological transformation.

12.3 Cell Fusion as an Initiating Catastrophic Event in Tumourigenesis

The proposal that cell fusion between diverse cancer cells with cells from the adjacent tissue can contribute to tumourigenesis is a long-standing concept and can be dated back to the German pathologist Otto Aichel in the early twentieth century. His revolutionary idea was based on the observation that leukocytes can fuse with cancer cells to inherit genetic background from both parental cells, necessity to combine cancer-imposed uncontrolled proliferation capacity with the natural migratory potential of leukocytes, endowing the newly formed hybrid cells to spread throughout the body and form metastases at distant sites (Otto Aichel 1911). In his pioneer article “About cell fusion with qualitatively abnormal chromosome distribution as cause for tumour formation”, he provided a potential link on how hybrid cells derived by cancer cell fusion become chromosomal instable as a consequence of the abnormal chromosomal distribution induced by atypical mitotic progression of the hybrid cells (Otto Aichel 1911; Pawelek 2005). Since then, a multitude of studies have given new insights into how cancer cells acquire malignant attributes when fusing with cells from the immune system, epithelium, stroma or even with stem cells (Lindström et al. 2017; Ding et al. 2012; Zhang et al. 2019b;

Melzer et al. 2021; Weiler und Dittmar 2019). The latter became the subject of interest as an alternative explanation for the cancer stem cell hypothesis, whereby cancer cells regain stemness properties by reactivating embryonic/stemness genes like Oct4, Nanog, ALDH1 or Sox2 after the fusion with embryonic or adult stem cells (Xu et al. 2014; Dittmar et al. 2009; Xue et al. 2015). However, recent findings revealed that reprogramming toward a stem cell-like phenotype does not necessarily require fusion with a stem cell. In this regard, Ding et al. unveiled that fusion between macrophages with MDA-MB-231 or MCF-7 breast epithelial cells gave rise to hybrid cells exhibiting a $CD44^{+}/CD24^{-}$/low stem cell-like phenotype with enhance mammosphere formation capability (Ding et al. 2012). Likewise, the study conducted by Uygur et al. demonstrated that cell fusion of prostate cancer cells with smooth muscle cells of the prostate can induce stemness in the newly formed hybrid cells by the upregulation of the stem cell marker CD133 coupled with an increase in tumour-sphere-forming ability of those hybrids (Uygur et al. 2019). Meanwhile, some studies indicated that the fusion of normal, non-transformed cells with themselves can directly drive oncogenesis by producing hybrid cells with cancer-initiating potential. Otto Aichel's hypothesis, in this context, delves deeper than previously considered, proposing that malignancy can be incrementally induced by the manifestation of genomic changes that ultimately provoke a malignant transformation. Consequently, the inherent selection pressure within the hybrid cell population elicits the expansion of genetically distinct hybrid subclones and the subsequent clonal succession or clonal convergence of hybrid subtypes exhibiting enhanced survival and fitness advantages. This, in turn, implies that traits are not solely inherited after the fusion process but rather evolve as a result of the genomic/epigenetic instability instigated by the fusion process itself. Hass and colleagues postulated that aberrant cell fusion events undergone by cancer cells with somatic cells or among each other underlie a post-hybrid selection process (PHSP), thereby enabling the acquisition of novel properties in a stepwise karyotypic fine-tuning process (Hass et al. 2021). Although many hybrid cells die during the PHSP due to severe and lethal genomic aberrations, a subset of these hybrid cells might develop a malignant phenotype linked with metastatic potential and self-renewal capabilities. First attempts to investigate whether cell–cell fusion of non-tumourigenic cells culminates in the formation of hybrids with tumour-initiating potential were provided by Duelli et al. using hybrid cells derived from virus-induced fusion among D551 fibroblasts (Duelli et al. 2007). They could show that, unlike the diploid D551 parental cells, hybrids are capable of forming tumours when injected into nude mice and attributed this to the highly unstable and heterogeneous karyotype that evolved in the hybrids. Similarly, the study of Zhou et al. showed that homotypic cell fusion of IEC-6 epithelial cells induced massive chromosomal damage linked with chromosomal translocations and rearrangements sufficient to induce tumour formation in vivo (Zhou et al. 2015). Consistent with the finding above, studies by Delespaul et al. and Merle et al. have indicated that homotypic cell fusion of immortalised, non-malignant cells resulted in hybrid cells that exhibited cancer-initiating properties (Delespaul et al. 2019; Merle et al. 2020). Both studies further explored that hybrid cells were chromosomally instable and developed a unique rearranged genome associated with the loss or gain of distinct chromosomal regions (copy number alterations) and complex structural chromosomal aberrations (structural variations). Even though all studies have shown that cell–cell fusion per se can initiate tumourigenesis and cause the formation of chromosomal instable hybrid cells, changes in cell ploidy and the manifestation of chromosomal aberrations, including numerical and structural variations, seem to be strictly hybrid-clone-dependent. Additionally, considering the PHSP theory, hybrid cells have to undergo a post-fusion reshuffling program in order to cope with the extra genomic material, as well as the epigenomic setup inherited from both parental cells. In the following, we will focus on the specific strategies that enable hybrid cells to overcome cell-cycle arrest and apoptosis and to fulfil the genomic/cellular demands indispensable for the viability of the hybrid cell population, but that simultaneously can act as a catalyst to fuel the trajectory of hybrid cells with malignant, tumourigenic alterations.

12.4 Chromosomal Instability as Driver for Genome Heterogeneity and Cellular Plasticity

Chromosomal instability (CIN) describes a condition in which cells experience frequent changes in chromosome number or structure that can contribute to the development of progeny exhibiting distinct genetic and phenotypic traits compared to their parent cells (Chen et al. 2025; Lukow und Sheltzer 2022). Under normal physiological processes, CIN typically remains at a low level, as high-grade CIN can result in unpredictable genomic chaos linked with numerical changes (gain/loss of whole chromosomes) or structural modifications (chromosomal-arm deletions, translocations, amplifications) (Chen et al. 2025). Beyond that, CIN in the course of aberrant mitotic progression can give rise to the formation of extrachromosomal DNA, which is either encapsulated in its own envelope to form small extranuclear bodies (micronuclei) or exists as circular DNA (double minutes) within the karyoplasm (Lukow und Sheltzer 2022). In contrast to the somatic mutation theory (Holmquist und Gao 1997), which includes that cellular mutations occur in a gradual, stepwise process driven by the accumulation of small alterations composed in single nucleotides (single nucleotide variation) or small nucleotide groups over time, large-scale genetic alterations triggered by CIN involve the sudden and rapid exchange, loss, or gain of entire chromosomes or chromosome segments (Tijhuis et al. 2019). Considering this, CIN can rapidly reshape the genome at the chromosomal level through large-scale alterations and further contributes to clonal heterogeneity by reproducing progenies that exhibit a unique and modified genome. In addition, CIN not only changes the architecture of the genome, but it can also cause epigenetic changes triggered by the up- or downregulation of epigenetic enzymes via chromosomal mis-segregation or fragmentation, induce the remodelling of the 3D chromatin structure and thereby the accessibility of regulatory genetic elements (enhancers, silencers, insulators), or indirectly transmit aberrant epigenetic patterns upon micronucleus reincorporation into the main nucleus (Chen et al. 2025; Baumann et al. 2025; Agustinus et al. 2023). Besides, the fusion of two heterotypic cell types itself involves the combination of two distinct epigenomic profiles, which can result in adoption, chimerism (cross-mixing) or remodelling of the hybrid epigenome. Thus, diverse studies have shown that cell fusion between cancer cells with stem and somatic cells gives rise to hybrid cells that confer increasing stress adaptability, drug resistance, invasiveness, stemness or immune tolerance (Melzer et al. 2018, 2021; Li et al. 2014a; Tajima et al. 2024; Liu et al. 2025; Sieler et al. 2021; Lindström et al. 2017). This is triggered by CIN-induced structural genomic rearrangements of the inherited parental genetic material, coupled with epigenetic reprogramming that overrides the original parental epigenomes. These changes empower hybrid cells to change their cellular identity, functionality or phenotype in response to various stressors, but also as a result of the ongoing selection process within the hybrid cell ecosystem (Hass et al. 2021). The capability of hybrid cells to modify their identity, exhibit chimeric phenotypes or switch between diverse cell or lineage states is referred to as cellular plasticity (Yuan et al. 2019). In the following section, we will elaborate on how aberrant fusion events drive genomic and epigenomic instability, fostering diversity at both levels, and endowing hybrid cells with enhanced plasticity, a key mechanism frequently implicated in cancer evolution.

12.5 Cell Fusion as a Source for Genomic Instability and Hybrid Genome Reorganisation

Genomic instability encompasses a broad spectrum of genomic alterations, including point mutations, sequence-specific structural rearrangements and numerical chromosome aberrations (Russo et al. 2015). In this regard, CIN represents a subtype of genomic instability, in which persistent segregation errors can contribute to genomic alterations, resulting in karyotypic diversity and ultimately causing cell or lineage heterogeneity (Bakhoum und Landau 2017). Chromosomal instability caused by illicit cell fusion can be attributed to certain determinants that apply to the hybrid cell, such as whole-genome doubling, centrosome overduplication and cellular enlargement, which collectively can compromise mitotic fidelity and promote aneuploidy and genomic instability. The simplest cell fusion scenario involves the fusion of two single cells (binary cell fusion) that, at least initially, exist as a binucleated entity with two discrete nuclei (heterokaryons). However, upon mitotic progression, the chromosomal content of both nuclei is commingled by the breakdown of the nuclear envelopes, which in general yields in the formation of two mononucleated daughter cells (synkaryons), each now equipped with four copies of each chromosome (tetraploid) (Sottile et al. 2016; Ho et al. 2010). Thus, tetraploidy, which in this context results from a whole-genome duplication (WGD) event under non-physiological circumstances, typically activates cell-cycle control mechanisms to preserve the cell from disastrous DNA replication linked with replication stress, DNA strand breaks and chromosomal damage (Manic et al. 2022). However, to protect the cell from polyploidy-associated genomic damage, tetraploid cells tend to arrest at the G1/S checkpoint via p53 stabilisation, resulting in either senescence or the induction of apoptosis (Andreassen et al. 2001; Senovilla et al. 2009). Decades before, the study of Harris et al. explored how spontaneous fusion between cancer cells and non-malignant cells negated the tumourigenic potential in the arising hybrid cells (Harris et al. 1969). In this study, cell fusion of diverse cancer cells with murine A9 fibroblast gave rise to hybrid cells that proliferated less progressively and failed to produce tumours in vivo. However, the authors also showed that weeks to months later, descendants of the original non-malignant hybrids revealed a significant loss of chromosome number that was associated with the reversion of malignancy. Although not yet known, Harris et al.'s discovery was seminal for further research and gave the first hints of a protective tumour-suppressive mechanism, either when cells illegally become tetraploid or when cells occasionally fuse together to share their genomic content. Harris and co-workers considered that the loss of one or more specific chromosomes provides the hybrid cell with the ability to regain tumour-forming capacity. Exactly one decade later, multiple research groups independently discovered a 53 kDa protein and named it p53 (Levine und Oren 2009). From there on, extensive research in this field has revealed that loss of TP53 tumour suppressor function enables tetraploid (cancer) cells to bypass senescence and re-enter the cell-cycle to generate aneuploid and CIN-positive progeny with enhanced metastatic potential (Fujiwara et al. 2005; Thompson und Compton 2010; Zhang et al. 2020; Andreassen et al. 2001). This has led to the assumption that TP53 (encodes for p53) protects the cell from accumulating genomic errors that, in the context of WGD, would otherwise elicit replication stress, chromosomal mis-segregation and structural lesions. Besides, the proposal of p53 as "guardian of the genome" (Anwar et al. 2022) is controversial, and nowadays diverse studies have demonstrated that tetraploidy can be tolerated under certain intrinsic or contextual circumstances, even without TP53 loss/mutation, including oncogene overexpression, compensation through other p53 family members or through deregulation of the Hippo pathway (Ganem et al. 2014; Potapova et al. 2016; Crockford et al. 2017; Talos et al. 2007; Nguyen et al. 2009). Nevertheless, subsequent studies have revealed that tetraploidy may act as an intermediate premalignant state, endowing cells with the capacity to rapidly generate aneuploid progeny through mitotic aberrations (Andreassen et al. 2001; Ganem et al. 2007). This assumption was based on the observation that solid tumours contain a certain fraction of tetraploid cancer cells ranging from 30% to 40%, and thus in some cancer types tetraploidy occurrs especially during the early stages of cancer onset (Olaharski et al. 2006; Reid et al. 1996). The Pan-cancer cohort study from over 9,000 sequenced cancer patients unveiled that WGD is a common hallmark in human cancer and arose early during cancer development (Bielski et al. 2018). In this study, time study analysis revealed that WGD was linked with large-scale copy number losses and occurred robustly prior to TP53 mutations in 97% of all cases. However, TP53 dysfunction was not obligate and could be replaced by other driver alterations like CCNE1 amplifications or RB1 loss-of-function. This and other Pan-cancer whole-genome studies demonstrated that tetraploid cancer cells are more prone to generate sub-tetraploid and aneuploid progeny and additionally tolerate genomic lesions accrued by large-scale single- or copy number alterations than diploid cancer cells (Zack et al. 2013; McPherson et al. 2025; Bielski et al. 2018; Campbell et al. 2020; Consortium 2020). Overall, all these studies underscore how WGD shapes the landscape of tumours as a powerful catalyst for macroevolutionary changes driven by large-scale, population-wide genomic alterations such as aneuploidy, copy number variation and chromosomal rearrangements. However, cell–cell fusion not only can contribute to WGD, but it might also be a mechanism that can dramatically accelerate intratumour heterogeneity in a parasexual manner. The study by Miroshnychenko et al. demonstrated that spontaneous fusion of cancer cells, whether derived from the same lineage or distinct tumour types, gives rise to hybrid subclones exhibiting marked phenotypic diversity compared to their parental cells, which in addition increases with time. The study impressively demonstrated that the mutation rate increased dynamically during fusion-mediated recombination and was substantially elevated compared to that in non-fused cancer cells (Miroshnychenko et al. 2021).

12.6 Tetraploidy as a Premalignant State Associated with Chromosomal Instability (CIN)

However, the acquisition of a post-fusion cell state includes not only genome doubling but also the exchange of cytoplasmic components, e.g. organelles, macromolecules and other diverse organic molecules from both fusion partners. Hence, cell–cell fusion automatically gives rise to the heritable transfer of supernumerary centrosomes, which then undergo S-phase replication and subsequently initiate the assembly of multipolar spindles, resulting in chaotic multipolar cell division upon mitotic entry (Dörnen et al. 2020). Microscopic studies have shown that post-fusion-mediated cell division is usually associated with tri- or tetrapolar cell division (Zhou et al. 2015; Dittmar et al. 2021; Frade et al. 2019; Chen et al. 2019), which typically leads to genomically unstable progeny highly susceptible to ongoing mitotic aberrations (Duncan et al. 2010; Potapova et al. 2013). Some studies, however, also showed that tetraploid (hybrid) cells can undergo controlled cell division resulting in genomic stable progenies, which either remain tetraploid or can undergo controlled ploidy reduction to reacquire a diploid karyotype (Frade et al. 2019; Ho et al. 2010; Guidotti et al. 2003). Whether controlled ploidy reduction represents a mechanism by which stem cells maintain genome integrity and evade oncogenesis has been proposed and needs further investigation in ongoing research. Nevertheless, in a cancer context, ploidy reduction following multipolar cell division is rarely controlled and is commonly coupled with the generation of genetically distinct subclones that drive intratumoural heterogeneity and phenotypic plasticity (Potapova et al. 2013). However, it has also been shown that tetraploid cancer cells harbouring supernumerary centrosomes can mimic normal bipolar mitosis by a mechanism known as centrosome coalescence or clustering, where two centrosome pairs aggregate to form a bipolar spindle (Bloomfield und Cimini 2023; Silkworth et al. 2009; Krämer et al. 2011). Although multipolar mitosis can expedite the generation of aneuploid daughter cells to facilitate cancer evolution by genetic diversity, it also increases the likelihood of severe, deleterious chromosomal imbalances, including monosomies (loss of one chromosome copy) and nullisomies (loss of both copies of a chromosome pair) (Bloomfield und Cimini 2023). One possibility to overcome this deleterious impact is that tetraploid cancer cells undergo centrosome clustering to re-establish mitotic polarity and proceed with cell division in a pseudo-bipolar fashion, effectively reducing the risk of essential chromosome loss (Krämer et al. 2011). A critical aspect of centrosome clustering is that all newly formed daughter cells carry two centrosomes, which can still lead to multipolar cell division in the next mitosis if all centrosomes remain separated during spindle pole formation. Alternatively, asymmetrical clustering of only one centrosome pair alongside two individual centrosomes may inevitably lead to tripolar mitosis (Bloomfield und Cimini 2023). The last option would be that three centrosomes cluster together, while one remains singular, leading to a bipolar cell division in which one daughter cell becomes corrected and bears only a single centrosome. In vitro studies dealing with this issue have shown that cells that became tetraploid quickly lost their extra centrosomes through asymmetrical clustering, which automatically favoured those cells that inherited the correct number of centrosomes, thereby enabling these cells to survive and outcompete other cells in a natural selection process (Baudoin et al. 2020; Galofré et al. 2020). Another critical consequence of centrosome amplification is that cells with more than the usual number of centrosomes are highly susceptible to microtubule-kinetochore attachment errors, a condition in which metaphase chromosomes become mis-attached to the spindle apparatus, leading to chromosome segregation errors, non-disjunction or the appearance of lagging chromosomes during anaphase (Ganem et al. 2009; Cimini et al. 2001; Gregan et al. 2011). In general, it has been evidenced that incorrect microtubule-kinetochore attachments significantly contribute to CIN, wherein especially merotelic kinetochore attachments (a single kinetochore bound to microtubules from both poles) can be found to be prevalent upon anaphase onset in tetraploid cells harbouring extra centrosomes. This merotelic attachment is strongly linked with the formation of so-called lagging chromosomes, in which the opposing pulling forces exerted by the spindles cause the mis-attached chromosome(s) to lag behind the other chromosomes, preventing the correct segregation to the spindle pole (Cimini et al. 2001, 2002; Ganem et al. 2009). If not corrected, lagging chromosomes can lead to aneuploidy or alternatively can cause massive chromosomal damage, fragmentation, and micronucleus formation if the lagging chromosome remains trapped in the cleavage furrow during cytokinesis (Gregan et al. 2011; He et al. 2019; Janssen et al. 2011; Crasta et al. 2012). The formation of a micronucleus is largely associated with chromothripsis, a catastrophic genomic event in which a single or a few chromosomes are shattered into dozens or hundreds of fragments, also termed chromosome pulverisation (Crasta et al. 2012; Stephens et al. 2011; Zhang et al. 2015). These fragments are then randomly rearranged, resulting in complex intrachromosomal translocations, inversions or the loss of chromosomal regions. Sequencing analysis of 2658 whole-cancer genomes has uncovered that chromothripsis is an early event in cancer progression, with chromothripsis detected in 22.3% of all cases and commonly more prevalent in cancers that have undergone WGD (Consortium 2020). Unlike punctual mutations that accumulate over time, chromothripsis can drastically reshape the genome in a single catastrophic event. If the cell survives this genome crisis, the resulting complex chromosomal rearrangements can significantly accelerate the dynamic transformation toward a malignant phenotype (Consortium 2020; Lin et al. 2023). Consequently, the observation that WGD-positive cells are several times more prone than diploid cancer cells to evolve a chromosomally unstable phenotype can be attributed to an interplay of interrelated factors, including centrosome overduplication, multipolar mitoses and kinetochore-microtubule attachment errors, which collectively increase the frequency of lagging chromosomes, micronucleus formation and chromothripsis. Lastly, cells that undergo unscheduled tetraploidisation often experience excessive replication stress, indicated by nucleotide pool depletion and insufficient availability of replication factors upon S-phase entry (Gemble et al. 2022; Bester et al. 2011). This imbalance in replication-licensing factors leads to replication fork stalling and collapse, resulting in DNA strand breaks and activation of the DNA damage response (Gemble et al. 2022; Zheng et al. 2012). If the DNA damage is not precisely repaired upon entry into the M-phase, this can further cause structural and numerical CIN. Several studies have shown that DNA damage instigated by replication stress can further be transmitted into the M-phase, subsequently giving rise to chromosomal aberrations upon mitosis, such as anaphase bridges (end-to-end chromosome fusions that stretch between spindle poles during anaphase), lagging chromosomes or acentric fragments (chromosomal fragments without a centromere) (Burrell et al. 2013; Wangsa et al. 2018). Strikingly, one study revealed that due to the shortage of replication factors, tetraploid cells that transit from G1 to S-phase underlie replication stress, where a compromise in DNA replication fidelity results in under- and over-replicated DNA regions, replication fork collapse and the arising of DNA damage (Gemble et al. 2022). This damage is transmitted to G2/M-phase, resulting in genomic instability, indicated by copy number alterations. Crucially, the study demonstrated that CIN in tetraploid cells arises during the first S-phase, prior to the structural chromosome segregation defects that emerge and accumulate in subsequent mitotic cycles.

12.7 Cell Size as a Metric of Mitotic Fidelity

Under normal physiological conditions, cell size scales almost proportionally to its ploidy level and its cytoplasm content, which is composed of organelles, total RNA, and proteins (Chadha et al. 2024; Rhind 2021). Cells that become excessively large without an accurate increase in nucleic acid are more prone to enter a permanent, non-dividing state known as cellular senescence due to cytoplasmic dilution of critical regulatory proteins like DNA replication factors, DNA repair proteins and cell-cycle regulators (Lanz et al. 2022; Neurohr et al. 2019). Therefore, cells may acquire a senescence-associated phenotype to avoid replication stress-dependent damage linked with mitotic errors, which can ultimately lead to the manifestation of genomic instability (Gemble et al. 2022). While some studies refer that uncoupling of the nuclear-to-cytoplasmic ratio (N/C ratio) decisively determines if a cell may become malignant, other studies revealed that other criteria like the cytoplasmic composition, nucleocytoplasmic transport and the transcriptional machinery can also substantially contribute to a malignant phenotype, independently of the N/C ratio (Cantwell und Nurse 2019). Another limiting factor to bypass a senescence-like condition is the total concentration of proteins, which, when scaled linearly with the cell size, can protect the cell from senescence, and thereby polyploidisation might be a crucial mechanism for organisms to bypass senescence, while concurrently enhancing protein synthesis and metabolic turnover (Lanz et al. 2022). Although polyploidy can positively contribute to the metabolic capacity within the growing cell and successively promote the biosynthetic rate, mitotic fidelity in polyploid cells is often compromised. Apart from the already discussed obstacles of kinetochore attachment issues and centrosome overduplication, cell-size dependent grow upon polyploidisation can also alter the mitotic spindle dynamics, resulting in incorrect or failed chromosome segregation and genomic instability. Studies indicated that even with mechanisms such as pseudo-bipolar division and centrosome loss to preserve mitotic fidelity in tetraploid cells, alterations in spindle size or architecture can still lead to mitotic errors and genomic instability (Bloomfield et al. 2020; Storchová et al. 2006; Quinton et al. 2021). Specifically, in the study by Bloomfield et al., tetraploid DLD-1 cells that differed in size were picked and analysed for their mitotic potential (Bloomfield et al. 2020). It was clearly observed that larger tetraploid cells exhibited a significantly prolonged mitotic duration compared to their smaller counterparts, which relied onto the variations in spindle pole size and microtubule abundance and commonly resulted in prolonged metaphase arrest. In addition, other studies reported that larger cells tend to exhibit a weakened spindle assembly checkpoint (SAC) response, which can compromise checkpoint robustness and increase the chance of premature anaphase onset, even in the presence of incorrect microtubule-kinetochore attachments (Galli und Morgan 2016; Gerhold et al. 2018). All in all, dilution of critical cell-cycle and/or SAC components during cell enlargement impairs SAC signalling and mitotic fidelity, potentially triggering senescence or, in some cases, when the metaphase-arrested cell escapes SAC-mediated silencing (mitotic slippage), can even lead to endoreplication

(DNA replication without mitosis or cytokinesis) (Khurana et al. 2023; Lok et al. 2020). A recent study has calculated a model to demonstrate that the spindle assembly dynamics increase proportionally to the ploidy level in yeast cells (Jelenić et al. 2018). Here, upon reaching the maximum mitotic arrest duration, cells enter anaphase even with unattached kinetochores, resulting in frequent loss of chromosomes that can result in a 1000-fold increase in chromosome loss compared to diploid or haploid cell strains. However, it remains to be investigated whether this model can be adapted to polyploid mammalian (tumour) cells, in which ploidy reduction is also a common event during cell division.

12.8 Cell Fusion as a Mechanism for Inducing High-Level Polyploidy

While most studies focus on binary cell fusion, the possibility of multiple rounds of fusion events (multinary cell fusion) after the initial fusion incidence might also occur in non-physiological processes, which may result in highly polyploid cells (> 4n) with exceptionally large cell dimensions. If the arising hybrid cell contains several spatially separated nuclei, we generally speak of cellular ploidy, while in hybrid cells exhibiting a single large polyploid nucleus, the term nuclear ploidy is used (Donne et al. 2020). As discussed in a previous section, cellular ploidy is a common phenomenon in physiological processes and is tightly regulated in a stepwise manner, characterised by fusion priming, fusion readiness, and the post-fusion state (Zhang et al. 2021a). For instance, myogenesis is orchestrated by the fusion of two myoblasts forming the nascent myotube, which can then fuse with more myoblasts or other myotubes to form larger, mature, multinucleated myotubes that ultimately drive myotube growth (Horsley et al. 2003; Hindi et al. 2013). In principle, cancer hybrid cells may retain fusogenic properties and fuse again with other cells from the surrounding environment to achieve a higher ploidy level. Although this possibility exists, and it is known that cancer cells can merge with various cell types, the experimental evidence to support this issue is scarcely represented. One study revealed that Hodgkin lymphoma cells are capable of continues fusion rounds after the primary fusion event. In this study, homotypic cell fusion of mononucleated Hodgkin lymphoma cells generated the formation of binucleated cells that were capable of re-fusion, which nearly exclusively occured among the daughter cells to produce giant multinucleated Hodgkin and Reed–Sternberg cells (Rengstl et al. 2013). Duelli et al. proposed in the mid-2000s that virus infection of host cells can induce the expression of fusogenic viral proteins on the host cell membrane, which subsequently can fuse with other cells via fusogen–receptor interactions to form hybrid cells that become cancerous, senescent or may form syncytia when fusogenic activity is conserved in the hybrids (Duelli und Lazebnik 2007). Furthermore, studies have shown that virus-induced aberrant expression of envelope fusogens, such as Syncytin-1, HIV-1 Env, SARS-CoV-2

Spike, or the HTLV-1 envelope protein gp46, can trigger syncytia formation and the emergence of giant multinucleated cancer cells, also termed polyploid giant cancer cells (PGCCs) (Kunihiro et al. 2016; Symeonides et al. 2015; Li et al. 2021; Buchrieser et al. 2020). Other fusion proteins, such as CD9 and CD47, have also been detected in PGCCs and are suggested to be involved in regulating PGCC formation via cell fusion (Fei et al. 2019). Additionally, PGCCs are more represented in metastasis lesions and are characterised by the abundant overexpression of the fusogen Syncytin-1, which suggests that cell fusion mainly contributes to metastasis by driving PGCC formation. However, cells that have undergone fusion may further increase their ploidy through non-canonical endoreplicative mechanisms. Thus, fusion and endoreplication are not mutually exclusive and may even be functionally coupled in specific biological contexts, as exemplified by syncytiotrophoblast development, osteoclast maturation or liver regeneration (Zybina und Zybina 2020; Sakaue-Sawano et al. 2013; Wang et al. 2021; Takegahara et al. 2016). Cancer cells may exploit mechanisms such as cell fusion, endoreplication, or both to rapidly acquire polyploidy while bypassing mitotic and DNA damage checkpoints, thereby avoiding apoptosis and the putative loss of essential chromosomes or chromosomal segments (Zhang et al. 2018; Hayashi et al. 2024). In this context, PGCCs can reside as dormant entities within the tumour microenvironment, which can be reactivated for cell division under stressful conditions and rapidly generate genetically heterogeneous daughter cells through asymmetric cell division (Zhang et al. 2014; Jiang et al. 2015).

12.9 Polyploid Giant Cancer Cells in the Context of Tumour Evolution

Polyploid giant cancer cells represent a distinct subclass of cancer cells commonly found within the tumour bulk. A characteristic morphological feature of PGCCs is their dramatically enlarged cell size compared to diploid cancer cells, typically three- to tenfold larger, but in extreme cases, reaching up to 20 times the size of their diploid counterparts (Zhang et al. 2013, 2014; Qu et al. 2013). Despite their enormous size, PGCCs are also marked by their extreme ploidy amount, which can extend from 4 N to over 16 N depending on the tumour type and the environmental contingencies (Xuan et al. 2018; Amend et al. 2019). Although PGCCs are typically underrepresented compared to diploid or aneuploid cancer cells, which account for less than 5% of the population in untreated tumours, their formation is dramatically elevated under stress conditions, such as hypoxia, nutrient deprivation, radiation, or cytotoxic drug treatment and frequently can climb to 20–30%, and, in some cancers, even become the dominant cell fraction with approximately 65–80% of the total cancer population (Zhang et al. 2021b; Niu et al. 2021; Xuan et al. 2020). Specifically, under stress induction, cancer cells are capable of fusing with each other to form PGCCs (Zhang et al. 2014; Yart et al. 2022; Li et al. 2021), which are then better equipped to survive harsh environmental changes through metabolic reprogramming and suspend energy-consuming processes such as mitosis until environmental conditions stabilise (You et al. 2022; Donovan et al. 2014; Sirois et al. 2019). As noted, PGCC enrichment is frequently associated with higher tumour grade and advanced stage (Fei et al. 2019), and consequently correlates with poor survival, therapy resistance, and increased cancer aggressiveness (You et al. 2022; Saini et al. 2022; Zhang et al. 2014, 2021b). It has long been assumed that PGCCs represent terminally senescent cancer cells and, because of that not substantially contribute to cancer progression or aggressiveness (Sikora et al. 2016; Saleh et al. 2020). This paradigm has shifted, as several studies have revealed that PGCCs are not exclusively senescent, even though they often exhibit slow-cycling or long-term dormant behaviour (Alhaddad et al. 2023; Jiao et al. 2024; Zhang et al. 2014). Instead, PGCCs exhibit a rather stem cell-like phenotype and progress through an alternative "giant cell cycle" enabling PGCCs to persist stress situations by undergoing multiple rounds of endoreplication to form highly polyploid giant mother cells. These mother cells can subsequently undergo de-polyploidisation and generate progenies via asymmetric (often multipolar) cell divisions or through a budding-like process known as neosis, thereby regenerating proliferative, genetically diverse daughter cells that contribute to tumour repopulation and therapy resistance (Niu et al. 2016; Jiang et al. 2015). Moreover, it would be suggested that PGCCs not only reactivate stem cell transcription factors like OCT3/4, NANOG, SOX-2, ABCG2 or SSEA1 but also reactivate key meiotic genes to regain embryonic-like stemness capacity, enabling PGCCs to form blastomere-like spheroids with the capacity for multilineage differentiation (Niu et al. 2017; Erenpreisa et al. 2015; Zhang et al. 2014). With this regard, polyploidisation in the cancer context, triggered by cell–cell fusion and/or endoreplication, could generate stem cell-like PGCCs that could withstand extreme stress-rich microenvironments (Zhang et al. 2014). Upon stress reduction, PGCCs can repopulate the tumour mass via asymmetric cell division or nuclear budding to reproduce low-ploid and aneuploid progenies with enhanced plasticity and lineage-dependent function (Zhang et al. 2014; Zhou et al. 2022; Yang et al. 2024). Thus, PGCCs, even at low levels in the early stages of cancer evolution, can significantly fuel intratumoural heterogeneity, confer drug resistance and drive tumour relapse.

12.10 Epigenetic Reprogramming During Cell–Cell Fusion

Epigenetic reprogramming after the initial cell fusion event involves the rewiring of the epigenomes of both parental cells without altering the underlying DNA sequence. Instead, epigenetic reprogramming relies on reshaping the chemical modification landscape and causes the remodelling of the chromatin structure by altering DNA methylation pattern, modifying histone marks and modulating the interplay of chromatin-associated proteins, such as transcription factors, histone-modifying enzymes or histone compaction factors (Bae et al. 2025; Yamanaka und

Blau 2010). Thereby, cell fusion can alter the overall gene expression profile by inducing the activation or silencing of diverse genes (Palermo et al. 2009), modulating chromatin accessibility at regulatory elements such as enhancers and promoters (Su et al. 2015) and enabling the co-expression of gene products derived from both parental cells (Gast et al. 2018). As such, cell fusion can serve as a powerful tool for somatic reprogramming, overriding somatic genomic imprints through epigenetic modifications and ultimately resetting the somatic genome to an undifferentiated, stem cell-like state. Considering this, cell–cell fusion between various somatic cell types and embryonic stem cells has long served, and continues to serve as, a valuable approach alongside induced pluripotent stem cell (iPSC) technology to drive nuclear reprogramming and generate hybrid cells capable of differentiating into all three germ layers (Yamanaka und Blau 2010). For instance, cell fusion between murine fibroblasts with embryonic stem cells from the mouse induced the generation of pluripotent ES cell/fibroblast hybrid cells with full pluripotency and developmental potential, giving rise to the development of viable, near-tetraploid adult chimeric mice when injected into mouse blastocysts (Battulin et al. 2012; Kruglova et al. 2008). Interestingly, the near-tetraploid karyotype of the ES cell/fibroblast remained stable throughout the development of the chimeras, with no evidence of genomic instability. However, which epigenome dominates in the hybrid cell depends on several factors, including the inherited ploidy contribution from each parental cell (Kruglova et al. 2010; Palermo et al. 2009), the relative abundance of epigenetic regulators (Pereira et al. 2008), the cell-cycle status upon fusion (Sullivan et al. 2006), the differentiation state (Pasque et al. 2011) and the transcriptional and chromatin states of the parental nuclei (Santos et al. 2010). Beyond that, studies have shown that the ability of embryonic stem cells to circumvent somatic epigenetic programs is dependent on a biphasic reprogramming mechanism (Foshay et al. 2012). This mechanism initially involves the rapid reactivation of trans-acting factors following the fusion process, and subsequently, the reactivation of cis-silenced pluripotency genes such as Oct4 and Nanog after the first DNA replication round. In this context, nuclear reprogramming toward pluripotency requires DNA demethylation and chromatin remodelling for the reactivation of critical pluripotency factors, a process that at least requires DNA replication (Foshay et al. 2012; Tsubouchi et al. 2013). Nevertheless, unlike genomic alterations, which involve sequence-specific structural changes, epigenetic reprogramming does not necessarily require the merging of the genomes to change gene expression patterns or drive cell fate conversion (Pomerantz et al. 2009; Tsubouchi et al. 2013). On the other hand, cell fusion also plays a pivotal role in tissue regeneration, whereby fusion between damaged somatic cells and adult stem cells resulted in the generation of hybrid cells capable of adopting the tissue-specific phenotype of the recipient cell to restore tissue function (Terada et al. 2002; Ferrand et al. 2011). However, other studies have demonstrated that cell fusion between bone marrow-derived cells (BMDCs) and somatic cells after bone marrow transplantation can also result in a mixed chimerism phenotype (Bonde et al. 2010; Quintana-Bustamante et al. 2012). Concurrently, the regenerative potential of adult stem cells, including hematopoietic stem cells (HSCs), BMDCs, and mesenchymal stem/stromal cells (MSCs) has been reported in numerous studies, wherein spontaneous or inflammation-induced cell fusion gives rise to hybrid cells that at least partially display phenotypic traits of the host cell lineage. (Nygren et al. 2004; Bonde et al. 2010; Quintana-Bustamante et al. 2012; Alvarez-Dolado et al. 2003; Davies et al. 2009; Skinner et al. 2012). Whether hybrid cells derived from stem cell fusion remain genetically stable is a pivotal question, as it hinges on their mitotic fidelity, specifically, their ability to divide in a controlled manner, regardless of whether they undergo ploidy reduction or maintain their tetraploid karyotype through symmetric cell division (Skinner et al. 2012; Battulin et al. 2012; Frade et al. 2019). Another critical aspect is that reprogramming must also occur in a controlled manner to prevent the epigenetic reset from deviating into a disruptive regulatory program, which could lead to aberrant gene expression and probably to cancer formation. The condition in which cells exhibit epigenetic dysregulation is referred to as epigenetic instability, and it can account for the phenotypic diversity of cancer cells, how they develop therapy resistance, and their high plasticity under certain stress situations (Guo et al. 2019; Choi und Lee 2013).

12.11 Epigenetic Instability and Cellular Plasticity: Key Drivers of Tumour Malignancy

Although cell fusion in the context of somatic reprogramming towards a more dedifferentiated, pluripotent-like cell state can substantially contribute to tissue regeneration and wound healing, it also carries the risk of generating hybrid cells with cancer-initiating potential (Dörnen et al. 2020). The earliest suggestion of this concept was given by Otto Aichel, who proposed that macrophage–cancer cell fusion could generate hybrid cells inheriting traits from both parental cells (Otto Aichel 1911). This suggests that such hybrids must have undergone transformative genomic and epigenomic changes, which ultimately result in the expression of a combined "chimeric" expression profile in the arising hybrid cell. Based on this hypothesis, cell fusion between cancer cells with stem cells can give rise to cancer stem cells (CSCs) that combine self-renewal capacity with the metastatic potential of both parental cells, a dangerous mixture, as those CSCs are capable of forming secondary tumours at distant sites, exhibit increased therapeutic resistance and can contribute to the recurrence of the tumour (Wang et al. 2016; Dittmar et al. 2009; Lu und Kang 2011; El-Tanani et al. 2025). In fact, a wealth of studies have now demonstrated that, across a wide range of cancer types, fusion of tumour cells with stem cells can give rise to CSCs, in which both malignancy and stemness can be enhanced further (Li et al. 2014a; Melzer et al. 2018; Xue et al. 2015). The studies provided clear evidence that cell fusion can actively promote epigenetic reprogramming, as indicated by the reactivation or overexpression of stem cell-specific factors such as Oct4, Nanog, or ALDH that foster pluripotency in the hybrid cell (Rubio et al. 2022; Merle et al. 2021; Xu et al. 2014). However, which phenotype dominates is highly context-dependent, and even if the pluripotent state typically overrides the differentiated state, a reversal toward the somatic, likely malignant, phenotype is still possible if conditions change. Yao et al. have reported that upon cell fusion between hepatoma cells and embryonic stem cells, reprogramming towards pluripotency was induced in the hybrid cells, indicated by the reactivation of stem cell markers Oct4, Nanog or Sox2 (Yao et al. 2010). However, after fusion-induced reprogramming, ES–Hepa hybrids rapidly reacquired tumourigenic potential upon induction of differentiation, accompanied by transcriptional silencing of the tumour suppressor $p16^{INK4a}$ and subsequent tumour formation in vivo. This study, together with others, indicates that epigenomic changes are not unilateral but are rather shaped by the differentiation state of the cell and the surrounding microenvironment, further underscoring the plasticity driven by the inherent epigenetic instability in those hybrids (Yao et al. 2010, 2016). In addition to the extrinsic impact on the phenotypic expression, cell fusion between distinct lineages leads to the mixing of lineage-specific epigenetic regulators (trans-acting factors), such as transcription factors, chromatin modifiers and non-coding RNAs, which altogether reshape the epigenome, initiate transcriptional reprogramming and may even give rise to novel phenotypic traits (Bure et al. 2022; Bonasio et al. 2010). For instance, in the study by Long et al., virus-mediated fusion between human fibroblasts and mouse myotubes successfully generated stable heterokaryons, in which acetylation of histone H3K9/K14 at the MyoD promoter in the fibroblast nucleus led to MyoD activation and myogenic reprogramming. Parlermo et al. revealed that nuclear reprogramming upon cell–cell fusion between distinct somatic and interspecies cells resulted in hybrid cells, in which reprogramming can occur in both directions, resulting either in the phenotypic conversion into a muscle or keratinocyte cell, whereby cell fate reprogramming was decisively dictated by the myocyte/keratinocyte nuclear ratio and the global gene expression pattern (Palermo et al. 2009). Somatic cell fusion among luminal and basal breast cancer subtypes has shown that through epigenetic mechanisms driven by the overexpression of the transcription factors TBX18, EN1, and TCF4, hybrid cells were generated in which the basal-like phenotype dominated over the luminal-type, concurrently resulting in the epigenetic repression of luminal transcription factors (Su et al. 2015). Despite the dominate nature of the basal-like state, the epigenetic landscape of the basal-like parental cells, as well as of the luminal/basal-like hybrids, was highly heterogeneous, suggesting a putative phenotypic plasticity, which may contribute to stress adaptation and therapy resistance. Another often emerging phenomenon upon cell fusion in a cancer context is that hybrid cells underwent epithelial-to-mesenchymal transition (EMT), which confers on hybrids with increased motility and invasiveness associated with the breakdown of cell junctions and the secretion of proteinases necessary to degrade the extracellular matrix and penetrate into foreign tissues (Xu et al. 2014; Zhang et al. 2019a; Melzer et al. 2018). It is known that upregulation of the EMT transcription factors (EMT-TFs) Snail, Slug, Twist1/2 or ZEB1/2 induces the epigenetic repression of E-cadherin, first by direct binding to its promoter region and secondly by the recruitment of chromatin modification enzymes that ultimately mediate epigenetic silencing of E-cadherin

(Dong et al. 2012; Bolós et al. 2003; Vesuna et al. 2008; Wang et al. 2013). Furthermore, these EMT-TFs are master regulators of the EMT program by downregulation of diverse epithelial markers, such as E-cadherin, Claudin-1 or Occludin, while at the same time stimulating the expression of mesenchymal markers such as N-cadherin or Vimentin (Bolós et al. 2003; Martínez-Estrada et al. 2006; Ohkubo und Ozawa 2004; Alexander et al. 2006; Debnath et al. 2022). In differentiated epithelial cells, EMT is typically suppressed through epigenetic silencing, such as DNA methylation at the promoter side of EMT-TFs (Lee und Kong 2016). Hence, when stem cells or cancer cells expressing active EMT-TFs undergo fusion, the resulting hybrid cells may adopt a mesenchymal or hybrid epithelial/mesenchymal (E/M) phenotype by partially or fully overriding the epithelial identity. One study exemplified that spontaneous fusion between MSCs and lung cancer cells gave rise to hybrids cells where the EMT transcription factor FOXF1 was abundantly overexpressed and decisively contributed to the EMT and stemness phenotype of the hybrids (Wei et al. 2014). However, hybrid cell formation upon MSC-cancer fusion can also lead to mesenchymal–epithelial transition (MET) or to an E/M intermediate state, revealing that further genetic and epigenetic factors play a role in the cell fate decision of the hybrid cell. For instance, miRNAs are critical involved in the EMT plasticity network, and so miRNAs such as miR-137, miR-34a, miR-720 or miR-22 can target EMT-TFs like Twist1 or Snail to suppress EMT progression. (Zhu et al. 2020; Li et al. 2014b; Dong et al. 2016) The ZEB/miR-200 feedback loop, for instance, is known to regulate each other through a double-negative feedback loop, suggesting that hybrid cells with an imbalanced ratio are more prone to undergo EMT or MET, respectively (Brabletz und Brabletz 2010). Merckens et al. demonstrated that cell fusion between M13SV1 human breast epithelial cells and the human breast cancer cell line HS578T resulted in hybrid cells exhibiting a comparable reduction in miRNA-200 levels as seen in the parental cancer cell line, suggesting that the low miRNA levels facilitate the EMT phenotype in those hybrids (Merckens et al. 2023). In fact, stemness and EMT are tightly intertwined through reciprocal regulatory circuits. For instance, upregulation of EMT-inducing transcription factors such as Twist1 or Snail1 can promote a cancer stem cell-like state by triggering pathways that upregulate core stemness factors, such as Oct4, Nanog, and KLF4 (Liu et al. 2014; Izadpanah und Forghanifard 2022; Izadpanah et al. 2017). Conversely, upregulation of the stemness transcription factors Oct4, Nanog or Sox2 can increase the expression of EMT-TFs by directly binding to their promoter regions or by activating signalling pathways that subsequently co-activate EMT (Gao et al. 2015; Palla et al. 2015; Xie et al. 2022; Mandal et al. 2018; Herreros-Villanueva et al. 2013; Chiou et al. 2010). The emergence of an intermediate E/M phenotype can significantly contribute to the malignancy of the tumour, as those E/M hybrid cells can dynamically switch between both phenotypes in relation to environmental cues and thereby can be either motile or sessile depending on the current condition (Yu et al. 2013; Pastushenko und Blanpain 2019). Moreover, a growing body of evidence shows that cells in a partial or hybrid E/M state exhibit enhanced stemness, tumour aggressiveness, and therapy resistance compared to cancer cells that have undergone a full EMT (Grosse-Wilde et al. 2015;

Lüönd et al. 2021; Andriani et al. 2016; Jolly et al. 2015). Considering this, cell fusion may serve as a rapid mechanism for generating hybrid E/M phenotypes, whereby the fusion of epithelial cells with mesenchymal cells gives rise to hybrid cells that co-express markers of both lineages and contribute to epithelial–mesenchymal/stemness plasticity. Studies have revealed that fusion events between diverse cancer cells in accordance with MSCs give rise to hybrid cells with E/M phenotypic attributes, confirming that cell fusion can promote the E/M hybrid state linked with enhanced tumour aggressiveness and cancer stem cell plasticity (Zhang et al. 2019b; Li et al. 2014a; Xue et al. 2015). Although it is difficult to directly demonstrate that epigenetic instability precedes genomic instability, except in heterokaryon-based experiments, structural genomic alterations resulting from genome merging after the fusion process can further reinforce epigenetic instability. Therefore, structural genomic rearrangements, such as translocations, amplifications, or deletions, can lead to loss- or gain-of-function alterations in genes encoding epigenetic regulators, thereby rewiring the epigenetic network by aberrant activation or silencing of regulatory elements or by the decompaction of heterochromatin (Wang et al. 2020). Moreover, genomic catastrophic events, like chromothripsis or micronucleus reincorporation can systematically change chromatin-associated patterns (Agustinus et al. 2023; Ijaz et al. 2024). In this context, CIN-induced by aberrant cell–cell fusion events can potentiate epigenomic instability, while abnormal DNA methylation, histone modification or chromatin remodelling can further fuel genomic instability, either by silencing genes associated with DNA damage repair or by promoting chromatin destabilisation that predisposes to DNA damage and structural rearrangements (Baumann et al. 2025; Biggs et al. 2019; Choi und Lee 2013). Even if it is impossible to ascertain whether an epigenetic modification originates from genetic or epigenetic mechanisms, those epigenetic alterations that confer a fitness advantage are subject to positive selection. This process subsequently drives the clonal succession of beneficial genetic and/or epigenetic traits within the hybrid cell tumour population (Shlush und Hershkovitz 2015). Recent studies have shown that epigenetic reprogramming upon cell fusion can lead to immunosuppression by upregulating PD-L1 in hybrid cells or by modulating macrophage M2 polarisation (Tajima et al. 2024; Liu et al. 2025). Collectively, aberrant fusion events can induce inherent genomic and epigenomic instability in the hybrid cell population, which reciprocally reinforces one another to drive intratumoural heterogeneity and phenotypic plasticity, a vicious cycle fuelling tumour evolution.

References

Agustinus AS, Al-Rawi D, Dameracharla B, Raviram R, Jones BSCL, Stransky S et al (2023) Epigenetic dysregulation from chromosomal transit in micronuclei. Nature 619(7968):176–183. https://doi.org/10.1038/s41586-023-06084-7

Ahodantin J, Bou-Nader M, Cordier C, Mégret J, Soussan P, Desdouets C, Kremsdorf D (2019) Hepatitis B virus X protein promotes DNA damage propagation through disruption of liver polyploidization and enhances hepatocellular carcinoma initiation. Oncogene 38(14): 2645–2657. https://doi.org/10.1038/s41388-018-0607-3

Alexander NR, Tran NL, Rekapally H, Summers CE, Glackin C, Heimark RL (2006) N-cadherin gene expression in prostate carcinoma is modulated by integrin-dependent nuclear translocation of Twist1. Cancer Res 66(7):3365–3369. https://doi.org/10.1158/0008-5472.CAN-05-3401

Alhaddad L, Chuprov-Netochin R, Pustovalova M, Osipov AN, Leonov S (2023) Polyploid/multinucleated giant and slow-cycling cancer cell enrichment in response to X-ray irradiation of human glioblastoma Multiforme cells differing in Radioresistance and TP53/PTEN status. Int J Mol Sci 24(2). https://doi.org/10.3390/ijms24021228

Alvarez-Dolado M, Pardal R, Garcia-Verdugo JM, Fike JR, Lee HO, Pfeffer K et al (2003) Fusion of bone-marrow-derived cells with Purkinje neurons, cardiomyocytes and hepatocytes. Nature 425(6961):968–973. https://doi.org/10.1038/nature02069

Amend SR, Torga G, Lin K-C, Kostecka LG, de Marzo A, Austin RH, Pienta KJ (2019) Polyploid giant cancer cells: unrecognized actuators of tumorigenesis, metastasis, and resistance. Prostate 79(13):1489–1497. https://doi.org/10.1002/pros.23877

Andreassen PR, Lohez OD, Lacroix FB, Margolis RL (2001) Tetraploid state induces p53-dependent arrest of nontransformed mammalian cells in G1. Mol Biol Cell 12(5): 1315–1328. https://doi.org/10.1091/mbc.12.5.1315

Andriani F, Bertolini G, Facchinetti F, Baldoli E, Moro M, Casalini P et al (2016) Conversion to stem-cell state in response to microenvironmental cues is regulated by balance between epithelial and mesenchymal features in lung cancer cells. Mol Oncol 10(2):253–271. https://doi.org/10.1016/j.molonc.2015.10.002

Anwar M, Farooq Z, Tauseef M, Avin Balaji Ragunathrao V (Hg.) (2022) p53. A guardian of the genome and beyond. Erscheinungsort nicht ermittelbar: IntechOpen. Online verfügbar unter https://directory.doabooks.org/handle/20.500.12854/97101

Bae W, Ra EA, Lee MH (2025) Epigenetic regulation of reprogramming and pluripotency: insights from histone modifications and their implications for cancer stem cell therapies. Front Cell Dev Biol 13:1559183. https://doi.org/10.3389/fcell.2025.1559183

Bakhoum SF, Landau DA (2017) Chromosomal instability as a driver of tumor heterogeneity and evolution. Cold Spring Harb Perspect Med 7(6). https://doi.org/10.1101/cshperspect.a029611

Battulin NR, Khabarova AA, Boyarskikh U'y A, Menzorov AG, Filipenko ML, Serov OL (2012) Reprogramming somatic cells by fusion with embryonic stem cells does not cause silencing of the Dlk1-Dio3 region in mice. World J Stem Cells 4(8):87–93. https://doi.org/10.4252/wjsc.v4.i8.87

Baudoin NC, Nicholson JM, Soto K, Martin O, Chen J, Cimini D (2020) Asymmetric clustering of centrosomes defines the early evolution of tetraploid cells. elife 9. https://doi.org/10.7554/eLife.54565

Baumann AA, Buribayev Z, Wolkenhauer O, Salybekov AA, Wolfien M (2025) Epigenomic echoes-decoding genomic and epigenetic instability to distinguish lung cancer types and predict relapse. Epigenomes 9(1). https://doi.org/10.3390/epigenomes9010005

Bester AC, Roniger M, Oren YS, Im MM, Sarni D, Chaoat M et al (2011) Nucleotide deficiency promotes genomic instability in early stages of cancer development. Cell 145(3):435–446. https://doi.org/10.1016/j.cell.2011.03.044

Bielski CM, Zehir A, Penson AV, Donoghue MTA, Chatila W, Armenia J et al (2018) Genome doubling shapes the evolution and prognosis of advanced cancers. Nat Genet 50(8):1189–1195. https://doi.org/10.1038/s41588-018-0165-1

Biggs R, Liu PZ, Stephens AD, Marko JF (2019) Effects of altering histone posttranslational modifications on mitotic chromosome structure and mechanics. Mol Biol Cell 30(7):820–827. https://doi.org/10.1091/mbc.E18-09-0592

Bloomfield M, Cimini D (2023) The fate of extra centrosomes in newly formed tetraploid cells: should I stay, or should I go? Front Cell Dev Biol 11:1210983. https://doi.org/10.3389/fcell.2023.1210983

Bloomfield M, Chen J, Cimini D (2020) Spindle architectural features must be considered along with cell size to explain the timing of mitotic checkpoint silencing. Front Physiol 11:596263. https://doi.org/10.3389/fphys.2020.596263

Bolós V, Peinado H, Pérez-Moreno MA, Fraga MF, Esteller M, Cano A (2003) The transcription factor slug represses E-cadherin expression and induces epithelial to mesenchymal transitions: a comparison with snail and E47 repressors. J Cell Sci 116(Pt 3):499–511. https://doi.org/10.1242/jcs.00224

Bonasio R, Tu S, Reinberg D (2010) Molecular signals of epigenetic states. Science (New York, NY) 330(6004):612–616. https://doi.org/10.1126/science.1191078

Bonde S, Pedram M, Stultz R, Zavazava N (2010) Cell fusion of bone marrow cells and somatic cell reprogramming by embryonic stem cells. FASEB J 24(2):364–373. https://doi.org/10.1096/fj.09-137141

Brabletz S, Brabletz T (2010) The ZEB/miR-200 feedback loop–a motor of cellular plasticity in development and cancer? EMBO Rep 11(9):670–677. https://doi.org/10.1038/embor.2010.117

Brukman NG, Uygur B, Podbilewicz B, Chernomordik LV (2019) How cells fuse. J Cell Biol 218(5):1436–1451. https://doi.org/10.1083/jcb.201901017

Buchrieser J, Dufloo J, Hubert M, Monel B, Planas D, Rajah MM et al (2020) Syncytia formation by SARS-CoV-2-infected cells. EMBO J 39(23):e106267. https://doi.org/10.15252/embj.2020106267

Bure IV, Nemtsova MV, Kuznetsova EB (2022) Histone modifications and non-coding RNAs: mutual epigenetic regulation and role in pathogenesis. Int J Mol Sci 23(10). https://doi.org/10.3390/ijms23105801

Burrell RA, McClelland SE, Endesfelder D, Groth P, Weller M-C, Shaikh N et al (2013) Replication stress links structural and numerical cancer chromosomal instability. Nature 494(7438):492–496. https://doi.org/10.1038/nature11935

Campbell PJ, Getz G, Korbel JO, Stuart JM, Jennings JL, Stein LD, Schlesner M (2020) Pan-cancer analysis of whole genomes. Universität Augsburg/Springer, Augsburg

Cantwell H, Nurse P (2019) Unravelling nuclear size control. Curr Genet 65(6):1281–1285. https://doi.org/10.1007/s00294-019-00999-3

Chadha Y, Khurana A, Schmoller KM (2024) Eukaryotic cell size regulation and its implications for cellular function and dysfunction. Physiol Rev 104(4):1679–1717. https://doi.org/10.1152/physrev.00046.2023

Chen S, Liu M, Huang H, Li B, Zhao H, Feng X-Q, Zhao H-P (2019) Heat stress-induced multiple multipolar divisions of human cancer cells. Cells 8(8). https://doi.org/10.3390/cells8080888

Chen B, You W, Wang Y, Shan T (2020) The regulatory role of Myomaker and Myomixer-Myomerger-Minion in muscle development and regeneration. Cell Mol Life Sci 77(8):1551–1569. https://doi.org/10.1007/s00018-019-03341-9

Chen X, Agustinus AS, Li J, DiBona M, Bakhoum SF (2025) Chromosomal instability as a driver of cancer progression. Nat Rev Genet 26(1):31–46. https://doi.org/10.1038/s41576-024-00761-7

Chiou S-H, Wang M-L, Chou Y-T, Chen C-J, Hong C-F, Hsieh W-J et al (2010) Coexpression of Oct4 and Nanog enhances malignancy in lung adenocarcinoma by inducing cancer stem cell-like properties and epithelial-mesenchymal transdifferentiation. Cancer Res 70(24):10433–10444. https://doi.org/10.1158/0008-5472.CAN-10-2638

Choi JD, Lee J-S (2013) Interplay between epigenetics and genetics in cancer. Genomics Inform 11(4):164–173. https://doi.org/10.5808/GI.2013.11.4.164

Ciccia A, Elledge SJ (2010) The DNA damage response: making it safe to play with knives. Mol Cell 40(2):179–204. https://doi.org/10.1016/j.molcel.2010.09.019

Cimini D, Howell B, Maddox P, Khodjakov A, Degrassi F, Salmon ED (2001) Merotelic kinetochore orientation is a major mechanism of aneuploidy in mitotic mammalian tissue cells. J Cell Biol 153(3):517–527. https://doi.org/10.1083/jcb.153.3.517

Cimini D, Fioravanti D, Salmon ED, Degrassi F (2002) Merotelic kinetochore orientation versus chromosome mono-orientation in the origin of lagging chromosomes in human primary cells. J Cell Sci 115(Pt 3):507–515. https://doi.org/10.1242/jcs.115.3.507

Consortium, The ICGC/TCGA Pan-Cancer Analysis of Whole Genomes (2020) Pan-cancer analysis of whole genomes. Nature (London) 578(7793):82–93. https://doi.org/10.1038/s41586-020-1969-6

Coticchio G, Cimadomo D, Cermisoni GC, Rienzi L, Papaleo E, Ubaldi FM et al (2023) The first mitotic division: a perilous bridge connecting the zygote and the early embryo. Hum Reprod 38(6):1019–1027. https://doi.org/10.1093/humrep/dead067

Crasta K, Ganem NJ, Dagher R, Lantermann AB, Ivanova EV, Pan Y et al (2012) DNA breaks and chromosome pulverization from errors in mitosis. Nature 482(7383):53–58. https://doi.org/10.1038/nature10802

Crockford A, Zalmas LP, Grönroos E, Dewhurst SM, McGranahan N, Cuomo ME et al (2017) Cyclin D mediates tolerance of genome-doubling in cancers with functional p53. Ann Oncol 28(1):149–156. https://doi.org/10.1093/annonc/mdw612

Darmasaputra GS, Geerlings CC, Chuva de Sousa Lopes SM, Clevers H, Galli M (2024) Binucleated human hepatocytes arise through late cytokinetic regression during endomitosis M phase. J Cell Biol 223(8). https://doi.org/10.1083/jcb.202403020

Davies PS, Powell AE, Swain JR, Wong MH (2009) Inflammation and proliferation act together to mediate intestinal cell fusion. PLoS One 4(8):e6530. https://doi.org/10.1371/journal.pone.0006530

Debnath P, Huirem RS, Dutta P, Palchaudhuri S (2022) Epithelial-mesenchymal transition and its transcription factors. Biosci Rep 42(1). https://doi.org/10.1042/BSR20211754

Delespaul L, Merle C, Lesluyes T, Lagarde P, Le Guellec S, Pérot G et al (2019) Fusion-mediated chromosomal instability promotes aneuploidy patterns that resemble human tumors. Oncogene 38(33):6083–6094. https://doi.org/10.1038/s41388-019-0859-6

Ding J, Jin W, Chen C, Shao Z, Wu J (2012) Tumor associated macrophage × cancer cell hybrids may acquire cancer stem cell properties in breast cancer. PLoS One 7(7):e41942. https://doi.org/10.1371/journal.pone.0041942

Dittmar T, Nagler C, Schwitalla S, Reith G, Niggemann B, Zänker KS (2009) Recurrence cancer stem cells–made by cell fusion? Med Hypotheses 73(4):542–547. https://doi.org/10.1016/j.mehy.2009.05.044

Dittmar T, Weiler J, Luo T, Hass R (2021) Cell-cell fusion mediated by viruses and HERV-derived Fusogens in cancer initiation and progression. Cancer 13(21). https://doi.org/10.3390/cancers13215363

Dong C, Wu Y, Yao J, Wang Y, Yu Y, Rychahou PG et al (2012) G9a interacts with snail and is critical for snail-mediated E-cadherin repression in human breast cancer. J Clin Invest 122(4): 1469–1486. https://doi.org/10.1172/JCI57349

Dong P, Xiong Y, Watari H, Hanley SJB, Konno Y, Ihira K et al (2016) MiR-137 and miR-34a directly target snail and inhibit EMT, invasion and sphere-forming ability of ovarian cancer cells. J Exp Clin Cancer Res 35(1):132. https://doi.org/10.1186/s13046-016-0415-y

Donne R, Saroul-Aïnama M, Cordier P, Celton-Morizur S, Desdouets C (2020) Polyploidy in liver development, homeostasis and disease. Nat Rev Gastroenterol Hepatol 17(7):391–405. https://doi.org/10.1038/s41575-020-0284-x

Donne R, Sangouard F, Celton-Morizur S, Desdouets C (2021) Hepatocyte polyploidy: driver or gatekeeper of chronic liver diseases. Cancer 13(20). https://doi.org/10.3390/cancers13205151

Donovan P, Cato K, Legaie R, Jayalath R, Olsson G, Hall B et al (2014) Hyperdiploid tumor cells increase phenotypic heterogeneity within glioblastoma tumors. Mol BioSyst 10(4):741–758. https://doi.org/10.1039/c3mb70484j

Dörnen J, Sieler M, Weiler J, Keil S, Dittmar T (2020) Cell fusion-mediated tissue regeneration as an inducer of polyploidy and aneuploidy. Int J Mol Sci 21(5). https://doi.org/10.3390/ijms21051811

Duelli D, Lazebnik Y (2007) Cell-to-cell fusion as a link between viruses and cancer. Nat Rev Cancer 7(12):968–976. https://doi.org/10.1038/nrc2272

Duelli DM, Padilla-Nash HM, Berman D, Murphy KM, Ried T, Lazebnik Y (2007) A virus causes cancer by inducing massive chromosomal instability through cell fusion. Curr Biol 17(5): 431–437. https://doi.org/10.1016/j.cub.2007.01.049

Duncan AW, Hickey RD, Paulk NK, Culberson AJ, Olson SB, Finegold MJ, Grompe M (2009) Ploidy reductions in murine fusion-derived hepatocytes. PLoS Genet 5(2):e1000385. https://doi.org/10.1371/journal.pgen.1000385

Duncan AW, Taylor MH, Hickey RD, Hanlon Newell AE, Lenzi ML, Olson SB et al (2010) The ploidy conveyor of mature hepatocytes as a source of genetic variation. Nature 467(7316): 707–710. https://doi.org/10.1038/nature09414

Duncan AW, Hanlon Newell AE, Smith L, Wilson EM, Olson SB, Thayer MJ et al (2012) Frequent aneuploidy among normal human hepatocytes. Gastroenterology 142(1):25–28. https://doi.org/10.1053/j.gastro.2011.10.029

El-Tanani M, Rabbani SA, Satyam SM, Rangraze IR, Wali AF, El-Tanani Y, Aljabali AAA (2025) Deciphering the role of cancer stem cells: drivers of tumor evolution, therapeutic resistance, and precision medicine strategies. Cancer 17(3). https://doi.org/10.3390/cancers17030382

Erenpreisa J, Salmina K, Huna A, Jackson TR, Vazquez-Martin A, Cragg MS (2015) The "virgin birth", polyploidy, and the origin of cancer. Oncoscience 2(1):3–14. https://doi.org/10.18632/oncoscience.108

Faggioli F, Sacco MG, Susani L, Montagna C, Vezzoni P (2008) Cell fusion is a physiological process in mouse liver. Hepatology (Baltimore, MD) 48(5):1655–1664. https://doi.org/10.1002/hep.22488

Fava LL, Schuler F, Sladky V, Haschka MD, Soratroi C, Eiterer L et al (2017) The PIDDosome activates p53 in response to supernumerary centrosomes. Genes Dev 31(1):34–45. https://doi.org/10.1101/gad.289728.116

Fei F, Li C, Wang X, Du J, Liu K, Li B et al (2019) Syncytin 1, CD9, and CD47 regulating cell fusion to form PGCCs associated with cAMP/PKA and JNK signaling pathway. Cancer Med 8(6):3047–3058. https://doi.org/10.1002/cam4.2173

Feliciano D, Ott CM, Espinosa-Medina I, Weigel AV, Benedetti L, Milano KM et al (2021) YAP1 nuclear efflux and transcriptional reprograming follow membrane diminution upon VSV-G-induced cell fusion. Nat Commun 12(1):4502. https://doi.org/10.1038/s41467-021-24708-2

Ferrand J, Noël D, Lehours P, Prochazkova-Carlotti M, Chambonnier L, Ménard A et al (2011) Human bone marrow-derived stem cells acquire epithelial characteristics through fusion with gastrointestinal epithelial cells. PLoS One 6(5):e19569. https://doi.org/10.1371/journal.pone.0019569

Foshay KM, Looney TJ, Chari S, Mao FF, Lee JH, Zhang L et al (2012) Embryonic stem cells induce pluripotency in somatic cell fusion through biphasic reprogramming. Mol Cell 46(2): 159–170. https://doi.org/10.1016/j.molcel.2012.02.013

Fox DT, Duronio RJ (2013) Endoreplication and polyploidy: insights into development and disease. Development 140(1):3–12. https://doi.org/10.1242/dev.080531

Frade J, Nakagawa S, Cortes P, Di Vicino U, Romo N, Lluis F, Cosma MP (2019) Controlled ploidy reduction of pluripotent 4n cells generates 2n cells during mouse embryo development. Sci Adv 5(10):eaax4199. https://doi.org/10.1126/sciadv.aax4199

Fujiwara T, Bandi M, Nitta M, Ivanova EV, Bronson RT, Pellman D (2005) Cytokinesis failure generating tetraploids promotes tumorigenesis in p53-null cells. Nature (London) 437(7061): 1043–1047. https://doi.org/10.1038/nature04217

Galli M, Morgan DO (2016) Cell size determines the strength of the spindle assembly checkpoint during embryonic development. Dev Cell 36(3):344–352. https://doi.org/10.1016/j.devcel.2016.01.003

Galofré C, Asensio E, Ubach M, Torres IM, Quintanilla I, Castells A, Camps J (2020) Centrosome reduction in newly-generated tetraploid cancer cells obtained by separase depletion. Sci Rep 10(1):9152. https://doi.org/10.1038/s41598-020-65975-1

Ganem NJ, Storchova Z, Pellman D (2007) Tetraploidy, aneuploidy and cancer. Curr Opin Genet Dev 17(2):157–162. https://doi.org/10.1016/j.gde.2007.02.011
Ganem NJ, Godinho SA, Pellman D (2009) A mechanism linking extra centrosomes to chromosomal instability. Nature 460(7252):278–282. https://doi.org/10.1038/nature08136
Ganem NJ, Cornils H, Chiu S-Y, O'Rourke KP, Arnaud J, Yimlamai D et al (2014) Cytokinesis failure triggers hippo tumor suppressor pathway activation. Cell 158(4):833–848. https://doi.org/10.1016/j.cell.2014.06.029
Gao H, Teng C, Huang W, Peng J, Wang C (2015) SOX2 promotes the epithelial to mesenchymal transition of esophageal squamous cells by modulating slug expression through the activation of STAT3/HIF-α signaling. Int J Mol Sci 16(9):21643–21657. https://doi.org/10.3390/ijms160921643
Gast CE, Silk AD, Zarour L, Riegler L, Burkhart JG, Gustafson KT et al (2018) Cell fusion potentiates tumor heterogeneity and reveals circulating hybrid cells that correlate with stage and survival. Sci Adv 4(9):eaat7828. https://doi.org/10.1126/sciadv.aat7828
Gemble S, Wardenaar R, Keuper K, Srivastava N, Nano M, Macé A-S et al (2022) Genetic instability from a single S phase after whole-genome duplication. Nature 604(7904):146–151. https://doi.org/10.1038/s41586-022-04578-4
Gentric G, Maillet V, Paradis V, Couton D, L'Hermitte A, Panasyuk G et al (2015) Oxidative stress promotes pathologic polyploidization in nonalcoholic fatty liver disease. J Clin Invest 125(3):981–992. https://doi.org/10.1172/JCI73957
Gerhold AR, Poupart V, Labbé J-C, Maddox PS (2018) Spindle assembly checkpoint strength is linked to cell fate in the Caenorhabditis elegans embryo. Mol Biol Cell 29(12):1435–1448. https://doi.org/10.1091/mbc.E18-04-0215
Goldman-Wohl D, Yagel S (2014) United we stand not dividing: the syncytiotrophoblast and cell senescence. Placenta 35(6):341–344. https://doi.org/10.1016/j.placenta.2014.03.012
Gregan J, Polakova S, Zhang L, Tolić-Nørrelykke IM, Cimini D (2011) Merotelic kinetochore attachment: causes and effects. Trends Cell Biol 21(6):374–381. https://doi.org/10.1016/j.tcb.2011.01.003
Grosse-Wilde A, Fouquier d'Hérouël A, McIntosh E, Ertaylan G, Skupin A, Kuestner RE et al (2015) Stemness of the hybrid epithelial/mesenchymal state in breast cancer and its association with poor survival. PLoS One 10(5):e0126522. https://doi.org/10.1371/journal.pone.0126522
Guidotti J-E, Brégerie O, Robert A, Debey P, Brechot C, Desdouets C (2003) Liver cell polyploidization: a pivotal role for binuclear hepatocytes. J Biol Chem 278(21):19095–19101. https://doi.org/10.1074/jbc.M300982200
Guo M, Peng Y, Gao A, Du C, Herman JG (2019) Epigenetic heterogeneity in cancer. Biomark Res 7:23. https://doi.org/10.1186/s40364-019-0174-y
Harris H, Miller OJ, Klein G, Worst P, Tachibana T (1969) Suppression of malignancy by cell fusion. Nature 223(5204):363–368. https://doi.org/10.1038/223363a0
Hass R, von der Ohe J, Dittmar T (2021) Cancer cell fusion and post-hybrid selection process (PHSP). Cancer 13(18). https://doi.org/10.3390/cancers13184636
Hayashi K, Horisaka K, Harada Y, Ogawa Y, Yamashita T, Kitano T et al (2024) Polyploidy mitigates the impact of DNA damage while simultaneously bearing its burden. Cell Death Discov 10(1):436. https://doi.org/10.1038/s41420-024-02206-w
He B, Gnawali N, Hinman AW, Mattingly AJ, Osimani A, Cimini D (2019) Chromosomes missegregated into micronuclei contribute to chromosomal instability by missegregating at the next division. Oncotarget 10(28):2660–2674. https://doi.org/10.18632/oncotarget.26853
Heinke P, Rost F, Rode J, Trus P, Simonova I, Lázár E et al (2022) Diploid hepatocytes drive physiological liver renewal in adult humans. Cell Syst 13(6):499–507.e12. https://doi.org/10.1016/j.cels.2022.05.001
Herreros-Villanueva M, Zhang J-S, Koenig A, Abel EV, Smyrk TC, Bamlet WR et al (2013) SOX2 promotes dedifferentiation and imparts stem cell-like features to pancreatic cancer cells. Oncogene 2(8):e61. https://doi.org/10.1038/oncsis.2013.23

Hindi SM, Tajrishi MM, Kumar A (2013) Signaling mechanisms in mammalian myoblast fusion. Sci Signal 6(272):re2. https://doi.org/10.1126/scisignal.2003832

Ho CC, Hau PM, Marxer M, Poon RYC (2010) The requirement of p53 for maintaining chromosomal stability during tetraploidization. Oncotarget 1(7):583–595. https://doi.org/10.18632/oncotarget.193

Holmquist GP, Gao S (1997) Somatic mutation theory, DNA repair rates, and the molecular epidemiology of p53 mutations. Mutat Res 386(1):69–101. https://doi.org/10.1016/s1383-5742(96)00045-2

Horsley V, Jansen KM, Mills ST, Pavlath GK (2003) IL-4 acts as a myoblast recruitment factor during mammalian muscle growth. Cell 113(4):483–494. https://doi.org/10.1016/s0092-8674(03)00319-2

Ijaz J, Harry E, Raine K, Menzies A, Beal K, Quail MA et al (2024) Haplotype-specific assembly of shattered chromosomes in esophageal adenocarcinomas. Cell Genomics 4(2):100484. https://doi.org/10.1016/j.xgen.2023.100484

Izadpanah MH, Forghanifard MM (2022) TWIST1 plays role in expression of Stemness state markers in ESCC. Genes 13(12). https://doi.org/10.3390/genes13122369

Izadpanah MH, Abbaszadegan MR, Fahim Y, Forghanifard MM (2017) Ectopic expression of TWIST1 upregulates the stemness marker OCT4 in the esophageal squamous cell carcinoma cell line KYSE30. Cell Mol Biol Lett 22:33. https://doi.org/10.1186/s11658-017-0065-x

Janic A, Abad E, Amelio I (2025) Decoding p53 tumor suppression: a crosstalk between genomic stability and epigenetic control? Cell Death Differ 32(1):1–8. https://doi.org/10.1038/s41418-024-01259-9

Janssen, Aniek; van der Burg, Marja; Szuhai, Karoly; Kops, Geert J. P. L.; Medema, René H. (2011): Chromosome segregation errors as a cause of DNA damage and structural chromosome aberrations Science 333 (6051), 1895–1898. https://doi.org/10.1126/science.1210214

Jelenić I, Selmecki A, Laan L, Pavin N (2018) Spindle dynamics model explains chromosome loss rates in yeast polyploid cells. Front Genet 9:296. https://doi.org/10.3389/fgene.2018.00296

Jiang Q, Zhang Q, Wang S, Xie S, Fang W, Liu Z et al (2015) A fraction of CD133+ CNE2 cells is made of giant cancer cells with morphological evidence of asymmetric mitosis. J Cancer 6(12):1236–1244. https://doi.org/10.7150/jca.12626

Jiao Y, Yu Y, Zheng M, Yan M, Wang J, Zhang Y, Zhang S (2024) Dormant cancer cells and polyploid giant cancer cells: the roots of cancer recurrence and metastasis. Clin Transl Med 14(2):e1567. https://doi.org/10.1002/ctm2.1567

Jolly MK, Boareto M, Huang B, Jia D, Lu M, Ben-Jacob E et al (2015) Implications of the hybrid epithelial/mesenchymal phenotype in metastasis. Front Oncol 5:155. https://doi.org/10.3389/fonc.2015.00155

Kaji K, Kudo A (2004) The mechanism of sperm-oocyte fusion in mammals. Reproduction 127(4):423–429. https://doi.org/10.1530/rep.1.00163

Kashofer K, Siapati EK, Bonnet D (2006) In vivo formation of unstable heterokaryons after liver damage and hematopoietic stem cell/progenitor transplantation. Stem Cells 24(4):1104–1112. https://doi.org/10.1634/stemcells.2005-0405

Khurana A, Chadha Y, Schmoller KM (2023) Too big not to fail: different paths lead to senescence of enlarged cells. Mol Cell 83(22):3946–3947. https://doi.org/10.1016/j.molcel.2023.10.024

Krämer A, Maier B, Bartek J (2011) Centrosome clustering and chromosomal (in)stability: a matter of life and death. Mol Oncol 5(4):324–335. https://doi.org/10.1016/j.molonc.2011.05.003

Kreutz C, MacNelly S, Follo M, Wäldin A, Binninger-Lacour P, Timmer J, Bartolomé-Rodríguez MM (2017) Hepatocyte ploidy is a diversity factor for liver homeostasis. Front Physiol 8:862. https://doi.org/10.3389/fphys.2017.00862

Kruglova AA, Kizilova EA, Zhelezova AI, Gridina MM, Golubitsa AN, Serov OL (2008) Embryonic stem cell/fibroblast hybrid cells with near-tetraploid karyotype provide high yield of chimeras. Cell Tissue Res 334(3):371–380. https://doi.org/10.1007/s00441-008-0702-9

Kruglova AA, Matveeva NM, Gridina MM, Battulin NR, Karpov A, Kiseleva EV et al (2010) Dominance of parental genomes in embryonic stem cell/fibroblast hybrid cells depends on the ploidy of the somatic partner. Cell Tissue Res 340(3):437–450. https://doi.org/10.1007/s00441-010-0987-3

Kunihiro M, Fujii H, Miyagi T, Takahashi Y, Tanaka R, Fukushima T et al (2016) Heat shock enhances the expression of the human T cell leukemia virus type-I (HTLV-I) trans-activator (tax) antigen in human HTLV-I infected primary and cultured T cells. Viruses 8(7). https://doi.org/10.3390/v8070191

Lanz MC, Zatulovskiy E, Swaffer MP, Zhang L, Ilerten I, Zhang S et al (2022) Increasing cell size remodels the proteome and promotes senescence. Mol Cell 82(17):3255–3269.e8. https://doi.org/10.1016/j.molcel.2022.07.017

Lee J-Y, Kong G (2016) Roles and epigenetic regulation of epithelial-mesenchymal transition and its transcription factors in cancer initiation and progression. Cell Mol Life Sci 73(24): 4643–4660. https://doi.org/10.1007/s00018-016-2313-z

Levine AJ, Oren M (2009) The first 30 years of p53: growing ever more complex. Nat Rev Cancer 9(10):749–758. https://doi.org/10.1038/nrc2723

Li H, Feng Z, Tsang TC, Tang T, Jia X, He X et al (2014a) Fusion of HepG2 cells with mesenchymal stem cells increases cancer-associated and malignant properties: an in vivo metastasis model. Oncol Rep 32(2):539–547. https://doi.org/10.3892/or.2014.3264

Li L-Z, Zhang CZ, Liu L-L, Yi C, Lu S-X, Zhou X et al (2014b) miR-720 inhibits tumor invasion and migration in breast cancer by targeting TWIST1. Carcinogenesis 35(2):469–478. https://doi.org/10.1093/carcin/bgt330

Li Z, Zheng M, Zhang H, Yang X, Fan L, Fu F et al (2021) Arsenic trioxide promotes tumor progression by inducing the formation of PGCCs and embryonic hemoglobin in colon cancer cells. Front Oncol 11:720814. https://doi.org/10.3389/fonc.2021.720814

Lin Y-F, Hu Q, Mazzagatti A, Valle-Inclán JE, Maurais EG, Dahiya R et al (2023) Mitotic clustering of pulverized chromosomes from micronuclei. Nature (London) 618(7967): 1041–1048. https://doi.org/10.1038/s41586-023-05974-0

Lindström A, Midtbö K, Arnesson L-G, Garvin S, Shabo I (2017) Fusion between M2-macrophages and cancer cells results in a subpopulation of radioresistant cells with enhanced DNA-repair capacity. Oncotarget 8(31):51370–51386. https://doi.org/10.18632/oncotarget.17986

Liu C-W, Li C-H, Peng Y-J, Cheng Y-W, Chen H-W, Liao P-L et al (2014) Snail regulates Nanog status during the epithelial-mesenchymal transition via the Smad1/Akt/GSK3β signaling pathway in non-small-cell lung cancer. Oncotarget 5(11):3880–3894. https://doi.org/10.18632/oncotarget.2006

Liu Z, Gu S, Peng Z, Wang Y, Li H, Zeng X et al (2025) Fusion of glioma-associated mesenchymal stem/stromal cells with glioma cells promotes macrophage recruitment and M2 polarization via m6A modification of CSF1. Cell Death Dis 16(1):345. https://doi.org/10.1038/s41419-025-07678-x

Lok TM, Wang Y, Xu WK, Xie S, Ma HT, Poon RYC (2020) Mitotic slippage is determined by p31comet and the weakening of the spindle-assembly checkpoint. Oncogene 39(13): 2819–2834. https://doi.org/10.1038/s41388-020-1187-6

López S, Lim EL, Horswell S, Haase K, Huebner A, Dietzen M et al (2020) Interplay between whole-genome doubling and the accumulation of deleterious alterations in cancer evolution. Nat Genet 52(3):283–293. https://doi.org/10.1038/s41588-020-0584-7

Lu X, Kang Y (2011) Cell fusion hypothesis of the cancer stem cell. Adv Exp Med Biol 714:129–140. https://doi.org/10.1007/978-94-007-0782-5_6

Lukow DA, Sheltzer JM (2022) Chromosomal instability and aneuploidy as causes of cancer drug resistance. Trends Cancer 8(1):43–53. https://doi.org/10.1016/j.trecan.2021.09.002

Lüönd F, Sugiyama N, Bill R, Bornes L, Hager C, Tang F et al (2021) Distinct contributions of partial and full EMT to breast cancer malignancy. Dev Cell 56(23):3203–3221.e11. https://doi.org/10.1016/j.devcel.2021.11.006

Machida K, Liu J-C, McNamara G, Levine A, Duan L, Lai MMC (2009) Hepatitis C virus causes uncoupling of mitotic checkpoint and chromosomal polyploidy through the Rb pathway. J Virol 83(23):12590–12600. https://doi.org/10.1128/JVI.02643-08

Mandal G, Biswas S, Roy Chowdhury S, Chatterjee A, Purohit S, Khamaru P et al (2018) Heterodimer formation by Oct4 and Smad3 differentially regulates epithelial-to-mesenchymal transition-associated factors in breast cancer progression. Biochim Biophys Acta Mol basis Dis 1864(6 Pt A):2053–2066. https://doi.org/10.1016/j.bbadis.2018.03.010

Manic G, Galluzzi L, Vitale I (2022) Catastrophic DNA replication in unscheduled tetraploid cells. Trends Genet 38(8):787–788. https://doi.org/10.1016/j.tig.2022.04.005

Martínez-Estrada OM, Culleres A, Soriano FX, Peinado H, Bolós V, Martínez FO et al (2006) The transcription factors slug and snail act as repressors of Claudin-1 expression in epithelial cells. Biochem J 394(Pt 2):449–457. https://doi.org/10.1042/BJ20050591

Matsumoto T (2022) Implications of polyploidy and ploidy alterations in hepatocytes in liver injuries and cancers. Int J Mol Sci 23(16). https://doi.org/10.3390/ijms23169409

Matsumoto T, Wakefield L, Tarlow BD, Grompe M (2020) In vivo lineage tracing of polyploid hepatocytes reveals extensive proliferation during liver regeneration. Cell Stem Cell 26(1): 34–47.e3. https://doi.org/10.1016/j.stem.2019.11.014

Matsumoto T, Wakefield L, Peters A, Peto M, Spellman P, Grompe M (2021) Proliferative polyploid cells give rise to tumors via ploidy reduction. Nat Commun 12(1):646. https://doi.org/10.1038/s41467-021-20916-y

McPherson A, Vázquez-García I, Myers MA, Al-Rawi DH, Zatzman M, Weiner AC et al (2025) Ongoing genome doubling shapes evolvability and immunity in ovarian cancer. Nature 644(8078):1078–1087. https://doi.org/10.1038/s41586-025-09240-3

Melzer C, von der Ohe J, Hass R (2018) Enhanced metastatic capacity of breast cancer cells after interaction and hybrid formation with mesenchymal stroma/stem cells (MSC). Cell Commun Signal 16(1):2. https://doi.org/10.1186/s12964-018-0215-4

Melzer C, von der Ohe J, Luo T, Hass R (2021) Spontaneous fusion of MSC with breast cancer cells can generate tumor dormancy. Int J Mol Sci 22(11). https://doi.org/10.3390/ijms22115930

Merckens A, Sieler M, Keil S, Dittmar T (2023) Altered phenotypes of breast epithelial × breast cancer hybrids after ZEB1 knock-out. Int J Mol Sci 24(24). https://doi.org/10.3390/ijms242417310

Merle C, Thébault N, LeGuellec S, Baud J, Pérot G, Lesluyes T et al (2020) Tetraploidization of immortalized myoblasts induced by cell fusion drives myogenic sarcoma development with DMD deletion. Cancer 12(5). https://doi.org/10.3390/cancers12051281

Merle C, Lagarde P, Lartigue L, Chibon F (2021) Acquisition of cancer stem cell capacities after spontaneous cell fusion. BMC Cancer 21(1):241. https://doi.org/10.1186/s12885-021-07979-2

Middelkamp S, van Tol HTA, Spierings DCJ, Boymans S, Guryev V, Roelen BAJ et al (2020) Sperm DNA damage causes genomic instability in early embryonic development. Sci Adv 6(16):eaaz7602. https://doi.org/10.1126/sciadv.aaz7602

Miroshnychenko D, Baratchart E, Ferrall-Fairbanks MC, Velde RV, Laurie MA, Bui MM et al (2021) Spontaneous cell fusions as a mechanism of parasexual recombination in tumour cell populations. Nat Ecol Evol 5(3):379–391. https://doi.org/10.1038/s41559-020-01367-y

Miyamoto T (2006) The dendritic cell-specific transmembrane protein DC-STAMP is essential for osteoclast fusion and osteoclast bone-resorbing activity. Modern Rheumatol 16(6):341–342. https://doi.org/10.1007/s10165-006-0524-0

Miyaoka Y, Ebato K, Kato H, Arakawa S, Shimizu S, Miyajima A (2012) Hypertrophy and unconventional cell division of hepatocytes underlie liver regeneration. Curr Biol 22(13): 1166–1175. https://doi.org/10.1016/j.cub.2012.05.016

Mukund K, Subramaniam S (2020) Skeletal muscle: a review of molecular structure and function, in health and disease. Wiley Interdiscip Rev Syst Biol Med 12(1):e1462. https://doi.org/10.1002/wsbm.1462

Müller M, May S, Bird TG (2021) Ploidy dynamics increase the risk of liver cancer initiation. Nat Commun 12(1):1896. https://doi.org/10.1038/s41467-021-21897-8

Myerson D, Parkin RK (2019) Donor-derived hepatocytes in human hematopoietic cell transplant recipients: evidence of fusion. Virchows Arch 474(3):365–374. https://doi.org/10.1007/s00428-018-2497-8

Nawroth JC, Petropolis DB, Manatakis DV, Maulana TI, Burchett G, Schlünder K et al (2021) Modeling alcohol-associated liver disease in a human liver-chip. Cell Rep 36(3):109393. https://doi.org/10.1016/j.celrep.2021.109393

Nern C, Wolff I, Macas J, von Randow J, Scharenberg C, Priller J, Momma S (2009) Fusion of hematopoietic cells with Purkinje neurons does not lead to stable heterokaryon formation under noninvasive conditions. J Neurosci 29(12):3799–3807. https://doi.org/10.1523/JNEUROSCI.5848-08.2009

Neurohr GE, Terry RL, Lengefeld J, Bonney M, Brittingham GP, Moretto F et al (2019) Excessive cell growth causes cytoplasm dilution and contributes to senescence. Cell 176(5):1083–1097.e18. https://doi.org/10.1016/j.cell.2019.01.018

Nguyen HG, Makitalo M, Yang D, Chinnappan D, St Hilaire C, Ravid K (2009) Deregulated Aurora-B induced tetraploidy promotes tumorigenesis. FASEB J 23(8):2741–2748. https://doi.org/10.1096/fj.09-130963

Niu N, Zhang J, Zhang N, Mercado-Uribe I, Tao F, Han Z et al (2016) Linking genomic reorganization to tumor initiation via the giant cell cycle. Oncogene 5(12):e281. https://doi.org/10.1038/oncsis.2016.75

Niu N, Mercado-Uribe I, Liu J (2017) Dedifferentiation into blastomere-like cancer stem cells via formation of polyploid giant cancer cells. Oncogene 36(34):4887–4900. https://doi.org/10.1038/onc.2017.72

Niu N, Yao J, Bast RC, Sood AK, Liu J (2021) IL-6 promotes drug resistance through formation of polyploid giant cancer cells and stromal fibroblast reprogramming. Oncogene 10(9):65. https://doi.org/10.1038/s41389-021-00349-4

Nygren JM, Jovinge S, Breitbach M, Säwén P, Röll W, Hescheler J et al (2004) Bone marrow-derived hematopoietic cells generate cardiomyocytes at a low frequency through cell fusion, but not transdifferentiation. Nat Med 10(5):494–501. https://doi.org/10.1038/nm1040

Ohkubo T, Ozawa M (2004) The transcription factor snail downregulates the tight junction components independently of E-cadherin downregulation. J Cell Sci 117(Pt 9):1675–1685. https://doi.org/10.1242/jcs.01004

Olaharski AJ, Sotelo R, Solorza-Luna G, Gonsebatt ME, Guzman P, Mohar A, Eastmond DA (2006) Tetraploidy and chromosomal instability are early events during cervical carcinogenesis. Carcinogenesis 27(2):337–343. https://doi.org/10.1093/carcin/bgi218

Otto A (Hg.) (1911) Vorträge und Aufsätze über Entwickelungsmechanik der Organismen: Über Zellverschmelzung mit qualitativ abnormer Chromosomenverteilung als Ursache der Geschwulstbildung

Palermo A, Doyonnas R, Bhutani N, Pomerantz J, Alkan O, Blau HM (2009) Nuclear reprogramming in heterokaryons is rapid, extensive, and bidirectional. FASEB J 23(5):1431–1440. https://doi.org/10.1096/fj.08-122903

Palla AR, Piazzolla D, Alcazar N, Cañamero M, Graña O, Gómez-López G et al (2015) The pluripotency factor NANOG promotes the formation of squamous cell carcinomas. Sci Rep 5:10205. https://doi.org/10.1038/srep10205

Pasque V, Jullien J, Miyamoto K, Halley-Stott RP, Gurdon JB (2011) Epigenetic factors influencing resistance to nuclear reprogramming. Trends Genet 27(12):516–525. https://doi.org/10.1016/j.tig.2011.08.002

Pastushenko I, Blanpain C (2019) EMT transition states during tumor progression and metastasis. Trends Cell Biol 29(3):212–226. https://doi.org/10.1016/j.tcb.2018.12.001

Pawelek JM (2005) Tumour-cell fusion as a source of myeloid traits in cancer. Lancet Oncol 6(12):988–993. https://doi.org/10.1016/S1470-2045(05)70466-6

Pereira CF, Terranova R, Ryan NK, Santos J, Morris KJ, Cui W et al (2008) Heterokaryon-based reprogramming of human B lymphocytes for pluripotency requires Oct4 but not Sox2. PLoS Genet 4(9):e1000170. https://doi.org/10.1371/journal.pgen.1000170

Pomerantz JH, Mukherjee S, Palermo AT, Blau HM (2009) Reprogramming to a muscle fate by fusion recapitulates differentiation. J Cell Sci 122(Pt 7):1045–1053. https://doi.org/10.1242/jcs.041376
Potapova TA, Zhu J, Li R (2013) Aneuploidy and chromosomal instability: a vicious cycle driving cellular evolution and cancer genome chaos. Cancer Metastasis Rev 32(3–4):377–389. https://doi.org/10.1007/s10555-013-9436-6
Potapova TA, Seidel CW, Box AC, Rancati G, Li R (2016) Transcriptome analysis of tetraploid cells identifies cyclin D2 as a facilitator of adaptation to genome doubling in the presence of p53. Mol Biol Cell 27(20):3065–3084. https://doi.org/10.1091/mbc.E16-05-0268
Qu Y, Zhang L, Rong Z, He T, Zhang S (2013) Number of glioma polyploid giant cancer cells (PGCCs) associated with vasculogenic mimicry formation and tumor grade in human glioma. J Exp Clin Cancer Res 32(1):75. https://doi.org/10.1186/1756-9966-32-75
Quintana-Bustamante O, Grueso E, Garcia-Escudero R, Arza E, Alvarez-Barrientos A, Fabregat I et al (2012) Cell fusion reprogramming leads to a specific hepatic expression pattern during mouse bone marrow derived hepatocyte formation in vivo. PLoS One 7(3):e33945. https://doi.org/10.1371/journal.pone.0033945
Quinton RJ, DiDomizio A, Vittoria MA, Kotýnková K, Ticas CJ, Patel S et al (2021) Whole-genome doubling confers unique genetic vulnerabilities on tumour cells. Nature 590(7846):492–497. https://doi.org/10.1038/s41586-020-03133-3
Reid BJ, Barrett MT, Galipeau PC, Sanchez CA, Neshat K, Cowan DS, Levine DS (1996) Barrett's esophagus: ordering the events that lead to cancer. Eur J Cancer Prev 5(Suppl 2):57–65. https://doi.org/10.1097/00008469-199612002-00009
Rengstl B, Newrzela S, Heinrich T, Weiser C, Thalheimer FB, Schmid F et al (2013) Incomplete cytokinesis and re-fusion of small mononucleated Hodgkin cells lead to giant multinucleated reed-Sternberg cells. Proc Natl Acad Sci USA 110(51):20729–20734. https://doi.org/10.1073/pnas.1312509110
Rhind N (2021) Cell-size control. Curr Biol 31(21):R1414–R1420. https://doi.org/10.1016/j.cub.2021.09.017
Rubio C, Avendaño-Ortiz J, Ruiz-Palomares R, Karaivanova V, Alberquilla O, Sánchez-Domínguez R et al (2022) Toward tumor fight and tumor microenvironment remodeling: PBA induces cell cycle arrest and reduces tumor hybrid cells' pluripotency in bladder cancer. Cancer 14(2). https://doi.org/10.3390/cancers14020287
Russo A, Pacchierotti F, Cimini D, Ganem NJ, Genescà A, Natarajan AT et al (2015) Genomic instability: crossing pathways at the origin of structural and numerical chromosome changes. Environ Mol Mutagen 56(7):563–580. https://doi.org/10.1002/em.21945
Sabe H, Yahara Y, Ishii M (2024) Cell fusion dynamics: mechanisms of multinucleation in osteoclasts and macrophages. Inflamm Regen 44(1):49. https://doi.org/10.1186/s41232-024-00360-3
Saini G, Joshi S, Garlapati C, Li H, Kong J, Krishnamurthy J et al (2022) Polyploid giant cancer cell characterization: new frontiers in predicting response to chemotherapy in breast cancer. Semin Cancer Biol 81:220–231. https://doi.org/10.1016/j.semcancer.2021.03.017
Sakaue-Sawano A, Hoshida T, Yo M, Takahashi R, Ohtawa K, Arai T et al (2013) Visualizing developmentally programmed endoreplication in mammals using ubiquitin oscillators. Development 140(22):4624–4632. https://doi.org/10.1242/dev.099226
Saleh T, Bloukh S, Carpenter VJ, Alwohoush E, Bakeer J, Darwish S et al (2020) Therapy-induced senescence: an "old" friend becomes the enemy. Cancer 12(4). https://doi.org/10.3390/cancers12040822
Santos J, Pereira CF, Di-Gregorio A, Spruce T, Alder O, Rodriguez T et al (2010) Differences in the epigenetic and reprogramming properties of pluripotent and extra-embryonic stem cells implicate chromatin remodelling as an important early event in the developing mouse embryo. Epigenetics Chromatin 3:1. https://doi.org/10.1186/1756-8935-3-1

Senovilla L, Vitale I, Galluzzi L, Vivet S, Joza N, Younes AB et al (2009) p53 represses the polyploidization of primary mammary epithelial cells by activating apoptosis. Cell Cycle (Georgetown, Tex) 8(9):1380–1385. https://doi.org/10.4161/cc.8.9.8305

Shlush LI, Hershkovitz D (2015) Clonal evolution models of tumor heterogeneity. In: American Society of Clinical Oncology educational book. American Society of Clinical Oncology. Annual Meeting, pp e662–e665. https://doi.org/10.14694/EdBook_AM.2015.35.e662

Shu Z, Row S, Deng W-M (2018) Endoreplication: the good, the bad, and the ugly. Trends Cell Biol 28(6):465–474. https://doi.org/10.1016/j.tcb.2018.02.006

Sieler M, Weiler J, Dittmar T (2021) Cell-cell fusion and the roads to novel properties of tumor hybrid cells. Cells 10(6). https://doi.org/10.3390/cells10061465

Sikora E, Mosieniak G, Sliwinska MA (2016) Morphological and functional characteristic of senescent cancer cells. Curr Drug Targets 17(4):377–387. https://doi.org/10.2174/1389450116666151019094724

Silkworth WT, Nardi IK, Scholl LM, Cimini D (2009) Multipolar spindle pole coalescence is a major source of kinetochore mis-attachment and chromosome mis-segregation in cancer cells. PLoS One 4(8):e6564. https://doi.org/10.1371/journal.pone.0006564

Sirois I, Aguilar-Mahecha A, Lafleur J, Fowler E, Vu V, Scriver M et al (2019) A unique morphological phenotype in chemoresistant triple-negative breast cancer reveals metabolic reprogramming and PLIN4 expression as a molecular vulnerability. Mol Cancer Res 17(12): 2492–2507. https://doi.org/10.1158/1541-7786.MCR-19-0264

Skinner AM, Grompe M, Kurre P (2012) Intra-hematopoietic cell fusion as a source of somatic variation in the hematopoietic system. J Cell Sci 125(Pt 12):2837–2843. https://doi.org/10.1242/jcs.100123

Sladky VC, Knapp K, Soratroi C, Heppke J, Eichin F, Rocamora-Reverte L et al (2020a) E2F-family members engage the PIDDosome to limit hepatocyte ploidy in liver development and regeneration. Dev Cell 52(3):335–349.e7. https://doi.org/10.1016/j.devcel.2019.12.016

Sladky VC, Knapp K, Szabo TG, Braun VZ, Bongiovanni L, van den Bos H et al (2020b) PIDDosome-induced p53-dependent ploidy restriction facilitates hepatocarcinogenesis. EMBO Rep 21(12):e50893. https://doi.org/10.15252/embr.202050893

Sottile F, Aulicino F, Theka I, Cosma MP (2016) Mesenchymal stem cells generate distinct functional hybrids in vitro via cell fusion or entosis. Sci Rep 6:36863. https://doi.org/10.1038/srep36863

Spees JL, Olson SD, Ylostalo J, Lynch PJ, Smith J, Perry A et al (2003) Differentiation, cell fusion, and nuclear fusion during ex vivo repair of epithelium by human adult stem cells from bone marrow stroma. Proc Natl Acad Sci USA 100(5):2397–2402. https://doi.org/10.1073/pnas.0437997100

Stephens PJ, Greenman CD, Fu B, Yang F, Bignell GR, Mudie LJ et al (2011) Massive genomic rearrangement acquired in a single catastrophic event during cancer development. Cell 144(1): 27–40. https://doi.org/10.1016/j.cell.2010.11.055

Storchová Z, Breneman A, Cande J, Dunn J, Burbank K, O'Toole E, Pellman D (2006) Genome-wide genetic analysis of polyploidy in yeast. Nature 443(7111):541–547. https://doi.org/10.1038/nature05178

Su Y, Subedee A, Bloushtain-Qimron N, Savova V, Krzystanek M, Li L et al (2015) Somatic cell fusions reveal extensive heterogeneity in basal-like breast cancer. Cell Rep 11(10):1549–1563. https://doi.org/10.1016/j.celrep.2015.05.011

Sullivan S, Pells S, Hooper M, Gallagher E, McWhir J (2006) Nuclear reprogramming of somatic cells by embryonic stem cells is affected by cell cycle stage. Cloning Stem Cells 8(3):174–188. https://doi.org/10.1089/clo.2006.8.174

Symeonides M, Murooka TT, Bellfy LN, Roy NH, Mempel TR, Thali M (2015) HIV-1-induced small T cell syncytia can transfer virus particles to target cells through transient contacts. Viruses 7(12):6590–6603. https://doi.org/10.3390/v7122959

Tajima Y, Shibasaki F, Masai H (2024) Cell fusion upregulates PD-L1 expression for evasion from immunosurveillance. Cancer Gene Ther 31(1):158–173. https://doi.org/10.1038/s41417-023-00693-0

Takegahara N, Kim H, Mizuno H, Sakaue-Sawano A, Miyawaki A, Tomura M et al (2016) Involvement of receptor activator of nuclear factor-κB ligand (RANKL)-induced incomplete cytokinesis in the polyploidization of osteoclasts. J Biol Chem 291(7):3439–3454. https://doi.org/10.1074/jbc.M115.677427

Talos F, Nemajerova A, Flores ER, Petrenko O, Moll UM (2007) p73 suppresses polyploidy and aneuploidy in the absence of functional p53. Mol Cell 27(4):647–659. https://doi.org/10.1016/j.molcel.2007.06.036

Terada N, Hamazaki T, Oka M, Hoki M, Mastalerz DM, Nakano Y et al (2002) Bone marrow cells adopt the phenotype of other cells by spontaneous cell fusion. Nature 416(6880):542–545. https://doi.org/10.1038/nature730

Thompson SL, Compton DA (2010) Proliferation of aneuploid human cells is limited by a p53-dependent mechanism. J Cell Biol 188(3):369–381. https://doi.org/10.1083/jcb.200905057

Tijhuis AE, Johnson SC, McClelland SE (2019) The emerging links between chromosomal instability (CIN), metastasis, inflammation and tumour immunity. Mol Cytogenet 12:17. https://doi.org/10.1186/s13039-019-0429-1

Trebichalská Z, Holubcová Z (2020) Perfect date-the review of current research into molecular bases of mammalian fertilization. J Assist Reprod Genet 37(2):243–256. https://doi.org/10.1007/s10815-019-01679-4

Tsubouchi T, Soza-Ried J, Brown K, Piccolo FM, Cantone I, Landeira D et al (2013) DNA synthesis is required for reprogramming mediated by stem cell fusion. Cell 152(4):873–883. https://doi.org/10.1016/j.cell.2013.01.012

Uygur B, Leikina E, Melikov K, Villasmil R, Verma SK, Vary CPH, Chernomordik LV (2019) Interactions with muscle cells boost fusion, stemness, and drug resistance of prostate cancer cells. Mol Cancer Res 17(3):806–820. https://doi.org/10.1158/1541-7786.MCR-18-0500

Vesuna F, van Diest P, Chen JH, Raman V (2008) Twist is a transcriptional repressor of E-cadherin gene expression in breast cancer. Biochem Biophys Res Commun 367(2):235–241. https://doi.org/10.1016/j.bbrc.2007.11.151

Wang Y, Shi J, Chai K, Ying X, Zhou BP (2013) The role of snail in EMT and tumorigenesis. Curr Cancer Drug Targets 13(9):963–972. https://doi.org/10.2174/15680096113136660102

Wang R, Chen S, Li C, Ng KTP, Kong C-w, Cheng J et al (2016) Fusion with stem cell makes the hepatocellular carcinoma cells similar to liver tumor-initiating cells. BMC Cancer 16:56. https://doi.org/10.1186/s12885-016-2094-7

Wang M-J, Chen F, Lau JTY, Hu Y-P (2017) Hepatocyte polyploidization and its association with pathophysiological processes. Cell Death Dis 8(5):e2805. https://doi.org/10.1038/cddis.2017.167

Wang W-J, Li L-Y, Cui J-W (2020) Chromosome structural variation in tumorigenesis: mechanisms of formation and carcinogenesis. Epigenetics Chromatin 13(1):49. https://doi.org/10.1186/s13072-020-00371-7

Wang N, Hao F, Shi Y, Wang J (2021) The controversial role of polyploidy in hepatocellular carcinoma. Onco Targets Ther 14:5335–5344. https://doi.org/10.2147/OTT.S340435

Wangsa D, Quintanilla I, Torabi K, Vila-Casadesús M, Ercilla A, Klus G et al (2018) Near-tetraploid cancer cells show chromosome instability triggered by replication stress and exhibit enhanced invasiveness. FASEB J 32(7):3502–3517. https://doi.org/10.1096/fj.201700247RR

Warmerdam DO, Kanaar R (2010) Dealing with DNA damage: relationships between checkpoint and repair pathways. Mutat Res 704(1–3):2–11. https://doi.org/10.1016/j.mrrev.2009.12.001

Wei H-J, Nickoloff JA, Chen W-H, Liu H-Y, Lo W-C, Chang Y-T et al (2014) FOXF1 mediates mesenchymal stem cell fusion-induced reprogramming of lung cancer cells. Oncotarget 5(19):9514–9529. https://doi.org/10.18632/oncotarget.2413

Weiler J, Dittmar T (2019) Minocycline impairs TNF-α-induced cell fusion of M13SV1-Cre cells with MDA-MB-435-pFDR1 cells by suppressing NF-κB transcriptional activity and its

induction of target-gene expression of fusion-relevant factors. Cell Commun Signal 17(1):1–20. https://doi.org/10.1186/s12964-019-0384-9
Wilkinson PD, Duncan AW (2021) Differential roles for diploid and polyploid hepatocytes in acute and chronic liver injury. Semin Liver Dis 41(1):42–49. https://doi.org/10.1055/s-0040-1719175
Wilkinson PD, Delgado ER, Alencastro F, Leek MP, Roy N, Weirich MP et al (2019) The polyploid state restricts hepatocyte proliferation and liver regeneration in mice. Hepatology 69(3):1242–1258. https://doi.org/10.1002/hep.30286
Wilson SR, Duncan AW (2023) The ploidy state as a determinant of hepatocyte proliferation. Semin Liver Dis 43(4):460–471. https://doi.org/10.1055/a-2211-2144
Xie W, Yu J, Yin Y, Zhang X, Zheng X, Wang X (2022) OCT4 induces EMT and promotes ovarian cancer progression by regulating the PI3K/AKT/mTOR pathway. Front Oncol 12:876257. https://doi.org/10.3389/fonc.2022.876257
Xu M-H, Gao X, Luo D, Zhou X-D, Xiong W, Liu G-X (2014) EMT and acquisition of stem cell-like properties are involved in spontaneous formation of tumorigenic hybrids between lung cancer and bone marrow-derived mesenchymal stem cells. PLoS One 9(2):e87893. https://doi.org/10.1371/journal.pone.0087893
Xuan B, Ghosh D, Cheney EM, Clifton EM, Dawson MR (2018) Dysregulation in actin cytoskeletal organization drives increased stiffness and migratory persistence in polyploidal giant cancer cells. Sci Rep 8(1):11935. https://doi.org/10.1038/s41598-018-29817-5
Xuan B, Ghosh D, Jiang J, Shao R, Dawson MR (2020) Vimentin filaments drive migratory persistence in polyploidal cancer cells. Proc Natl Acad Sci USA 117(43):26756–26765. https://doi.org/10.1073/pnas.2011912117
Xue J, Zhu Y, Sun Z, Ji R, Zhang X, Xu W et al (2015) Tumorigenic hybrids between mesenchymal stem cells and gastric cancer cells enhanced cancer proliferation, migration and stemness. BMC Cancer 15:793. https://doi.org/10.1186/s12885-015-1780-1
Yamanaka S, Blau HM (2010) Nuclear reprogramming to a pluripotent state by three approaches. Nature 465(7299):704–712. https://doi.org/10.1038/nature09229
Yang X, Sun J, Ning Y, Wang J, Xu J, Zhang S (2024) Role of the CTCF/p300 axis in osteochondrogenic-like differentiation of polyploid giant cancer cells with daughter cells. Cell Commun Signal 22(1):546. https://doi.org/10.1186/s12964-024-01933-y
Yao J-Y, Zhang L, Zhang X, He Z-Y, Ma Y, Hui L-J et al (2010) H3K27 trimethylation is an early epigenetic event of p16INK4a silencing for regaining tumorigenesis in fusion reprogrammed hepatoma cells. J Biol Chem 285(24):18828–18837. https://doi.org/10.1074/jbc.M109.077974
Yao J, Zhang L, Hu L, Guo B, Hu X, Borjigin U et al (2016) Tumorigenic potential is restored during differentiation in fusion-reprogrammed cancer cells. Cell Death Dis 7(7):e2314. https://doi.org/10.1038/cddis.2016.189
Yart L, Bastida-Ruiz D, Allard M, Dietrich P-Y, Petignat P, Cohen M (2022) Linking unfolded protein response to ovarian cancer cell fusion. BMC Cancer 22(1):622. https://doi.org/10.1186/s12885-022-09648-4
You B, Xia T, Gu M, Zhang Z, Zhang Q, Shen J et al (2022) AMPK-mTOR-mediated activation of autophagy promotes formation of dormant Polyploid giant cancer cells. Cancer Res 82(5): 846–858. https://doi.org/10.1158/0008-5472.CAN-21-2342
Yu M, Bardia A, Wittner BS, Stott SL, Smas ME, Ting DT et al (2013) Circulating breast tumor cells exhibit dynamic changes in epithelial and mesenchymal composition. Science (New York, NY) 339(6119):580–584. https://doi.org/10.1126/science.1228522
Yuan S, Norgard RJ, Stanger BZ (2019) Cellular plasticity in cancer. Cancer Discov 9(7):837–851. https://doi.org/10.1158/2159-8290.CD-19-0015
Zack TI, Schumacher SE, Carter SL, Cherniack AD, Saksena G, Tabak B et al (2013) Pan-cancer patterns of somatic copy number alteration. Nat Genet 45(10):1134–1140. https://doi.org/10.1038/ng.2760
Zhang S, Mercado-Uribe I, Liu J (2013) Generation of erythroid cells from fibroblasts and cancer cells in vitro and in vivo. Cancer Lett 333(2):205–212. https://doi.org/10.1016/j.canlet.2013.01.037

Zhang S, Mercado-Uribe I, Xing Z, Sun B, Kuang J, Liu J (2014) Generation of cancer stem-like cells through the formation of polyploid giant cancer cells. Oncogene 33(1):116–128. https://doi.org/10.1038/onc.2013.96

Zhang C-Z, Spektor A, Cornils H, Francis JM, Jackson EK, Liu S et al (2015) Chromothripsis from DNA damage in micronuclei. Nature 522(7555):179–184. https://doi.org/10.1038/nature14493

Zhang S, Zhou K, Luo X, Li L, Tu H-C, Sehgal A et al (2018) The Polyploid state plays a tumor-suppressive role in the liver. Dev Cell 47(3):390. https://doi.org/10.1016/j.devcel.2018.10.020

Zhang L-N, Huang Y-H, Zhao L (2019a) Fusion of macrophages promotes breast cancer cell proliferation, migration and invasion through activating epithelial-mesenchymal transition and Wnt/β-catenin signaling pathway. Arch Biochem Biophys 676:108137. https://doi.org/10.1016/j.abb.2019.108137

Zhang L-N, Kong C-F, Zhao D, Cong X-L, Wang S-S, Ma L, Huang Y-H (2019b) Fusion with mesenchymal stem cells differentially affects tumorigenic and metastatic abilities of lung cancer cells. J Cell Physiol 234(4):3570–3582. https://doi.org/10.1002/jcp.27011

Zhang C, Liu J, Xu D, Zhang T, Hu W, Feng Z (2020) Gain-of-function mutant p53 in cancer progression and therapy. J Mol Cell Biol 12(9):674–687. https://doi.org/10.1093/jmcb/mjaa040

Zhang H, Ma H, Yang X, Fan L, Tian S, Niu R et al (2021a) Cell fusion-related proteins and signaling pathways, and their roles in the development and progression of cancer. Front Cell Dev Biol 9:809668. https://doi.org/10.3389/fcell.2021.809668

Zhang Z, Feng X, Deng Z, Cheng J, Wang Y, Zhao M et al (2021b) Irradiation-induced polyploid giant cancer cells are involved in tumor cell repopulation via neosis. Mol Oncol 15(8): 2219–2234. https://doi.org/10.1002/1878-0261.12913

Zheng L, Dai H, Zhou M, Li X, Liu C, Guo Z et al (2012) Polyploid cells rewire DNA damage response networks to overcome replication stress-induced barriers for tumour progression. Nat Commun 3:815. https://doi.org/10.1038/ncomms1825

Zhou X, Platt JL (2011) Molecular and cellular mechanisms of mammalian cell fusion. Adv Exp Med Biol 713:33–64. https://doi.org/10.1007/978-94-007-0763-4_4

Zhou P, Wirthlin L, McGee J, Annett G, Nolta J (2009) Contribution of human hematopoietic stem cells to liver repair. Semin Immunopathol 31(3):411–419. https://doi.org/10.1007/s00281-009-0166-3

Zhou X, Merchak K, Lee W, Grande JP, Cascalho M, Platt JL (2015) Cell fusion connects oncogenesis with tumor evolution. Am J Pathol 185(7):2049–2060. https://doi.org/10.1016/j.ajpath.2015.03.014

Zhou X, Zhou M, Zheng M, Tian S, Yang X, Ning Y et al (2022) Polyploid giant cancer cells and cancer progression. Front Cell Dev Biol 10:1017588. https://doi.org/10.3389/fcell.2022.1017588

Zhu S-T, Wang X, Wang J-Y, Xi G-H, Liu Y (2020) Downregulation of miR-22 contributes to epithelial-mesenchymal transition in osteosarcoma by targeting Twist1. Front Oncol 10:406. https://doi.org/10.3389/fonc.2020.00406

Zybina TG, Zybina EV (2020) Role of cell cycling and polyploidy in placental trophoblast of different mammalian species. Reprod Domest Anim Zuchthygiene 55(8):895–904. https://doi.org/10.1111/rda.13732

Chapter 13
Cell–Cell Fusion as a Mechanism of Immune Evasion and Metastatic Progression

Eduardo López-Collazo, Laura Hurtado-Navarro, and Gloria C. Bonel-Pérez

Abstract The cancer cell fusion theory has emerged in the last few years as a critical yet under-recognised mechanism contributing to tumour progression and metastasis. This chapter revisits its historical foundations and explores the cellular and molecular mechanisms that underpin fusion events. The resulting tumour hybrid cells (THCs) display enhanced migratory and invasive properties and chemoresistant properties, together with profound immunomodulatory capacities. These hybrids often express immune cell markers and stem-like features, enabling them to evade immune surveillance and adapt to hostile microenvironments. Recognition of cell fusion as a driver of tumour evolution challenges classical models of metastasis and highlights the urgent need to identify biomarkers and therapeutic targets associated with fusion-derived hybrids.

Authors Eduardo López-Collazo and Laura Hurtado-Navarro have contributed equally to this chapter.

E. López-Collazo (✉)
The Innate Immune Response Group, IdiPAZ, La Paz University Hospital, Madrid, Spain

Tumour Immunology Laboratory, IdiPAZ, La Paz University Hospital, Madrid, Spain

CIBER of Respiratory Diseases (CIBERES), Madrid, Spain

UNIE University, Madrid, Spain
e-mail: elopezc@salud.madrid.org

L. Hurtado-Navarro (✉)
The Innate Immune Response Group, IdiPAZ, La Paz University Hospital, Madrid, Spain

Tumour Immunology Laboratory, IdiPAZ, La Paz University Hospital, Madrid, Spain

UNIE University, Madrid, Spain
e-mail: laura.hurtado@idipaz.es

G. C. Bonel-Pérez
The Innate Immune Response Group, IdiPAZ, La Paz University Hospital, Madrid, Spain

Tumour Immunology Laboratory, IdiPAZ, La Paz University Hospital, Madrid, Spain

© The Author(s), under exclusive license to Springer Nature Switzerland AG 2026
J. Weiler et al. (eds.), *Cell Fusion*, Advances in Experimental Medicine and Biology 1509, https://doi.org/10.1007/978-3-032-22637-2_13

Keywords Cell fusion · Macrophages · Tumour hybrid cell · Immune evasion · Metastatic traits

Abbreviations

BMDC	Bone marrow-derived cells
CSC	Cancer stem cell
CTC	Circulating tumour cell
EMT	Epithelial-to-mesenchymal transition
HST	Heterokaryon-to-synkaryon transition
IC	Immune checkpoint
MDSC	Myeloid-derived suppressor cells
MET	Mesenchymal–epithelial transition
NK	Natural killer
TAM	Tumour-associated macrophages
THC	Tumour hybrid cells
VEGF	Vascular endothelial growth factor

13.1 Introduction

13.1.1 Historical Context and Metastasis Theories

Cell fusion is a biologically significant process that occurs under both physiological and pathological conditions. During normal development, it is essential for fertilisation, placentation, myogenesis, wound healing, and tissue repair (Hernández and Podbilewicz 2017; Ogle et al. 2005). Such events result in multinucleated syncytia or polyploid cells that perform specialised functions (Camargo et al. 2003; Shabo et al. 2020; Anatskaya and Vinogradov 2007). Osteoclasts, for example, arise through the fusion of monocyte/macrophage precursors and are indispensable for bone resorption, underscoring the direct role of immune cells in physiological fusion processes (Gambari et al. 2020).

Historically, multinucleated cells produced by fusion were first recognised in viral infections such as measles and smallpox (Forkner 1930). These discoveries broadened the understanding of fusion beyond infectious pathology, suggesting a potential role in tumourigenesis and cancer progression.

Metastasis represents a complex multistep process encompassing the detachment of tumour cells from the primary site, invasion of surrounding tissue, intravasation, survival in circulation, extravasation, and colonisation of distant organs (Chaffer and Weinberg 2011; Chambers et al. 2002). To succeed, tumour cells must evade immune detection, resist anoikis, tolerate mechanical stress, and adapt metabolically to unfavourable environments (López-Collazo and Hurtado-Navarro 2025; Hunter et al. 2008). Despite major advances, the molecular mechanisms governing metastasis remain incompletely understood, and early, non-invasive detection of metastatic spread continues to be a major clinical challenge (Rosenkrantz et al. 2018; O'Connor et al. 2017).

Over the past century, several models have sought to explain metastatic behaviour. Paget's seminal *seed-and-soil* hypothesis (1889) proposed that dissemination depends on the interaction between disseminated tumour "seeds" and a receptive organ "soil", thereby introducing the concept of organotropism (Paget 1989; Akhtar et al. 2019). More recently, the epithelial–mesenchymal transition (EMT) model suggested that epithelial tumour cells undergo transcriptional reprogramming, acquiring mesenchymal features that enhance motility and invasiveness (Thiery 2002; Kalluri 2009). However, EMT is inconsistently observed in clinical samples, and many metastatic lesions retain epithelial characteristics (Zheng et al. 2015; Garber 2008). Furthermore, tumour cells frequently manifest a range of EMT phenotypes, conferring reversible phenotypic plasticity through mesenchymal–epithelial transition (MET) (Chaffer and Weinberg 2011).

Additional hypotheses highlighted the role of tumour-associated macrophages (TAMs) in the remodelling of the tumour microenvironment and promoting immune evasion (Bingle et al. 2002; Pollard 2004). Yet, none fully explain how tumour cells acquire the metastatic phenotype. In 1911, the German pathologist Otto Aichel proposed that the fusion of malignant cells with leukocytes could give rise to hybrids with enhanced migratory and immune-evasive capacities (Aichel 1911). This hypothesis, initially speculative, gained support decades later through the work of Mekler and Goldenberg (Mekler 1968; Goldenberg et al. 1967). In 1974, Goldenberg and colleagues provided karyotypic evidence of in vivo fusion between transplanted human tumour cells and hamster host cells, generating highly aggressive hybrid neoplasms (Goldenberg et al. 1974). Subsequent studies confirmed that spontaneous or experimentally induced fusion between tumour cells and bone marrow-derived cells, particularly macrophages, produces hybrids with increased capacities for dissemination, immune evasion, and adaptation to distant tissues (Levin et al. 2010; Sutton et al. 2022; Montalbán-Hernández et al. 2022; Aguirre et al. 2020; Brito et al. 2021; Gast et al. 2018; Powell et al. 2011).

13.1.2 Cancer Cell Fusion: Definition, Mechanisms, and Macrophage Involvement

Cell–cell fusion, otherwise referred to as somatic hybridisation, is defined as the process by which two or more distinct cells combine to form a single hybrid cell (López-Collazo and Hurtado-Navarro 2025). This process is characterised by a series of sequential steps that are meticulously regulated (López-Collazo and Hurtado-Navarro 2025): these include initial contact and adhesion, the fusion of both plasma membranes with the consequent mixing of cytoplasmic contents, and, in certain cases, the fusion of nuclei (Brukman et al. 2019; Chernomordik and Kozlov 2008). The immediate outcome of this process is frequently a heterokaryon, defined as a multinucleated cell containing separate parental nuclei. These can either persist in this state or undergo a heterokaryon-to-synkaryon transition (HST), leading to a mononucleated hybrid (synkaryon) in which chromosomal material from both fusion partners is combined (Dörnen et al. 2020; Sieler et al. 2021). This process frequently results in polyploidy, which, while physiologically relevant in certain contexts, contributes to genomic instability when dysregulated—particularly in tumours.

At the molecular level, cell fusion is not a spontaneous property; rather, it requires that cells transition into a pro-fusogenic state, a process orchestrated by multiple molecular players. At the core of this process are fusogens, a specialised class of proteins that facilitate the merging of lipid bilayers by overcoming membrane repulsion and catalysing the formation of fusion intermediates. These proteins have been demonstrated to facilitate the alignment, dehydration, and destabilisation of adjacent membranes, thereby opening a fusion pore (Brukman et al. 2019; Segev et al. 2018). In mammals, only a few fusogens have been clearly identified: syncytins in placental development (Oren-Suissa and Podbilewicz 2007; Muir et al. 2006), myomaker and myomerger in muscle differentiation (Millay et al. 2013; Leikina et al. 2018), and the IZUMO1–JUNO complex in gamete fusion (Inoue et al. 2005; Bianchi and Wright 2015). Despite the established association of these proteins with physiological cell–cell fusion, the specific fusogens responsible for pathological fusion events, such as those observed in cancer, remain to be fully elucidated. However, it is important to note that inflammation, hypoxia, and tissue injury—all of which are characteristic of the tumour microenvironment—have been shown to enhance fusogenic signalling and membrane fusion competence (Noubissi et al. 2015; Melzer et al. 2021; Johansson et al. 2008; Davies et al. 2009).

Within the context of tumours, macrophages have been identified as the primary immune cells capable of engaging in cell–cell fusion with cancer cells (Powell et al. 2011; Huysentruyt et al. 2008). In comparison with other haematopoietic lineages, such as B and T lymphocytes, macrophages exhibit a markedly elevated fusogenic capacity (Sutton et al. 2022; Powell et al. 2011). The fusion of macrophages with tumour cells gives rise to the formation of tumour hybrid cells (THCs), which exhibit characteristics derived from both their parental lineages (Gast et al. 2018; Clawson et al. 2017). These hybrids have been shown to gain enhanced migratory and invasive abilities, increased plasticity, and even immune-evasive features derived from the macrophage lineage, such as the expression of leukocyte adhesion molecules and chemokine receptors (Montalbán-Hernández et al. 2022; Aguirre et al. 2020). Thus, cell fusion provides a non-mutational, rapid mechanism for the acquisition of metastatic and immune-resistant phenotypes.

13.2 Metastatic Potential Conferred to Cellular Fusion

Cell fusion involving cancer and immune cells is an unpredictable and highly variable biological process, whose outcome—whether the resultant hybrid is viable, functional, or detrimental—cannot be anticipated beforehand. As a result, the population of THCs that arises from this process comprises multiple clonal populations, each displaying unique phenotypic and functional properties. While certain hybrids may acquire enhanced malignant traits, others can display impaired metastatic potential or lower viability compared to their parental tumour cells (Sieler et al. 2021). Nevertheless, this diversity increases the likelihood of generating variants with clinically significant properties.

THCs have been shown to adopt myeloid-like functions such as chemotactic migration, invasiveness, resistance to chemotherapy, and mechanical resilience (Pawelek and Chakraborty 2008; Pawelek et al. 2006). Fusion events thus influence not only tumour cell behaviour but also the wider tumour microenvironment, modulating immune responses, angiogenesis, and stromal interactions.

13.2.1 Acquisition of Migratory and Invasive Phenotypes Through Fusion

Multiple experimental studies have demonstrated that the fusion of tumour cells with bone marrow-derived cells (BMDCs), particularly macrophages, gives rise to hybrid cells that exhibit markedly elevated migratory activity and invasive behaviour (Aguirre et al. 2020; Gast et al. 2018; Montalbán-Hernández et al. 2022; Ramakrishnan et al. 2013). One potential explanation for this phenomenon is that it is related to the intrinsic characteristics of macrophages. These cells are programmed to move dynamically through dense extracellular structures and inflamed tissue compartments. When fused with tumour cells, this migratory programme may be partially transferred, enabling the tumour hybrids to display more flexible cytoskeletal architecture, amoeboid-like motility, and an improved capacity to respond to chemotactic signals (Pawelek and Chakraborty 2008; Cendrowicz et al. 2021). These characteristics offer a selective advantage for tissue infiltration and may facilitate pivotal steps in the metastatic cascade, including intravasation and colonisation of distant organs.

This phenotype has been observed in several tumour types, including colorectal, pulmonary, melanoma, and oral squamous cell carcinomas (Aguirre et al. 2020; Montalbán-Hernández et al. 2022; Ramakrishnan et al. 2013; Xia et al. 2024; Dietz et al. 2021). For example, THCs derived from colon cancer–macrophage fusion has been shown to result in a marked increase in invasive capabilities (Gast et al. 2018). Furthermore, in models of lung, melanoma, and ovarian cancer, the fusion products exhibit accelerated proliferation and enhanced chemotactic movement (Aguirre et al. 2020). In such cases, the hybrids frequently exhibit overexpression of CXCR4, a chemokine receptor that is associated with augmented migratory capacity (Ramakrishnan et al. 2013; Anderson et al. 2024). Likewise, in vitro studies utilising Boyden chamber assays have supported that hybrids derived from MC38 colon cancer cells demonstrate enhanced motility in comparison to non-fused cells (Gast et al. 2018).

At the molecular level, the process of fusion might result in the upregulation of adhesion molecules and integrins, including ICAM-1 and VCAM-1 (Song et al. 2014). These molecules contribute to dynamic interactions with the tumour stroma and endothelial surfaces. The structural features that are characteristic of macrophages, such as lamellipodia and podosome-like projections, have also been observed in tumour hybrids (Faust et al. 2019). Furthermore, melanoma-derived fusion hybrids exhibit elevated expression of GnT-V, a glycosyltransferase known to modify surface receptors in ways that promote migration and facilitate metastatic spread (Nagae et al. 2018). This further supports their ability to migrate directionally and transmigrate through tissue barriers.

Additionally, THCs may remodel the tumour microenvironment by secreting signalling molecules such as CCL12, recruiting myeloid-derived suppressor cells (MDSCs), and preparing distant pre-metastatic niches (Liu et al. 2021; Shang et al. 2022).

13.2.2 Pro-angiogenic Properties

The fusion of tumour cells with myeloid-derived cells, particularly macrophages and dendritic cells, has been demonstrated to contribute to tumour angiogenesis, which is a critical factor in the progression of metastasis.

For instance, the fusion of dendritic cells with endothelial progenitor cells has been observed to yield hybrid cells that exhibit accelerated proliferation and a significantly augmented capacity to form microvessels (Shen et al. 2019). In addition, in murine sarcoma models, hybrids generated from MethA tumour cells and macrophages exhibited increased intratumoural microvessel density and vessel maturation (Busund et al. 2002). These hybrids secreted elevated levels of angiogenic factors, such as vascular endothelial growth factor (VEGF), which exceeded the levels produced by either parental cell type (Busund et al. 2002). Thus, THCs may acquire the pro-angiogenic machinery of their immune precursors, actively promoting vascular growth and remodelling.

13.3 Immune Evasion Through Fusion-Derived Cellular Plasticity

The capacity of tumour cells to evade the detection by the immune system is a defining factor in the progression and metastasis of cancer. While numerous mechanisms have been described in the literature, including alterations in antigen loss, immune checkpoint expression, and changes in the tumour microenviroment (TME), recent evidence suggests that cancer cell fusion may represent an additional, non-mutational mechanism of immune escape.

Fusion-derived THCs frequently acquire phenotypic and functional traits from their immune cell fusion partners, thereby enhancing their plasticity. This observed adaptability may facilitate the adoption of immune-like behaviours by THCs, the modulation of surface marker expression, and the resistance of immunological clearance, amongst other potential outcomes. The subsequent sections will examine the strategies by which THCs circumvent the immune system.

13.3.1 Mechanisms of Immune Escape in Fusion-Derived Tumour Cells

A critical challenge in cancer progression is the ability of tumour cells to avoid recognition and elimination by the immune system. The fusion of tumour cells and immune cells, specifically macrophages and monocytes, has emerged as a promising avenue for achieving this objective.

As demonstrated by experimental models, hybrids derived from cell fusion present an augmented capacity to withstand immune attack. Concretely, the THCs exhibit the ability to evade destruction by cytotoxic T cells, to suppress the perforin production of natural killer (NK) cells and to mitigate the proliferative responses of both $CD4^{+}$ and $CD8^{+}$ lymphocytes (López-Collazo and Hurtado-Navarro 2025; Cozzo et al. 2023). These processes might be associated with the upregulation of immune checkpoints (ICs), including CTLA-4, PD-1, and SIGLEC5, all of which have been identified as inhibitors of immune effector function (Aguirre et al. 2020; Tanjak et al. 2024). Likewise, in lung cancer cell–monocyte hybrids, the upregulation of HLA-B and HLA-E by THCs may also contribute to the inhibition of NK cell function (Aguirre et al. 2020; Beelen et al. 2023).

Also, tumour hybrids display resistance to inflammatory cytokines, such as TNF-α or TGF-β, which may render them more likely to survive in hostile immune environments (Aguirre et al. 2020; Gast et al. 2018).

Moreover, in circulation, clusters of hybrids are frequently co-localised with circulating tumour cells (CTCs). The identification of these complexes is typically facilitated by their characteristic multinucleated structure and expression of myeloid markers. They are known to benefit from anoikis-suppressing signals, including downregulation of major histocompatibility complex (MHC) class II expression, which affords them protection against apoptosis and immune attack (Cheung and Ewald 2016; Cheung et al. 2016).

13.3.2 Immune Identity Transfer Through Cancer Cell Fusion

If the fusion process is to be understood as a mechanism of immune evasion by tumour cells, it is logical to hypothesise that these hybrids must acquire molecular traits typical of the immune cells with which they fuse. Supporting this, several studies have identified THCs through the co-expression of membrane markers characteristic of both tumour cells and macrophage/monocyte lineages (López-Collazo and Hurtado-Navarro 2025; Cozzo et al. 2023).

A well-documented example of this is the expression of CD163, a scavenger receptor commonly associated with M2-polarised macrophages and tumour-promoting immunosuppressive activity (Shabo et al. 2009). Clinically, elevated CD163 expression has been linked to a poor prognosis, likely due to its role in promoting tumour oxygenation and vascularisation, thereby facilitating tumour growth and immune escape (Plevriti et al. 2024; Ma et al. 2018).

In addition to CD163, THC has also been demonstrated to express other myeloid-lineage markers, including CD14 (Aguirre et al. 2020; Montalbán-Hernández et al. 2022; Calvo et al. 1987). It is noteworthy that Aguirre et al. reported the presence of THCs in lung cancer composed of fused macrophages and tumour cells co-expressing CD36, CD14, and pan-cytokeratins (Aguirre et al. 2020). Others have defined them as $CD45^{+}CD14^{+}EpCAM^{+}$ (Montalbán-Hernández et al. 2022; Kaigorodova et al. 2022). This phenotypic profile lends support to the hypothesis that fusion results in the acquisition of immune-like traits by tumour cells.

The current body of evidence indicates that the fusion of malignant cells with BMDCs results in the formation of hybrid cells that exhibit enhanced migratory capacity, immune evasiveness, and metastatic potential.

13.3.3 Hybrid Cells and the Emergence of Stem Cell Features

A growing body of evidence suggests that cell fusion results in the acquisition of stem-like properties by tumour cells. For instance, fusion-derived hybrids from human breast cancer display enhanced colony-forming capacity in comparison to their non-fused counterparts (Gauck et al. 2017). Moreover, studies utilising murine and human embryonal carcinoma cells have shown that fusion-derived hybrids present activated key transcription factors associated with pluripotency, including OCT4, SOX2, and Nanog (Flasza et al. 2003).

Beyond the link to pluripotency, cell fusion has also been associated with the promotion of EMT, a process which is closely related to stemness and metastatic potential (Celià-Terrassa and Jolly 2020). The fusion of breast cancer cells (e.g. MCF-7 and MDA-MB-231) with macrophages has been observed to result in the acquisition of mesenchymal traits. In detail, the hybrid cells displayed a decrease in E-cadherin expression and an increase in mesenchymal markers, such as Snail1 and Vimentin. Furthermore, these cells have a $CD44^{+}/CD24^{-}$ phenotype, which is widely recognised as a cancer stem cell (CSC)-associated signature (Ding et al. 2012).

Taken together, these observations suggest that cell fusion contributes to phenotypic plasticity and the generation of tumour subpopulations with enhanced stemness, metastatic potential, and immune evasiveness, all of which have been linked to worse clinical outcomes.

13.3.4 Acquired Drug Resistance in Fusion-Generated Tumour Hybrid Cells

Accumulating evidence from both experimental models and clinical observations suggests that cell fusion contributes to the emergence of tumour subpopulations capable of withstanding chemotherapeutic and radiotherapeutic stress.

For instance, in a metastatic colon carcinoma model, cells derived from fusion exhibited resistance to 5-fluorouracil and oxaliplatin—two cornerstone agents in colorectal cancer treatment (Carloni et al. 2013). Furthermore, the fusion of tumour subclones in murine systems resulted in a greater degree of resistance to melphalan and methotrexate in comparison to that observed in either parental population (Carloni et al. 2013).

Clinical studies of breast cancer and primary invasive carcinoma have also shown that hybrid cancer cells co-expressing epithelial markers and mesenchymal stem cell traits, particularly fibroblast activation protein-α and pan-cytokeratin, were presented within chemoresistant metastatic lesions. They concluded that these hybrids appeared in primary tumours, contributing not only to tumour dissemination but also to therapeutic failure (Augimeri et al. 2023). Notably, in glioblastoma, radiation-induced stress has been shown to trigger the formation of resistant multinucleated giant cells, which likely arise through fusion between progenitor tumour cells (Kaur et al. 2015).

Furthermore, cancer cell fusion has been proposed as a plausible origin for recurrent cancer stem cells (rCSCs)—a subset associated with enhanced malignancy and treatment resistance (López-Collazo and Hurtado-Navarro 2025; Dittmar et al. 2009). Conventional models describe the emergence of rCSC to the selection of genetically altered clones under therapeutic resistance. However, fusion-induced genetic recombination and aneuploidy present an alternative mechanism through which such malignant traits may be acquired.

13.4 Concluding Remarks and Future Perspectives

Over the past two decades, our understanding of cancer–cell fusion has undergone a profound transformation. Once regarded as a biological curiosity, it is now recognised as a plausible and consequential mechanism in tumour evolution. This conceptual shift stems not only from advances in experimental modelling—particularly in vivo systems that enable the visualisation and functional tracking of hybrid cells—but also from the increasing detection of fusion-derived cells in human malignancies. The hypothesis that even rare fusion events can profoundly amplify clonal diversity has been further strengthened by mathematical modelling and genomic evidence, positioning cell fusion as a potentially underestimated source of tumour heterogeneity and adaptability (Miroshnychenko et al. 2021).

Emerging studies indicate that THCs may endow malignant clones with enhanced metastatic competence, immune evasion, and resistance to conventional therapies (Carloni et al. 2013; Augimeri et al. 2023). If corroborated in larger, longitudinal cohorts, these findings would establish cancer cell fusion as a central mechanism underpinning therapeutic failure and disease progression, rather than a peripheral or incidental phenomenon.

Nevertheless, significant challenges remain. Current detection methods depend largely on the co-expression of lineage-specific markers derived from both parental cells, an approach that inevitably overlooks subpopulations of "cryptic" or phenotypically unstable hybrids. The true prevalence and functional diversity of THCs in vivo, therefore, remain uncertain (López-Collazo and Hurtado-Navarro 2025; Weiler and Dittmar 2019). Furthermore, it is equally imperative to decipher the molecular circuitry that governs the fusion process itself. This includes the identity of fusogenic proteins, the role of inflammatory and hypoxic signals in promoting hybridisation, and the regulatory checkpoints that may restrain or reverse it.

In the forthcoming years, the integration of single-cell multi-omics, advanced imaging, and functional lineage tracing will be essential to elucidate the biology of fusion-derived hybrids. The acquisition of such knowledge may ultimately facilitate the development of diagnostic tools and therapeutic strategies that are targeted at the hybrid state. This concept has the potential to redefine our approach to metastasis, immune escape, and therapeutic resistance in cancer.

In understanding how cells merge, we gain insight into how tumours adapt, survive, and evolve. The study of cell fusion thus exceeds the limits of molecular biology, providing a new perspective on the fundamental logic of cancer—its ability to utilise disorder to ensure its own survival. The mapping of the hybrid frontier may offer insights into the emerging paradigm within the field of oncology.

Acknowledgements We would like to thank the entire research team of the Immune Response group and the Tumour Immunology laboratory at IdiPAZ (La Paz University Hospital) for their work over more than two decades on the topic addressed here, and their comments and suggestions in the development of this project.

Funding The author(s) declare that financial support was received for the research, authorship, and/or publication of this article. They have been supported by funds from the Instituto de Salud Carlos III (ISCiii, PI21/00869), Fundación para la Investigación Biomédica del Hospital Universitario La Paz (FIBHULP), and Fundación Familia Alonso.

References

Aguirre LA, Montalbán-Hernández K, Avendaño-Ortiz J, Marín E, Lozano R, Toledano V et al (2020) Tumor stem cells fuse with monocytes to form highly invasive tumor-hybrid cells. Onco Targets Ther 9(1):1773204

Aichel O (1911) Vorträge und Aufsätze über Entwickelungsmechanik der Organismen: Über Zellverschmelzung mit qualitativ abnormer Chromosomenverteilung als Ursache der Geschwulstbildung, 1–115p

Akhtar M, Haider A, Rashid S, Al-Nabet ADMH (2019) Paget's "seed and soil" theory of cancer metastasis: an idea whose time has come. Adv Anat Pathol 26(1):69–74

Anatskaya OV, Vinogradov AE (2007) Genome multiplication as adaptation to tissue survival: evidence from gene expression in mammalian heart and liver. Genomics 89(1):70–80

Anderson AN, Conley P, Klocke CD, Sengupta SK, Pang A, Farley HC et al (2024) Detection of neoplastic-immune hybrid cells with metastatic properties in uveal melanoma. Biomark Res 12(1):67

Augimeri G, Gonzalez ME, Paolì A, Eido A, Choi Y, Burman B et al (2023) A hybrid breast cancer/ mesenchymal stem cell population enhances chemoresistance and metastasis. JCI Insight 8-(18) [Internet] [Cited 2025 July 16]. Available from: https://insight.jci.org/articles/view/164216

Beelen NA, Ehlers FAI, Bos GMJ, Wieten L (2023) Inhibitory receptors for HLA class I as immune checkpoints for natural killer cell-mediated antibody-dependent cellular cytotoxicity in cancer immunotherapy. Cancer Immunol Immunother 72(4):797–804

Bianchi E, Wright GJ (2015) Cross-species fertilization: the hamster egg receptor, Juno, binds the human sperm ligand, Izumo1. Philos Trans R Soc B 370(1661):20140101

Bingle L, Brown NJ, Lewis CE (2002) The role of tumour-associated macrophages in tumour progression: implications for new anticancer therapies. J Pathol 196(3):254–265

Brito A, Merle C, Lagarde P, Faustin B, Devin A, Lartigue L et al (2021) Cell fusion enhances energy metabolism of mesenchymal tumor hybrid cells to sustain their proliferation and invasion. BMC Cancer 21(1):863

Brukman NG, Uygur B, Podbilewicz B, Chernomordik LV (2019) How cells fuse. J Cell Biol 218(5):1436–1451

Busund LTR, Killie MK, Bartnes K, Seljelid R (2002) Spontaneously formed tumorigenic hybrids of Meth A sarcoma and macrophages grow faster and are better vascularized than the parental tumor. Int J Cancer 100(4):407–413

Calvo F, Martin P, Jabrane N, De Cremoux P, Magdelenat H (1987) Human breast cancer cells share antigens with the myeloid monocyte lineage. Br J Cancer 56(1):15–19

Camargo FD, Green R, Capetenaki Y, Jackson KA, Goodell MA (2003) Single hematopoietic stem cells generate skeletal muscle through myeloid intermediates. Nat Med 9(12):1520–1527

Carloni V, Mazzocca A, Mello T, Galli A, Capaccioli S (2013) Cell fusion promotes chemoresistance in metastatic colon carcinoma. Oncogene 32(21):2649–2660

Celià-Terrassa T, Jolly MK (2020) Cancer stem cells and epithelial-to-mesenchymal transition in cancer metastasis. Cold Spring Harb Perspect Med 10(7):a036905

Cendrowicz E, Sas Z, Bremer E, Rygiel TP (2021) The role of macrophages in cancer development and therapy. Cancers (Basel) 13(8):1946

Chaffer CL, Weinberg RA (2011) A perspective on cancer cell metastasis. Science 331(6024): 1559–1564

Chambers AF, Groom AC, MacDonald IC (2002) Dissemination and growth of cancer cells in metastatic sites. Nat Rev Cancer 2(8):563–572

Chernomordik LV, Kozlov MM (2008) Mechanics of membrane fusion. Nat Struct Mol Biol 15(7): 675–683

Cheung KJ, Ewald AJ (2016) A collective route to metastasis: seeding by tumor cell clusters. Science 352(6282):167–169

Cheung KJ, Padmanaban V, Silvestri V, Schipper K, Cohen JD, Fairchild AN et al (2016) Polyclonal breast cancer metastases arise from collective dissemination of keratin

14-expressing tumor cell clusters. Proc Natl Acad Sci USA 113(7). Available from: https://pnas.org/doi/full/10.1073/pnas.1508541113
Clawson GA, Matters GL, Xin P, McGovern C, Wafula E, dePamphilis C et al (2017) "Stealth dissemination" of macrophage-tumor cell fusions cultured from blood of patients with pancreatic ductal adenocarcinoma. Batra SK, editor. PLoS One 12(9):e0184451
Cozzo AJ, Coleman MF, Hursting SD (2023) You complete me: tumor cell-myeloid cell nuclear fusion as a facilitator of organ-specific metastasis. Front Oncol 13:1191332
Davies PS, Powell AE, Swain JR, Wong MH (2009) Inflammation and proliferation act together to mediate intestinal cell fusion. Bereswill S, editor. PLoS One 4(8):e6530
Dietz MS, Sutton TL, Walker BS, Gast CE, Zarour L, Sengupta SK et al (2021) Relevance of circulating hybrid cells as a non-invasive biomarker for myriad solid tumors. Sci Rep 11(1):13630
Ding J, Jin W, Chen C, Shao Z, Wu J et al (2012) PLoS One 7(7):e41942
Dittmar T, Nagler C, Schwitalla S, Reith G, Niggemann B, Zänker KS (2009) Recurrence cancer stem cells—made by cell fusion? Med Hypotheses 73(4):542–547
Dörnen J, Sieler M, Weiler J, Keil S, Dittmar T (2020) Cell fusion-mediated tissue regeneration as an inducer of polyploidy and aneuploidy. IJMS 21(5):1811
Faust JJ, Balabiyev A, Heddleston JM, Podolnikova NP, Baluch DP, Chew TL et al (2019) An actin-based protrusion originating from a podosome-enriched region initiates macrophage fusion. Mol Biol Cell 30(17):2254–2267
Flasza M, Shering AF, Smith K, Andrews PW, Talley P, Johnson PA (2003) Reprogramming in inter-species embryonal carcinoma–somatic cell hybrids induces expression of pluripotency and differentiation markers. Cloning Stem Cells 5(4):339–354
Forkner CE (1930) The origin and fate of two types of multi-nucleated giant cells in the circulating blood. J Exp Med 52(2):279–297
Gambari L, Grassi F, Roseti L, Grigolo B, Desando G (2020) Learning from monocyte-macrophage fusion and multinucleation: potential therapeutic targets for osteoporosis and rheumatoid arthritis. Int J Mol Sci 21(17):6001
Garber K (2008) Epithelial-to-mesenchymal transition is important to metastasis, but questions remain. J Natl Cancer Inst 100(4):232–233, 239
Gast CE, Silk AD, Zarour L, Riegler L, Burkhart JG, Gustafson KT et al (2018) Cell fusion potentiates tumor heterogeneity and reveals circulating hybrid cells that correlate with stage and survival. Sci Adv 4(9):eaat7828
Gauck D, Keil S, Niggemann B, Zänker KS, Dittmar T (2017) Hybrid clone cells derived from human breast epithelial cells and human breast cancer cells exhibit properties of cancer stem/initiating cells. BMC Cancer 17(1):515
Goldenberg DM, Müller E, Witte S (1967) In vivo proliferation of heterotransplanted human cancer cells. Eur J Cancer (1965) 3(4–5):315–319
Goldenberg DM, Pavia RA, Tsao MC (1974) In vivo hybridisation of human tumour and normal hamster cells. Nature 250(5468):649–651
Hernández JM, Podbilewicz B (2017) The hallmarks of cell-cell fusion. Development 144(24):4481–4495
Hunter KW, Crawford NP, Alsarraj J (2008) Mechanisms of metastasis. Breast Cancer Res 10(S1). [Internet] [Cited 2025 July 11]. Available from: http://breast-cancer-research.biomedcentral.com/articles/10.1186/bcr1988
Huysentruyt LC, Mukherjee P, Banerjee D, Shelton LM, Seyfried TN (2008) Metastatic cancer cells with macrophage properties: evidence from a new murine tumor model. Int J Cancer 123(1):73–84
Inoue N, Ikawa M, Isotani A, Okabe M (2005) The immunoglobulin superfamily protein Izumo is required for sperm to fuse with eggs. Nature 434(7030):234–238
Johansson CB, Youssef S, Koleckar K, Holbrook C, Doyonnas R, Corbel SY et al (2008) Extensive fusion of haematopoietic cells with Purkinje neurons in response to chronic inflammation. Nat Cell Biol 10(5):575–583

Kaigorodova EV, Kozik AV, Zavaruev IS, Grishchenko MY (2022) Hybrid/atypical forms of circulating tumor cells: current state of the art. Biochem Mosc 87(4):380–390
Kalluri R (2009) EMT: when epithelial cells decide to become mesenchymal-like cells. J Clin Invest 119(6):1417–1419
Kaur E, Rajendra J, Jadhav S, Shridhar E, Goda JS, Moiyadi A et al (2015) Radiation-induced homotypic cell fusions of innately resistant glioblastoma cells mediate their sustained survival and recurrence. Carcinogenesis 36(6):685–695
Leikina E, Gamage DG, Prasad V, Goykhberg J, Crowe M, Diao J et al (2018) Myomaker and Myomerger work independently to control distinct steps of membrane remodeling during myoblast fusion. Dev Cell 46(6):767–780.e7
Levin TG, Powell AE, Davies PS, Silk AD, Dismuke AD, Anderson EC et al (2010) Characterization of the intestinal cancer stem cell marker CD166 in the human and mouse gastrointestinal tract. Gastroenterology 139(6):2072–2082.e5
Liu H, Ling CC, Yeung WHO, Pang L, Liu J, Zhou J et al (2021) Monocytic MDSC mobilization promotes tumor recurrence after liver transplantation via CXCL10/TLR4/MMP14 signaling. Cell Death Dis 12(5):489
López-Collazo E, Hurtado-Navarro L (2025) Cell fusion as a driver of metastasis: re-evaluating an old hypothesis in the age of cancer heterogeneity. Front Immunol 16:1524781
Ma C, Horlad H, Ohnishi K, Nakagawa T, Yamada S, Kitada S et al (2018) CD163-positive cancer cells are potentially associated with high malignant potential in clear cell renal cell carcinoma. Med Mol Morphol 51(1):13–20
Mekler LB (1968) Hybridization of transformed cells with lymphocytes as 1 of the probable causes of the progression leading to the development of metastatic malignant cells. Vestn Akad Med Nauk SSSR 26(8):80–89
Melzer C, Von Der Ohe J, Luo T, Hass R (2021) Spontaneous fusion of MSC with breast cancer cells can generate tumor dormancy. IJMS 22(11):5930
Millay DP, O'Rourke JR, Sutherland LB, Bezprozvannaya S, Shelton JM, Bassel-Duby R et al (2013) Myomaker is a membrane activator of myoblast fusion and muscle formation. Nature 499(7458):301–305
Miroshnychenko D, Baratchart E, Ferrall-Fairbanks MC, Velde RV, Laurie MA, Bui MM et al (2021) Spontaneous cell fusions as a mechanism of parasexual recombination in tumour cell populations. Nat Ecol Evol 5(3):379–391
Montalbán-Hernández K, Cantero-Cid R, Casalvilla-Dueñas JC, Avendaño-Ortiz J, Marín E, Lozano-Rodríguez R et al (2022) Colorectal cancer stem cells fuse with monocytes to form tumour hybrid cells with the ability to migrate and evade the immune system. Cancer 14(14): 3445
Muir A, Lever AML, Moffett A (2006) Human endogenous retrovirus-W envelope (syncytin) is expressed in both villous and extravillous trophoblast populations. J Gen Virol 87(7): 2067–2071
Nagae M, Kizuka Y, Mihara E, Kitago Y, Hanashima S, Ito Y et al (2018) Structure and mechanism of cancer-associated N-acetylglucosaminyltransferase-V. Nat Commun 9(1). [Internet]. [Cited 2025 Jul 16] Available from: https://www.nature.com/articles/s41467-018-05931-w
Noubissi FK, Harkness T, Alexander CM, Ogle BM (2015) Apoptosis-induced cancer cell fusion: a mechanism of breast cancer metastasis. FASEB J 29(9):4036–4045
O'Connor JPB, Aboagye EO, Adams JE, Aerts HJWL, Barrington SF, Beer AJ et al (2017) Imaging biomarker roadmap for cancer studies. Nat Rev Clin Oncol 14(3):169–186
Ogle BM, Cascalho M, Platt JL (2005) Biological implications of cell fusion. Nat Rev Mol Cell Biol 6(7):567–575
Oren-Suissa M, Podbilewicz B (2007) Cell fusion during development. Trends Cell Biol 17(11): 537–546
Paget S (1989) The distribution of secondary growths in cancer of the breast. 1889. Cancer Metastasis Rev 8(2):98–101

Pawelek JM, Chakraborty AK (2008) The cancer cell—leukocyte fusion theory of metastasis. Adv Cancer Res 101:397–444

Pawelek J, Chakraborty A, Lazova R, Yilmaz Y, Cooper D, Brash D et al (2006) Co-opting macrophage traits in cancer progression: a consequence of tumor cell fusion? In: Dittmar T, Zaenker KS, Schmidt A (eds) Contributions to microbiology [internet]. KARGER, Basel, pp 138–155. [Cited 2024 July 26]. Available from: https://karger.com/books/book/2535/chapter/5741821

Plevriti A, Lamprou M, Mourkogianni E, Skoulas N, Giannakopoulou M, Sajib MS et al (2024) The role of soluble CD163 (sCD163) in human physiology and pathophysiology. Cells 13(20):1679

Pollard JW (2004) Tumour-educated macrophages promote tumour progression and metastasis. Nat Rev Cancer 4(1):71–78

Powell AE, Anderson EC, Davies PS, Silk AD, Pelz C, Impey S et al (2011) Fusion between intestinal epithelial cells and macrophages in a cancer context results in nuclear reprogramming. Cancer Res 71(4):1497–1505

Ramakrishnan M, Mathur SR, Mukhopadhyay A (2013) Fusion-derived epithelial cancer cells express hematopoietic markers and contribute to stem cell and migratory phenotype in ovarian carcinoma. Cancer Res 73(17):5360–5370

Rosenkrantz AB, Duszak R, Babb JS, Glover M, Kang SK (2018) Discrepancy rates and clinical impact of imaging secondary interpretations: a systematic review and meta-analysis. J Am Coll Radiol 15(9):1222–1231

Segev N, Avinoam O, Podbilewicz B (2018) Fusogens. Curr Biol 28(8):R378–R380

Shabo I, Olsson H, Sun X, Svanvik J (2009) Expression of the macrophage antigen CD163 in rectal cancer cells is associated with early local recurrence and reduced survival time. Int J Cancer 125(8):1826–1831

Shabo I, Svanvik J, Lindström A, Lechertier T, Trabulo S, Hulit J et al (2020) Roles of cell fusion, hybridization and polyploid cell formation in cancer metastasis. World J Clin Oncol 11(3):121–135

Shang C, Sun Y, Wang Y, Shi H, Han X, Mo Y et al (2022) CXCL10 conditions alveolar macrophages within the premetastatic niche to promote metastasis. Cancer Lett 537:215667

Shen Z, Gu X, Mao W, Cao H, Zhang R, Zhou Y et al (2019) Dendritic cells fused with endothelial progenitor cells play immunosuppressive effects on angiogenesis in acute myeloid leukemia mice. Am J Transl Res 11(5):2816–2829

Sieler M, Weiler J, Dittmar T (2021) Cell–cell fusion and the roads to novel properties of tumor hybrid cells. Cells 10(6):1465

Song K, Song Y, Zhao XP, Shen H, Wang M, Yan TL et al (2014) Oral cancer/endothelial cell fusion experiences nuclear fusion and acquisition of enhanced survival potential. Exp Cell Res 328(1):156–163

Sutton TL, Patel RK, Anderson AN, Bowden SG, Whalen R, Giske NR et al (2022) Circulating cells with macrophage-like characteristics in cancer: the importance of circulating neoplastic-immune hybrid cells in cancer. Cancer 14(16):3871

Tanjak P, Chaiboonchoe A, Suwatthanarak T, Thanormjit K, Acharayothin O, Chanthercrob J et al (2024) Tumor-immune hybrid cells evade the immune response and potentiate colorectal cancer metastasis through CTLA4. Clin Exp Med 25(1):2

Thiery JP (2002) Epithelial–mesenchymal transitions in tumour progression. Nat Rev Cancer 2(6):442–454

Weiler J, Dittmar T (2019) Cell fusion in human cancer: the dark matter hypothesis. Cells 8(2):132

Xia C, Zhang Q, Pu Y, Hu Q, Wang Y (2024) Cell fusion between tumor cells and macrophages promotes the metastasis of OSCC patient through the activation of the chemokine signaling pathway. Cancer Med 13(4). [Internet] [Cited 2025 July 16]. Available from: https://onlinelibrary.wiley.com/doi/10.1002/cam4.6940

Zheng X, Carstens JL, Kim J, Scheible M, Kaye J, Sugimoto H et al (2015) Epithelial-to-mesenchymal transition is dispensable for metastasis but induces chemoresistance in pancreatic cancer. Nature 527(7579):525–530

Chapter 14
Implications of Heterotypic Cell Fusion in Cancer

Kathryn L. Fowler, Ashley N. Anderson, Taelor J. Ekstrom, Nicole R. Giske, Abigail Moore, Ranish K. Patel, Amanda Zucker, and Melissa H. Wong

Abstract The interplay between discrete neoplastic cells and cells within the tumor microenvironment represents a critical component of cancer evolution that ultimately impacts patient outcomes. A newly identified neoplastic cell that is a product of cell fusion between neoplastic and immune cells has emerged as a key player in metastatic spread. Indeed, these hybrid cells are detected across the metastatic cascade: within the primary tumor, disseminated into the peripheral blood, and in metastatic sites. While multiple mechanisms of hybrid cells generation have been posited, cell fusion has the most rigorous prior study. Most commonly between tumor cells and macrophages, fusion progeny are identified by the co-expression of parental cell type features (e.g., epithelial proteins and immune cell antigens). After fusion, they become reprogrammed to be distinct from other neoplastic cells and demonstrate properties that enhance their ability to successfully travel through the metastatic cascade. As such, cell fusion hybrids have more recently gained traction as a means of tumorigenesis, metastatic spread, and treatment resistance. Importantly, investigators have demonstrated clinical utility for disseminated tumor-immune hybrids, called circulating hybrid cells (CHCs) in peripheral blood, for early detection, metastatic risk stratification, treatment resistance, and disease progression. Herein, we review the study of tumor hybrid cells as a new frontier of cancer biology and as a developing biomarker to enhance metrics for cancer patient care and survival.

K. L. Fowler · R. K. Patel
Department of Surgery, Oregon Health & Science University (OHSU), Portland, OR, USA
e-mail: fowlekat@ohsu.edu; patera@ohsu.edu

A. N. Anderson · T. J. Ekstrom · N. R. Giske · A. Moore · A. Zucker
Department of Cell, Developmental and Cancer Biology, OHSU, Portland, OR, USA
e-mail: barling@ohsu.edu; ekstromt@ohsu.edu; giske@ohsu.edu; mooreab@ohsu.edu; zuckeram@ohsu.edu

M. H. Wong (✉)
Department of Cell, Developmental and Cancer Biology, OHSU, Portland, OR, USA

Knight Cancer Institute, OHSU, Portland, OR, USA
e-mail: wongme@ohsu.edu

© The Author(s), under exclusive license to Springer Nature Switzerland AG 2026
J. Weiler et al. (eds.), *Cell Fusion*, Advances in Experimental Medicine and Biology 1509, https://doi.org/10.1007/978-3-032-22637-2_14

Keywords Circulating hybrid cells · Tumor-immune hybrid cells · Cell fusion · Translational research

Abbreviations

CA 19-9	Carbohydrate antigen 19-9
CAMLs	Cancer-associated macrophage-like cells
CCC	Circulating cancer cell
CEA	Carcinoembryonic antigen
CHC	Circulating hybrid cell
CRC	Colorectal cancer
CTC	Circulating tumor cell
EMT	Epithelial-to-mesenchymal transition
GFP	Green fluorescent protein
PDAC	Pancreatic ductal adenocarcinoma
PGCC	Polyploid giant cancer cell
RFP	Red fluorescent protein
TME	Tumor microenvironment

14.1 Introduction

Tumor-immune cell fusion hybrids are a recently identified neoplastic cell type with emerging data supporting their unique biology and clinical utility for disease monitoring (Sutton et al. 2019, 2022; Dietz et al. 2021; Gast et al. 2018). Categorized as heterotypic cell fusion, tumor-immune cell fusion in oncology is considered similar, but related to, the dysregulation of other physiologic types of cell fusion (Noubissi et al. 2024). Indeed, heterotypic cell fusion is fundamental for all life, with the union of egg and sperm giving rise to diverse organisms. Cell fusion is an established and fundamental mechanism in embryonic development and is conserved in adults to drive functional muscle and bone formation (Lehka and Redowicz 2020; Hindi et al. 2013; Horsley and Pavlath 2004; Yahara et al. 2022; Jansen et al. 2012; Soe 2020). Cell fusion also functions in tumor progression, where developmental mechanisms are hijacked by the tumor to evade cellular restraints. Tumor cell fusion produces a novel cell population, and research has demonstrated that this population has an important role in tumor progression and metastatic spread of cancer (Sutton et al. 2019, 2022; Patel et al. 2024b). While predominantly described as a heterotypic fusion between a neoplastic cell and an immune cell, homotypic fusion between two neoplastic cells has also been described (Demin et al. 2022; Dittmar et al. 2021; Hass et al. 2021a, b; Manjunath et al. 2020b; Wang et al. 2021). Tumor-immune hybrid cells (Fig. 14.1) are detected across the metastatic cascade in the primary tumor, peripheral blood, and metastatic sites, and thus have biologic significance in driving malignant progression. These cells, once disseminated into peripheral blood, where

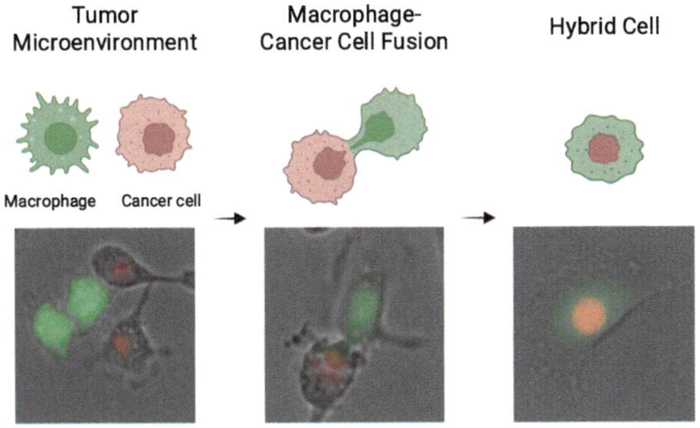

Fig. 14.1 Macrophage–cancer cell fusion results in hybrid cell identity. Within the tumor microenvironment, macrophages expressing green fluorescent protein (GFP, green) and tumor cells expressing red fluorescent protein (RFP, red) spontaneously fuse in cell culture. The resulting hybrid cell co-expresses GFP and RFP, demonstrating dual tumor and immune identity

they are termed circulating hybrid cells (CHCs), have promise to provide prognostic and predictive value for patients with cancer. Herein, we review the characteristics of tumor-immune cell hybrids, known mechanisms of cell fusion, the presence of this unique cell population across the metastatic cascade, and their clinical utility. This tumor cell population demonstrates great promise as a new frontier in elucidating cancer biology, discovering novel therapeutic targets, and improving early oncology detection and monitoring diagnostics.

14.2 Shared Mechanisms of Cell Fusion in Development and Cancer

Cell fusion plays essential roles in developmental biology and homeostasis to enhance diversity and provide essential cellular functions. In addition, cell fusion has an emerging impact in oncology, contributing to tumor progression, heterogeneity, and metastasis. Mechanistically, cell fusion involves the joining of two separate cells, including their membranes and cellular contents (i.e., nuclei, mitochondria, and lysosomes). Overall, whether it is egg–sperm, myoblast, or tumor-immune cell fusion, the process of cell fusion imparts the resulting progeny the ability to gain enhanced functional properties, ultimately gaining the ability to adapt to diverse microenvironments.

14.2.1 Physiologic Cell Fusion

Cell fusion can be broadly categorized based on the parent cell types involved and post-fusion nuclear features. Homotypic fusion constitutes fusion between two like cells, while heterotypic fusion occurs between different cell types. Fusion results in a synkaryon—a cell with a single nucleus—or a heterokaryon—a cell with multiple nuclei (Sabe et al. 2024). The most well-recognized example of physiologic cell fusion is a heterotypic fusion between egg and a sperm during fertilization to form a single cell called the zygote (Hernandez and Podbilewicz 2017). Throughout development and in tissue regeneration, cell fusion continues to be an essential process within the body. During placental development, the syncytiotrophoblast—the barrier between the maternal and fetal blood supply—is formed by the fusion of cytotrophoblast cells (Potgens et al. 2002; Oren-Suissa and Podbilewicz 2007; Renaud and Jeyarajah 2022). Additionally, skeletal muscles form their fibers when myoblast cells undergo homotypic fusion to form myotubes (Lehka and Redowicz 2020; Hindi et al. 2013; Horsley and Pavlath 2004). Newly formed myotubes then continue to fuse with surrounding myoblasts. In a similar process, bone formation, repair, and resorption also occur by the fusion of mononuclear osteoclasts, forming multinucleated osteoclasts that are key to bone homeostasis (Yahara et al. 2022; Jansen et al. 2012; Soe 2020).

On the cellular level, fusion between two cells involves the reconstruction of the cell membrane and the merging of cellular contents. This process is highly regulated by both internal and external factors via the expression of specific fusogens, including syncytins, phosphatidylserine, inflammatory mediators, and microenvironmental factors, such as hypoxia (Grandi and Tramontano 2018; Blond et al. 2000; Hamoud et al. 2014; Hochreiter-Hufford et al. 2013; Noubissi and Ogle 2016; Noubissi et al. 2015; Davies et al. 2009; Yan et al. 2017; Song et al. 2012). Dysregulation of this process contributes to diseases, including osteoporosis, muscle malformation, infertility, fetal growth restriction, preeclampsia, and cancer (Zhou et al. 2015b; Buckingham 2006; Horsley and Pavlath 2004; Chen et al. 2006; Langbein et al. 2008; Lee et al. 2001; Bastida-Ruiz et al. 2016; Clark and Naz 2013; Yu et al. 2018; Wang et al. 2022; Ruebner et al. 2013; Huang et al. 2014; Knerr et al. 2004). In general, the steps occurring during fusion, regardless of cell type, involve priming, chemotaxis, membrane adhesion, membrane fusion, and post-fusion resetting (Zhou and Platt 2011). In addition, a pre-hybrid preparation and post-hybrid selection process has been identified, in which post-fusion selection is believed to address the aneuploidy–genomic instability that results from cell–cell fusion (Hass 2020; Melzer et al. 2020; Hass et al. 2021a; Hernandez and Podbilewicz 2017; Chunduri and Storchova 2019; Holland and Cleveland 2009). This is the mechanism by which hybrid cells stabilize chromosomal imbalances, and thus this process plays an important role in the fate of the hybrid cell. It is thought that the segregation of chromosomes after fusion results in multiple centrosomes, chromosomal missegregation, aneuploidy, and chromothripsis (Hass et al. 2021a; Song et al. 2021; Chunduri and Storchova 2019; Holland and Cleveland 2009; Duncan et al. 2009; Ganem et al. 2009; Godinho et al. 2009; Passerini et al. 2016; Duncan et al. 2010). Thus, the redistribution of chromosomes and genetic material might be stochastic and unpredictable, and therefore most hybrid cells do not survive the resulting genetic rearrangement (Do et al. 2017; Miroshnychenko et al. 2021). When investigating cancer cell fusion, important conserved features have been demonstrated between physiology and oncology, including the upregulation of similar fusogens and the regulation of fusion via conserved pathways. However, given the inherent heterogeneity associated with neoplastic states, including tumor cell plasticity and differences in tumor microenvironmental cues, there are mechanistic complexities that have yet to be fully elucidated.

14.2.2 Oncologic Cell Fusion: Physiology and Mechanisms

Cell fusion in oncology has long been purported as a mechanism of neoplastic initiation, cancer progression, or metastatic spread and, more recently, has been investigated as impacting treatment resistance, genetic instability, and tumor heterogeneity (Kerbel et al. 1983; Lu and Kang 2009; Seyfried and Huysentruyt 2013). In 1911, Otto Aichel, a German pathologist, proposed that cancer cells fuse with leukocytes to contribute to tumor initiation and enable the acquisition of migratory properties from an immune cell, enhancing the cancer cell's metastatic potential (Jiang and Wu 2024). He reasoned that cell fusion may explain the observed chromosomal rearrangements in tumor cells, and that by acquiring leukocytic characteristics through fusion, tumor cells could migrate throughout the body, resulting in the metastatic spread of disease. Since then, leveraging both in vitro and in vivo assays as well as human specimens, many studies have demonstrated fusion between primary tumor cells and nearby non-neoplastic cells (including macrophages, fibroblasts, and mesenchymal stem cells) resulting in tumor-fusion hybrids (Fig. 14.1) that have increased metastatic potential (Chakraborty et al. 2004; Mandel et al. 2013; Pawelek 2014; Pawelek and Chakraborty 2008b; Rachkovsky et al. 1998; Yang et al. 2015; Matsumoto et al. 2021; Bielski et al. 2018; Ben-David and Amon 2020; Dornen et al. 2020; Powell et al. 2011).

The mechanisms underlying cell fusion in oncology are not fully elucidated, given the inherent heterogeneity associated with cancer itself and the complexity of the fusion process. However, there are conserved patterns identified in the mechanisms of physiologic cell fusion and oncologic cell fusion. Syncytin-1 and -2, two human endogenous retrovirus envelope genes, are known to be involved in physiologic cell fusion (Grandi and Tramontano 2018; Blond et al. 2000). These genes are upregulated or overexpressed along with their receptors (ASCT1, ASCT2) in cancer and cancer cell lines (Bastida-Ruiz et al. 2016; Fei et al. 2019), including endometrial carcinoma (Liu et al. 2019; Strick et al. 2007; Strissel et al. 2012), breast cancer (Bjerregaard et al. 2006; Larsson et al. 2007), T-cell lymphoma (Maliniemi et al. 2013), leukemia (Sun et al. 2010), urothelial cell carcinoma (Yu et al. 2014), non-small cell lung cancer (Fu et al. 2021; Li et al. 2019), colorectal cancer (Montalban-Hernandez et al. 2022; Larsen et al. 2009), prostate cancer (Uygur et al. 2019), and hepatocellular carcinoma (Zhou et al. 2021). Inhibition of syncytins or their receptors using an antisense oligonucleotide or inhibitory peptides prevents fusion in cancer cells (Strick et al. 2007; Wang et al. 2015). In addition, phosphatidylserine, normally a component of the inner leaflet of eukaryotic plasma membranes, is associated with cell fusion when translocated to the outer leaflet of the membrane, activating the ELMO-Dock pathway (Hamoud et al. 2014; Hochreiter-Hufford et al. 2013; Noubissi et al. 2015; Noubissi and Ogle 2016; Li et al. 2013; Abu-Thuraia et al. 2015) in myoblast, macrophage, and egg–sperm fusion (Van Den Eijnde et al. 2001; Jeong and Conboy 2011; Faust et al. 2019; Adler et al. 1995; Das et al. 2004; Zigon et al. 2015; Rival et al. 2019). In cancer, inhibition of phosphatidylserine prevented fusion between breast cancer cells and mesenchymal/multipotent stem/stromal cells (Noubissi et al. 2015; Noubissi and Ogle 2016), highlighting conserved mechanisms.

Extracellular or microenvironmental factors, including inflammation, inflammatory cytokines, and cell stress, including hypoxia, also contribute to fusion in physiologic and neoplastic states (Davies et al. 2009; Melzer et al. 2018; Weiler and Dittmar 2019; Yan et al. 2017; Song et al. 2012; Mohr et al. 2015; Skokos et al. 2011; Hotokezaka et al. 2007). Inflammation, hypoxia, and oxidative stress, all features of the tumor microenvironment (TME) of most cancers (Dvorak 1986; Ribatti and Tamma 2018; Balkwill and Mantovani 2001), promote phosphatidylserine externalization (Brukman et al. 2019). Sites of inflammation have long been associated with the development of cancer, and macrophages recruited to inflamed sites have proven to be the primary fusion partners with intestinal epithelial cells (Powell et al. 2011). Tumor necrosis factor alpha (TNFα), a key inflammatory mediator, and other cytokines were shown to enhance fusion in squamous cell carcinoma (Song et al. 2012; Yan et al. 2017) and in breast cancer cells (Mohr et al. 2015; Weiler et al. 2018; Weiler and Dittmar 2019). In addition to inflammation, sites of high epithelial proliferation were shown to promote cell fusion (Davies et al. 2009). Stimulation of the Wnt/β-catenin signaling pathway, which regulates cell proliferation, was shown to contribute to cell fusion in sperm–oocyte adhesion, cytotrophoblast fusion, and human choriocarcinoma cell fusion (Matsuura et al. 2011; Takezawa et al. 2011; Aghababaei et al. 2015). Additionally, in a murine model of colorectal cancer, mice with constitutive activation of Wnt/β-catenin signaling supported enhanced cell fusion and increased tumor burden compared with wild-type mice (Rizvi et al. 2006; Hamilton et al. 2013).

In physiologic conditions, hypoxia promotes fusion between cell types, including bone-marrow–derived mesenchymal stem cells and cardiomyocytes or myoblasts (Haneef et al. 2014; Archacka et al. 2021) as well as urine-derived stem cells and liver cells (Hu et al. 2021). In cancer, hypoxic conditions were shown to increase fusion in breast cancer cells and oral squamous carcinoma cells (Noubissi et al. 2015; Noubissi and Ogle 2016; Huang et al. 2018; Mohr et al. 2015). However, the effect of hypoxia is known to have a differential impact depending on which cell type is present (Yart et al. 2022; Archacka et al. 2021; Huang et al. 2018; Kudo et al. 2003), highlighting the complexity of elucidating the complete mechanism of oncologic cell fusion. While there is an incomplete understanding of neoplastic fusogenic mechanisms, it is clear that conserved mechanistic patterns exist between physiologic and oncologic fusion. Given the presence of fusion cells across the metastatic cascade, increased understanding of how neoplastic cells initiate and undergo fusion may inform future cancer management. Tumor-immune hybrid cells (Fig. 14.1) and their peripheral blood counterparts, CHCs, have garnered increasing interest as a readout for the noninvasive diagnosis and monitoring of oncologic progression as well as potential therapeutic targets (Gast et al. 2018; Lazova et al. 2013; Laberge et al. 2017; Shabo et al. 2009; Lopez-Collazo and Hurtado-Navarro 2025).

14.3 Cellular Fusion Across the Metastatic Cascade

Hybrid cells have been analyzed across the metastatic cascade, from the primary tumor to the peripheral circulation to metastatic sites. Genetic advantages gained through cell fusion can allow for adaptation to the tumor microenvironments. Cell fusion can result in cellular senescence or growth arrest, an important part of tumor suppression and tissue repair (Gal and Krizhanovsky 2014; Chuprin et al. 2013; Bojko et al. 2020), but hybrid clones that survive and escape senescence may have phenotypic advantages that increase their oncologic potential (Rachkovsky et al. 1998; He et al. 2015; Melzer et al. 2019; Gast et al. 2018; Lartigue et al. 2020; Merle et al. 2021; Miroshnychenko et al. 2021; Zhou et al. 2015a; Delespaul et al. 2019, 2020; Hass et al. 2019; Xu et al. 2014; Xue et al. 2015; Zhang et al. 2019b; Yan et al. 2015; Sun et al. 2019; Sodi et al. 1998; Rappa et al. 2012; Miller et al. 1989; Mi et al. 2012; Luo et al. 2016; Duelli and Lazebnik 2007; Duelli et al. 2007; Ding et al. 2012). Many have demonstrated that fusion hybrids in cancer demonstrate increased tumorigenicity, motility, phenotypic heterogeneity, and evasion of immune surveillance—all of which are hallmarks of metastatic spread in cancer (Fig. 14.2) (Hanahan and Weinberg 2000, 2011). Hybrid cells that escape from the primary tumor into the peripheral circulation are even more primed to promote metastasis.

14.3.1 Cell Fusion Within the Primary Tumor

Neoplastic-immune hybrids are identified by co-expression of at least one tumor-specific protein and the pan-leukocyte antigen CD45, but are reported to express a number of different tumor and immune antigens (Table 14.1). Cell fusion may represent a more common process than previously appreciated and may result in aneuploidy and altered chromosomal content, which could contribute to metastatic potential (Matsumoto et al. 2021; Zack et al. 2013; Bielski et al. 2018; Quinton et al. 2021; Dornen et al. 2020; Ben-David and Amon 2020). However, rates of cell fusion have varied across in vivo analyses (0.5–51%) (Miroshnychenko et al. 2021; Yan et al. 2016; Powell et al. 2011; Gast et al. 2018; Melzer et al. 2019; Ramakrishnan et al. 2013; Rizvi et al. 2006) and have not been demonstrated to be a direct cause of widespread tumor aneuploidy. Regardless of the rate of cell fusion within the primary tumor, hybrid cells have been detected using immunofluorescence-based analyses of tumor tissue and in postanalyses of scRNA-sequencing datasets, as well as in murine models of primary tumors (Akhter et al. 2018; Cho et al. 2012; Yang

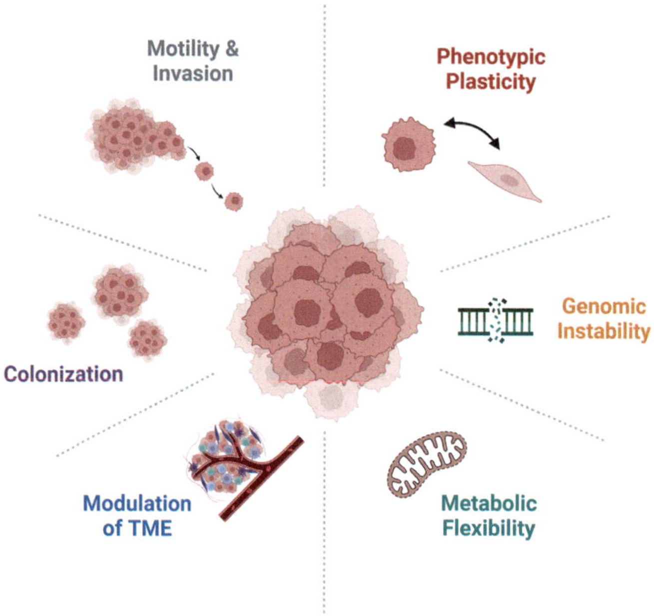

Fig. 14.2 Hallmarks of metastasis progression. Identified features that contribute to the development of metastasis include genomic instability, metabolic flexibility, the ability to modulate the tumor microenvironment (TME), colonization of distant sites, increased motility and invasion, and phenotypic plasticity

Table 14.1 Protein identifiers for circulating hybrid cells across cancer types

Cancer site	CHC protein identifier	References
Ampullary carcinoma	*CD45*, **panCK**	Dietz et al. (2021)
Breast cancer	*CD45*, **CK7, CK19, ECAD**	Allan et al. (2005)
		Dietz et al. (2021)
		Lustberg et al. (2014)
Cholangiocarcinoma	*CD45*, **panCK**	Dietz et al. (2021)
		Patel et al. (2024)
Colon cancer	*CD45*, **EpCAM**, **ECAD**, **panCK**	Dietz et al. (2021)
		Whalen et al. (2024)
Esophageal carcinoma	*CD45*, **panCK**	Walker et al. (2021)
Gastric cancer	*CD45*, **EpCAM**	Toyoshima et al. (2015)
Gastrointestinal stromal tumor	*CD14*, *CD68*, **CSV**	Li et al. (2018)
Glioma (adult & pediatric)	*CD45*, **GFAP**	Dietz et al. (2021)
Head & neck squamous cell carcinoma	*CD45*, **panCK**	Dietz et al. (2021)
		Henn et al. (2021)

(continued)

Table 14.1 (continued)

Cancer site	CHC protein identifier	References
Lung cancer	*CD14*, *CD45*, *CD133*, **CK8**, **CK18**, **CK19**, **EpCAM**, **NCAD**, **panCK**	Dietz et al. (2021)
		Manjunath et al. (2020a, b)
		Nel et al. (2014)
		Stott et al. (2010)
Melanoma (cutaneous)	*CD14*, *CD45*, *CD68*, *CD163*, *CD204*, *CD206*, **ALCAM**, **EpCAM**, **MLANA**, **panCK**	Clawson et al. (2015)
Ovarian cancer	*CD45*, **EpCAM**, **panCK**	Akhter et al. (2018)
		Dietz et al. (2021)
Pancreatic adenocarcinoma	*CD45*, *CD68*, *CD163*, *CD204*, *CD206*, **ALDH1A1**, **CD44**, **CEP8**, **CXCR4**, **EpCAM**, **ECAD**, **MUC4**, **MIF**, **panCK**, **S100PBP**, **ZG16B**	Clawson et al. (2017)
		Dietz et al. (2021)
		Patel et al. (2024)
		Whalen et al. (2024)
		Zhang et al. (2015)
Pancreatic neuroendocrine carcinoma	*CD45*, **CHGA**, **SYP**	Dietz et al. (2021)
Prostate cancer	*CD45*, **EpCAM**, **PSA**	Dietz et al. (2021)
		Stott et al. (2010)
Rectal cancer	*CD45*, **EpCAM**, **ECAD**, **panCK**	Dietz et al. (2021)
		Walker et al. (2021)
Uveal melanoma	*CD45*, **gp100**, **HTR2B**, **MITF**, **MLANA**, **TYR**	Anderson et al. (2024)
		Dietz et al. (2021)
		Parappilly et al. (2022)

Circulating hybrid cells (CHCs) have been studied in a myriad of cancer types. CHCs, by definition, express both a tumor (bold text) and immune (italic text) identifier, and multiple groups have identified expression patterns specific to solid organ sites (Dietz et al. 2021; Allan et al. 2005; Lustberg et al. 2014; Patel et al. 2024c; Whalen et al. 2024; Walker et al. 2021; Toyoshima et al. 2015; Li et al. 2018; Henn et al. 2021; Manjunath et al. 2020a; Nel et al. 2014; Stott et al. 2010; Clawson et al. 2015; 2017; Akhter et al. 2018; Patel et al. 2024a; Anderson et al. 2024; Parappilly et al. 2022; Zhang et al. 2015)

et al. 2024; Aguirre et al. 2020; Reduzzi et al. 2020; Li et al. 2018; Lazova et al. 2013; Laberge et al. 2017, 2019; Luo et al. 2016; Ogle et al. 2005; Duelli and Lazebnik 2003; Lustberg et al. 2014; Xu et al. 2014; Xue et al. 2015; Yilmaz et al. 2005; Goldenberg et al. 2012; Lartigue et al. 2020; Fan and Lu 2014; Zhang et al. 2019a; Cao et al. 2019; Ding et al. 2012; Rappa et al. 2012; Lizier et al. 2016; Manjunath et al. 2020a; Clawson et al. 2015; Clawson et al. 2017; Pawelek 2000, 2005, 2014; Pawelek and Chakraborty 2008b; Pawelek and Chakraborty 2008a; Ramakrishnan et al. 2013; Adams et al. 2014; Gast et al. 2018; Powell et al. 2011).

Oncologic cell fusion has been definitively demonstrated in a variety of cancer models. Heterotypic fusion between neoplastic cells and adjacent immune cells has been demonstrated in murine models of tumorigenesis (Gast et al. 2018). Red fluorescent protein (RFP)-labeled tumor cells injected into actin–green fluorescent protein (GFP) transgenic mice develop tumor hybrids that co-express GFP and RFP. When hybrid cells were isolated by fluorescence-activated cell sorting and orthotopically injected into wild-type mice, rapid tumor growth was seen (Dietz et al. 2021), demonstrating the tumorigenic nature of tumor-immune cell hybrids. Neoplastic cells can fuse with a variety of cell types in the TME. Prostate cancer cells have been shown to fuse with adjacent stromal and skeletal muscle cells, while breast cancer cells fuse with normal mammary gland cells and endothelial cells and result in increased metastatic potential, proliferation, and drug resistance (Frost et al. 1983; Kerbel et al. 1983). In vitro, fusion between oncologic cell lines and other cell types has been shown to increase malignant potential. Breast cancer cells can fuse with adipose-derived stem cells, forming hybrid breast cancer stem cells ($CD44^{+}$/$CD24^{-/low}$/$EpCAM^{+}$) with higher tumorigenic potential than the unfused mammary cell lines (Chan et al. 2020). Human hepatocellular carcinoma cells have demonstrated fusion with mesenchymal stromal cells, resulting in increased heterogeneity and metastatic potential (Li et al. 2014). Multiple myeloma cells and bone marrow–derived mesenchymal stromal cells have been shown to fuse, and the resultant progeny had increased heterogeneity, stemness markers, and drug resistance (Wang et al. 2018). Therefore, studies of cells from myriad organ sites have demonstrated the malignant properties of fusion cells reflective of those within the primary tumor sites.

Clearly, cell fusion between neoplastic cells and a variety of different types of neighboring cells is a common event found across solid malignancies. In a murine parabiosis model, macrophages were identified as the most common hematopoietic fusion partners of tumor cells, whereas B and T cells underwent fusion significantly less frequently (Powell et al. 2011). In patients, fusion has been demonstrated in women who received sex-mismatched bone marrow transplants and later developed secondary solid tumors. Evaluation of tumors with fluorescent in situ hybridization probes demonstrated that the Y chromosome from a male donor was present in neoplastic cells and premalignant lesions in female patients with pancreatic ductal adenocarcinoma, renal cell carcinoma, head and neck squamous cell carcinoma, and lung adenocarcinoma (Gast et al. 2018). A wide array of solid malignancies harbor a population of tumor-immune hybrids, and the range of tumor types and their fusion partners dictate how they are identified. The proteins utilized for the identification of tumor-immune hybrids and CHCs are summarized in Table 14.1.

Tumor-immune hybrid cells may contribute to mechanisms of treatment resistance, tumor heterogeneity, and metastatic potential, likely due in part to features gained from their macrophage parent cells. It has been shown that tumor-associated macrophages (TAMs) in glioblastoma utilize CCL21/CCR7 signaling to promote tumorigenesis, leading to therapy resistance (Geraldo et al. 2023). In lung cancer, P2X7, which promotes proliferation, invasion, and metastasis, is upregulated in TAMs, facilitating the progression of disease despite treatment with immunotherapy and chemotherapy (Qin et al. 2020). Other mechanisms of drug resistance include the expression of drug efflux membrane transporters, which are prominently expressed by macrophages (He et al. 2018; Singh and Gautam 2023; Emran et al. 2022). In colorectal cancer (CRC), TAMs contribute to 5-fluorouracil resistance

(Zhang et al. 2023). Given the preponderance of tumor–macrophage hybrid cell fusion in primary tumors, it is feasible that these cells convey treatment resistance and therapeutic challenges in cancer.

Tumor progression, invasion, and metastasis are complex, multistage processes. One key step involves epithelial-to-mesenchymal transition (EMT), which enables neoplastic cells to exit the primary tumor and enter the circulation, while the reverse process—mesenchymal-to-epithelial transition—facilitates their colonization and growth at distant sites (Pawelek and Chakraborty 2008b; Lopez-Collazo and Hurtado-Navarro 2025; May et al. 2023). The genes expressed during EMT are similar to those expressed in migratory immune cells such as macrophages (Fernandes et al. 2019; Brukman et al. 2019; Pawelek and Chakraborty 2008b; Lopez-Collazo and Hurtado-Navarro 2025). The fusion of tumor cells with bone marrow–derived cells allows resultant tumor-immune hybrids to express genes associated with migration, invasion, survival, and metastasis, including SPARC (Robert et al. 2006; Chakraborty and Yamaga 2003), MCR1 (Chakraborty et al. 2003; Garraway et al. 2005), and MET (Pawelek and Chakraborty 2008b). Further, hybrid cells resulting from the fusion of lung cancer cells and mesenchymal stem cells exhibited increased expression of EMT transcription factors such as Snail1, Slug, Twist1, and Zeb1 and Zeb2 (Zhang et al. 2019b). Co-expression of epithelial and mesenchymal markers has been seen in the tumor cells of breast, colorectal, head and neck squamous cell carcinoma, lung, pancreatic, ovarian, prostate, and esophageal cancers (Pastushenko et al. 2018; Puram et al. 2017; Fan et al. 2012; Rasheed et al. 2010; Long et al. 2015; Kong et al. 2010), and these characteristics have been associated with metastatic spread, including enhanced invasive and migratory capacity and more effective seeding of secondary tumor sites (Fan et al. 2012; Rasheed et al. 2010; Long et al. 2015; Kong et al. 2010; Mego et al. 2019; Bulfoni et al. 2016; Yu et al. 2013). The phenotypic qualities conferred to cells after fusion, therefore, likely promote features of tumor spread, intravasation, colonization, and progression.

14.3.2 Cell Fusion Hybrids in Circulation

Tumor-immune hybrid cells are detected in the peripheral blood, where they are termed circulating hybrid cells (CHCs). These cells are detected in a myriad of cancer types, including melanoma (Dietz et al. 2021), oral cavity carcinoma (Henn et al. 2021), pancreatic cancer (Patel et al. 2024a; Zhang et al. 2015), colorectal cancer (Tanjak et al. 2024), cholangiocarcinoma (Patel et al. 2024c), esophageal adenocarcinoma (Walker et al. 2021), glioma (Dietz et al. 2021), head and neck squamous cell carcinoma (Henn et al. 2021), breast cancer (Menyailo et al. 2023; Allan et al. 2005), lung cancer (Dietz et al. 2021; Nel et al. 2014; Stott et al. 2010), gastrointestinal stromal tumors (Toyoshima et al. 2015), ampullary cancer, ovarian cancer, and prostate cancer (Dietz et al. 2021). Their dissemination from the primary tumor is likely due to selection pressure and stochastic chromosomal retention, causing a phenotypic advantage that supports extravasation into the bloodstream. One of many disseminated tumor cell populations in the peripheral blood of patients with cancer, CHCs fall under the umbrella of circulating cancer cells (CCCs), which are key metastatic propagators and thus are the focus of understanding metastatic progression as well as the development of a noninvasive biomarker to evaluate tumor stage and patient prognosis. CCCs encompass a variety of neoplastic cells, including polyploid giant cancer cells (PGCCs), cancer-associated macrophage-like cells (CAMLs), circulating tumor cells (CTCs), and the focus of this chapter, CHCs. Of these populations, CHCs, CTCs, and PGCCs have reported tumor growth properties. PGCCs exhibit stem properties, increased migration, and increased invasiveness in cancer (Liu et al. 2024; Amend et al. 2019). Although multiple mechanisms have been proposed for their origin, cell fusion is one recognized contributor. CAMLs are cells of myeloid origin that have phagocytosed components of tumor cells (Pirrello et al. 2024). These cells have dissemination properties and track with poor prognosis. They likely represent an immune reaction to tumors, therefore increasing in number while CHCs decrease with good tumor response to therapies (Sutton et al. 2022; Adams et al. 2014; Pirrello et al. 2024). CTCs, perhaps the most well-studied CCC, are shed from the primary tumor and disseminate into the peripheral blood (Alix-Panabieres and Pantel 2025); however, they have been found in only extremely low numbers—approximately 1–2 cells per 7.5 mL of blood—in late-stage CRC (Gast et al. 2018; Cohen et al. 2009; Tol et al. 2010; Krebs et al. 2010). CHCs are phenotypically distinct from all other disseminated cells in that they harbor a dual immune-tumor cell identity and, moreover, possess functional properties of both cell types. It is hypothesized that the features conferred by both parent cells enable CHCs to adapt to the complex process of metastasis, including phenotypic plasticity, motility, and colonization of distant sites, microenvironmental modulation, metabolic flexibility, and genomic instability (Fig. 14.2): the tumor cell identity promotes CHC expansion in the primary tumor and growth at metastatic sites, while the immune functions support intravasation into circulation and survival in the peripheral blood (Anderson et al. 2024; Gast et al. 2018).

Tumor-immune hybrids are highly motile and invasive (Gast et al. 2018; Ye et al. 2023; Uygur et al. 2019; Luo et al. 2016; Wang et al. 2021; Ma et al. 2014). By the nature of their dual cell type parentage, tissue-based tumor-immune hybrids, as well as blood-based CHCs, demonstrate great heterogeneity and acquire features of both parent cell types (immune and oncologic), aiding in the evasion of normal immune functions or potential recognition as an immune cell themselves. $EpCAM^{+}/CD45^{+}$ hybrid cells found in the ascites of patients with ovarian cancer have been shown to overexpress major histocompatibility complex I antigens, allowing them to evade normal natural killer cell immune-mediated destruction (Akhter et al. 2018). By definition, CHCs, like their tumor-resident counterparts, co-express features of the primary tumor as well as an immune marker, commonly the pan-leukocyte antigen CD45 (Table 14.1). Phenotypic plasticity is well-documented in macrophages—the most prominent nonneoplastic fusion partners generating tumor-immune hybrids (Powell et al. 2011). Macrophages demonstrate plasticity in response to the TME, where local chemotactic factors can affect the activation of the classical pathway with a resultant M1 macrophage or the alternative pathway with a more M2 macrophage phenotype (Leopold Wager and Wormley Jr. 2014; Sutton et al. 2022; Malyshev and Malyshev 2015). M1 macrophages release pro-inflammatory cytokines such as TNFα, IL-6, and IFNγ, thought to contribute to tumor destruction, while M2 macrophages perform actions permissive to cell motility and promote tumor invasion, such as angiogenesis and basement membrane breakdown (Yunna et al. 2020; Sutton et al. 2022). With a spectrum of phenotypes available between this binary nomenclature, macrophages offer great plasticity based on the local TME, likely contributing to the discrete hybrid subpopulations seen in the primary tumor compared to the blood and metastasis (Whalen et al. 2024). There is some evidence to suggest that M2 macrophages have higher rates of fusion with tumor cells compared to M1-polarized macrophages, as exhibited in human lung cancer cells (Aguirre et al. 2020) and melanoma CHCs (Clawson et al. 2015). Though their phenotypic expression patterns are heterogeneous, CHCs retain features of their macrophage parent cells (Patel et al. 2024b; Powell et al. 2011). It is therefore plausible that they utilize mechanisms similar to those of the parent macrophage to promote survival and avoid immune destruction. For example, colony-stimulating factor-1 (CSF-1) contributes to the migratory capacity (Wyckoff et al. 2004) and immune identity of macrophages, and likewise the CSF-1 receptor is expressed on CHCs and has been demonstrated to facilitate hybrid cell migration (Gast et al. 2018). It is feasible that CSF1R and other immune identity proteins convey immune evasion to CHCs, contributing to oncologic progression.

Marked heterogeneity among the CHC subpopulations has been reported, likely due to their dual cell type parentage and post-fusion reprogramming. Protein expression, evaluated by cyclic immunofluorescence, of hybrid cells within the tumor and disseminated into the peripheral blood, demonstrated that discrete subpopulations exist within the tumor versus the blood (Whalen et al. 2024). As hybrid cells escape from the primary tumor, they retain conserved features of the primary tumor but also develop discrete changes permitting their dissemination. It is possible that this hybrid cell heterogeneity may be due to stochastic chromosomal rearrangements, missegregation, and chromothripsis (Miroshnychenko et al. 2021; Hass et al. 2021a; Song et al. 2021; Chunduri and Storchova 2019; Holland and Cleveland 2009; Duncan et al. 2009, 2010; Ganem et al. 2009; Godinho et al. 2009; Passerini et al. 2016), and the survival of CHCs that can adapt to their diverse microenvironment increases their chances of metastatic progression. Further work into the granularity of these chromosomal rearrangements may determine the genetic makeup of the daughter cells and could identify important mechanisms of therapeutic resistance and oncologic progression.

14.3.3 Cell Fusion Hybrids in Metastatic Sites

While globally defined, the comprehensive understanding of the cellular mechanisms that underlie the metastatic cascade remains incomplete. In 1889, Dr. Stephen Paget posited the “Seed and Soil” theory of metastasis, proposing that both tumor cells with potential to metastasize (the "seeds") and an environment appropriate for metastatic colonization (the "soil") needed to exist in concert to promote the distant spread of cancer (Paget 1989; Akhtar et al. 2019). Organotropism, or the preference for metastasis to a specific organ site, is not well understood but has been associated with circulatory patterns affecting access to a given organ, tumor-specific genetics/epigenetics, and microenvironmental factors impacting pre-metastatic tumor cells (Smith et al. 2018; Gao et al. 2019). It is believed that tumor-immune hybrid status may also contribute to a preference for metastasis to certain organ sites. Liver, lung, and bone are common sites of metastasis, and they are tissues that contain large numbers of bone-marrow–derived cells (Fidler 2003). The genomic and phenotypic alterations associated with fusion between tumor and immune cells may confer advantages to seed a different microenvironment than that of the primary tumor (Seyfried and Huysentruyt 2013). CHCs from patients with uveal melanoma were found to express genes highly expressed in the liver, including glutathione peroxidase 1 (GPX1) and selenoprotein P1 (SEPP1), which may contribute to the selective metastasis of uveal melanoma to the liver (Anderson et al. 2024). Interestingly, CHCs from patients with pancreatic ductal adenocarcinoma (PDAC) transplanted orthotopically in the pancreas of nude mice were shown to seed well-differentiated islands of cells in multiple organ sites, but did not readily form tumors in the implanted location (Clawson et al. 2017). Similarly, CHCs extracted from patients with melanoma implanted in the subcutaneous space of nude mice formed metastatic lesions but formed no local subcutaneous tumors (Clawson et al. 2015). CHCs derived from a murine melanoma B16F10 cell line and bone-marrow–derived macrophages had a markedly increased propensity to promote metastatic burden compared with unfused tumor cells (Gast et al. 2018). Clearly, tumor-immune hybrids and CHCs are corresponding populations that inform the biology of oncology and contribute to both tumor progression, disease heterogeneity, and metastatic spread. The translational utility of this population has great potential and is an area of ongoing study.

14.4 Clinical Applications

As more is understood about CHC biology, specifically their role in metastatic progression, there is a growing potential for their translation to the clinic. With the ability to evade the immune system and proliferate in metastatic sites, CHCs outnumber other neoplastic cell populations in numerous cancer sites, including, but not limited to the prostate, breast, skin, uveal melanoma, pancreas, colon, and lung (Chai et al. 2022; Whalen et al. 2024; Dietz et al. 2021; Tanjak et al. 2024; Anderson et al. 2024; Henn et al. 2021; Akhter et al. 2018; Ishizawa et al. 2019), highlighting their critical clinical relevance. CHCs thus represent a population of interest within translational research and warrant further assessment as a therapeutic target to improve patient outcomes. Leveraging the disease- and site-specific biomarkers harbored by CHCs remains an area of ongoing investigation and significant clinical opportunity, as does the potential to prognosticate from CHC gene signature patterns within and across cancers.

14.4.1 Early Detection and Biomarker Development

The correlation of CHCs with tumor burden and disease stage presents an area of significant clinical opportunity. It is well-known that, across cancer types, early diagnosis correlates with improved survival and outcomes in patients (Crosby et al. 2022). Those with lung, pancreatic, esophageal, and stomach tumors have been shown to die more commonly due to metastatic disease, while breast and prostate cancer patients are less likely to die from metastatic cancer (Mani et al. 2024), likely due to improvements in early diagnosis and treatment of the latter two cancers. On the other hand, the diagnosis of pancreatic cancer, which is the third leading cause of cancer-related death in the United States (Siegel et al. 2024; Park et al. 2021), is limited by a lack of early symptoms and the current inefficacy of screening modalities. Cross-sectional imaging under-stages patients up to 70% of the time (Baugh et al. 2019) and fails to capture occult metastatic disease in 8–19% of patients (Suker et al. 2019; Gemenetzis et al. 2018)—and this barrier is not limited to cancer of the pancreas (Crosby et al. 2022). Further, biopsies at any organ site are plagued by issues of sampling error and can underdiagnose cancer.

The development of a biomarker that can detect early-stage cancer or premalignancy as well as reliably monitor disease course would revolutionize oncology care. Blood-based biomarkers, due to their noninvasive nature and procedural feasibility in any clinical setting, are the focus of the future in oncology care. Currently available options fall short, but CHCs have shown great promise in early diagnostics across various cancer types (Fig. 14.3). Carbohydrate antigen 19-9 (CA 19-9), carcinoembryonic antigen (CEA), cancer antigen 125, and prostate-specific antigen are the current gold standard serum biomarkers for pancreatic, colorectal, ovarian, and prostate cancers, respectively. Though FDA-approved, these markers have not been validated as independent screening or diagnostic tools for the general population, and are primarily used for disease surveillance and assessment of treatment response. Even in this setting, their effectiveness is limited because serum levels can be influenced by confounding clinical conditions. CA 19-9 can be elevated in nonmalignant hepatic, pulmonary, gynecologic, and endocrine disorders (Kim et al. 2020), and notably, 5–10% of the general population are Lewis antigen–negative and do not synthesize CA 19-9 (Parra-Robert et al. 2018). Given their tumor-specific origin, CHCs may serve as more reliable biomarkers for diagnosis, prognosis, and treatment response (Fig. 14.3). Data comparing CHC levels to CA 19-9 in patients with biliary tract and pancreatic cancer suggest their potential superiority as oncologic biomarkers (Patel et al. 2024c). In patients with PDAC, CRC, esophageal adenocarcinoma, and uveal melanoma (Gast et al. 2018; Walker et al. 2021; Parappilly et al. 2022; Patel et al. 2024a), increasing numbers of CHCs

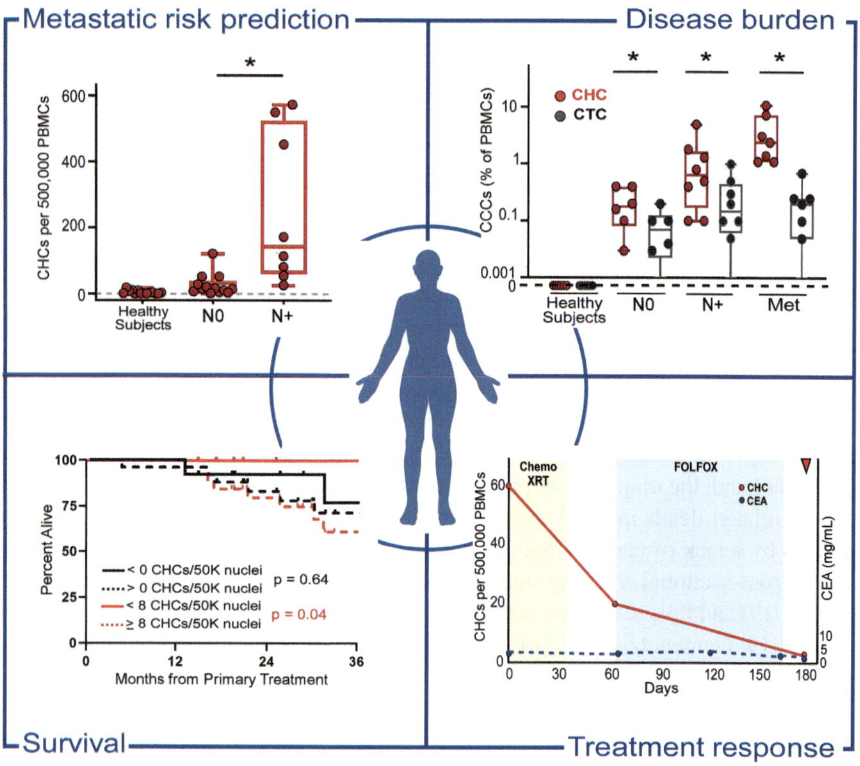

Fig. 14.3 Circulating hybrid cells have promising clinical utility. Increasing numbers of circulating hybrid cells (CHCs) correlate with metastatic risk progression (top left), disease burden (top right), survival (bottom left), and treatment response (bottom right). (Top Left) In oral cavity carcinoma, patients who underwent surgical resection had significantly higher numbers of CHCs in node-positive disease compared to patients with node-negative disease (Modified from Henn et al. 2021). (Top right) In patients with pancreatic ductal adenocarcinoma (PDAC), increased numbers of CHCs correlated with disease burden, while circulating tumor cells (CTCs) failed to correlate with increasing disease stage (Modified from Gast et al. 2018). (Bottom left) Identification of CHCs in uveal melanoma patients showed that, while increased numbers of CTCs did not correlate with survival outcomes, increased numbers of CHCs corresponded to significantly worse overall survival (Modified from Parappilly et al. 2022). (Bottom right) In colorectal cancer, CHC numbers corresponded to therapeutic response to neoadjuvant chemoradiation, while the protein marker carcinogenic embryonic antigen (CEA) did not change with response to therapies. (Modified from Walker et al. 2021, reprinted with permission)

correlate with worse disease stage and decreased survival. Additionally, CHCs have the potential to differentiate between the presence of benign polyps, premalignant polyps, and cancers that are biopsied or removed during standard-of-care colonoscopies—a feat not yet reliably achieved by other blood-based biomarker platforms (Patel et al. 2024d).

In addition to CHCs, CTCs and CAMLs also represent CCC populations that may serve as biomarkers of malignancy. These cellular populations are, advantageously, intact cells. This provides a more comprehensive representation of tumor biology compared to noncellular biomarkers, as these populations may inherently select from clonal populations that are malignant. Their analysis enables the assessment of key oncologic characteristics, including mutational alterations, protein expression, signaling pathway activation, and proliferative status. For example, in PDAC, CHCs have been shown to reflect treatment-induced phenotypic changes in metastatic lesions (Patel et al. 2024a). Furthermore, CTCs co-expressing mesenchymal and epithelial markers have been associated with more aggressive disease biology and worse overall survival (Gemenetzis et al. 2018; Zhao et al. 2019), while CAML levels in breast cancer patients have been correlated with both survival outcomes and responses to cytotoxic chemotherapy (Adams et al. 2014; Mu et al. 2017). Importantly, CTCs and CAMLs benefit from standardized detection platforms, such as the CellSearch® system (Menarini Silicon Biosystems, Castel Maggiore, BO, Italy) and the CellSieve™ microfiltration system (Creatv MicroTech, Inc., Potomac, MD), which enhance their clinical applicability. In contrast, CHC detection lacks similarly standardized platforms, and high-throughput automated detection is not yet possible, although it is an area of ongoing study (Heussner et al. 2024; Cheng et al. 2020). However, CHCs offer the distinct advantage of being more reliably detectable, particularly in earlier stages of malignancy. For instance, a 7.5 mL peripheral blood sample may yield only 1–2 CTCs or CAMLs in advanced PDAC and often none in early-stage cases (Court et al. 2018; Nitschke et al. 2022), whereas CHCs are detected at magnitudes that are higher across all disease stages (Gast et al. 2018).

Beyond emerging circulating cellular biomarkers, significant advancements have been made in the development of noncellular circulating biomarkers. Interest in circulating tumor DNA (ctDNA) and cell-free DNA (cfDNA)—fragments of DNA released into circulation through cell death—has surged in recent years. The recent prospective, nationwide ECLIPSE study represented a major breakthrough that demonstrated that a cfDNA-based blood test could serve as a colorectal cancer screening tool in an average-risk population, achieving high sensitivity and specificity. However, its performance was limited in detecting advanced precancerous lesions (Chung et al. 2024). Furthermore, persistent ctDNA following colorectal cancer resection has been associated with worse disease-free and overall survival, as demonstrated in the observational CIRCULATE-Japan GALAXY study and the BESPOKE CRC study (Nakamura et al. 2024; Kasi et al. 2024). Despite these promising applications, several limitations must be considered. ctDNA is likely shed from both indolent and aggressive clones, which may reduce its reliability as a direct reflection of tumor cell biology. Additionally, ctDNA is not uniformly detectable across all malignancies, is often scarce in early-stage disease, and current detection platforms require further refinement to enhance sensitivity and accuracy. Beyond ctDNA, protease-based assays have also garnered significant interest as potential biomarkers of malignancy given the established role of protease activation in tumor progression. However, protease detection assays have historically faced challenges in demonstrating clinical applicability and practicality, largely due to the absence of accessible platforms for widespread evaluation (Garcia-Pardo et al. 2022). Recent advancements in this field have shown promise, particularly with the development of a rapid, magnetic nanosensor–based protease assay ("PAC-MANN-1"), which successfully differentiated patients with PDAC from healthy controls and those with precancerous lesions with a specificity of 98% and a sensitivity of 73%. When combined with CA 19-9, this diagnostic test achieved an 85% sensitivity and 96% specificity for stage I PDAC, representing a significant step toward an accessible, blood-based screening tool (Montoya Mira et al. 2025). However, the influence of confounding inflammatory conditions on protease-based assays, their tumor specificity, and their potential applicability in treatment response monitoring and disease surveillance—capabilities that early data suggest CHCs may offer—remain areas of ongoing investigation.

14.4.2 Prognostication and Disease Monitoring

Clear trends have been established between the prevalence of CHCs and disease progression or recurrence that warrant further consideration of this population in clinical settings. When used in combination, imaging studies, endoscopic procedures, and biopsies are the current gold standard for determining pathologic complete response in gastrointestinal malignancies, although the sensitivity of these methods is limited (Hiotis et al. 2002). Response to preoperative chemotherapy and the lack of a visible or palpable tumor do not always indicate complete disease remission, as in some cases, recurrence goes unnoticed. CHC levels have also been shown to reliably correlate with disease course, even in the event that other biomarker analyses and imaging studies fall short (Walker et al. 2021; Patel et al. 2024a). For example, the number of CHCs increased in correlation with disease progression in patients with intrahepatic cholangiocarcinoma, while cross-sectional imaging showed no change in tumor size or disease progression (Patel et al. 2024c). The prevalence of CHCs in rectal adenocarcinoma and esophageal adenocarcinoma has been demonstrated to directly correlate with treatment response to neoadjuvant therapy and in addition to residual nodal involvement in esophageal adenocarcinoma (Walker et al. 2021). Additionally, a subset of patients enrolled in the HELIX-1 phase II clinical trial (NCT04251715) demonstrated increased CHC levels that correlated with treatment resistance by the end of the trial, reflecting disease progression or recurrence. In contrast, CHCs remained at undetectable levels in patients enrolled in the trial who demonstrated clinical evidence of therapeutic response (Patel et al. 2024c).

The detection of increased numbers of CHCs has been shown to correlate with worse survival in uveal melanoma, esophageal adenocarcinoma, and PDAC (Fig. 14.3) (Walker et al. 2021; Parappilly et al. 2022; Patel et al. 2024a). Further, a population of CHCs lacking vimentin expression reliably classified patients with lymph node metastasis and recurrent disease, whereas CHCs harboring vimentin expression consistently aligned with less severe disease (Cheng et al. 2020), indicating that expression patterns in CHCs can provide a readout for oncologic prognostication and diagnosis. In contrast, although the detection of CTCs has been approved for monitoring certain cancers, including colorectal, breast, and prostate cancers, CTC enumeration has not reliably correlated with patient survival, though it has been used to inform the need for chemotherapeutic regimen alterations. The SWOG S0500 clinical trial evaluated the use of CTCs to inform treatment effectiveness in breast cancer but failed to show a survival difference for breast cancer patients whose chemotherapeutic regimen was informed by CTC monitoring (Smerage et al. 2014). Noninvasive disease monitoring through CHC detection has the potential to impact patient survival by informing personalized oncology treatment decisions and remains an unmet frontier of ongoing research.

These findings present a call to design more comprehensive care that personalizes and optimizes therapies to address current gaps in oncology care. There is significant translational potential to leverage CHCs as a noninvasive biomarker for early detection, monitoring disease progression, and prediction of survival outcomes. As CHC gene expression is heterogeneous across malignancies, this population could serve as a cancer-specific therapeutic target and a window into the development of personalized oncology therapies (Dietz et al. 2021; Walker et al. 2021). Early data provide a strong foundation supporting the potential role of CHCs in this capacity; however, further validation through large-scale prospective studies is essential to determine how they may be integrated with other biomarkers and effectively utilized to enhance clinical care and tailor therapeutic strategies to the unique biology of individual tumors.

14.5 Prospectus

The biological importance of cell fusion is indisputable. Cell fusion is a mechanism underlying both life and disease. Tumor-immune hybrid cells in primary neoplasms and peripheral blood are drivers of metastatic progression in solid malignancies. Found in greater numbers than other CCCs, CHCs are a promising tool for noninvasive blood-based biomarker assays for cancer diagnostics, prognostication, and surveillance. Further distilling the biology of CHCs (e.g., their mechanism of formation, epigenome, and regulation of gene expression), their role in the metastatic cascade, and their impact on patient outcomes is necessary to solidify the clinical utility of CHCs in the management and surveillance of disease burden. This understanding will guide the implementation of clinical trials evaluating CHCs in patient care settings, providing physicians with knowledge that has the potential to revolutionize patient-centered approaches to treating solid malignancies.

Acknowledgments KLF is supported by the Norton Fellowship from the OHSU Brenden-Colson Center for Pancreatic Care.

References

Abu-Thuraia A, Gauthier R, Chidiac R, Fukui Y, Screaton RA, Gratton JP, Cote JF (2015) Axl phosphorylates Elmo scaffold proteins to promote Rac activation and cell invasion. Mol Cell Biol 35:76–87

Adams DL, Martin SS, Alpaugh RK, Charpentier M, Tsai S, Bergan RC, Ogden IM, Catalona W, Chumsri S, Tang CM, Cristofanilli M (2014) Circulating giant macrophages as a potential biomarker of solid tumors. Proc Natl Acad Sci USA 111:3514–3519

Adler RR, Ng AK, Rote NS (1995) Monoclonal antiphosphatidylserine antibody inhibits intercellular fusion of the choriocarcinoma line, JAR. Biol Reprod 53:905–910

Aghababaei M, Hogg K, Perdu S, Robinson WP, Beristain AG (2015) ADAM12-directed ectodomain shedding of E-cadherin potentiates trophoblast fusion. Cell Death Differ 22: 1970–1984

Aguirre LA, Montalban-Hernandez K, Avendano-Ortiz J, Marin E, Lozano R, Toledano V, Sanchez-Maroto L, Terron V, Valentin J, Pulido E, Casalvilla JC, Rubio C, Diekhorst L, Laso-Garcia F, Del Fresno C, Collazo-Lorduy A, Jimenez-Munarriz B, Gomez-Campelo P, Llanos-Gonzalez E, Fernandez-Velasco M, Rodriguez-Antolin C, Perez De Diego R, Cantero-Cid R, Hernadez-Jimenez E, Alvarez E, Rosas R, Dies Lopez-Ayllon B, De Castro J, Wculek SK, Cubillos-Zapata C, Ibanez De Caceres I, Diaz-Agero P, Gutierrez Fernandez M, Paz De Miguel M, Sancho D, Schulte L, Perona R, Belda-Iniesta C, Bosca L, Lopez-Collazo E (2020) Tumor stem cells fuse with monocytes to form highly invasive tumor-hybrid cells. Onco Targets Ther 9:1773204

Akhtar M, Haider A, Rashid S, Al-Nabet A (2019) Paget's "seed and soil" theory of cancer metastasis: An idea whose time has come. Adv Anat Pathol 26:69–74

Akhter MZ, Sharawat SK, Kumar V, Kochat V, Equbal Z, Ramakrishnan M, Kumar U, Mathur S, Kumar L, Mukhopadhyay A (2018) Aggressive serous epithelial ovarian cancer is potentially propagated by EpCAM(+)CD45(+) phenotype. Oncogene 37:2089–2103

Alix-Panabieres C, Pantel K (2025) From discovery to diagnosis: a perspective for circulating tumor cells in personalized oncology. Cancer Discov 15:OF1–OF17

Allan AL, Vantyghem SA, Tuck AB, Chambers AF, Chin-Yee IH, Keeney M (2005) Detection and quantification of circulating tumor cells in mouse models of human breast cancer using immunomagnetic enrichment and multiparameter flow cytometry. Cytometry A 65:4–14

Amend SR, Torga G, Lin KC, Kostecka LG, De Marzo A, Austin RH, Pienta KJ (2019) Polyploid giant cancer cells: unrecognized actuators of tumorigenesis, metastasis, and resistance. Prostate 79:1489–1497

Anderson AN, Conley P, Klocke CD, Sengupta SK, Pang A, Farley HC, Gillingham AR, Dawson AD, Fan Y, Jones JA, Gibbs SL, Skalet AH, Wu G, Wong MH (2024) Detection of neoplastic-immune hybrid cells with metastatic properties in uveal melanoma. Biomark Res 12:67

Archacka K, Grabowska I, Mierzejewski B, Graffstein J, Gorzynska A, Krawczyk M, Rozycka AM, Kalaszczynska I, Muras G, Streminska W, Janczyk-Ilach K, Walczak P, Janowski M, Ciemerych MA, Brzoska E (2021) Hypoxia preconditioned bone marrow-derived mesenchymal stromal/stem cells enhance myoblast fusion and skeletal muscle regeneration. Stem Cell Res Ther 12:448

Balkwill F, Mantovani A (2001) Inflammation and cancer: back to Virchow? Lancet 357:539–545

Bastida-Ruiz D, Van Hoesen K, Cohen M (2016) The dark side of cell fusion. Int J Mol Sci 17:638

Baugh KA, Tran Cao HS, Van Buren G 2nd, Silberfein EJ, Hsu C, Chai C, Barakat O, Fisher WE, Massarweh NN (2019) Understaging of clinical stage I pancreatic cancer and the impact of multimodality therapy. Surgery 165:307–314

Ben-David U, Amon A (2020) Context is everything: aneuploidy in cancer. Nat Rev Genet 21:44–62

Bielski CM, Zehir A, Penson AV, Donoghue MTA, Chatila W, Armenia J, Chang MT, Schram AM, Jonsson P, Bandlamudi C, Razavi P, Iyer G, Robson ME, Stadler ZK, Schultz N, Baselga J, Solit DB, Hyman DM, Berger MF, Taylor BS (2018) Genome doubling shapes the evolution and prognosis of advanced cancers. Nat Genet 50:1189–1195

Bjerregaard B, Holck S, Christensen IJ, Larsson LI (2006) Syncytin is involved in breast cancer-endothelial cell fusions. Cell Mol Life Sci 63:1906–1911
Blond JL, Lavillette D, Cheynet V, Bouton O, Oriol G, Chapel-Fernandes S, Mandrand B, Mallet F, Cosset FL (2000) An envelope glycoprotein of the human endogenous retrovirus HERV-W is expressed in the human placenta and fuses cells expressing the type D mammalian retrovirus receptor. J Virol 74:3321–3329
Bojko A, Staniak K, Czarnecka-Herok J, Sunderland P, Dudkowska M, Sliwinska MA, Salmina K, Sikora E (2020) Improved Autophagic flux in escapers from doxorubicin-induced senescence/polyploidy of breast cancer cells. Int J Mol Sci 21:6084
Brukman NG, Uygur B, Podbilewicz B, Chernomordik LV (2019) How cells fuse. J Cell Biol 218:1436–1451
Buckingham M (2006) Myogenic progenitor cells and skeletal myogenesis in vertebrates. Curr Opin Genet Dev 16:525–532
Bulfoni M, Gerratana L, Del Ben F, Marzinotto S, Sorrentino M, Turetta M, Scoles G, Toffoletto B, Isola M, Beltrami CA, Di Loreto C, Beltrami AP, Puglisi F, Cesselli D (2016) In patients with metastatic breast cancer the identification of circulating tumor cells in epithelial-to-mesenchymal transition is associated with a poor prognosis. Breast Cancer Res 18:30
Cao MF, Chen L, Dang WQ, Zhang XC, Zhang X, Shi Y, Yao XH, Li Q, Zhu J, Lin Y, Liu S, Chen Q, Cui YH, Zhang X, Bian XW (2019) Hybrids by tumor-associated macrophages x glioblastoma cells entail nuclear reprogramming and glioblastoma invasion. Cancer Lett 442:445–452
Chai S, Ruiz-Velasco C, Naghdloo A, Pore M, Singh M, Matsumoto N, Kolatkar A, Xu L, Shishido S, Aparicio A, Zurita AJ, Hicks J, Kuhn P (2022) Identification of epithelial and mesenchymal circulating tumor cells in clonal lineage of an aggressive prostate cancer case. NPJ Precis Oncol 6:41
Chakraborty AK, Yamaga S (2003) Differential gene expression in genetically matched mouse melanoma cells with different metastatic potential. Gene 315:165–175
Chakraborty AK, Kolesnikova N, Sousa Jde F, Espreafico EM, Peronni KC, Pawelek J (2003) Expression of c-met proto-oncogene in metastatic macrophage x melanoma fusion hybrids: implication of its possible role in MSH-induced motility. Oncol Res 14:163–174
Chakraborty A, Lazova R, Davies S, Backvall H, Ponten F, Brash D, Pawelek J (2004) Donor DNA in a renal cell carcinoma metastasis from a bone marrow transplant recipient. Bone Marrow Transplant 34:183–186
Chan YW, So C, Yau KL, Chiu KC, Wang X, Chan FL, Tsang SY (2020) Adipose-derived stem cells and cancer cells fuse to generate cancer stem cell-like cells with increased tumorigenicity. J Cell Physiol 235:6794–6807
Chen CP, Wang KG, Chen CY, Yu C, Chuang HC, Chen H (2006) Altered placental syncytin and its receptor ASCT2 expression in placental development and pre-eclampsia. BJOG 113:152–158
Cheng KS, Pan R, Pan H, Li B, Meena SS, Xing H, Ng YJ, Qin K, Liao X, Kosgei BK, Wang Z, Han RPS (2020) Alice: a hybrid AI paradigm with enhanced connectivity and cybersecurity for a serendipitous encounter with circulating hybrid cells. Theranostics 10:11026–11048
Cho EH, Wendel M, Luttgen M, Yoshioka C, Marrinucci D, Lazar D, Schram E, Nieva J, Bazhenova L, Morgan A, Ko AH, Korn WM, Kolatkar A, Bethel K, Kuhn P (2012) Characterization of circulating tumor cell aggregates identified in patients with epithelial tumors. Phys Biol 9:016001
Chunduri NK, Storchova Z (2019) The diverse consequences of aneuploidy. Nat Cell Biol 21:54–62
Chung DC, Gray DM 2nd, Singh H, Issaka RB, Raymond VM, Eagle C, Hu S, Chudova DI, Talasaz A, Greenson JK, Sinicrope FA, Gupta S, Grady WM (2024) A cell-free DNA blood-based test for colorectal cancer screening. N Engl J Med 390:973–983
Chuprin A, Gal H, Biron-Shental T, Biran A, Amiel A, Rozenblatt S, Krizhanovsky V (2013) Cell fusion induced by ERVWE1 or measles virus causes cellular senescence. Genes Dev 27:2356–2366

Clark S, Naz RK (2013) Presence and incidence of Izumo antibodies in sera of immunoinfertile women and men. Am J Reprod Immunol 69:256–263

Clawson GA, Matters GL, Xin P, Imamura-Kawasawa Y, Du Z, Thiboutot DM, Helm KF, Neves RI, Abraham T (2015) Macrophage-tumor cell fusions from peripheral blood of melanoma patients. PLoS One 10:e0134320

Clawson GA, Matters GL, Xin P, Mcgovern C, Wafula E, Depamphilis C, Meckley M, Wong J, Stewart L, D'jamoos C, Altman N, Imamura Kawasawa Y, Du Z, Honaas L, Abraham T (2017) "Stealth dissemination" of macrophage-tumor cell fusions cultured from blood of patients with pancreatic ductal adenocarcinoma. PLoS One 12:e0184451

Cohen SJ, Punt CJ, Iannotti N, Saidman BH, Sabbath KD, Gabrail NY, Picus J, Morse MA, Mitchell E, Miller MC, Doyle GV, Tissing H, Terstappen LW, Meropol NJ (2009) Prognostic significance of circulating tumor cells in patients with metastatic colorectal cancer. Ann Oncol 20:1223–1229

Court CM, Ankeny JS, Sho S, Winograd P, Hou S, Song M, Wainberg ZA, Girgis MD, Graeber TG, Agopian VG, Tseng HR, Tomlinson JS (2018) Circulating tumor cells predict occult metastatic disease and prognosis in pancreatic cancer. Ann Surg Oncol 25:1000–1008

Crosby D, Bhatia S, Brindle KM, Coussens LM, Dive C, Emberton M, Esener S, Fitzgerald RC, Gambhir SS, Kuhn P, Rebbeck TR, Balasubramanian S (2022) Early detection of cancer. Science 375:eaay9040

Das M, Xu B, Lin L, Chakrabarti S, Shivaswamy V, Rote NS (2004) Phosphatidylserine efflux and intercellular fusion in a BeWo model of human villous cytotrophoblast. Placenta 25:396–407

Davies PS, Powell AE, Swain JR, Wong MH (2009) Inflammation and proliferation act together to mediate intestinal cell fusion. PLoS One 4:e6530

Delespaul L, Merle C, Lesluyes T, Lagarde P, Le Guellec S, Perot G, Baud J, Carlotti M, Danet C, Fevre M, Rousseau B, Durrieu S, Teichmann M, Coindre JM, Lartigue L, Chibon F (2019) Fusion-mediated chromosomal instability promotes aneuploidy patterns that resemble human tumors. Oncogene 38:6083–6094

Delespaul L, Gelabert C, Lesluyes T, Le Guellec S, Perot G, Leroy L, Baud J, Merle C, Lartigue L, Chibon F (2020) Cell-cell fusion of mesenchymal cells with distinct differentiations triggers genomic and transcriptomic remodelling toward tumour aggressiveness. Sci Rep 10:21634

Demin S, Berdieva M, Goodkov A (2022) Cell-cell fusions and cell-in-cell phenomena in healthy cells and cancer: lessons from protists and invertebrates. Semin Cancer Biol 81:96–105

Dietz MS, Sutton TL, Walker BS, Gast CE, Zarour L, Sengupta SK, Swain JR, Eng J, Parappilly M, Limbach K, Sattler A, Burlingame E, Chin Y, Gower A, Mira JLM, Sapre A, Chiu YJ, Clayburgh DR, Pommier SJ, Cetnar JP, Fischer JM, Jaboin JJ, Pommier RF, Sheppard BC, Tsikitis VL, Skalet AH, Mayo SC, Lopez CD, Gray JW, Mills GB, Mitri Z, Chang YH, Chin K, Wong MH (2021) Relevance of circulating hybrid cells as a non-invasive biomarker for myriad solid tumors. Sci Rep 11:13630

Ding J, Jin W, Chen C, Shao Z, Wu J (2012) Tumor associated macrophage x cancer cell hybrids may acquire cancer stem cell properties in breast cancer. PLoS One 7:e41942

Dittmar T, Weiler J, Luo T, Hass R (2021) Cell-cell fusion mediated by viruses and HERV-derived fusogens in cancer initiation and progression. Cancers (Basel) 13:5363

Do M, Kwak IH, Ahn JH, Lee IJ, Lee JH (2017) Survivin protects fused cancer cells from cell death. BMB Rep 50:361–366

Dornen J, Sieler M, Weiler J, Keil S, Dittmar T (2020) Cell fusion-mediated tissue regeneration as an inducer of polyploidy and aneuploidy. Int J Mol Sci 21:1811

Duelli D, Lazebnik Y (2003) Cell fusion: a hidden enemy? Cancer Cell 3:445–448

Duelli D, Lazebnik Y (2007) Cell-to-cell fusion as a link between viruses and cancer. Nat Rev Cancer 7:968–976

Duelli DM, Padilla-Nash HM, Berman D, Murphy KM, Ried T, Lazebnik Y (2007) A virus causes cancer by inducing massive chromosomal instability through cell fusion. Curr Biol 17:431–437

Duncan AW, Hickey RD, Paulk NK, Culberson AJ, Olson SB, Finegold MJ, Grompe M (2009) Ploidy reductions in murine fusion-derived hepatocytes. PLoS Genet 5:e1000385

Duncan AW, Taylor MH, Hickey RD, Hanlon Newell AE, Lenzi ML, Olson SB, Finegold MJ, Grompe M (2010) The ploidy conveyor of mature hepatocytes as a source of genetic variation. Nature 467:707–710

Dvorak HF (1986) Tumors: wounds that do not heal. Similarities between tumor stroma generation and wound healing. N Engl J Med 315:1650–1659

Emran TB, Shahriar A, Mahmud AR, Rahman T, Abir MH, Siddiquee MF, Ahmed H, Rahman N, Nainu F, Wahyudin E, Mitra S, Dhama K, Habiballah MM, Haque S, Islam A, Hassan MM (2022) Multidrug resistance in cancer: understanding molecular mechanisms, Immunoprevention and Therapeutic Approaches. Front Oncol 12:891652

Fan H, Lu S (2014) Fusion of human bone hemopoietic stem cell with esophageal carcinoma cells didn't generate esophageal cancer stem cell. Neoplasma 61:540–545

Fan F, Samuel S, Evans KW, Lu J, Xia L, Zhou Y, Sceusi E, Tozzi F, Ye XC, Mani SA, Ellis LM (2012) Overexpression of snail induces epithelial-mesenchymal transition and a cancer stem cell-like phenotype in human colorectal cancer cells. Cancer Med 1:5–16

Faust JJ, Balabiyev A, Heddleston JM, Podolnikova NP, Baluch DP, Chew TL, Ugarova TP (2019) An actin-based protrusion originating from a podosome-enriched region initiates macrophage fusion. Mol Biol Cell 30:2254–2267

Fei F, Li C, Wang X, Du J, Liu K, Li B, Yao P, Li Y, Zhang S (2019) Syncytin 1, CD9, and CD47 regulating cell fusion to form PGCCs associated with cAMP/PKA and JNK signaling pathway. Cancer Med 8:3047–3058

Fernandes C, Prabhu P, Juvale K, Suares D, Yc M (2019) Cancer cell fusion: a potential target to tackle drug-resistant and metastatic cancer cells. Drug Discov Today 24:1836–1844

Fidler IJ (2003) The pathogenesis of cancer metastasis: the 'seed and soil' hypothesis revisited. Nat Rev Cancer 3:453–458

Frost P, Kerbel RS, Bauer E, Tartamella-Biondo R, Cefalu W (1983) Mutagen treatment as a means for selecting immunogenic variants from otherwise poorly immunogenic malignant murine tumors. Cancer Res 43:125–132

Fu Y, Zhuang X, Xia X, Li X, Xiao K, Liu X (2021) Correlation between promoter hypomethylation and increased expression of syncytin-1 in non-small cell lung cancer. Int J Gen Med 14:957–965

Gal H, Krizhanovsky V (2014) Cell fusion induced senescence. Aging (Albany NY) 6:353–354

Ganem NJ, Godinho SA, Pellman D (2009) A mechanism linking extra centrosomes to chromosomal instability. Nature 460:278–282

Gao Y, Bado I, Wang H, Zhang W, Rosen JM, Zhang XH (2019) Metastasis organotropism: redefining the congenial soil. Dev Cell 49:375–391

Garcia-Pardo M, Makarem M, Li JJN, Kelly D, Leighl NB (2022) Integrating circulating-free DNA (cfDNA) analysis into clinical practice: opportunities and challenges. Br J Cancer 127:592–602

Garraway LA, Widlund HR, Rubin MA, Getz G, Berger AJ, Ramaswamy S, Beroukhim R, Milner DA, Granter SR, Du J, Lee C, Wagner SN, Li C, Golub TR, Rimm DL, Meyerson ML, Fisher DE, Sellers WR (2005) Integrative genomic analyses identify MITF as a lineage survival oncogene amplified in malignant melanoma. Nature 436:117–122

Gast CE, Silk AD, Zarour L, Riegler L, Burkhart JG, Gustafson KT, Parappilly MS, Roh-Johnson M, Goodman JR, Olson B, Schmidt M, Swain JR, Davies PS, Shasthri V, Iizuka S, Flynn P, Watson S, Korkola J, Courtneidge SA, Fischer JM, Jaboin J, Billingsley KG, Lopez CD, Burchard J, Gray J, Coussens LM, Sheppard BC, Wong MH (2018) Cell fusion potentiates tumor heterogeneity and reveals circulating hybrid cells that correlate with stage and survival. Sci Adv 4:eaat7828

Gemenetzis G, Groot VP, Blair AB, Ding D, Thakker SS, Fishman EK, Cameron JL, Makary MA, Weiss MJ, Wolfgang CL, He J (2018) Incidence and risk factors for abdominal occult metastatic disease in patients with pancreatic adenocarcinoma. J Surg Oncol 118:1277–1284

Geraldo LH, Garcia C, Xu Y, Leser FS, Grimaldi I, De Camargo Magalhaes ES, Dejaegher J, Solie L, Pereira CM, Correia AH, De Vleeschouwer S, Tavitian B, Canedo NHS, Mathivet T, Thomas JL, Eichmann A, Lima FRS (2023) CCL21-CCR7 signaling promotes microglia/macrophage recruitment and chemotherapy resistance in glioblastoma. Cell Mol Life Sci 80:179

Godinho SA, Kwon M, Pellman D (2009) Centrosomes and cancer: how cancer cells divide with too many centrosomes. Cancer Metastasis Rev 28:85–98

Goldenberg DM, Zagzag D, Heselmeyer-Haddad KM, Berroa Garcia LY, Ried T, Loo M, Chang CH, Gold DV (2012) Horizontal transmission and retention of malignancy, as well as functional human genes, after spontaneous fusion of human glioblastoma and hamster host cells in vivo. Int J Cancer 131:49–58

Grandi N, Tramontano E (2018) HERV envelope proteins: physiological role and pathogenic potential in cancer and autoimmunity. Front Microbiol 9:462

Hamilton KE, Noubissi FK, Katti PS, Hahn CM, Davey SR, Lundsmith ET, Klein-Szanto AJ, Rhim AD, Spiegelman VS, Rustgi AK (2013) IMP1 promotes tumor growth, dissemination and a tumor-initiating cell phenotype in colorectal cancer cell xenografts. Carcinogenesis 34:2647–2654

Hamoud N, Tran V, Croteau LP, Kania A, Cote JF (2014) G-protein coupled receptor BAI3 promotes myoblast fusion in vertebrates. Proc Natl Acad Sci USA 111:3745–3750

Hanahan D, Weinberg RA (2000) The hallmarks of cancer. Cell 100:57–70

Hanahan D, Weinberg RA (2011) Hallmarks of cancer: the next generation. Cell 144:646–674

Haneef K, Naeem N, Khan I, Iqbal H, Kabir N, Jamall S, Zahid M, Salim A (2014) Conditioned medium enhances the fusion capability of rat bone marrow mesenchymal stem cells and cardiomyocytes. Mol Biol Rep 41:3099–3112

Hass R (2020) Role of MSC in the tumor microenvironment. Cancers (Basel) 12:2107

Hass R, Von Der Ohe J, Ungefroren H (2019) Potential role of MSC/cancer cell fusion and EMT for breast cancer stem cell formation. Cancers (Basel) 11:1432

Hass R, Von Der Ohe J, Dittmar T (2021a) Cancer cell fusion and post-hybrid selection process (PHSP). Cancers (Basel) 13:4636

Hass R, Von Der Ohe J, Dittmar T (2021b) Hybrid formation and fusion of cancer cells in vitro and in vivo. Cancers (Basel) 13:4496

He X, Li B, Shao Y, Zhao N, Hsu Y, Zhang Z, Zhu L (2015) Cell fusion between gastric epithelial cells and mesenchymal stem cells results in epithelial-to-mesenchymal transition and malignant transformation. BMC Cancer 15:24

He H, Buckley M, Britton B, Mu Y, Warner K, Kumar S, Cory TJ (2018) Polarized macrophage subsets differentially express the drug efflux transporters MRP1 and BCRP, resulting in altered HIV production. Antivir Chem Chemother 26:2040206617745168

Henn TE, Anderson AN, Hollett YR, Sutton TL, Walker BS, Swain JR, Sauer DA, Clayburgh DR, Wong MH (2021) Circulating hybrid cells predict presence of occult nodal metastases in oral cavity carcinoma. Head Neck 43:2193–2201

Hernandez JM, Podbilewicz B (2017) The hallmarks of cell-cell fusion. Development 144:4481–4495

Heussner RT, Whalen RM, Anderson A, Theison H, Baik J, Gibbs S, Wong MH, Chang YH (2024) Quantitative image analysis pipeline for detecting circulating hybrid cells in immunofluorescence images with human-level accuracy. Cytometry A 105:345–355

Hindi SM, Tajrishi MM, Kumar A (2013) Signaling mechanisms in mammalian myoblast fusion. Sci Signal 6:re2

Hiotis SP, Weber SM, Cohen AM, Minsky BD, Paty PB, Guillem JG, Wagman R, Saltz LB, Wong WD (2002) Assessing the predictive value of clinical complete response to neoadjuvant therapy for rectal cancer: an analysis of 488 patients. J Am Coll Surg 194:131–135. discussion 135-6

Hochreiter-Hufford AE, Lee CS, Kinchen JM, Sokolowski JD, Arandjelovic S, Call JA, Klibanov AL, Yan Z, Mandell JW, Ravichandran KS (2013) Phosphatidylserine receptor BAI1 and apoptotic cells as new promoters of myoblast fusion. Nature 497:263–267

Holland AJ, Cleveland DW (2009) Boveri revisited: chromosomal instability, aneuploidy and tumorigenesis. Nat Rev Mol Cell Biol 10:478–487

Horsley V, Pavlath GK (2004) Forming a multinucleated cell: molecules that regulate myoblast fusion. Cells Tissues Organs 176:67–78

Hotokezaka H, Sakai E, Ohara N, Hotokezaka Y, Gonzales C, Matsuo K, Fujimura Y, Yoshida N, Nakayama K (2007) Molecular analysis of RANKL-independent cell fusion of osteoclast-like cells induced by TNF-alpha, lipopolysaccharide, or peptidoglycan. J Cell Biochem 101:122–134

Hu C, He Y, Liu D, Zhao L, Fang S, Tan B, Dong S, Wang Y, He T, Bi Y (2021) Hypoxia preconditioning promotes the proliferation and migration of human urine-derived stem cells in chronically injured liver of mice by upregulating CXCR4. Stem Cells Dev 30:526–536

Huang Q, Chen H, Wang F, Brost BC, Li J, Gao Y, Li Z, Gao Y, Jiang SW (2014) Reduced syncytin-1 expression in choriocarcinoma BeWo cells activates the calpain1-AIF-mediated apoptosis, implication for preeclampsia. Cell Mol Life Sci 71:3151–3164

Huang CM, Yan TL, Xu Z, Wang M, Zhou XC, Jiang EH, Liu K, Shao Z, Shang ZJ (2018) Hypoxia enhances fusion of oral squamous carcinoma cells and epithelial cells partly via the epithelial-mesenchymal transition of epithelial cells. Biomed Res Int 2018:5015203

Ishizawa K, Yamanaka M, Saiki Y, Miyauchi E, Fukushige S, Akaishi T, Asao A, Mimori T, Saito R, Tojo Y, Yamashita R, Abe M, Sakurada A, Pham NA, Li M, Okada Y, Ishii T, Ishii N, Kobayashi S, Nagasaki M, Ichinose M, Tsao MS, Horii A (2019) CD45(+)CD326(+) cells are predictive of poor prognosis in non-small cell lung cancer patients. Clin Cancer Res 25:6756–6763

Jansen ID, Vermeer JA, Bloemen V, Stap J, Everts V (2012) Osteoclast fusion and fission. Calcif Tissue Int 90:515–522

Jeong J, Conboy IM (2011) Phosphatidylserine directly and positively regulates fusion of myoblasts into myotubes. Biochem Biophys Res Commun 414:9–13

Jiang C, Wu J (2024) Hypothesis: hematogenous metastatic cancer cells of solid tumors may disguise themselves as memory macrophages for metastasis. Front Oncol 14:1412296

Kasi PM, Aushev VN, Ensor J, Langer N, Wang C, Cannon T, Berim L, Feinstein T, Grothey A, Mccollom J, Kalmadi S, Zakari A, Dayyani F, Gravenor D, Meyer J, Sharif S, Jurdi A, Liu M, Aleshin A, Kopetz S (2024) Circulating tumor DNA (ctDNA) for informing adjuvant chemotherapy (ACT) in stage II/III colorectal cancer (CRC): interim analysis of BESPOKE CRC study. J Clin Oncol 42:267127519

Kerbel RS, Lagarde AE, Dennis JW, Donaghue TP (1983) Spontaneous fusion in vivo between normal host and tumor cells: possible contribution to tumor progression and metastasis studied with a lectin-resistant mutant tumor. Mol Cell Biol 3:523–538

Kim S, Park BK, Seo JH, Choi J, Choi JW, Lee CK, Chung JB, Park Y, Kim DW (2020) Carbohydrate antigen 19-9 elevation without evidence of malignant or pancreatobiliary diseases. Sci Rep 10:8820

Knerr I, Huppertz B, Weigel C, Dotsch J, Wich C, Schild RL, Beckmann MW, Rascher W (2004) Endogenous retroviral syncytin: compilation of experimental research on syncytin and its possible role in normal and disturbed human placentogenesis. Mol Hum Reprod 10:581–588

Kong D, Banerjee S, Ahmad A, Li Y, Wang Z, Sethi S, Sarkar FH (2010) Epithelial to mesenchymal transition is mechanistically linked with stem cell signatures in prostate cancer cells. PLoS One 5:e12445

Krebs MG, Hou JM, Ward TH, Blackhall FH, Dive C (2010) Circulating tumour cells: their utility in cancer management and predicting outcomes. Ther Adv Med Oncol 2:351–365

Kudo Y, Boyd CA, Sargent IL, Redman CW (2003) Hypoxia alters expression and function of syncytin and its receptor during trophoblast cell fusion of human placental BeWo cells: implications for impaired trophoblast syncytialisation in pre-eclampsia. Biochim Biophys Acta 1638:63–71

Laberge GS, Duvall E, Grasmick Z, Haedicke K, Pawelek J (2017) A melanoma lymph node metastasis with a donor-patient hybrid genome following bone marrow transplantation: a second case of leucocyte-tumor cell hybridization in cancer metastasis. PLoS One 12:e0168581

Laberge GS, Duvall E, Haedicke K, Pawelek J (2019) Leukocyte(−)cancer cell fusion-genesis of a deadly journey. Cells 8:170

Langbein M, Strick R, Strissel PL, Vogt N, Parsch H, Beckmann MW, Schild RL (2008) Impaired cytotrophoblast cell-cell fusion is associated with reduced Syncytin and increased apoptosis in patients with placental dysfunction. Mol Reprod Dev 75:175–183
Larsen JM, Christensen IJ, Nielsen HJ, Hansen U, Bjerregaard B, Talts JF, Larsson LI (2009) Syncytin immunoreactivity in colorectal cancer: potential prognostic impact. Cancer Lett 280: 44–49
Larsson LI, Holck S, Christensen IJ (2007) Prognostic role of syncytin expression in breast cancer. Hum Pathol 38:726–731
Lartigue L, Merle C, Lagarde P, Delespaul L, Lesluyes T, Le Guellec S, Perot G, Leroy L, Coindre JM, Chibon F (2020) Genome remodeling upon mesenchymal tumor cell fusion contributes to tumor progression and metastatic spread. Oncogene 39:4198–4211
Lazova R, Laberge GS, Duvall E, Spoelstra N, Klump V, Sznol M, Cooper D, Spritz RA, Chang JT, Pawelek JM (2013) A melanoma brain metastasis with a donor-patient hybrid genome following bone marrow transplantation: first evidence for fusion in human cancer. PLoS One 8:e66731
Lee X, Keith JC Jr, Stumm N, Moutsatsos I, Mccoy JM, Crum CP, Genest D, Chin D, Ehrenfels C, Pijnenborg R, Van Assche FA, Mi S (2001) Downregulation of placental syncytin expression and abnormal protein localization in pre-eclampsia. Placenta 22:808–812
Lehka L, Redowicz MJ (2020) Mechanisms regulating myoblast fusion: a multilevel interplay. Semin Cell Dev Biol 104:81–92
Leopold Wager CM, Wormley FL, Jr (2014) Classical versus alternative macrophage activation: the Ying and the Yang in host defense against pulmonary fungal infections. Mucosal Immunol 7: 1023–1035
Li H, Yang L, Fu H, Yan J, Wang Y, Guo H, Hao X, Xu X, Jin T, Zhang N (2013) Association between Galphai2 and Elmo1/Dock180 connects chemokine signalling with Rac activation and metastasis. Nat Commun 4:1706
Li H, Feng Z, Tsang TC, Tang T, Jia X, He X, Pennington ME, Badowski MS, Liu AK, Chen D, Harris DT, Martinez J, Meade-Tollin LC (2014) Fusion of HepG2 cells with mesenchymal stem cells increases cancer-associated and malignant properties: an in vivo metastasis model. Oncol Rep 32:539–547
Li H, Meng QH, Noh H, Somaiah N, Torres KE, Xia X, Batth IS, Joseph CP, Liu M, Wang R, Li S (2018) Cell-surface vimentin-positive macrophage-like circulating tumor cells as a novel biomarker of metastatic gastrointestinal stromal tumors. Onco Targets Ther 7:e1420450
Li X, Fu Y, Xia X, Zhang X, Xiao K, Zhuang X, Zhang Y (2019) Knockdown of SP1/Syncytin1 axis inhibits the proliferation and metastasis through the AKT and ERK1/2 signaling pathways in non-small cell lung cancer. Cancer Med 8:5750–5759
Liu C, Xu J, Wen F, Yang F, Li X, Geng D, Li L, Chen J, Zheng J (2019) Upregulation of syncytin-1 promotes invasion and metastasis by activating epithelial-mesenchymal transition-related pathway in endometrial carcinoma. Onco Targets Ther 12:31–40
Liu P, Wang L, Yu H (2024) Polyploid giant cancer cells: origin, possible pathways of formation, characteristics, and mechanisms of regulation. Front Cell Dev Biol 12:1410637
Lizier M, Anselmo A, Mantero S, Ficara F, Paulis M, Vezzoni P, Lucchini F, Pacchiana G (2016) Fusion between cancer cells and macrophages occurs in a murine model of spontaneous neu+ breast cancer without increasing its metastatic potential. Oncotarget 7:60793–60806
Long H, Xiang T, Qi W, Huang J, Chen J, He L, Liang Z, Guo B, Li Y, Xie R, Zhu B (2015) CD133+ ovarian cancer stem-like cells promote non-stem cancer cell metastasis via CCL5 induced epithelial-mesenchymal transition. Oncotarget 6:5846–5859
Lopez-Collazo E, Hurtado-Navarro L (2025) Cell fusion as a driver of metastasis: re-evaluating an old hypothesis in the age of cancer heterogeneity. Front Immunol 16:1524781
Lu X, Kang Y (2009) Efficient acquisition of dual metastasis organotropism to bone and lung through stable spontaneous fusion between MDA-MB-231 variants. Proc Natl Acad Sci USA 106:9385–9390
Luo F, Liu T, Wang J, Li J, Ma P, Ding H, Feng G, Lin D, Xu Y, Yang K (2016) Bone marrow mesenchymal stem cells participate in prostate carcinogenesis and promote growth of prostate cancer by cell fusion in vivo. Oncotarget 7:30924–30934

Lustberg MB, Balasubramanian P, Miller B, Garcia-Villa A, Deighan C, Wu Y, Carothers S, Berger M, Ramaswamy B, Macrae ER, Wesolowski R, Layman RM, Mrozek E, Pan X, Summers TA, Shapiro CL, Chalmers JJ (2014) Heterogeneous atypical cell populations are present in blood of metastatic breast cancer patients. Breast Cancer Res 16:R23

Ma X, Xiao Z, Li X, Wang F, Zhang J, Zhou R, Wang J, Liu L (2014) Prognostic role of circulating tumor cells and disseminated tumor cells in patients with prostate cancer: a systematic review and meta-analysis. Tumour Biol 35:5551–5560

Maliniemi P, Vincendeau M, Mayer J, Frank O, Hahtola S, Karenko L, Carlsson E, Mallet F, Seifarth W, Leib-Mosch C, Ranki A (2013) Expression of human endogenous retrovirus-w including syncytin-1 in cutaneous T-cell lymphoma. PLoS One 8:e76281

Malyshev I, Malyshev Y (2015) Current concept and update of the macrophage plasticity concept: intracellular mechanisms of reprogramming and M3 macrophage "switch" phenotype. Biomed Res Int 2015:341308

Mandel K, Yang Y, Schambach A, Glage S, Otte A, Hass R (2013) Mesenchymal stem cells directly interact with breast cancer cells and promote tumor cell growth in vitro and in vivo. Stem Cells Dev 22:3114–3127

Mani K, Deng D, Lin C, Wang M, Hsu ML, Zaorsky NG (2024) Causes of death among people living with metastatic cancer. Nat Commun 15:1519

Manjunath Y, Mitchem JB, Suvilesh KN, Avella DM, Kimchi ET, Staveley-O'carroll KF, Deroche CB, Pantel K, Li G, Kaifi JT (2020a) Circulating giant tumor-macrophage fusion cells are independent prognosticators in patients with NSCLC. J Thorac Oncol 15:1460–1471

Manjunath Y, Porciani D, Mitchem JB, Suvilesh KN, Avella DM, Kimchi ET, Staveley-O'carroll KF, Burke DH, Li G, Kaifi JT (2020b) Tumor-cell-macrophage fusion cells as liquid biomarkers and tumor enhancers in cancer. Int J Mol Sci 21:1872

Matsumoto T, Wakefield L, Peters A, Peto M, Spellman P, Grompe M (2021) Proliferative polyploid cells give rise to tumors via ploidy reduction. Nat Commun 12:646

Matsuura K, Jigami T, Taniue K, Morishita Y, Adachi S, Senda T, Nonaka A, Aburatani H, Nakamura T, Akiyama T (2011) Identification of a link between Wnt/beta-catenin signalling and the cell fusion pathway. Nat Commun 2:548

May AM, Batoon L, Mccauley LK, Keller ET (2023) The role of tumor epithelial-mesenchymal transition and macrophage crosstalk in cancer progression. Curr Osteoporos Rep 21:117–127

Mego M, Karaba M, Minarik G, Benca J, Silvia J, Sedlackova T, Manasova D, Kalavska K, Pindak D, Cristofanilli M, Reuben JM, Mardiak J (2019) Circulating tumor cells with epithelial-to-mesenchymal transition phenotypes associated with inferior outcomes in primary breast cancer. Anticancer Res 39:1829–1837

Melzer C, Von Der Ohe J, Hass R (2018) In vitro fusion of normal and neoplastic breast epithelial cells with human mesenchymal stroma/stem cells partially involves tumor necrosis factor receptor signaling. Stem Cells 36:977–989

Melzer C, Von Der Ohe J, Hass R (2019) In vivo cell fusion between mesenchymal stroma/stem-like cells and breast cancer cells. Cancers (Basel) 11:185

Melzer C, Ohe JV, Hass R (2020) Altered tumor plasticity after different cancer cell fusions with MSC. Int J Mol Sci 21:8347

Menyailo ME, Zainullina VR, Khozyainova AA, Tashireva LA, Zolotareva SY, Gerashchenko TS, Alifanov VV, Savelieva OE, Grigoryeva ES, Tarabanovskaya NA, Popova NO, Choinzonov EL, Cherdyntseva NV, Perelmuter VM, Denisov EV (2023) Heterogeneity of circulating epithelial cells in breast cancer at single-cell resolution: identifying tumor and hybrid cells. Adv Biol (Weinh) 7:e2200206

Merle C, Lagarde P, Lartigue L, Chibon F (2021) Acquisition of cancer stem cell capacities after spontaneous cell fusion. BMC Cancer 21:241

Mi R, Pan C, Bian X, Song L, Tian W, Cao F, Yin J, Peng H, Ma J (2012) Fusion between tumor cells enhances melanoma metastatic potential. J Cancer Res Clin Oncol 138:1651–1658

Miller FR, Mohamed AN, Mceachern D (1989) Production of a more aggressive tumor cell variant by spontaneous fusion of two mouse tumor subpopulations. Cancer Res 49:4316–4321

Miroshnychenko D, Baratchart E, Ferrall-Fairbanks MC, Velde RV, Laurie MA, Bui MM, Tan AC, Altrock PM, Basanta D, Marusyk A (2021) Spontaneous cell fusions as a mechanism of parasexual recombination in tumour cell populations. Nat Ecol Evol 5:379–391

Mohr M, Tosun S, Arnold WH, Edenhofer F, Zanker KS, Dittmar T (2015) Quantification of cell fusion events human breast cancer cells and breast epithelial cells using a Cre-LoxP-based double fluorescence reporter system. Cell Mol Life Sci 72:3769–3782

Montalban-Hernandez K, Cantero-Cid R, Casalvilla-Duenas JC, Avendano-Ortiz J, Marin E, Lozano-Rodriguez R, Terron-Arcos V, Vicario-Bravo M, Marcano C, Saavedra-Ambrosy J, Prado-Montero J, Valentin J, Perez De Diego R, Cordoba L, Pulido E, Del Fresno C, Duenas M, Lopez-Collazo E (2022) Colorectal cancer stem cells fuse with monocytes to form tumour hybrid cells with the ability to migrate and evade the immune system. Cancers (Basel) 14:3445

Montoya Mira JL, Quentel A, Patel RK, Keith D, Sousa M, Minnier J, Kingston BR, David L, Esener SC, Sears RC, Lopez CD, Sheppard BC, Demirci U, Wong MH, Fischer JM (2025) Early detection of pancreatic cancer by a high-throughput protease-activated nanosensor assay. Sci Transl Med 17:eadq3110

Mu Z, Wang C, Ye Z, Rossi G, Sun C, Li L, Zhu Z, Yang H, Cristofanilli M (2017) Prognostic values of cancer associated macrophage-like cells (CAML) enumeration in metastatic breast cancer. Breast Cancer Res Treat 165:733–741

Nakamura Y, Watanabe J, Akazawa N, Hirata K, Kataoka K, Yokota M, Kato K, Kotaka M, Kagawa Y, Yeh KH, Mishima S, Yukami H, Ando K, Miyo M, Misumi T, Yamazaki K, Ebi H, Okita K, Hamabe A, Sokuoka H, Kobayashi S, Laliotis G, Aushev VN, Sharma S, Jurdi A, Liu MC, Aleshin A, Rabinowitz M, Bando H, Taniguchi H, Takemasa I, Kato T, Kotani D, Mori M, Yoshino T, Oki E (2024) ctDNA-based molecular residual disease and survival in resectable colorectal cancer. Nat Med 30:3272–3283

Nel I, Jehn U, Gauler T, Hoffmann AC (2014) Individual profiling of circulating tumor cell composition in patients with non-small cell lung cancer receiving platinum based treatment. Transl Lung Cancer Res 3:100–106

Nitschke C, Markmann B, Konczalla L, Kropidlowski J, Pereira-Veiga T, Scognamiglio P, Schonrock M, Sinn M, Tolle M, Izbicki J, Pantel K, Uzunoglu FG, Wikman H (2022) Circulating cancer associated macrophage-like cells as a potential new prognostic marker in pancreatic ductal adenocarcinoma. Biomedicine 10:2955

Noubissi FK, Ogle BM (2016) Cancer cell fusion: mechanisms slowly unravel. Int J Mol Sci 17: 1587

Noubissi FK, Harkness T, Alexander CM, Ogle BM (2015) Apoptosis-induced cancer cell fusion: a mechanism of breast cancer metastasis. FASEB J 29:4036–4045

Noubissi FK, Odubanjo OV, Ogle BM, Tchounwou PB (2024) Mechanisms of cell fusion in cancer. Results Probl Cell Differ 71:407–432

Ogle BM, Cascalho M, Platt JL (2005) Biological implications of cell fusion. Nat Rev Mol Cell Biol 6:567–575

Oren-Suissa M, Podbilewicz B (2007) Cell fusion during development. Trends Cell Biol 17:537–546

Paget S (1989) The distribution of secondary growths in cancer of the breast. 1889. Cancer Metastasis Rev 8:98–101

Parappilly MS, Chin Y, Whalen RM, Anderson AN, Robinson TS, Strgar L, Sutton TL, Conley P, Klocke C, Gibbs SL, Chang YH, Wu G, Wong MH, Skalet AH (2022) Circulating neoplastic-immune hybrid cells predict metastatic progression in uveal melanoma. Cancers (Basel) 14: 4617

Park W, Chawla A, O'reilly EM (2021) Pancreatic cancer: a review. JAMA 326:851–862

Parra-Robert M, Santos VM, Canis SM, Pla XF, Fradera JMA, Porto RM (2018) Relationship between ca 19.9 and the Lewis phenotype: options to improve diagnostic efficiency. Anticancer Res 38:5883–5888

Passerini V, Ozeri-Galai E, De Pagter MS, Donnelly N, Schmalbrock S, Kloosterman WP, Kerem B, Storchova Z (2016) The presence of extra chromosomes leads to genomic instability. Nat Commun 7:10754

Pastushenko I, Brisebarre A, Sifrim A, Fioramonti M, Revenco T, Boumahdi S, Van Keymeulen A, Brown D, Moers V, Lemaire S, De Clercq S, Minguijon E, Balsat C, Sokolow Y, Dubois C, De Cock F, Scozzaro S, Sopena F, Lanas A, D'haene N, Salmon I, Marine JC, Voet T, Sotiropoulou PA, Blanpain C (2018) Identification of the tumour transition states occurring during EMT. Nature 556:463–468

Patel RK, Parappilly M, Farley HC, Latour EJ, Wang LG, Nair AM, Lu ES, Sims Z, Park B, Nelson K, Mayo SC, Mills GB, Sheppard BC, Chang YH, Gibbs SL, Kardosh A, Lopez CD, Wong MH (2024a) Circulating neoplastic-immune hybrid cells are biomarkers of occult metastasis and treatment response in pancreatic cancer. Cancers (Basel) 16:3650

Patel RK, Parappilly M, Rahman S, Schwantes IR, Sewell M, Giske NR, Whalen RM, Durmus NG, Wong MH (2024b) The hallmarks of circulating hybrid cells. Results Probl Cell Differ 71:467–485

Patel RK, Parappilly MS, Walker BS, Heussner RT, Fung A, Chang YH, Kardosh A, Lopez CD, Mayo SC, Wong MH (2024c) Exploratory analyses of circulating neoplastic-immune hybrid cells as prognostic biomarkers in advanced intrahepatic cholangiocarcinoma. Int J Mol Sci 25:9198

Patel RK, Parappilly MS, Rahman S, Lu E, Nair A, Latour E, Mayo S, Cote G, Tsikitis V, Wong M (2024d) Circulating hybrid cells as a blood based screening tool for colorectal neoplasia. Ann Surg Oncol 31:S13–S14

Pawelek JM (2000) Tumour cell hybridization and metastasis revisited. Melanoma Res 10:507–514

Pawelek JM (2005) Tumour-cell fusion as a source of myeloid traits in cancer. Lancet Oncol 6:988–993

Pawelek JM (2014) Fusion of bone marrow-derived cells with cancer cells: metastasis as a secondary disease in cancer. Chin J Cancer 33:133–139

Pawelek JM, Chakraborty AK (2008a) The cancer cell–leukocyte fusion theory of metastasis. Adv Cancer Res 101:397–444

Pawelek JM, Chakraborty AK (2008b) Fusion of tumour cells with bone marrow-derived cells: a unifying explanation for metastasis. Nat Rev Cancer 8:377–386

Pirrello A, Killingsworth M, Spring K, Rasko JEJ, Yeo D (2024) Cancer-associated macrophage-like cells as a prognostic biomarker in solid tumors. J Liq Biopsy 6:100275

Potgens AJ, Schmitz U, Bose P, Versmold A, Kaufmann P, Frank HG (2002) Mechanisms of syncytial fusion: a review. Placenta 23(Suppl A):S107–S113

Powell AE, Anderson EC, Davies PS, Silk AD, Pelz C, Impey S, Wong MH (2011) Fusion between intestinal epithelial cells and macrophages in a cancer context results in nuclear reprogramming. Cancer Res 71:1497–1505

Puram SV, Tirosh I, Parikh AS, Patel AP, Yizhak K, Gillespie S, Rodman C, Luo CL, Mroz EA, Emerick KS, Deschler DG, Varvares MA, Mylvaganam R, Rozenblatt-Rosen O, Rocco JW, Faquin WC, Lin DT, Regev A, Bernstein BE (2017) Single-cell transcriptomic analysis of primary and metastatic tumor ecosystems in head and neck cancer. Cell 171:1611–1624. e24

Qin J, Zhang X, Tan B, Zhang S, Yin C, Xue Q, Zhang Z, Ren H, Chen J, Liu M, Qian M, Du B (2020) Blocking P2X7-mediated macrophage polarization overcomes treatment resistance in lung cancer. Cancer Immunol Res 8:1426–1439

Quinton RJ, Didomizio A, Vittoria MA, Kotynkova K, Ticas CJ, Patel S, Koga Y, Vakhshoorzadeh J, Hermance N, Kuroda TS, Parulekar N, Taylor AM, Manning AL, Campbell JD, Ganem NJ (2021) Whole-genome doubling confers unique genetic vulnerabilities on tumour cells. Nature 590:492–497

Rachkovsky M, Sodi S, Chakraborty A, Avissar Y, Bolognia J, Mcniff JM, Platt J, Bermudes D, Pawelek J (1998) Melanoma x macrophage hybrids with enhanced metastatic potential. Clin Exp Metastasis 16:299–312

Ramakrishnan M, Mathur SR, Mukhopadhyay A (2013) Fusion-derived epithelial cancer cells express hematopoietic markers and contribute to stem cell and migratory phenotype in ovarian carcinoma. Cancer Res 73:5360–5370

Rappa G, Mercapide J, Lorico A (2012) Spontaneous formation of tumorigenic hybrids between breast cancer and multipotent stromal cells is a source of tumor heterogeneity. Am J Pathol 180: 2504–2515

Rasheed ZA, Yang J, Wang Q, Kowalski J, Freed I, Murter C, Hong SM, Koorstra JB, Rajeshkumar NV, He X, Goggins M, Iacobuzio-Donahue C, Berman DM, Laheru D, Jimeno A, Hidalgo M, Maitra A, Matsui W (2010) Prognostic significance of tumorigenic cells with mesenchymal features in pancreatic adenocarcinoma. J Natl Cancer Inst 102:340–351

Reduzzi C, Vismara M, Gerratana L, Silvestri M, De Braud F, Raspagliesi F, Verzoni E, Di Cosimo S, Locati LD, Cristofanilli M, Daidone MG, Cappelletti V (2020) The curious phenomenon of dual-positive circulating cells: longtime overlooked tumor cells. Semin Cancer Biol 60:344–350

Renaud SJ, Jeyarajah MJ (2022) How trophoblasts fuse: an in-depth look into placental syncytiotrophoblast formation. Cell Mol Life Sci 79:433

Ribatti D, Tamma R (2018) A revisited concept. Tumors: wounds that do not heal. Crit Rev Oncol Hematol 128:65–69

Rival CM, Xu W, Shankman LS, Morioka S, Arandjelovic S, Lee CS, Wheeler KM, Smith RP, Haney LB, Isakson BE, Purcell S, Lysiak JJ, Ravichandran KS (2019) Phosphatidylserine on viable sperm and phagocytic machinery in oocytes regulate mammalian fertilization. Nat Commun 10:4456

Rizvi AZ, Swain JR, Davies PS, Bailey AS, Decker AD, Willenbring H, Grompe M, Fleming WH, Wong MH (2006) Bone marrow-derived cells fuse with normal and transformed intestinal stem cells. Proc Natl Acad Sci USA 103:6321–6325

Robert G, Gaggioli C, Bailet O, Chavey C, Abbe P, Aberdam E, Sabatie E, Cano A, Garcia De Herreros A, Ballotti R, Tartare-Deckert S (2006) Sparc represses E-cadherin and induces mesenchymal transition during melanoma development. Cancer Res 66:7516–7523

Ruebner M, Strissel PL, Ekici AB, Stiegler E, Dammer U, Goecke TW, Faschingbauer F, Fahlbusch FB, Beckmann MW, Strick R (2013) Reduced syncytin-1 expression levels in placental syndromes correlates with epigenetic hypermethylation of the ERVW-1 promoter region. PLoS One 8:e56145

Sabe H, Yahara Y, Ishii M (2024) Cell fusion dynamics: mechanisms of multinucleation in osteoclasts and macrophages. Inflamm Regen 44:49

Seyfried TN, Huysentruyt LC (2013) On the origin of cancer metastasis. Crit Rev Oncog 18:43–73

Shabo I, Olsson H, Sun XF, Svanvik J (2009) Expression of the macrophage antigen CD163 in rectal cancer cells is associated with early local recurrence and reduced survival time. Int J Cancer 125:1826–1831

Siegel RL, Giaquinto AN, Jemal A (2024) Cancer statistics, 2024. CA Cancer J Clin 74:12–49

Singh K, Gautam PK (2023) Macrophage infiltration in 3D cancer spheroids to recapitulate the TME and unveil interactions within cancer cells and macrophages to modulate chemotherapeutic drug efficacy. BMC Cancer 23:1201

Skokos EA, Charokopos A, Khan K, Wanjala J, Kyriakides TR (2011) Lack of TNF-alpha-induced MMP-9 production and abnormal E-cadherin redistribution associated with compromised fusion in MCP-1-null macrophages. Am J Pathol 178:2311–2321

Smerage JB, Barlow WE, Hortobagyi GN, Winer EP, Leyland-Jones B, Srkalovic G, Tejwani S, Schott AF, O'rourke MA, Lew DL, Doyle GV, Gralow JR, Livingston RB, Hayes DF (2014) Circulating tumor cells and response to chemotherapy in metastatic breast cancer: SWOG S0500. J Clin Oncol 32:3483–3489

Smith BA, Balanis NG, Nanjundiah A, Sheu KM, Tsai BL, Zhang Q, Park JW, Thompson M, Huang J, Witte ON, Graeber TG (2018) A human adult stem cell signature Marks aggressive variants across epithelial cancers. Cell Rep 24:3353–3366. e5

Sodi SA, Chakraborty AK, Platt JT, Kolesnikova N, Rosemblat S, Keh-Yen A, Bolognia JL, Rachkovsky ML, Orlow SJ, Pawelek JM (1998) Melanoma x macrophage fusion hybrids acquire increased melanogenesis and metastatic potential: altered N-glycosylation as an underlying mechanism. Pigment Cell Res 11:299–309

Soe K (2020) Osteoclast fusion: physiological regulation of multinucleation through heterogeneity-potential implications for drug sensitivity. Int J Mol Sci 21:7717

Song K, Zhu F, Zhang HZ, Shang ZJ (2012) Tumor necrosis factor-alpha enhanced fusions between oral squamous cell carcinoma cells and endothelial cells via VCAM-1/VLA-4 pathway. Exp Cell Res 318:1707–1715

Song Y, Zhao Y, Deng Z, Zhao R, Huang Q (2021) Stress-induced polyploid giant cancer cells: unique way of formation and non-negligible characteristics. Front Oncol 11:724781

Stott SL, Hsu CH, Tsukrov DI, Yu M, Miyamoto DT, Waltman BA, Rothenberg SM, Shah AM, Smas ME, Korir GK, Floyd FP Jr, Gilman AJ, Lord JB, Winokur D, Springer S, Irimia D, Nagrath S, Sequist LV, Lee RJ, Isselbacher KJ, Maheswaran S, Haber DA, Toner M (2010) Isolation of circulating tumor cells using a microvortex-generating herringbone-chip. Proc Natl Acad Sci USA 107:18392–18397

Strick R, Ackermann S, Langbein M, Swiatek J, Schubert SW, Hashemolhosseini S, Koscheck T, Fasching PA, Schild RL, Beckmann MW, Strissel PL (2007) Proliferation and cell-cell fusion of endometrial carcinoma are induced by the human endogenous retroviral Syncytin-1 and regulated by TGF-beta. J Mol Med (Berl) 85:23–38

Strissel PL, Ruebner M, Thiel F, Wachter D, Ekici AB, Wolf F, Thieme F, Ruprecht K, Beckmann MW, Strick R (2012) Reactivation of codogenic endogenous retroviral (ERV) envelope genes in human endometrial carcinoma and prestages: emergence of new molecular targets. Oncotarget 3:1204–1219

Suker M, Koerkamp BG, Coene PP, Van Der Harst E, Bonsing BA, Vahrmeijer AL, Mieog JSD, Swijnenburg RJ, Dwarkasing RS, Roos D, Van Eijck CHJ (2019) Yield of staging laparoscopy before treatment of locally advanced pancreatic cancer to detect occult metastases. Eur J Surg Oncol 45:1906–1911

Sun Y, Ouyang DY, Pang W, Tu YQ, Li YY, Shen XM, Tam SC, Yang HY, Zheng YT (2010) Expression of syncytin in leukemia and lymphoma cells. Leuk Res 34:1195–1202

Sun C, Dai X, Zhao D, Wang H, Rong X, Huang Q, Lan Q (2019) Mesenchymal stem cells promote glioma neovascularization in vivo by fusing with cancer stem cells. BMC Cancer 19:1240

Sutton TL, Walker BS, Wong MH (2019) Circulating hybrid cells join the fray of circulating cellular biomarkers. Cell Mol Gastroenterol Hepatol 8:595–607

Sutton TL, Patel RK, Anderson AN, Bowden SG, Whalen R, Giske NR, Wong MH (2022) Circulating cells with macrophage-like characteristics in cancer: the importance of circulating neoplastic-immune hybrid cells in cancer. Cancers (Basel) 14:3871

Takezawa Y, Yoshida K, Miyado K, Sato M, Nakamura A, Kawano N, Sakakibara K, Kondo T, Harada Y, Ohnami N, Kanai S, Miyado M, Saito H, Takahashi Y, Akutsu H, Umezawa A (2011) Beta-catenin is a molecular switch that regulates transition of cell-cell adhesion to fusion. Sci Rep 1:68

Tanjak P, Chaiboonchoe A, Suwatthanarak T, Thanormjit K, Acharayothin O, Chanthercrob J, Parakonthun T, Methasate A, Fischer JM, Wong MH, Chinswangwatanakul V (2024) Tumor-immune hybrid cells evade the immune response and potentiate colorectal cancer metastasis through CTLA4. Clin Exp Med 25:2

Tol J, Koopman M, Miller MC, Tibbe A, Cats A, Creemers GJ, Vos AH, Nagtegaal ID, Terstappen LW, Punt CJ (2010) Circulating tumour cells early predict progression-free and overall survival in advanced colorectal cancer patients treated with chemotherapy and targeted agents. Ann Oncol 21:1006–1012

Toyoshima K, Hayashi A, Kashiwagi M, Hayashi N, Iwatsuki M, Ishimoto T, Baba Y, Baba H, Ohta Y (2015) Analysis of circulating tumor cells derived from advanced gastric cancer. Int J Cancer 137:991–998

Uygur B, Leikina E, Melikov K, Villasmil R, Verma SK, Vary CPH, Chernomordik LV (2019) Interactions with muscle cells boost fusion, stemness, and drug resistance of prostate cancer cells. Mol Cancer Res 17:806–820

Van Den Eijnde SM, Van Den Hoff MJ, Reutelingsperger CP, Van Heerde WL, Henfling ME, Vermeij-Keers C, Schutte B, Borgers M, Ramaekers FC (2001) Transient expression of phosphatidylserine at cell-cell contact areas is required for myotube formation. J Cell Sci 114: 3631–3642

Walker BS, Sutton TL, Zarour L, Hunter JG, Wood SG, Tsikitis VL, Herzig DO, Lopez CD, Chen EY, Mayo SC, Wong MH (2021) Circulating hybrid cells: a novel liquid biomarker of treatment response in gastrointestinal cancers. Ann Surg Oncol 28:8567–8578

Wang Q, Hardie RA, Hoy AJ, Van Geldermalsen M, Gao D, Fazli L, Sadowski MC, Balaban S, Schreuder M, Nagarajah R, Wong JJ, Metierre C, Pinello N, Otte NJ, Lehman ML, Gleave M, Nelson CC, Bailey CG, Ritchie W, Rasko JE, Holst J (2015) Targeting Asct2-mediated glutamine uptake blocks prostate cancer growth and tumour development. J Pathol 236:278–289

Wang Z, Yuan Y, Zhang L, Min Z, Zhou D, Yu S, Wang P, Ju S, Jun L, Fu J (2018) Impact of cell fusion in myeloma marrow microenvironment on tumor progression. Oncotarget 9:30997–31006

Wang HF, Xiang W, Xue BZ, Wang YH, Yi DY, Jiang XB, Zhao HY, Fu P (2021) Cell fusion in cancer hallmarks: current research status and future indications. Oncol Lett 22:530

Wang YN, Ye Y, Zhou D, Guo ZW, Xiong Z, Gong XX, Jiang SW, Chen H (2022) The role of syncytin in placental angiogenesis and fetal growth. Front Cell Dev Biol 10:852561

Weiler J, Dittmar T (2019) Minocycline impairs TNF-alpha-induced cell fusion of M13SV1-Cre cells with MDA-MB-435-PFDR1 cells by suppressing NF-kappaB transcriptional activity and its induction of target-gene expression of fusion-relevant factors. Cell Commun Signal 17:71

Weiler J, Mohr M, Zanker KS, Dittmar T (2018) Matrix metalloproteinase-9 (MMP9) is involved in the TNF-alpha-induced fusion of human M13SV1-Cre breast epithelial cells and human MDA-MB-435-PFDR1 cancer cells. Cell Commun Signal 16:14

Whalen RM, Anderson AN, Jones JA, Sims Z, Chang YH, Nederlof MA, Wong MH, Gibbs SL (2024) Ultra high content analyses of circulating and tumor associated hybrid cells reveal phenotypic heterogeneity. Sci Rep 14:7350

Wyckoff J, Wang W, Lin EY, Wang Y, Pixley F, Stanley ER, Graf T, Pollard JW, Segall J, Condeelis J (2004) A paracrine loop between tumor cells and macrophages is required for tumor cell migration in mammary tumors. Cancer Res 64:7022–7029

Xu MH, Gao X, Luo D, Zhou XD, Xiong W, Liu GX (2014) Emt and acquisition of stem cell-like properties are involved in spontaneous formation of tumorigenic hybrids between lung cancer and bone marrow-derived mesenchymal stem cells. PLoS One 9:e87893

Xue J, Zhu Y, Sun Z, Ji R, Zhang X, Xu W, Yuan X, Zhang B, Yan Y, Yin L, Xu H, Zhang L, Zhu W, Qian H (2015) Tumorigenic hybrids between mesenchymal stem cells and gastric cancer cells enhanced cancer proliferation, migration and stemness. BMC Cancer 15:793

Yahara Y, Nguyen T, Ishikawa K, Kamei K, Alman BA (2022) The origins and roles of osteoclasts in bone development, homeostasis and repair. Development 149:dev199908

Yan Y, Hsu Y, He X, Lu N, Wei W, Zhang Z, Zhu L (2015) Evidence of cell fusion in carcinogen-induced mice gastric carcinoma. Tumour Biol 36:5089–5094

Yan B, Wang J, Liu L (2016) Chemotherapy promotes tumour cell hybridization in vivo. Tumour Biol 37:5025–5030

Yan TL, Wang M, Xu Z, Huang CM, Zhou XC, Jiang EH, Zhao XP, Song Y, Song K, Shao Z, Liu K, Shang ZJ (2017) Up-regulation of syncytin-1 contributes to Tnf-alpha-enhanced fusion between OSCC and HUVECS partly via Wnt/beta-catenin-dependent pathway. Sci Rep 7: 40983

Yang Y, Otte A, Hass R (2015) Human mesenchymal stroma/stem cells exchange membrane proteins and alter functionality during interaction with different tumor cell lines. Stem Cells Dev 24:1205–1222

Yang C, Wang X, To KKW, Cui C, Luo M, Wu S, Huang L, Fu K, Pan C, Liu Z, Fan T, Yang C, Wang F, Fu L (2024) Circulating tumor cells shielded with extracellular vesicle-derived CD45 evade T cell attack to enable metastasis. Signal Transduct Target Ther 9:e84

Yart L, Bastida-Ruiz D, Allard M, Dietrich PY, Petignat P, Cohen M (2022) Linking unfolded protein response to ovarian cancer cell fusion. BMC Cancer 22:622

Ye X, Huang X, Fu X, Zhang X, Lin R, Zhang W, Zhang J, Lu Y (2023) Myeloid-like tumor hybrid cells in bone marrow promote progression of prostate cancer bone metastasis. J Hematol Oncol 16:46

Yilmaz Y, Lazova R, Qumsiyeh M, Cooper D, Pawelek J (2005) Donor Y chromosome in renal carcinoma cells of a female BMT recipient: visualization of putative BMT-tumor hybrids by FISH. Bone Marrow Transplant 35:1021–1024

Yu M, Bardia A, Wittner BS, Stott SL, Smas ME, Ting DT, Isakoff SJ, Ciciliano JC, Wells MN, Shah AM, Concannon KF, Donaldson MC, Sequist LV, Brachtel E, Sgroi D, Baselga J, Ramaswamy S, Toner M, Haber DA, Maheswaran S (2013) Circulating breast tumor cells exhibit dynamic changes in epithelial and mesenchymal composition. Science 339:580–584

Yu H, Liu T, Zhao Z, Chen Y, Zeng J, Liu S, Zhu F (2014) Mutations in 3-long terminal repeat of Herv-W family in chromosome 7 upregulate syncytin-1 expression in urothelial cell carcinoma of the bladder through interacting with c-Myb. Oncogene 33:3947–3958

Yu M, Zhao H, Chen T, Tian Y, Li M, Wu K, Bian Y, Su S, Cao Y, Ning Y, Liu H, Chen ZJ (2018) Mutational analysis of IZUMO1R in women with fertilization failure and polyspermy after in vitro fertilization. J Assist Reprod Genet 35:539–544

Yunna C, Mengru H, Lei W, Weidong C (2020) Macrophage M1/M2 polarization. Eur J Pharmacol 877:173090

Zack TI, Schumacher SE, Carter SL, Cherniack AD, Saksena G, Tabak B, Lawrence MS, Zhsng CZ, Wala J, Mermel CH, Sougnez C, Gabriel SB, Hernandez B, Shen H, Laird PW, Getz G, Meyerson M, Beroukhim R (2013) Pan-cancer patterns of somatic copy number alteration. Nat Genet 45:1134–1140

Zhang Y, Wang F, Ning N, Chen Q, Yang Z, Guo Y, Xu D, Zhang D, Zhan T, Cui W (2015) Patterns of circulating tumor cells identified by CEP8, CK and CD45 in pancreatic cancer. Int J Cancer 136:1228–1233

Zhang LN, Huang YH, Zhao L (2019a) Fusion of macrophages promotes breast cancer cell proliferation, migration and invasion through activating epithelial-mesenchymal transition and Wnt/beta-catenin signaling pathway. Arch Biochem Biophys 676:108137

Zhang LN, Kong CF, Zhao D, Cong XL, Wang SS, Ma L, Huang YH (2019b) Fusion with mesenchymal stem cells differentially affects tumorigenic and metastatic abilities of lung cancer cells. J Cell Physiol 234:3570–3582

Zhang L, Lu X, Xu Y, La X, Tian J, Li A, Li H, Wu C, Xi Y, Song G, Zhou Z, Bai W, An L, Li Z (2023) Tumor-associated macrophages confer colorectal cancer 5-fluorouracil resistance by promoting MRP1 membrane translocation via an intercellular CXCL17/CXCL22-CCR4-ATF6-GRP78 axis. Cell Death Dis 14:582

Zhao XH, Wang ZR, Chen CL, Di L, Bi ZF, Li ZH, Liu YM (2019) Molecular detection of epithelial-mesenchymal transition markers in circulating tumor cells from pancreatic cancer patients: potential role in clinical practice. World J Gastroenterol 25:138–150

Zhou X, Platt JL (2011) Molecular and cellular mechanisms of mammalian cell fusion. Adv Exp Med Biol 713:33–64

Zhou X, Merchak K, Lee W, Grande JP, Cascalho M, Platt JL (2015a) Cell fusion connects oncogenesis with tumor evolution. Am J Pathol 185:2049–2060

Zhou Y, Mohan A, Moore DC, Lin L, Zhou FL, Cao J, Wu Q, Qin YX, Reginato AM, Ehrlich MG, Yang W (2015b) Shp2 regulates osteoclastogenesis by promoting preosteoclast fusion. FASEB J 29:1635–1645

Zhou Y, Liu L, Liu Y, Zhou P, Yan Q, Yu H, Chen X, Zhu F (2021) Implication of human endogenous retrovirus W family envelope in hepatocellular carcinoma promotes Mek/Erk-mediated metastatic invasiveness and doxorubicin resistance. Cell Death Discov 7:177

Zigon P, Perdan Pirkmajer K, Tomsic M, Kveder T, Bozic B, Sodin Semrl S, Cucnik S, Ambrozic A (2015) Anti-phosphatidylserine/prothrombin antibodies are associated with adverse pregnancy outcomes. J Immunol Res 2015:975704

Chapter 15
Investigating the Consequences of Aneuploidy in Cancer and Normal Tissue

Joanna N. Modi and Alison M. Taylor

Abstract Aneuploidy is an unbalanced number of chromosomes arising from errors during cell division. Under normal conditions, cells use highly precise mechanisms to accurately segregate chromosomes during mitosis and meiosis. Errors in this process lead to chromosome missegregation and result in aneuploid daughter cells. Aneuploidy causes dosage imbalances of hundreds of different genes, disrupting homeostasis and causing adverse effects like genomic instability, metabolic and proteotoxic stress, cell cycle defects, and apoptosis (Hosea et al. 2024). Aneuploidy is the leading cause of miscarriage and congenital birth defects (Nagaoka et al. 2012). Somatic aneuploidy is found at low levels in neuronal cells and has also been observed post-mortem in brains with neurological pathologies such as Alzheimer's. Tetraploidy, octaploidy, and some aneuploidies are widely observed in hepatocytes, though their significance is not clearly understood (Simonetti et al. 2019). Additionally, aneuploidy is a hallmark of cancer, as 90% of solid tumors are aneuploid with whole chromosome or chromosome arm alterations (Taylor et al. 2018). This chapter will explore the causes and consequences of aneuploidy and its potential therapeutic implications.

Keywords Aneuploidy · Genomic instability · Cancer

15.1 Causes of Aneuploidy

15.1.1 Chromosomal Instability and Genomic Instability

Genomic instability is the increased tendency for the accumulation of genetic alterations in cells, spanning from point mutations to chromosomal gains or losses or rearrangements (Langie et al. 2015). Chromosomal Instability (CIN), one of the most common types of genomic instability, is a high rate of chromosomal gain or

J. N. Modi · A. M. Taylor (✉)
Department of Pathology and Cell Biology, Herbert Irving Comprehensive Cancer Center, Columbia University, New York, NY, USA
e-mail: at3488@cumc.columbia.edu

© The Author(s), under exclusive license to Springer Nature Switzerland AG 2026

J. Weiler et al. (eds.), *Cell Fusion*, Advances in Experimental Medicine and Biology 1509, https://doi.org/10.1007/978-3-032-22637-2_15

loss that often results in aneuploidy (Negrini et al. 2010). Aneuploidy was first discovered over a hundred years ago in tumors and more recently even in precancerous lesions (Li et al. 1997; Gorgoulis et al. 2005). CIN is linked to unfavorable outcomes in cancer patients and may play a key role in tumor progression by improving metastatic potential and chemotherapeutic resistance. CIN can cause two types of chromosomal changes: (1) numerical, which are gains or losses of whole chromosomes, and (2) structural, which are gains or losses of chromosomal segments, such as deletions, translocations, and amplifications (Thompson et al. 2010; van Jaarsveld and Kops 2016; Bakhoum et al. 2018; Hosea et al. 2024). CIN and aneuploidy are closely related, but aneuploidy refers to a state of having an imbalanced karyotype, while CIN is the rate at which chromosomes are missegregated during cell division. CIN often leads to aneuploidy, but aneuploid cells do not necessarily have CIN and can be chromosomally stable (Levine and Holland 2018). CIN is beneficial for tumors because it creates genetic diversity over time, increasing the possibility of metastasis and also resistance to chemotherapy.

15.1.2 Spindle Assembly Checkpoint Errors

The spindle assembly checkpoint (SAC) is a highly organized surveillance network that promotes proper chromosome segregation, mediating the interaction between kinetochores on duplicated chromosomes and microtubules (Lara-Gonzalez et al. 2021). If the microtubules are not properly attached to the kinetochores, the SAC is activated. Its effector molecule, the mitotic checkpoint complex (MCC), delays the onset of anaphase by inhibiting the anaphase-promoting complex/cyclosome (APC/C) (Malmanche et al. 2006; Chen et al. 2023). Once chromosomes are properly separated, the SAC is silenced, the MCC is disassembled, and the APC/C complex drives the ubiquitination and proteolytic degradation of cyclin B1 and securin. This leads to mitotic exit and sister chromatid separation, continuing cell division (Simonetti et al. 2019). An improperly functioning SAC can lead cells to enter anaphase prematurely. Additionally, errors in kinetochore-microtubule attachment can cause chromosome missegregation—for instance, if one kinetochore is attached by microtubules from opposite spindle poles (Kitajima 2018). If the SAC is faulty and fails to detect these errors, this can lead to gains and losses in chromosomes. Furthermore, mutations in MCC proteins, like BUBR1 and MAD1, are also involved in chromosomal instability and cancer development (Dobles et al. 2000; Bolanos-Garcia and Blundell 2011; Ryan et al. 2012).

15.1.3 Improper Cohesin Function

Chromosomal cohesin plays a significant role in preserving genomic stability. Cohesin works to keep newly synthesized sister chromatids from S phase together until metaphase. This tightly regulated complex allows precise genetic material to be distributed to each daughter cell. After proper kinetochore attachment, cohesin is cleaved by the enzyme separase, and the sister chromatids are moved to opposite poles. A weakened cohesin complex can lead to lagging chromosomes and uneven separation (Chiang et al. 2010), as well as chromosome bridges—structures that arise when chromosomes in daughter cells remain connected to each other. Chromosome bridges are formed due to telomere defects or DNA damage, resulting in the fusion of chromosome ends (Jiang and Chan 2024). These bridges can cause DNA damage and genomic instability if not properly resolved. Overexpression of separase has also been shown to lead to chromosomal abnormalities (Pati 2024; Musacchio 2015). A pan-cancer TCGA study implicated mutations in the cohesin complex across multiple cancer types, particularly in blood cancers (Leiserson et al. 2015; Boucher et al. 2024).

15.1.4 Centrosome Amplification

Centrosomes regulate signal transduction, cell division, polarity, and movement. They are essential for assembling a bipolar spindle, ensuring accurate chromosome alignment and segregation (Mittal et al. 2021). Centrosome amplification (CA) is a numerical excess of centrosomes that arises from centrosome duplication, mitotic failure, or cytokinesis defects. Studies have shown that CA leads to chromosomal instability and aneuploidy by altering the microtubule cytoskeleton (Lingle et al. 2002). Overamplification of centrosomes can lead to cells with multipolar spindles, and when anaphase occurs, there is a missegregation of chromosomes, resulting in aneuploid cells.

15.1.5 Replication Stress and Defective DNA Repair

During DNA replication, stress can cause the stalling of replication forks, leading to aneuploidy. It may arise from oncogenes, insufficient nucleotides, or DNA sequences and structural features that are challenging to replicate (Shaikh et al. 2022). In response, the DNA damage response (DDR) is activated to mitigate the effects of replication stress (Hosea et al. 2024). If left unattended, a stalled fork could collapse and cause double-stranded breaks that can also form translocations or deletions in chromosomes. Additionally, lagging chromosomes may arise, as well as micronuclei, which are hotspots for catastrophic genomic events (Burrell et al. 2013). In addition to the DDR, the Fanconi anemia pathway also works to repair interstrand cross-links, which prevent normal DNA strand separation (Naim and Rosselli 2009). Defects in this pathway have been demonstrated to cause chromosomal translocations and aneuploidy (Hosea et al. 2024).

15.1.6 Polyploidy

Polyploidy, multiple sets of chromosomes in a cell, can result from cell fusion, endoreplication, mitotic slippage, or cytokinesis failure (Zhang et al. 2022). Polyploidy is both a consequence and a cause of CIN, leading to further genetic chaos due to an excess of genetic material.

Cell fusion, when the membranes of two cells combine and merge their contents, often occurs between tumor cells and cells of the same origin, inflammatory cells, stromal cells, or stem cells (Zhang et al. 2022). Cell fusion forms hybrid cells with novel genetics and phenotypic traits, which can facilitate malignancy-related phenotypes, such as the reprogramming of somatic cells into induced pluripotent stem cells and epithelial-to-mesenchymal transition (Shabo et al. 2020). The formation of polyploid cells through cell fusion can confer survival advantages and contribute to cancer pathogenesis (Shultes et al. 2024).

Cytokinesis failure can occur due to defects in the contractile ring, which prevents the two daughter cells from separating. Moreover, chromosome bridges can also lead to cytokinesis failure. When this happens, the daughter cells remain as a single cell and have double the genetic content. Mitotic slippage occurs when cells exit mitosis without proper chromosome separation (Lok et al. 2020; Sanz-Gómez et al. 2023). Lastly, endoreplication is when cells undergo multiple rounds of DNA replication without mitosis (Hosea et al. 2024; Lee et al. 2009).

Whole genome doubling (WGD) is characterized by the duplication of the entire set of chromosomes, resulting in tetraploid cells instead of the normal diploid state. It is present in 30% of all solid cancers and has been associated with poorer prognosis and higher rates of metastasis (McPherson et al. 2025; Kikutake and Suyama 2023). Cells that undergo WGD are tumorigenic because they have cell cycle dysregulation and replication stress, which leads to chromosomal abnormalities. Despite the deleterious effects of WGD, cells often persist due to compensatory mechanisms of the extra genes, which may buffer against the negative effects of chromosomal instability (Quinton et al. 2021; McPherson et al. 2025).

15.2 Aneuploidy in Cancer Versus Normal Tissue

Aneuploidy can act as both a tumor promoter and a tumor suppressor. Under normal physiological conditions, cells employ a series of checkpoints and DNA damage-repair mechanisms to prevent genetic alterations and maintain genomic stability. If the damage is irreparable, apoptosis is activated to prevent the cell from replicating (Hosea et al. 2024). As a result, aneuploidy is not well tolerated in noncancerous cells, as it reduces organismal fitness and triggers cell cycle delays, proteotoxic stress, and metabolic alterations (Zhu et al. 2018). Aneuploid cells also grow more slowly than their diploid counterparts under normal conditions (Torres et al. 2007; Williams et al. 2008; Oromendia and Amon 2014; Pfau et al. 2016). On the other hand, aneuploidy enhances genetic diversity, allowing cancer cells to adapt to genetic imbalances and gene mutations, thereby giving them a survival advantage under stress (Torres et al. 2008). Additionally, the gain or loss of a chromosome leads to gene imbalances, resulting in disproportionate levels of mRNA and protein. Research in human trisomic cells found that although most protein levels increase with gene dosage, around 25% of proteins from these chromosomes remain near diploid levels, suggesting that aneuploid cells adapt to maintain protein stoichiometry, as described in more detail below (Stingele et al. 2012).

15.3 Models of Aneuploidy

Modeling aneuploidy is crucial for understanding the contribution of specific chromosomal abnormalities to the development of diseases and cancer. Traditionally, molecular geneticists have used gene overexpression, knockdown, and mutagenesis as the conventional tools for studying the role of individual tumor suppressors or oncogenes. However, single-gene manipulations are not sufficient to study large chromosomal alterations (Girish et al. 2023). Here, we will briefly summarize the currently successful strategies used to model aneuploidy in mouse, yeast, and human models, recently reviewed by others (Truong et al. 2023; Lakhani et al. 2023). Each model has its own pros and cons but offers a unique approach to uncover how aneuploidy drives dysregulation. Comparing these systems can offer insights into the context-dependent effects of chromosomal imbalances.

15.3.1 Yeast

Saccharomyces cerevisiae (budding yeast) has been instrumental in modeling aneuploidy due to its small genome and tolerance of chromosomal changes. Yeast has 16 chromosomes and can be haploid or diploid, which allows it to tolerate aneuploidy more easily than a haploid background. Additionally, yeast is an eukaryotic organism that shares many biological pathways with humans (Mulla et al. 2014), making it a great model system due to genetic tractability, rapid growth, and tolerance to aneuploidy (Gaikani et al. 2024).

Generation of random aneuploid events in yeast can occur through sporulation of polyploid strains with odd ploidy. When cells undergo meiosis, they segregate unevenly and produce spores with an abnormal number of chromosomes. A second popular method is treating cells with a kinetochore disruptor like radicicol, which causes missegregation and is a simple way to generate a diverse karyotype (Chen et al. 2012, 2015). The third method involves genome engineering, where the conditional *GAL1* promoter is inserted near the centromere. When induced, this promoter causes improper centromere function, resulting in nondisjunction of chromosomes. Using this method, monosomies and disomies have been made for every chromosome (Anders et al. 2009). Lastly, chromoduction utilizes chromosome transfer between two haploid strains during abortive mating (Torres et al. 2007). Overall, these systems have been proven useful for modeling the effects of the aneuploid state.

15.3.2 Mouse

Mouse models are generally thought to more closely mirror human biology. Although closer to human genomes, mouse chromosomes have little synteny with human chromosomes. For example, human trisomy 18 gene orthologues are syntenic to five different regions across three different mouse chromosomes (Tosh et al. 2022). Nevertheless, much knowledge has been gained by studying aneuploidy in mouse models, and some regions do exhibit useful synteny.

- *Robertsonian translocations*

Robertsonian translocations were first discovered in grasshoppers by Dr. W.R.-B. Robertson in 1916. The term refers to two different chromosomes that are attached to each other at their centromeres, forming a large metacentric chromosome and losing the shorter arm, which typically contains nonessential ribosomal DNA (Robertson 1916). This phenomenon naturally occurs in the house mouse (*Mus musculus domesticus M. m. domesticus*) as its chromosomes are acrocentric (Gerton 2024). Robertsonian translocations are frequently used to study whole chromosome aneuploidies like trisomy or monosomy.

Alexander Gropp and colleagues developed a protocol to create viable trisomic mouse embryos with Robertsonian translocations (Gropp et al. 1975). For example, mouse strain A might carry a Robertsonian translocation between chromosomes 6 and 16 (6.16), and mouse strain B a fusion between chromosomes 16 and 17 (16.17). In this case, since chromosome 16 is present as a translocation in both strains, it can be used to model trisomy 16. Mouse A and B are then bred to form compound heterozygotes because they contain both translocations 6.16 and 16.17. The male offspring of these mice are then bred with wild-type female mice that do not have any fused chromosomes. A small portion of the resulting offspring now contain one normal copy of chromosome 16 from the mother and fused chromosomes 6.16 and 16.17 from the father, resulting in a trisomy-16 mouse embryo. Cells are typically harvested from embryos around day 10.5 and used for research (Williams et al. 2008). This strategy successfully generates cell lines used to study gene dosage and chromosomal imbalance. However, this is limited to the chromosomes affected by a translocation. In addition, since most trisomies are embryonically lethal, these studies are often restricted to derived cell lines or early developmental stages.

- *Spindle Assembly Checkpoint Mutants*

Spindle Assembly Checkpoint (SAC) impairment is a powerful strategy to model aneuploidy in mice, as defective SAC activity significantly increases the chance of chromosome missegregation (Thompson et al. 2010). Mad2 plays a critical role in the SAC by preventing the separation of sister chromatids until the microtubules are attached to kinetochores. One strategy is to generate a mouse model carrying an inducible *Mad2* gene. Its overexpression has been shown to result in neoplasia, chromosome breakage, anaphase bridges, aneuploidy, and lymphomagenesis (Sotillo et al. 2007).

Mosaic aneuploidy is a condition in which an individual has both aneuploid and euploid cells. It often occurs during early embryonic development from mitotic errors (Malumbres and Villarroya-Beltri 2024). To model mosaicism, mouse embryos are treated with spindle assembly checkpoint inhibitors, such as reversine, during the four- to eight-cell division. This generates aneuploid cells, and the blastocysts are then implanted into recipient foster mothers. The mosaic embryos can be extracted for analysis or allowed to develop into pups (Bolton et al. 2016; Mashiko et al. 2020). SAC inhibition is a valuable strategy for studying aneuploidy, especially when investigating early embryonic development responses to CIN (Singla et al. 2020). However, because reversine induces random missegregation events, there is a lack of specificity within this model.

- *Engineering*

The Cre-*loxP*-based method is a popular strategy to introduce deletions, duplications, and inversions into the mouse genome. *Lox*P sites are inserted to flank a gene or chromosomal region of interest (Nagy 2000; Sauer and Henderson 1989). The engineered cells are then injected into blastocysts, which are implanted into a female mouse to create chimeras. These chimeric mice are bred to establish germline transmission of the floxed region. Offspring are then crossed with mice expressing CRE recombinase, which results in recombination at the *lox*P sites, resulting in deletions, duplications, or rearrangements (Duchon et al. 2008). This method has been used to model human chromosomal disorders such as Smith-Magenis syndrome, Down syndrome, and Charcot-Marie-Tooth type 1A (Hérault et al. 2010; Zheng et al. 2000). A key advantage of the Cre-*loxP* system is the ability to control spatial and temporal activity by using tissue-specific promoters and timing by using inducible promoters, such as tamoxifen. Additionally, this system allows for the precise study of specific targets and the effects of gene dosage.

Zuo et al. generated a mouse model of Turner syndrome by eliminating the Y chromosome. To do this, Cas9 mRNA and two Y chromosome-specific sgRNAs were injected into individual mouse zygotes and cultured to the blastocytes. DNA-FISH analysis revealed that a large portion of blastomeres of the male embryos were missing a green fluorescence signal for the Y chromosome probe, suggesting the Y chromosome was successfully lost. This was one of the first studies demonstrating CRISPR-Cas9 as a powerful and efficient tool for chromosomal engineering and modeling chromosomal disorders like Turner syndrome (Zuo et al. 2017).

Scott Lowe and his group developed a novel engineering strategy called MACHETE (molecular alteration of chromosomes with engineered tandem elements) to model deletions in mouse models of pancreatic cancer. With an initial focus on the deletion of chromosome 9p21.3, a common homozygous deletion across human cancers, MACHETE uses a bicistronic cassette with an antibiotic resistance gene and an inducible suicide gene inserted into a region of interest by CRISPR-Cas9 homology-directed repair (Barriga et al. 2022).

15.3.3 Human Cells

Human cells serve as a crucial model for studying the consequences of aneuploidy. In recent years, with the development of CRISPR-Cas9, significant strides have been made to genetically engineer targeted aneuploidies in these cells.

- *Microcell-mediated transfer*

Microcell-mediated transfer (MMCT), first developed in 1970 by Fournier and Ruddle (1977), uses a chromosome or a megabase-sized chromosome fragment that is transferred from a donor to a recipient cell. Usually, Chinese hamster ovary (CHO) or mouse A9 cells are treated with mitotic inhibitors such as Colcemid to induce hyperploidization (Liskovykh et al. 2016) and micronuclei formation. Micronuclei are then extruded from the cell using a centrifuge in the presence of a cytoskeletal disruptor, leading to the formation of microcells that contain chromosomes or chromosome fragments. These microcells are fused with recipient cells using agents such as polyethylene glycol or viral envelopes (Truong et al. 2023). MMCT allows for the transfer of many genes at once and is commonly used to study trisomies such as Down syndrome (Passerini et al. 2016). However, MMCT is very labor intensive and requires a lot of screening, as specific aneuploidies are created in a stochastic manner.

- *CRISPR-Cas9*

CRISPR-Cas9 has played a pivotal role in modeling aneuploidy in cells by enabling precise cutting and genome engineering, as highlighted in recent reviews (Li et al. 2023; Pacesa et al. 2024). Several studies have used CRISPR-Cas9 to generate targeted aneuploidies. One such method involves an artificial telomere recombining in place of the cut chromosome arm (Uno et al. 2017; Taylor et al. 2018). KaryoCreate (karyotype CRISPR-engineered aneuploidy technology) uses sgRNAs to target chromosome-specific CENPA-binding α-satellite repeats, co-expressed with dCAS9 fused to a mutant form of *Kinetochore Scaffold 1* (KNL1). This construct recruits defective kinetochore proteins, which impair normal mitotic function, resulting in missegregation and chromosome-specific aneuploidies (Bosco et al. 2023). Restoring disomy in aneuploid cells using CRISPR targeting (ReDACT) is another tool that makes use of negative selection (Girish et al. 2023; O'Brien and Berger 2023).

To study the fate of missegregated chromosomes, a strategy was developed by Ly et al. to selectively inactivate the centromere of the Y chromosome by exploiting a

CENP-A/histone H3 chimera. Following centromere inactivation, the Y chromosome underwent chromosome fragmentation and was later rejoined through nonhomologous end joining, leading to chromothripsis and genomic instability (Ly et al. 2017). CEN-SELECT combines the centromere-inactivation-induced chromosome fragmentation method (mentioned above) with a drug selection marker engineered into the Y chromosome for better tracking of genomic rearrangements after shattering (Ly et al. 2019). Additional genome editing tools have been developed to investigate the consequences of CRISPR-Cas9-induced double-stranded breaks and chromothripsis (Leibowitz et al. 2021).

15.4 Consequences of Aneuploidy and Modes of Adaptation

Aneuploidy leads to imbalances in protein abundance and cell cycle dysregulation, resulting in impaired cell growth. However, it also increases genetic diversity and promotes rapid cell adaptation. This "aneuploidy paradox," the coexistence of tumor-promoting and -suppressing effects, has been a focal point of studies for years. High levels of aneuploidy are associated with poorer overall survival, increased recurrence rates, and treatment resistance. For example, a meta-analysis of 141,163 breast cancer cases revealed a significantly higher presence of aneuploid cells in advanced tumors, larger tumors, and lymph node metastases, with patients experiencing worse clinical outcomes than those with diploid tumors (Xu et al. 2016). In stage III non-small cell lung cancer, tumors with a high level of aneuploidy have a greater incidence of distant metastasis (Alessi et al. 2023). Similarly, gastric cancer patients with aneuploid tumors have a greater risk of being late-stage and having lymph node metastasis, correlating with overall worse survival (Xu et al. 2019). Lastly, late-stage colorectal cancers are more aneuploid than early-stage, implying increased genomic instability with the cancer's progression (Laubert et al. 2015). This section will discuss the consequences of aneuploidy, in particular, its effects on cell proliferation, drug resistance, adaptation, evasion of immune response, and metabolic reprogramming.

15.4.1 Cell Proliferation and Drug Resistance

Aneuploidy is detrimental to noncancerous cells, causing a proliferative disadvantage. This proliferation defect is due to an accumulation of cyclins and a delay in G1 and S phases (Chunduri and Storchová 2019; Stingele et al. 2012) as errors in chromosome segregation result in p53 activation, which in turn leads to G1 arrest (Santaguida and Amon 2015). Fibroblasts isolated from a patient with Down syndrome showed delayed cell division compared to euploid cells from the same age (Segal and McCoy 1974). Aneuploidy modeled in human and mouse cells has also resulted in decreased proliferative capacities regardless of whether it was a gain or a loss (Williams et al. 2008; Taylor et al. 2018; Stingele et al. 2012). Similarly, yeast strains that had an extra copy of one or more chromosomes exhibited defects in cell cycle progression and glucose uptake (Torres et al. 2007; Niwa et al. 2006). On the other hand, although aneuploidy may slow cell proliferation, this delay in the G1 phase of the cell cycle can help cancer cells resist chemotherapeutic drugs' ability to damage their DNA (Replogle et al. 2020).

An analysis of triple-negative breast cancer tumor genomic data revealed two main patterns of aneuploidy: clonal *extinction* and clonal *persistence*. *Extinction* is the elimination of tumor cells, with only normal diploid cells present after treatment. The *persistent* clones harbor residual tumor cells with altered genotypes after treatment. Single-cell DNA and RNA sequencing revealed that the copy number alterations that emerged after treatment were present before therapy and were selected for, while the gene expression changes were acquired through transcriptional reprogramming. Moreover, some persistent clones expressed resistance-related genes even before treatment (Kim et al. 2018). These analyses demonstrate a critical role for aneuploidy in tumor evolution and the emergence of persistent subpopulations that confer resistance and promote metastasis.

The consequences of aneuploidy may also be tissue-type-specific. Chromosome 21, for instance, tends to be gained in leukemias but deleted in solid tumors (Duijf et al. 2013). In colorectal cancer, chromosome 5 gain increases invasiveness; however, the opposite is observed in retinal epithelial cells (Vasudevan et al. 2020).

15.4.2 Adaptation Under Stress

Aneuploidy provides cells with genetic variability, which offers avenues for cancer cells to resist treatment, circumvent oncogene addiction, and survive under stress (Salgueiro et al. 2020). One way cells adapt is by gaining or losing whole chromosomes or chromosome arms and selecting for the amplification of oncogenes and the loss of tumor suppressor genes (Holland and Cleveland 2012).

Aneuploid yeast strains have shown nongenetic heterogeneity in karyotypically identical populations, exhibiting variations in cell cycle duration and gene expression responses to stimuli. This heterogeneity enabled the aneuploid strains to have greater phenotypic diversity and a fitness advantage under stressful conditions such as nutrient scarcity, immune responses, extreme pH, and chemotherapeutic treatment (Beach et al. 2017).

Similar patterns are observed in human cells. For example, trisomic colorectal cancer cell lines have a proliferative advantage under hypoxic conditions, chemotherapeutic pressure, and serum starvation when compared to diploid cells (Pavelka et al. 2010). One study even found that trisomy 7 was acquired after serum-free passage of diploid immortalized human colonic epithelial cells (Ly et al. 2011). Taken together, hypoxic and nutrient-deprived conditions are commonly found in both solid and hematologic tumors, suggesting aneuploidy plays a key role in driving the evolution of malignant traits that help cells survive under stress.

15.4.3 Evasion of the Immune Response

Tumors avoid immune system detection and destruction by altering neoantigens, impairing antigen presentation, inhibiting immune cell infiltration, and suppressing cytotoxic activity (Davoli et al. 2017). Aneuploidy overall has been shown to inversely correlate with immune activity, with a particular reduction of cytotoxic CD8+ T cells and natural killer cells. A potential cause of this immune evasion is aneuploidy-induced protein imbalance, which may impair signals needed to promote cytotoxic immune infiltration. Aneuploidy could also weaken antigen presentation on major histocompatibility complexes (MHC), as more peptides are competing for space (Khong and Restifo 2002). This will hinder the immune system's ability to recognize and attack the tumor due to an imbalance of protein production, heightened competition for MHC protein binding, and reduced neoantigen presentation (Davoli et al. 2017). Tumor aneuploidy can also serve as an independent prognostic factor for patients with a low mutational burden and can predict responses to immunotherapy (Spurr et al. 2022b).

15.4.4 Metabolic Rewiring

Aneuploid cells often show metabolic reprogramming characterized by altered reactive oxygen species, glucose uptake, nucleotide, and carbohydrate metabolism, all of which are likely due to gene and protein imbalances (Zhakula-Kostadinova and Taylor 2024). Mammalian aneuploid cells exhibit higher glucose levels, nitrogen uptake, and lactate/ammonium byproducts (Vasudevan et al. 2021), suggesting that they are similar to cancer cells, which have an increased reliance on glucose and glutamine to meet their metabolic needs (Sheltzer 2013). These types of metabolic changes reshape cellular physiology, enabling survival in stressful conditions, facilitating rapid proliferation, and resistance to environmental factors such as drug treatment and nutrient deprivation.

Lipid biosynthesis is also impacted by the gain or loss of chromosomes, as aneuploid cells have elevated levels of ceramide, a sphingolipid that regulates apoptosis, senescence, differentiation, proliferation, and inflammation (Hannun and Obeid 2008). Increased ceramides can slow cell proliferation in aneuploid cells and help them endure metabolic stress. However, treating aneuploid cells with DL-PDMP, a ceramide glucosyltransferase antagonist, increased ceramide levels beyond a tolerable threshold, leading to apoptosis and cell proliferation inhibition (Tang et al. 2017). This vulnerability suggests that ceramide metabolism may be a potential therapeutic avenue for targeting aneuploid cells.

Aneuploidy often results in metabolic stress due to the increased reactive oxygen species (ROS) in the mitochondria. ROS are free radicals derived from oxygen that, in excess, can chemically alter and damage proteins (through aggregation and denaturation), lipids (via degradation), carbohydrates, and nucleotides (by changes in DNA) (Jakubczyk et al. 2020). ROS can promote single/double-stranded breaks in DNA, point mutations, or chromosomal breaks, which can activate oncogenes like RAS or MYC (Wang et al. 2025). ROS has also been shown to stimulate pathways such as PI3K/mTOR, NF-κB, and hypoxia signaling, which promote cell survival, proliferation, and angiogenesis (Wang et al. 2025). Aneuploid yeast strains, as well as a CIN model of *Drosophila,* have increased levels of ROS and sensitivity to oxidative stress (Dephoure et al. 2014; Shaukat et al. 2015). Increased ROS levels and oxidative DNA damage have been detected in several aneuploid mouse embryonic fibroblasts. The more severe the aneuploidy, the higher the levels of ROS and the stronger the activation of checkpoint responses such as p53 (Li et al. 2010). Taken together, the accumulation of ROS in aneuploid cells not only exacerbates oxidative stress but also activates several pathways that drive tumorigenesis, highlighting the complexity of the aneuploidy paradox.

15.4.5 Disruption of Protein Homeostasis

An unbalanced number of chromosomes causes proteotoxic stress in cells due to an excess of misfolded proteins and impaired protein degradation pathways. Aneuploid cells often have flawed protein quality-control pathways such as chaperone systems and degradation machinery, which lead to an accumulation of misfolded, damaged proteins, resulting in decreased fitness (Oromendia et al. 2012). A recent study using cells with extra chromosomes revealed an enhanced RNA degradation pathway to regulate transcriptional burden. At a protein level, cells may suppress protein translation and increase protein degradation to try to alleviate proteotoxic stress (Ippolito et al. 2024). In cells with chromosome losses, there can be a reduction of protein activity, haploinsufficiency, due to the reduced number of genes. In the case of chromosome gain, protein stoichiometric imbalances are responsible for cellular defects, as more demand is placed on the DNA replication machinery, chromosome maintenance, and segregation pathways to upkeep the extra chromosomal DNA (Torres et al. 2008). Similarly, it is important to note that differences might not be observed with copy number changes of one or two genes, but an imbalance of a large number of genes and proteins may have an additive effect, contributing to a significant decrease in fitness. Cells attempt to correct for this dosage imbalance by upregulating protein production in losses and eliminating proteins that are in excess (Schukken and Sheltzer 2022).

15.5 Therapeutic Implications and Conclusions

There is a significant effort underway to leverage aneuploidy for therapeutic benefit. Similar to oncogene addiction, certain cancers may be "addicted" to the aneuploidy found in their genome (Weinstein 2002; Girish et al. 2023). This phenomenon was observed in the case of trisomy of chromosome 1q, as its elimination completely stopped anchorage-dependent growth and xenograft formation. Furthermore, cells with 1q trisomy had an increased dependence on triploid *UCK2*, which could be targeted with cytotoxic nucleotide analogs (Girish et al. 2023). These findings underscore the potential of using aneuploidy addiction to develop targeted therapies for cancer.

Aneuploid cells adjust for their imbalanced DNA content by regulating their RNA and protein levels. Importantly, aneuploid cells have been found to be more sensitive to the perturbation of RNA degradation and proteasome inhibition (Ippolito et al. 2024). In addition to disrupting overall protein load, aneuploidy disrupts proper protein folding by impairing HSP90 function, leading to enhanced proteotoxic stress (Manchado and Malumbres 2011). Therefore, HSP90 inhibitors have been studied as potential therapeutic options to push proteotoxic stress beyond manageable limits in aneuploid tumors (Tang et al. 2011).

Aneuploid cells often upregulate the autophagy pathway to clear out damaged or excess proteins and organelles that have accumulated (Zhou et al. 2020). In human trisomic and tetrasomic cells, p62-dependent selective autophagy is much more active compared to normal cells (Stingele et al. 2012). Chloroquine, a drug that inhibits the late stages of autophagy, can induce p53-mediated apoptosis in trisomic mouse embryonic fibroblasts (Tang et al. 2011). Similarly, impaired autophagy in trisomic 13 MEFs leads to reduced proliferation (Tang et al. 2011). These findings suggest that suppressing autophagy could be a promising strategy for targeting aneuploid cancers.

Aneuploid cancer cells have been found to be more vulnerable to inhibitors targeting the spindle assembly checkpoint (SAC), such as MPS1 inhibitors. SAC inhibition combined with CDC20 overexpression leads to accelerated cell division and lethal levels of genetic instability (Zheng et al. 2025). KIF18A has been identified as a vulnerability in aneuploidy and genome-doubled cells (Cohen-Sharir et al. 2021; Quinton et al. 2021), and the success of preclinical studies with KIF18A inhibitors has boosted clinical development (Phillips et al. 2025).

Patients with highly metastatic non-small cell lung cancer who were treated simultaneously with radiotherapy and immune checkpoint blockade instead of sequentially had reduced tumors and improved overall survival. This benefit was observed specifically in patients with immunologically cold, highly aneuploid tumors, as opposed to less aneuploid tumors (Spurr and Pitroda 2024). Highly aneuploid tumors often have poor immune infiltration, but the dual treatment upregulated key immune pathways through increased cytokine production, activating the cGAS/STING pathway, and promoting CD8+ T cells (Weichselbaum et al. 2017). Radiotherapy and chemotherapy can stimulate immune infiltration and improve the response to immune checkpoint blockade. These findings suggest that tumor aneuploidy could serve as a biomarker for personalizing treatment for non-small cell lung cancer patients (Spurr et al. 2022a).

Clinical sequencing tools have simplified our ability to quantify aneuploidy. Using tools such as arm-level somatic copy number events in targeted sequencing (ASCETS) allows for aneuploidy to be a biomarker that can be applied in clinical settings and used to tailor patient therapies. As sequencing becomes increasingly widespread, harnessing aneuploidy as a biomarker has the potential to drive significant improvements in patient outcomes (Spurr et al. 2021). Furthermore, liquid biopsy is emerging as a noninvasive method to monitor aneuploidy through analysis of circulating tumor DNA (ctDNA) in patients. TRACERx and CIRCULATE-Japan have both been recently used to detect chromosomal gains and losses in patients over time using whole genome sequencing of ctDNA (Taniguchi et al. 2021; Abbosh et al. 2023).

Taken together, the emergence of novel sequencing technologies and the growing availability of aneuploidy models present a robust strategy for investigating the role of aneuploidy in various biological contexts. These methods have the ability to identify biomarkers, predict drug response, and discover new targets and treatment options for cancer patients. Gaining a deeper understanding of how chromosomal imbalances contribute to cancer development will be crucial in shifting aneuploidy from being a hallmark of cancer to a targeted focus of cancer therapy.

References

Abbosh C, Frankell AM, Harrison T et al (2023) Tracking early lung cancer metastatic dissemination in TRACERx using ctDNA. Nature 616:553–562. https://doi.org/10.1038/s41586-023-05776-4

Alessi JV, Price A, Richards AL et al (2023) Multi-institutional analysis of aneuploidy and outcomes to chemoradiation and durvalumab in stage III non-small cell lung cancer. J Immunother Cancer 11:e007618. https://doi.org/10.1136/jitc-2023-007618

Anders KR, Kudrna JR, Keller KE et al (2009) A strategy for constructing aneuploid yeast strains by transient nondisjunction of a target chromosome. BMC Genet 10:36. https://doi.org/10.1186/1471-2156-10-36

Bakhoum SF, Ngo B, Laughney AM et al (2018) Chromosomal instability drives metastasis through a cytosolic DNA response. Nature 553:467–472. https://doi.org/10.1038/nature25432

Barriga FM, Tsanov KM, Ho Y-J et al (2022) MACHETE identifies interferon-encompassing chromosome 9p21.3 deletions as mediators of immune evasion and metastasis. Nat Cancer 3:1367–1385. https://doi.org/10.1038/s43018-022-00443-5

Beach RR, Ricci-Tam C, Brennan CM et al (2017) Aneuploidy causes non-genetic individuality. Cell 169:229–242. https://doi.org/10.1016/j.cell.2017.03.021

Bolanos-Garcia VM, Blundell TL (2011) BUB1 and BUBR1: multifaceted kinases of the cell cycle. Trends Biochem Sci 36:141–150. https://doi.org/10.1016/j.tibs.2010.08.004

Bolton H, Graham SJL, Van der Aa N et al (2016) Mouse model of chromosome mosaicism reveals lineage-specific depletion of aneuploid cells and normal developmental potential. Nat Commun 7:11165. https://doi.org/10.1038/ncomms11165

Bosco N, Goldberg A, Zhao X et al (2023) KaryoCreate: a CRISPR-based technology to study chromosome-specific aneuploidy by targeting human centromeres. Cell 186:1985–2001. https://doi.org/10.1016/j.cell.2023.03.029

Boucher A, Murray J, Rao S (2024) Cohesin mutations in acute myeloid leukemia. Leukemia 38:2318–2328. https://doi.org/10.1038/s41375-024-02406-4

Burrell RA, McClelland SE, Endesfelder D et al (2013) Replication stress links structural and numerical cancer chromosomal instability. Nature 494:492–496. https://doi.org/10.1038/nature11935

Chen G, Bradford WD, Seidel CW, Li R (2012) Hsp90 stress potentiates rapid cellular adaptation through induction of aneuploidy. Nature 482:246–250. https://doi.org/10.1038/nature10795

Chen G, Mulla WA, Kucharavy A et al (2015) Targeting the adaptability of heterogeneous Aneuploids. Cell 160:771–784. https://doi.org/10.1016/j.cell.2015.01.026

Chen C, Piano V, Alex A et al (2023) The structural flexibility of MAD1 facilitates the assembly of the mitotic checkpoint complex. Nat Commun 14:1529. https://doi.org/10.1038/s41467-023-37235-z

Chiang T, Duncan FE, Schindler K et al (2010) Evidence that weakened centromere cohesion is a leading cause of age-related aneuploidy in oocytes. Curr Biol CB 20:1522–1528. https://doi.org/10.1016/j.cub.2010.06.069

Chunduri NK, Storchová Z (2019) The diverse consequences of aneuploidy. Nat Cell Biol 21:54–62. https://doi.org/10.1038/s41556-018-0243-8

Cohen-Sharir Y, McFarland JM, Abdusamad M et al (2021) Aneuploidy renders cancer cells vulnerable to mitotic checkpoint inhibition. Nature 590:486–491. https://doi.org/10.1038/s41586-020-03114-6

Davoli T, Uno H, Wooten EC, Elledge SJ (2017) Tumor aneuploidy correlates with markers of immune evasion and with reduced response to immunotherapy. Science 355:eaaf8399. https://doi.org/10.1126/science.aaf8399

Dephoure N, Hwang S, O'Sullivan C et al (2014) Quantitative proteomic analysis reveals post-translational responses to aneuploidy in yeast. elife 3:e03023. https://doi.org/10.7554/eLife.03023

Dobles M, Liberal V, Scott ML et al (2000) Chromosome missegregation and apoptosis in mice lacking the mitotic checkpoint protein Mad2. Cell 101:635–645. https://doi.org/10.1016/s0092-8674(00)80875-2

Duchon A, Besson V, Pereira PL et al (2008) Inducing segmental Aneuploid mosaicism in the mouse through targeted asymmetric sister chromatid event of recombination. Genetics 180:51–59. https://doi.org/10.1534/genetics.108.092312

Duijf PHG, Schultz N, Benezra R (2013) Cancer cells preferentially lose small chromosomes. Int J Cancer J Int Cancer 132:2316–2326. https://doi.org/10.1002/ijc.27924

Fournier RE, Ruddle FH (1977) Microcell-mediated transfer of murine chromosomes into mouse, Chinese hamster, and human somatic cells. Proc Natl Acad Sci USA 74:319–323. https://doi.org/10.1073/pnas.74.1.319

Gaikani HK, Stolar M, Kriti D et al (2024) From beer to breadboards: yeast as a force for biological innovation. Genome Biol 25:10. https://doi.org/10.1186/s13059-023-03156-9

Gerton JL (2024) A working model for the formation of Robertsonian chromosomes. J Cell Sci 137:jcs261912. https://doi.org/10.1242/jcs.261912

Girish V, Lakhani AA, Thompson SL et al (2023) Oncogene-like addiction to aneuploidy in human cancers. Science 381:eadg4521. https://doi.org/10.1126/science.adg4521

Gorgoulis VG, Vassiliou L-VF, Karakaidos P et al (2005) Activation of the DNA damage checkpoint and genomic instability in human precancerous lesions. Nature 434:907–913. https://doi.org/10.1038/nature03485

Gropp A, Kolbus U, Giers D (1975) Systematic approach to the study of trisomy in the mouse. II. Cytogenet Cell Genet 14:42–62. https://doi.org/10.1159/000130318

Hannun YA, Obeid LM (2008) Principles of bioactive lipid signalling: lessons from sphingolipids. Nat Rev Mol Cell Biol 9:139–150. https://doi.org/10.1038/nrm2329

Hérault Y, Duchon A, Maréchal D et al (2010) Controlled somatic and germline copy number variation in the mouse model. Curr Genomics 11:470–480. https://doi.org/10.2174/138920210793176038

Holland AJ, Cleveland DW (2012) Losing balance: the origin and impact of aneuploidy in cancer. EMBO Rep 13:501–514. https://doi.org/10.1038/embor.2012.55

Hosea R, Hillary S, Naqvi S et al (2024) The two sides of chromosomal instability: drivers and brakes in cancer. Signal Transduct Target Ther 9:75. https://doi.org/10.1038/s41392-024-01767-7

Ippolito MR, Zerbib J, Eliezer Y et al (2024) Increased RNA and protein degradation is required for counteracting transcriptional burden and Proteotoxic stress in human Aneuploid cells. Cancer Discov 14:2532–2553. https://doi.org/10.1158/2159-8290.CD-23-0309

Jakubczyk K, Dec K, Kałduńska J et al (2020) Reactive oxygen species - sources, functions, oxidative damage. Pol Merkur Lek Organ Pol Tow Lek 48:124–127

Jiang H, Chan YW (2024) Chromatin bridges: stochastic breakage or regulated resolution? Trends Genet 40:69–82. https://doi.org/10.1016/j.tig.2023.10.004

Khong HT, Restifo NP (2002) Natural selection of tumor variants in the generation of "tumor escape" phenotypes. Nat Immunol 3:999–1005. https://doi.org/10.1038/ni1102-999

Kikutake C, Suyama M (2023) Pan-cancer analysis of whole-genome doubling and its association with patient prognosis. BMC Cancer 23:619. https://doi.org/10.1186/s12885-023-11132-6

Kim C, Gao R, Sei E et al (2018) Chemoresistance evolution in triple-negative breast cancer delineated by single cell sequencing. Cell 173:879–893. https://doi.org/10.1016/j.cell.2018.03.041

Kitajima TS (2018) Mechanisms of kinetochore-microtubule attachment errors in mammalian oocytes. Develop Growth Differ 60:33–43. https://doi.org/10.1111/dgd.12410

Lakhani AA, Thompson SL, Sheltzer JM (2023) Aneuploidy in human cancer: new tools and perspectives. Trends Genet TIG 39:968–980. https://doi.org/10.1016/j.tig.2023.09.002

Langie SAS, Koppen G, Desaulniers D et al (2015) Causes of genome instability: the effect of low dose chemical exposures in modern society. Carcinogenesis 36:S61–S88. https://doi.org/10.1093/carcin/bgv031

Lara-Gonzalez P, Pines J, Desai A (2021) Spindle assembly checkpoint activation and silencing at kinetochores. Semin Cell Dev Biol 117:86–98. https://doi.org/10.1016/j.semcdb.2021.06.009

Laubert T, Freitag-Wolf S, Linnebacher M et al (2015) Stage-specific frequency and prognostic significance of aneuploidy in patients with sporadic colorectal cancer – a meta-analysis and current overview. Int J Color Dis 30:1015–1028. https://doi.org/10.1007/s00384-015-2259-x

Lee HO, Davidson JM, Duronio RJ (2009) Endoreplication: polyploidy with purpose. Genes Dev 23:2461–2477. https://doi.org/10.1101/gad.1829209

Leibowitz ML, Papathanasiou S, Doerfler PA et al (2021) Chromothripsis as an on-target consequence of CRISPR–Cas9 genome editing. Nat Genet 53:895–905. https://doi.org/10.1038/s41588-021-00838-7

Leiserson MDM, Vandin F, Wu H-T et al (2015) Pan-cancer network analysis identifies combinations of rare somatic mutations across pathways and protein complexes. Nat Genet 47:106–114. https://doi.org/10.1038/ng.3168

Levine MS, Holland AJ (2018) The impact of mitotic errors on cell proliferation and tumorigenesis. Genes Dev 32:620–638. https://doi.org/10.1101/gad.314351.118

Li R, Yerganian G, Duesberg P et al (1997) Aneuploidy correlated 100% with chemical transformation of Chinese hamster cells. Proc Natl Acad Sci USA 94:14506–14511. https://doi.org/10.1073/pnas.94.26.14506

Li M, Fang X, Baker DJ et al (2010) The ATM–p53 pathway suppresses aneuploidy-induced tumorigenesis. Proc Natl Acad Sci 107:14188–14193. https://doi.org/10.1073/pnas.1005960107

Li T, Yang Y, Qi H et al (2023) CRISPR/Cas9 therapeutics: progress and prospects. Signal Transduct Target Ther 8:36. https://doi.org/10.1038/s41392-023-01309-7

Lingle WL, Barrett SL, Negron VC et al (2002) Centrosome amplification drives chromosomal instability in breast tumor development. Proc Natl Acad Sci USA 99:1978–1983. https://doi.org/10.1073/pnas.032479999

Liskovykh M, Lee NC, Larionov V, Kouprina N (2016) Moving toward a higher efficiency of microcell-mediated chromosome transfer. Mol Ther Methods Clin Dev 3:16043. https://doi.org/10.1038/mtm.2016.43

Lok TM, Wang Y, Xu WK et al (2020) Mitotic slippage is determined by p31comet and the weakening of the spindle-assembly checkpoint. Oncogene 39:2819–2834. https://doi.org/10.1038/s41388-020-1187-6

Ly P, Eskiocak U, Kim SB et al (2011) Characterization of aneuploid populations with trisomy 7 and 20 derived from diploid human colonic epithelial cells. Neoplasia 13:348–357. https://doi.org/10.1593/neo.101580
Ly P, Teitz LS, Kim DH et al (2017) Selective Y centromere inactivation triggers chromosome shattering in micronuclei and repair by non-homologous end joining. Nat Cell Biol 19:68–75. https://doi.org/10.1038/ncb3450
Ly P, Brunner SF, Shoshani O et al (2019) Chromosome segregation errors generate a diverse spectrum of simple and complex genomic rearrangements. Nat Genet 51:705–715. https://doi.org/10.1038/s41588-019-0360-8
Malmanche N, Maia A, Sunkel CE (2006) The spindle assembly checkpoint: preventing chromosome mis-segregation during mitosis and meiosis. FEBS Lett 580:2888–2895. https://doi.org/10.1016/j.febslet.2006.03.081
Malumbres M, Villarroya-Beltri C (2024) Mosaic variegated aneuploidy in development, ageing and cancer. Nat Rev Genet 25:864–878. https://doi.org/10.1038/s41576-024-00762-6
Manchado E, Malumbres M (2011) Targeting aneuploidy for cancer therapy. Cell 144:465–466. https://doi.org/10.1016/j.cell.2011.01.037
Mashiko D, Ikeda Z, Yao T et al (2020) Chromosome segregation error during early cleavage in mouse pre-implantation embryo does not necessarily cause developmental failure after blastocyst stage. Sci Rep 10:854. https://doi.org/10.1038/s41598-020-57817-x
McPherson A, Vázquez-García I, Myers MA et al (2025) Ongoing genome doubling shapes evolvability and immunity in ovarian cancer. Nature:1–10. https://doi.org/10.1038/s41586-025-09240-3
Mittal K, Kaur J, Jaczko M et al (2021) Centrosome amplification: a quantifiable cancer cell trait with prognostic value in solid malignancies. Cancer Metastasis Rev 40:319–339. https://doi.org/10.1007/s10555-020-09937-z
Mulla W, Zhu J, Li R (2014) Yeast: a simple model system to study complex phenomena of aneuploidy. FEMS Microbiol Rev 38:201–212. https://doi.org/10.1111/1574-6976.12048
Musacchio A (2015) The molecular biology of spindle assembly checkpoint signaling dynamics. Curr Biol CB 25:R1002–R1018. https://doi.org/10.1016/j.cub.2015.08.051
Nagaoka SI, Hassold TJ, Hunt PA (2012) Human aneuploidy: mechanisms and new insights into an age-old problem. Nat Rev Genet 13:493–504. https://doi.org/10.1038/nrg3245
Nagy A (2000) Cre recombinase: the universal reagent for genome tailoring. Genesis 26:99–109
Naim V, Rosselli F (2009) The FANC pathway and BLM collaborate during mitosis to prevent micro-nucleation and chromosome abnormalities. Nat Cell Biol 11:761–768. https://doi.org/10.1038/ncb1883
Negrini S, Gorgoulis VG, Halazonetis TD (2010) Genomic instability – an evolving hallmark of cancer. Nat Rev Mol Cell Biol 11:220–228. https://doi.org/10.1038/nrm2858
Niwa O, Tange Y, Kurabayashi A (2006) Growth arrest and chromosome instability in aneuploid yeast. Yeast 23:937–950. https://doi.org/10.1002/yea.1411
O'Brien S, Berger AH (2023) Chromosome-level genome editing uncovers aneuploidy addiction in cancer. Cell Rep Methods 3:100618. https://doi.org/10.1016/j.crmeth.2023.100618
Oromendia AB, Amon A (2014) Aneuploidy: implications for protein homeostasis and disease. Dis Model Mech 7:15–20. https://doi.org/10.1242/dmm.013391
Oromendia AB, Dodgson SE, Amon A (2012) Aneuploidy causes proteotoxic stress in yeast. Genes Dev 26:2696–2708. https://doi.org/10.1101/gad.207407.112
Pacesa M, Pelea O, Jinek M (2024) Past, present, and future of CRISPR genome editing technologies. Cell 187:1076–1100. https://doi.org/10.1016/j.cell.2024.01.042
Passerini V, Ozeri-Galai E, de Pagter MS et al (2016) The presence of extra chromosomes leads to genomic instability. Nat Commun 7:10754. https://doi.org/10.1038/ncomms10754
Pati D (2024) Role of chromosomal cohesion and separation in aneuploidy and tumorigenesis. Cell Mol Life Sci CMLS 81:100. https://doi.org/10.1007/s00018-024-05122-5

Pavelka N, Rancati G, Zhu J et al (2010) Aneuploidy confers quantitative proteome changes and phenotypic variation in budding yeast. Nature 468:321–325. https://doi.org/10.1038/nature09529

Pfau SJ, Silberman RE, Knouse KA, Amon A (2016) Aneuploidy impairs hematopoietic stem cell fitness and is selected against in regenerating tissues in vivo. Genes Dev 30:1395–1408. https://doi.org/10.1101/gad.278820.116

Phillips AF, Zhang R, Jaffe M et al (2025) Targeting chromosomally unstable tumors with a selective KIF18A inhibitor. Nat Commun 16:307. https://doi.org/10.1038/s41467-024-55300-z

Quinton RJ et al (2021) Whole genome doubling confers unique genetic vulnerabilities on tumor cells. Nature 590:492–497. https://doi.org/10.1038/s41586-020-03133-3

Replogle JM, Zhou W, Amaro AE et al (2020) Aneuploidy increases resistance to chemotherapeutics by antagonizing cell division. Proc Natl Acad Sci USA 117:30566–30576. https://doi.org/10.1073/pnas.2009506117

Robertson WRB (1916) Chromosome studies. I. Taxonomic relationships shown in the chromosomes of tettigidae and acrididae: V-shaped chromosomes and their significance in acrididae, locustidae, and gryllidae: chromosomes and variation. J Morphol 27:179–331. https://doi.org/10.1002/jmor.1050270202

Ryan SD, Britigan EMC, Zasadil LM et al (2012) Up-regulation of the mitotic checkpoint component Mad1 causes chromosomal instability and resistance to microtubule poisons. Proc Natl Acad Sci 109:E2205–E2214. https://doi.org/10.1073/pnas.1201911109

Salgueiro L, Buccitelli C, Rowald K et al (2020) Acquisition of chromosome instability is a mechanism to evade oncogene addiction. EMBO Mol Med 12:e10941. https://doi.org/10.15252/emmm.201910941

Santaguida S, Amon A (2015) Short- and long-term effects of chromosome mis-segregation and aneuploidy. Nat Rev Mol Cell Biol 16:473–485. https://doi.org/10.1038/nrm4025

Sanz-Gómez N, González-Álvarez M, De Las Rivas J, de Cárcer G (2023) Whole-genome doubling as a source of cancer: how, when, where, and why? Front Cell Dev Biol 11. https://doi.org/10.3389/fcell.2023.1209136

Sauer B, Henderson N (1989) Cre-stimulated recombination at loxP-containing DNA sequences placed into the mammalian genome. Nucleic Acids Res 17:147–161. https://doi.org/10.1093/nar/17.1.147

Schukken KM, Sheltzer JM (2022) Extensive protein dosage compensation in aneuploid human cancers. Genome Res 32:1254–1270. https://doi.org/10.1101/gr.276378.121

Segal DJ, McCoy EE (1974) Studies on down's syndrome in tissue culture. I. Growth rates protein contents of fibroblast cultures. J Cell Physiol 83:85–90. https://doi.org/10.1002/jcp.1040830112

Shabo I, Svanvik J, Lindström A et al (2020) Roles of cell fusion, hybridization and polyploid cell formation in cancer metastasis. World J Clin Oncol 11:121–135. https://doi.org/10.5306/wjco.v11.i3.121

Shaikh N, Mazzagatti A, De Angelis S et al (2022) Replication stress generates distinctive landscapes of DNA copy number alterations and chromosome scale losses. Genome Biol 23:223. https://doi.org/10.1186/s13059-022-02781-0

Shaukat Z, Liu D, Choo A et al (2015) Chromosomal instability causes sensitivity to metabolic stress. Oncogene 34:4044–4055. https://doi.org/10.1038/onc.2014.344

Sheltzer JM (2013) A transcriptional and metabolic signature of primary aneuploidy is present in chromosomally unstable cancer cells and informs clinical prognosis. Cancer Res 73:6401–6412. https://doi.org/10.1158/0008-5472.CAN-13-0749

Shultes PV, Weaver DT, Tadele DS et al (2024) Cell-cell fusion in cancer: the next cancer hallmark? Int J Biochem Cell Biol 175:106649. https://doi.org/10.1016/j.biocel.2024.106649

Simonetti G, Bruno S, Padella A et al (2019) Aneuploidy: cancer strength or vulnerability? Int J Cancer 144:8–25. https://doi.org/10.1002/ijc.31718

Singla S, Iwamoto-Stohl LK, Zhu M, Zernicka-Goetz M (2020) Autophagy-mediated apoptosis eliminates aneuploid cells in a mouse model of chromosome mosaicism. Nat Commun 11:2958. https://doi.org/10.1038/s41467-020-16796-3

Sotillo R, Hernando E, Díaz-Rodríguez E et al (2007) Mad2 overexpression promotes aneuploidy and tumorigenesis in mice. Cancer Cell 11:9–23. https://doi.org/10.1016/j.ccr.2006.10.019

Spurr LF, Pitroda SP (2024) Exploiting tumor aneuploidy as a biomarker and therapeutic target in patients treated with immune checkpoint blockade. NPJ Precis Oncol 8:1. https://doi.org/10.1038/s41698-023-00492-8

Spurr LF, Touat M, Taylor AM et al (2021) Quantification of aneuploidy in targeted sequencing data using ASCETS. Bioinforma Oxf Engl 37:2461–2463. https://doi.org/10.1093/bioinformatics/btaa980

Spurr LF, Martinez CA, Kang W et al (2022a) Highly aneuploid non-small cell lung cancer shows enhanced responsiveness to concurrent radiation and immune checkpoint blockade. Nat Cancer 3:1498–1512. https://doi.org/10.1038/s43018-022-00467-x

Spurr LF, Weichselbaum RR, Pitroda SP (2022b) Tumor aneuploidy predicts survival following immunotherapy across multiple cancers. Nat Genet 54:1782–1785. https://doi.org/10.1038/s41588-022-01235-4

Stingele S, Stoehr G, Peplowska K et al (2012) Global analysis of genome, transcriptome and proteome reveals the response to aneuploidy in human cells. Mol Syst Biol 8:608. https://doi.org/10.1038/msb.2012.40

Tang Y-C, Williams BR, Siegel JJ, Amon A (2011) Identification of aneuploidy-selective antiproliferation compounds. Cell 144:499–512. https://doi.org/10.1016/j.cell.2011.01.017

Tang Y-C, Yuwen H, Wang K et al (2017) Aneuploid cell survival relies upon sphingolipid homeostasis. Cancer Res 77:5272–5286. https://doi.org/10.1158/0008-5472.CAN-17-0049

Taniguchi H, Nakamura Y, Kotani D et al (2021) CIRCULATE-Japan: circulating tumor DNA-guided adaptive platform trials to refine adjuvant therapy for colorectal cancer. Cancer Sci 112:2915–2920. https://doi.org/10.1111/cas.14926

Taylor AM, Shih J, Ha G et al (2018) Genomic and functional approaches to understanding cancer aneuploidy. Cancer Cell 33:676–689.e3. https://doi.org/10.1016/j.ccell.2018.03.007

Thompson SL, Bakhoum SF, Compton DA (2010) Mechanisms of chromosomal instability. Curr Biol CB 20:R285–R295. https://doi.org/10.1016/j.cub.2010.01.034

Torres EM, Sokolsky T, Tucker CM et al (2007) Effects of aneuploidy on cellular physiology and cell division in haploid yeast. Science 317:916–924. https://doi.org/10.1126/science.1142210

Torres EM, Williams BR, Amon A (2008) Aneuploidy: cells losing their balance. Genetics 179:737–746. https://doi.org/10.1534/genetics.108.090878

Tosh J, Tybulewicz V, Fisher EMC (2022) Mouse models of aneuploidy to understand chromosome disorders. Mamm Genome 33:157–168. https://doi.org/10.1007/s00335-021-09930-z

Truong MA, Cané-Gasull P, Lens SMA (2023) Modeling specific aneuploidies: from karyotype manipulations to biological insights. Chromosom Res 31:25. https://doi.org/10.1007/s10577-023-09735-7

Uno N, Hiramatsu K, Uno K et al (2017) CRISPR/Cas9-induced transgene insertion and telomere-associated truncation of a single human chromosome for chromosome engineering in CHO and A9 cells. Sci Rep 7:12739. https://doi.org/10.1038/s41598-017-10418-7

van Jaarsveld RH, Kops GJPL (2016) Difference makers: chromosomal instability versus aneuploidy in cancer. Trends Cancer 2:561–571. https://doi.org/10.1016/j.trecan.2016.09.003

Vasudevan A, Baruah PS, Smith JC et al (2020) Single-chromosomal gains can function as metastasis suppressors and promoters in colon cancer. Dev Cell 52:413–428.e6. https://doi.org/10.1016/j.devcel.2020.01.034

Vasudevan A, Schukken KM, Sausville EL et al (2021) Aneuploidy as a promoter and suppressor of malignant growth. Nat Rev Cancer 21:89–103. https://doi.org/10.1038/s41568-020-00321-1

Wang M, Xiao Y, Miao J et al (2025) Oxidative stress and inflammation: drivers of tumorigenesis and therapeutic opportunities. Antioxidants 14:735. https://doi.org/10.3390/antiox14060735

Weichselbaum RR, Liang H, Deng L, Fu Y-X (2017) Radiotherapy and immunotherapy: a beneficial liaison? Nat Rev Clin Oncol 14:365–379. https://doi.org/10.1038/nrclinonc.2016.211

Weinstein IB (2002) Addiction to oncogenes – the Achilles heal of cancer. Science 297:63–64. https://doi.org/10.1126/science.1073096

Williams BR, Prabhu VR, Hunter KE et al (2008) Aneuploidy affects proliferation and spontaneous immortalization in mammalian cells. Science 322:703–709. https://doi.org/10.1126/science.1160058

Xu J, Huang L, Li J (2016) DNA aneuploidy and breast cancer: a meta-analysis of 141,163 cases. Oncotarget 7:60218–60229. https://doi.org/10.18632/oncotarget.11130

Xu J, Zhu R, Fan L et al (2019) Prognostic value of DNA aneuploidy in gastric cancer: a meta-analysis of 3449 cases. BMC Cancer 19:650. https://doi.org/10.1186/s12885-019-5869-9

Zhakula-Kostadinova N, Taylor AM (2024) Patterns of aneuploidy and signaling consequences in cancer. Cancer Res 84:2575–2587. https://doi.org/10.1158/0008-5472.CAN-24-0169

Zhang J, Qiao Q, Xu H et al (2022) Human cell polyploidization: the good and the evil. Semin Cancer Biol 81:54–63. https://doi.org/10.1016/j.semcancer.2021.04.005

Zheng B, Sage M, Sheppeard EA et al (2000) Engineering mouse chromosomes with Cre-loxP: range, efficiency, and somatic applications. Mol Cell Biol 20:648–655. https://doi.org/10.1128/mcb.20.2.648-655.2000

Zheng S, Raz L, Zhou L et al (2025) High CDC20 levels increase sensitivity of cancer cells to MPS1 inhibitors. EMBO Rep 26:1036–1061. https://doi.org/10.1038/s44319-024-00363-8

Zhou L, Jilderda LJ, Foijer F (2020) Exploiting aneuploidy-imposed stresses and coping mechanisms to battle cancer. Open Biol 10:200148. https://doi.org/10.1098/rsob.200148

Zhu J, Tsai H-J, Gordon MR, Li R (2018) Cellular stress associated with aneuploidy. Dev Cell 44:420–431. https://doi.org/10.1016/j.devcel.2018.02.002

Zuo E, Huo X, Yao X et al (2017) CRISPR/Cas9-mediated targeted chromosome elimination. Genome Biol 18:224. https://doi.org/10.1186/s13059-017-1354-4

Chapter 16
Nano- and Microtubes: Tiny Fusion Events but Not Overall Plasma Membrane Merger for Exchanging Intracellular Components

Xianghe Qiao and Bo Li

Abstract Membrane nanotubes, including tunneling nanotubes (TNTs), tumor microtubes (TMs), cytonemes, and related structures, constitute open-ended conduits that directly connect cells. They enable the directed exchange of organelles, nucleic acids, proteins, and ions across tens to hundreds of micrometers while preserving cellular identities. Specifically, this chapter synthesizes the current understanding of mechanisms, delineates cargo selectivity and transfer modes, and integrates cancer-relevant functions, such as mitochondrial transfer, calcium-wave propagation, and therapy adaptation. We emphasize rigorous criteria and multimodal workflows to distinguish conduit-mediated exchange from gap junction coupling, extracellular vesicles (EVs), trogocytosis, entosis, and bona fide cell–cell fusion. Across carcinomas and malignancies, tubes organize multicellular networks that distribute stress, shape metabolic plasticity, and seed resistant subclones. We outline vulnerabilities in tube biogenesis and coupling, propose a communication–fusion continuum, and set near-term priorities for in vivo quantification and translational targeting.

Keywords Tunneling nanotubes (TNTs) · Tumor microtubes (TMs) · Mitochondria transfer · Intercellular cargo transfer · Non-fusogenic cell–cell communication

---

X. Qiao
State Key Laboratory of Oral Diseases, National Center for Stomatology, National Clinical Research Center for Oral Diseases, West China Hospital of Stomatology, Sichuan University, Chengdu, China

Department of Oral and Maxillofacial Surgery, The First Affiliated Hospital of Zhengzhou University, Zhengzhou, China

Bo Li (✉)
State Key Laboratory of Oral Diseases, National Center for Stomatology, National Clinical Research Center for Oral Diseases, West China Hospital of Stomatology, Sichuan University, Chengdu, China

Department of Orthodontics, West China Hospital of Stomatology, Sichuan University, Chengdu, China

© The Author(s), under exclusive license to Springer Nature Switzerland AG 2026

J. Weiler et al. (eds.), *Cell Fusion*, Advances in Experimental Medicine and Biology 1509, https://doi.org/10.1007/978-3-032-22637-2_16

16.1 Introduction: Scope and Positioning

Nano- and microtubes are open-ended membranous bridges that connect two or more cells and allow directional transport of cellular cargo without merging of the plasma membrane or mixing of nuclei, constituting a non-fusogenic mode of communication (Rustom et al. 2004; Sartori-Rupp et al. 2019; Roehlecke and Schmidt 2020). This definition distinguishes conduit-mediated exchange from bona fide fusion events that create syncytia or stable hybrids. The concept emerged from early microscopy that visualized long, thin, actin-rich tubes spanning tens to hundreds of micrometers and permitting the transfer of organelles and macromolecules (Rustom et al. 2004; Davis and Sowinski 2008). Subsequent cryo-electron microscopy showed that these tubes are continuous with the donor and recipient membranes and often contain distinct luminal tracks, supporting directed flow rather than wholesale cytoplasmic mixing (Sartori-Rupp et al. 2019; Gerdes and Carvalho 2008). Reviews in cancer and developmental biology have converged on the view that these conduits represent a dedicated communication mode with rules distinct from those governing vesicle secretion or junctional pores (Roehlecke and Schmidt 2020; Davis and Sowinski 2008; Gerdes and Carvalho 2008; Cordero Cervantes and Zurzolo 2021).

Functionally, tubes act as architectural motifs within the tumor microenvironment (TME). In glioblastoma, thick microtubes interconnect malignant cells into a conductive network that propagates signals and redistributes resources, thereby buffering focal injury and facilitating tumor regrowth after resection or irradiation (Weil et al. 2017; Wang et al. 2022; Osswald et al. 2015). These tumor microtubes (TMs) are longer lived and thicker than many classical nanotubes, frequently couple through connexins, and form a resilient network that supports invasion and therapy resistance (Weil et al. 2017; Osswald et al. 2015; Osswald et al. 2016). This model reframes glioma resilience as an emergent property of connectivity rather than solely a function of cell autonomous resistance (Weil et al. 2017; Osswald et al. 2015; Lou 2017).

Beyond the brain, thinner tunneling nanotubes (TNTs) occur in epithelial and hematologic malignancies that traffic mitochondria, lysosomes, protein complexes, and RNA (Ady et al. 2014; Marlein et al. 2017). Heterotypic exchanges between stromal and cancer cells can remodel metabolism and stress tolerance; for example, endothelial-to-tumor mitochondria transfer increases chemoresistance, and bone marrow stromal-to-leukemia mitochondria transfer enhances oxidative phosphorylation in blasts (Marlein et al. 2017; Pasquier et al. 2013a; Li et al. 2025a). TNT-mediated drug redistribution has also been observed, highlighting a noncanonical route for acquired resistance (Desir et al. 2018). In solid tumors, nanotube connectivity supports invasion and cross talk with immune cells, including macrophage-assisted streaming. Collectively, these observations position tubes as regulators of plasticity that operate across tumor compartments (Ady et al. 2014; Marlein et al. 2017; Pasquier et al. 2013a; Li et al. 2025a; Desir et al. 2018; Valdebenito et al. 2021).

A clear scope is essential because several cell–cell contact phenomena can be confused with fusion (Cordero Cervantes and Zurzolo 2021). In this chapter, we restrict the analysis to non-fusogenic exchanges via tubes. When cytoplasmic bulk mixing, nuclear coalescence, or stable karyotypic merger occurs, the event is categorized as fusion and is treated in other chapters. We also distinguish tubes from related structures and mechanisms, including cytonemes that specialize in morphogen presentation, gap junction channels that conduct ions and small metabolites without a lumen, and trogocytosis or entosis that involve material nibbling or cell-within-cell events (Sartori-Rupp et al. 2019; Davis and Sowinski 2008; Cordero Cervantes and Zurzolo 2021; Kornberg and Roy 2014). Methodological work has emphasized minimal reporting items to avoid misinterpretation, such as geometry, open-endedness, cytoskeletal composition, and effective connectivity (Sartori-Rupp et al. 2019; Cordero Cervantes and Zurzolo 2021).

Finally, this chapter positions biogenesis modules and network logic as the mechanistic backbone for understanding function. A canonical pathway centers on M-Sec, which collaborates with RalA and the exocyst to deform the membrane and initiate tube outgrowth, with defined domains mediating lipid binding and localization (Davis and Sowinski 2008; Kimura et al. 2016; Hase et al. 2009). Stress signals and receptor tyrosine kinase (RTK) programs may bias tube formation in cancer contexts. These mechanisms frame a testable path from molecular control to network-level phenotypes that influence therapeutic response. Subsequent sections develop standardized terminology, cargo selectivity, readouts, model systems, and intervention strategies.

16.2 Nomenclature and Taxonomy

TNTs are membranous conduits that typically exhibit diameters of approximately 50–700 nm and are rich in F-actin, although some contain microtubules (Rustom et al. 2004; Sartori-Rupp et al. 2019). These tubes form continuous bridges between cells and mediate long-distance transport of cellular cargoes, including organelles and macromolecules, without the merging of the plasma membranes or mixing of the cytoplasm. TNTs are often distinguished from TMs by their thinner structure and shorter lifespan. TMs, as frequently observed in glioma, are much thicker, enriched in microtubules, and remarkably long-lived, forming complex multicellular networks (Weil et al. 2017; Osswald et al. 2015). While TNTs are more commonly observed in various cell types, TMs play a critical role in glioblastoma by creating a durable network that helps the tumor survive and regenerate after injury or treatment. The key difference between TNTs and TMs lies not only in their morphology but also in their function. TMs are more frequently coupled through connexins, forming stable electrical and metabolic networks, while TNTs serve as transient connections that facilitate the movement of cellular components over shorter distances (Weil et al. 2017; Osswald et al. 2015).

Another important concept is the cytoneme, which was initially coined in the context of development. Cytonemes are thin, actin-based filopodia that mediate signaling between cells during tissue development and morphogenesis (Kornberg and Roy 2014). In cancer, similar structures can enable long-range communication between tumor and stromal cells. Unlike TNTs, which are open-ended conduits that allow direct cytoplasmic exchange, cytonemes are specialized, actin-based filopodia that transport morphogens along their length and transfer them at synapse-like contacts with target cells, typically without establishing open-ended cytoplasmic continuity (Kornberg and Roy 2014; Sherer and Mothes 2008). The critical distinction between TNTs and other similar structures is often subtle and can lead to misinterpretation if the definitions are not explicit. TNTs are open-ended conduits capable of transferring various cellular components, whereas other cell-contact phenomena, such as gap junctions, extracellular vesicles (EVs), trogocytosis, and entosis, have fundamentally different functions and structures.

For instance, gap junctions are composed of connexins and allow the diffusion of ions and small molecules but lack a lumen for larger cellular material (Li et al. 2025a; Goodenough and Paul 2009; Harris 2007). EVs are involved in the uptake and release of vesicles but follow discontinuous trajectories and do not form stable intercellular bridges (Welsh et al. 2024a; Weng et al. 2021; Théry et al. 2018). Trogocytosis and entosis transfer material from one cell to another via cell surface contact, typically by engulfment or "nibbling," not long-distance trafficking (Önfelt et al. 2006; Overholtzer et al. 2007). Moreover, it is crucial to distinguish TNTs from hemifusion, which is the transient merging of lipid bilayers without full membrane fusion (Kephart et al. 2024). Filopodia may resemble TNTs morphologically, but filopodia are usually much shorter, do not support directed transport, and are not open-ended in the same manner (Kornberg and Roy 2014).

Reporting standards for TNTs should specify key physical characteristics such as geometry (length, diameter), suspension versus substrate adherence, and the presence of open ends. The cytoskeletal composition, including actin and microtubules, should also be documented, as well as the functional properties such as connectivity metrics and the presence of gap junction coupling (Cordero Cervantes and Zurzolo 2021). To avoid confusion with filopodia or retraction fibers, it is essential to include these minimal items so that TNTs are identified consistently across studies. Given the challenges in distinguishing between TNTs and similar cellular structures, a negative set of criteria is also recommended. These should focus on structures that do not meet the definition of TNTs, such as closed filopodia (cytonemes), retraction fibers, and gap junctions. TNTs must have a continuous lumen capable of supporting the transport of organelles or macromolecules, which is a key feature that distinguishes them from other contact structures like trogocytosis, entosis, and hemifusion (Davis and Sowinski 2008; Kornberg and Roy 2014).

To standardize TNT reporting, we recommend, at a minimum, explicit measurements of length and diameter, identification of cytoskeletal content, evidence of open-ended continuity, and functional validation via intercellular transport assays (Sartori-Rupp et al. 2019; Cordero Cervantes and Zurzolo 2021). Adhering to these items, researchers can ensure consistent and reproducible studies of TNTs, ultimately advancing our understanding of these unique cellular conduits and their roles in cancer and other diseases.

16.3 Biogenesis and Molecular Machinery

Cells can generate long membranous bridges by at least three routes that are not mutually exclusive, including the mutual convergence and "handshake" of filopodia from two cells, the persistence and thinning of transient intercellular bridges, and membrane pulling under mechanical tension that draws long tethers between distant cells (Sartori-Rupp et al. 2019; Cordero Cervantes and Zurzolo 2021; Korenkova et al. 2020). Correlative light microscopy, focused ion beam scanning electron microscopy, and cryo-electron microscopy demonstrate that many conduits are open-ended, continuous with both the donor and recipient membranes, and contain distinct luminal tracks rather than broad cytoplasmic continuity (Sartori-Rupp et al. 2019). Mechanical deformation alone can generate close- or open-ended tunnels, implying that membrane tension and curvature thresholds are sufficient to nucleate tubes in permissive contexts (Chang et al. 2022). Classic membrane-physics studies further show that nanoscale tethers can be extracted by motor- or bead-based pulling, providing a physical basis for the long tubular bridges seen in cells (Leduc et al. 2004). Together, these data establish multiple entry points into the tubular state, unified by membrane continuity and lumen patency.

Membrane-curvature generators and actin nucleators cooperate to shape and stabilize tubes. In general, the BAR-domain superfamilies scaffolds and sense curvature, with I-BAR proteins such as IRSp53 enriching on negatively curved membranes typical of thin tubes (Mim and Unger 2012). In vitro reconstitution studies have shown that the I-BAR domain of IRSp53 can sense negative membrane curvature along artificial membrane tubules (Prévost et al. 2015), and live-cell imaging reveals that IRSp53 clusters recruit VASP to drive local actin elongation at the tips of filopodial protrusions (Tsai et al. 2022). Filopodia regulators and TNT programs differ, neuronal researchers show that the CDC42–IRSp53–VASP module that promotes filopodia can suppress TNT formation, whereas the actin bundler Eps8 promotes functional TNTs (Delage et al. 2016). Refinements using cryo-electron tomography and perturbations of branched actin nucleation indicate that reduced Arp2/3 activity favors long, linearly bundled filaments compatible with tube elongation (Henderson et al. 2023). These observations and findings support a unifying view that TNTs are molecularly distinct from filopodia despite superficial morphological similarity (Ljubojevic et al. 2021).

Motor proteins and trafficking modules help maintain lumenal patency and move cargo once a tube forms. Myosin-10 (Myo10), a canonical motor for tip-directed filopodial growth, supports long protrusions that can mature into stable bridges in appropriate signaling contexts (Bohil et al. 2006). Small GTPase pathways also tune tube competence: a Rab35-GTP–ACAP2–ARF6-GDP–EHD1 cascade promotes TNT formation and vesicle transfer in neuronal cells, highlighting endocytic recycling nodes as permissive hubs (Bhat et al. 2020). At the organelle level, mitochondrial motion along microtubules is mediated by the Miro-TRAK adapter axis, which recruits both kinesin and dynein to coordinate bidirectional transport (Sheng and Cai 2012). Ca^{2+} microdomains can propagate along thin bridges when connexins are present, coupling ionic signals to trafficking and survival pathways (Smith et al. 2011). Membrane tension further regulates curvature protein assembly, providing a physical rheostat for tube stability during cargo flow (Simunovic and Voth 2015).

A conserved initiation module centers on M-Sec, also known as TNFAIP2, which collaborates with RalA and the exocyst complex to deform the plasma membrane and promote TNT outgrowth (Hase et al. 2009). Structure–function studies show that the N-terminal polybasic region of M-Sec binds phosphatidylinositol 4,5-bisphosphate for membrane targeting, while a positively charged surface in the C-terminal half mediates interaction with active RalA to extend long protrusions (Kimura et al. 2016). M-Sec serves as a central factor and a practical marker for TNT biology, anchoring the field to a genetically tractable pathway (Ohno et al. 2010). Leukocyte-specific transcript one can scaffold RalA–exocyst at the membrane, reinforcing the idea that exocytic tethering complexes are repurposed to build intercellular bridges (Schiller et al. 2013). Research across diabetic nephropathy and cancer systems further connects M-Sec expression with TNT prevalence and stress resilience, underscoring the therapeutic interest in the module (Barutta et al. 2022; Qiao et al. 2024).

In glioma, TMs constitute a specialized, microtubule-rich variant that integrates growth-associated protein 43 (GAP-43) and connexin 43 (Cx43) to produce long-lived, electrically and metabolically coupled networks, linking developmental neurite programs to malignant connectivity (Osswald et al. 2015, 2016). TMs convey resistance to focal injury and chemoradiation by redistributing stress and enabling synchronized responses across connected cells (Weil et al. 2017). Cx43 localizes at tube contacts and gap junctions within the network, and pharmacologic modulation of connexin channels alters network integrity and the therapy response in preclinical systems (Wang et al. 2018). Contemporary reviews consolidate the structural and functional hallmarks of TMs, highlighting their length, thickness, microtubule content, and resilience relative to thinner TNTs (Wang et al. 2022). Moreover, researchers summarize pan-cancer evidence for TNTs and TMs, situating glioma networks as an extreme on the conduit spectrum (Roehlecke and Schmidt 2020; Osswald et al. 2015, 2016).

Notably, extrinsic signals bias tube formation and maturation in tumors by coupling stress sensors to the cytoskeleton and membrane remodeling. Hypoxia drives high-mobility group box 1 (HMGB1) release and signaling that upregulates RHOT1 and RAC1 to enhance mitochondria transfer through TNTs in hepatocellular carcinoma, linking metabolic stress to nanotube biogenesis (Jing et al. 2024). In non-small cell lung cancer (NSCLC), hepatocyte growth factor (HGF) acting through c-Met and integrin beta-1 (ITGB1), also known as CD29, promotes robust TNT formation and functional cargo exchange, illustrating how RTKs connect to the protrusive machinery (Awanis et al. 2023). Macrophage cues activate mitogen-activated protein kinase (MAPK) signaling and permit TNT assembly in ovarian cancer models, indicating that stromal programs can instruct nanotube emergence in malignant epithelia (Cole et al. 2021). More broadly, irradiation, chemotherapeutics, and inflammatory mediators increase TNT frequency in diverse systems, consistent with a stress-induced plasticity program (Kretschmer et al. 2019). These findings converge on the biogenesis modules above, providing a mechanistic bridge between TME pressures and the appearance of intercellular tubes.

Finally, several principles emerge across models that guide the rest of the chapter. First, tube formation is a state of membrane and cytoskeletal organization that multiple pathways can reach, but once formed, patency and cargo flow depend on specific motors and adapters (Sheng and Cai 2012). Second, tumor contexts co-opt conserved modules such as M-Sec–RalA–exocyst and BAR domain curvature systems while adding disease-specific modifiers like Cx43 in glioma (Osswald et al. 2015; Hase et al. 2009). Third, stress and growth factor circuits act as upstream dials that determine when and where tubes appear, ensuring that conduit biology is tightly coupled to the microenvironmental demands (Li et al. 2025a; Qiao et al. 2024; Jing et al. 2024; Awanis et al. 2023).

16.4 Cargo and Transfer Modes

Nano- and microtubes accommodate a broad cargo repertoire that spans mitochondrial, lysosomes, and autophagosomes, endosomes and segments of endoplasmic reticulum, protein complexes, diverse RNAs, and small ions, with transport occurring through an open lumen. When connexins concentrate at tube–cell interfaces, channel-based coupling can operate alongside luminal flow. In glioma networks, Cx43-enriched contacts at tube termini enable electrical and chemical coupling that supports the propagation of Ca^{2+} waves alongside luminal trafficking (Weil et al. 2017; Osswald et al. 2015, 2016; Goodenough and Paul 2009; Smith et al. 2011).

Among cargoes, mitochondrial transfer has emerged as the most consequential because donation from metabolically competent donors can restore respiration, reset redox poise, and increase stress tolerance in compromised recipients. Spees et al. first showed that intercellular mitochondria transfer rescues aerobic respiration in cells lacking functional organelles, launching a field that now spans cancer and immunity (Spees et al. 2006). In bladder cancer, nanotube-mediated donation increases invasiveness and oxidative phosphorylation in recipient cells (Lu et al. 2017). In acute myeloid leukemia, bone marrow stromal cells supply mitochondria to blasts to boost respiration, with NOX2-dependent signaling as a driver (Marlein et al. 2017). Endothelial-to-tumor transfer enhances chemoresistance and illustrates heterotypic exchange within solid tumors (Pasquier et al. 2013a). Recent in vivo and ex vivo work in ovarian cancer shows that cancer-associated mesenchymal stromal cells deliver mitochondria that expand tumor heterogeneity and metastatic fitness (Frisbie et al. 2024). Similarly, fibroblasts deliver mitochondria to rescue metabolically dysfunctional cancer cells in adenoid cystic carcinoma (Qiao et al. 2024). Mechanistically, oncogenic information can hitchhike with organelle flow, as seen with the transfer of mutant KRAS and its associated signaling capacity between colon cancer cells (Desir et al. 2019). In the brain, bidirectional mitochondrial exchange between glioblastoma cells and astrocytes remodels metabolic plasticity within the niche (Li et al. 2025a; Valdebenito et al. 2021).

Lysosomes, autophagosomes, and endocytic vesicles also traffic through nanotubes, linking degradation and stress responses across cells. In head and neck squamous cell carcinoma (HNSCC), long-range autophagosome transfer has been imaged and functionally tied to survival under stress (Sáenz-de-Santa-María et al. 2017). Reviews and primary imaging studies further document vesicle exchange and lysosomal movement, consolidating endolysosomal cargo as a consistent nanotube payload in cancer and stromal systems (Turos-Korgul et al. 2022; Thayanithy et al. 2014; Kolba et al. 2019). Beyond cancer, macrophage-mediated lysosomal cross-correction and vesicle delivery have been implicated in inherited metabolic disorders (Naphade et al. 2015). Segments of endoplasmic reticulum and the inositol trisphosphate receptor can traverse tubes, providing a route to coordinate calcium handling and protein synthesis domains between cells (Wang et al. 2011; Marzo et al. 2012). Pathogens and aggregation-prone proteins can exploit the same routes, highlighting the risk along with physiological function (Chakraborty et al. 2023; Pepe et al. 2022; Jansens et al. 2020).

RNA- and protein-based information flow through tubes adds a second tier of regulation atop organelle exchange. Full-length messenger RNA (mRNA) can move between mammalian cells through membrane nanotube-like extensions, with transfer requiring direct contact rather than diffusion (Haimovich et al. 2017). Cross-species experiments further confirm contact-dependent mRNA movement in vivo (Dasgupta et al. 2023). In tumors, microRNAs traverse nanotubes to reprogram gene expression and alter the drug response, as shown in osteosarcoma and ovarian models (Thayanithy et al. 2014; Lu et al. 2019). Comprehensive reviews synthesize nucleic-acid transfer routes and provide methodological guidance to distinguish true conduit-mediated exchange from confounders (Driscoll et al. 2022; Pinto et al. 2020).

Selectivity and directionality derive from the cytoskeletal and motor apparatus that shape the tube geometry and power cargo movement, together with metabolic gradients and organelle potential that bias donation. Myo10 supports filopodial protrusion and promotes nanotube formation in immune and other cells, providing plus-end-directed force along F-actin cores (Ljubojevic et al. 2021; Uhl et al. 2018). Long-range movement of mitochondria and large vesicles within thicker, microtubule-containing tubes aligns with kinesin- and dynein-based transport paradigms established in axons and other polarized cells (Pilling et al. 2006; Austefjord et al. 2014). The mitochondrial outer membrane adapter Miro-1 links organelles to motors and enhances intercellular mitochondrial transfer when overexpressed, implying that the intracellular trafficking logic extends into intercellular bridges (Ahmad et al. 2014; Babenko et al. 2018). Reviews of metabolic control in nanotube-mediated transfer further argue that the organelle membrane potential and redox state help set donor and recipient roles (Dong et al. 2023).

Methodologically, photoconvertible mitochondrial reporters and lineage separation controls enable direct visualization of the cargo handoff while excluding cell fusion. The PhAM mouse strains that express mitochondrially targeted Dendra2 (mito-Dendra2) permit precise photo-marking of donor organelles and tracking into recipients in living tissues, a strategy now widely adopted alongside time-lapse imaging and ultrastructural validation (Pham et al. 2012; Baldwin et al. 2024; Gkerküçük et al. 2020). In parallel, chemotherapeutic agents can redistribute through nanotubes, offering a visible readout of the luminal flow and revealing a noncanonical path to resistance (Desir et al. 2018). Altogether, these findings delineate two cooperating modes of transfer within the same architecture, with luminal shuttling of multiple cargoes and junction-mediated ions and small metabolites, which act together to reshape the tumor cell state in a temporospatial manner.

16.5 Cancer-Relevant Functions

In aggressive primary brain tumors, TMs organize malignant cells into a connected lattice that conducts ions, shares the metabolic load, and buffers focal damage, reframing resilience as a property of connectivity rather than only of cell-intrinsic resistance (Wang et al. 2022; Lou et al. 2024). Beyond structural conduction, glioma cells remap local neural circuits and exploit activity-dependent trophic signals, with circuit remodeling and activity coupling captured in patients and in mouse xenografts (Wang et al. 2022; Krishna et al. 2023). Across tumor types, conduit-connected meshes coordinate collective responses, such as calcium-wave propagation, synchronized polarization, and rapid rerouting around damaged nodes, which together raise the threshold for treatment-induced collapse (Wang et al. 2022; Lou et al. 2024).

Multiple groups place Cx43 within this architecture as both a structural and signaling hub that supports long-range coupling and can be pharmacologically perturbed with brain-penetrant modulators such as tonabersat, thereby weakening network integrity in preclinical models (Schneider et al. 2024; Zoteva et al. 2024). New mechanistic work dissects Cx43 interactions with cytoskeletal tracks that stabilize protrusions and influence the treatment response, pointing to additional druggable interfaces on the connexin scaffold (Smyth et al. 2025; Schmidt et al. 2025). As therapeutic concepts evolve, several groups test whether blocking Cx43 or modulating its noncanonical signaling sensitizes gliomas to alkylating chemotherapy, offering a way to degrade network benefits while preserving the standard of care (Zoteva et al. 2024; Che et al. 2023). Meanwhile, the same physical highways also create opportunities for therapeutic interception, since network function can be reduced by weakening junctional coupling, interrupting guidance cues that stabilize protrusions, or exploiting periods of fragility after a cytotoxic insult (Schneider et al. 2024; Zoteva et al. 2024).

Across epithelial tumors, stressors such as chemotherapy and hypoxia increase TNT abundance and bias traffic through these conduits, which is a phenomenon linked to drug adaptation and relapse in preclinical systems (Sarkari and Lou 2024; Chen and Zhao 2024). In lung adenocarcinoma models, RTK signaling via HGF and c-Met, together with ITGB1, enhances protrusive contacts that meet morphological and functional criteria for TNTs, strengthening the link between oncogenic programs and conduit assembly (Awanis et al. 2023; Manzano et al. 2024). Drug redistribution through conduits has emerged as a complementary route of resistance alongside enzyme-based inactivation and transporter-mediated efflux, with contemporary reviews summarizing the intercellular movement of small molecules and nanoparticles during treatment (Sarkari and Lou 2024; Chen and Zhao 2024). Prostate cancer and other solid tumors illustrate how conduit traffic fosters heterogeneity by disseminating oncogenic proteins, metabolic enzymes, and stress signals between clones, creating mosaics of tolerance within a single lesion (Chen and Zhao 2024; Heatlie et al. 2025).

As aforementioned, the most consequential cargo is the mitochondrion, whose donation restores respiration and redox control and thereby reshapes stress tolerance, invasion, and metastatic competence across tumor types (Guan et al. 2024; Marabitti et al. 2024). In high-grade serous ovarian cancer, carcinoma-associated mesenchymal stem cells preferentially donate their mitochondria to tumor cells with low endogenous organelle content, boosting oxidative phosphorylation, clonal diversity, and metastatic spread in vivo (Wang et al. 2022; Frisbie et al. 2024). Breast cancer models show that CAFs deliver mitochondria to malignant cells through open conduits, increasing three-dimensional migration and metabolic fitness, which ties the stromal support directly to invasive behavior (Chen and Zhao 2024; Goliwas et al. 2023). Astrocyte-to-glioma mitochondria transfer accelerates tumorigenesis and drug resistance, integrating with the preexisting TM network to enable rapid recovery after a focal insult (Wang et al. 2022; Krishna et al. 2023). Within the immune compartment, mitochondrial delivery also shapes antitumor efficacy, since cytotoxic T lymphocytes with higher organelle content persist and function better, and engineered gains in mitochondrial fitness improve their intratumoral performance (Baldwin et al. 2024).

Conduit biology can likewise dampen immunity, because mitochondrial or metabolite delivery from the stroma to leukemia or solid tumor cells amplifies oxidative phosphorylation and reduces susceptibility to programmed cell death, while reviews now outline how such exchanges promote immune escape (Guan et al. 2024; Liu et al. 2025). Hematologic malignancies offer clear examples of conduit-mediated metabolic rescue, including mitochondrial transfer that sustains stem cell-like features in leukemia models (Marlein et al. 2017) and emerging work in myeloma showing transmitophagy supported by stromal networks (Solimando et al. 2025). In multiple myeloma, recent preclinical data reveal stromal networks that accept damaged organelles from cancer cells and return survival signals, with conduit-like structures implicated in this recycling loop and linked to resistance to apoptosis (Solimando et al. 2025).

These behaviors are not generic features of any protrusion; they depend on the open-ended lumen continuity, motor-driven cargo transport, and frequent coupling through connexin-lined junctions at tube-to-cell interfaces (Manzano et al. 2024; Virtuoso et al. 2024). Reviews comparing protrusion-based communication with EVs and classical junctions conclude that conduits uniquely support the long-range movement of large cargo and organelles, explaining their outsized influence on plasticity in complex tissues (Hu et al. 2024; Kotarba et al. 2024). Methodological convergence on live imaging, correlative ultrastructure, and lineage separation has strengthened confidence that these are non-fusion exchanges, which are essential for interpreting functional readouts such as metabolic rescue or synchronized calcium signaling (Manzano et al. 2024; Hu et al. 2024). The mechanistic palette of conduit formation continues to widen, from proteome-defined membrane modules to growth-factor programs and integrin cues, which clarifies why tumors under stress often gain connectivity and why interventions that blunt these inputs reduce the conduit prevalence (Awanis et al. 2023; Manzano et al. 2024).

Within the brain TME, astrocyte–glioma interactions repeatedly map to Cx43-rich interfaces that enable both electrical and metabolic cooperation, providing a mechanistic bridge between junctional pores and lumen-based conduits (Paunikar and Tamagnone 2024; Catalano et al. 2023). Functional synthesis across cancers now supports a common view that open conduits integrate with junctional coupling to distribute organelles, second messengers, and protective proteins, thereby coordinating collective survival during chemoradiation and after surgery (Wang et al. 2022; Sarkari and Lou 2024). Finally, conceptual models now treat conduit networks as mesoscale targets whose failure points may be revealed by graph metrics and perturbed by junction blockers or therapies that disrupt protrusion stability, offering testable strategies to convert connectivity from an ally of the tumor into a liability during multimodal treatment (Lou et al. 2024; Schneider et al. 2024).

16.6 Methods and Evidence Grading

Investigating TNTs and TMs requires integrating complementary methods to visualize tube formation, verify cargo motion, and assess function. Live-cell time-lapse microscopy is central to observing the emergence of TNTs and TMs and to tracking cargo over time. This technique allows real-time monitoring of nanotube dynamics, which is crucial for understanding their behavior under different experimental conditions. Recent advances have enabled more precise tracking of specific cargo, such as mitochondria or RNA, using photoconvertible reporters, such as mito-Dendra2 and PhAM. These reporters allow researchers to track specific organelles or molecules as they traverse through nanotubes, ensuring that the observed transport occurs through the lumen and not along the cell surface (Pham et al. 2012).

In addition to live-cell imaging, super-resolution microscopy techniques like stimulated emission depletion (STED) and structured illumination microscopy (SIM) provide detailed insights into the structure and composition of TNTs. These methods permit accurate quantification of the tube diameter and the identification of actin filaments and microtubules within the tubes, which are essential for their structural integrity and cargo transport function. These approaches are especially useful for imaging thicker TMs, which are enriched with microtubules, in contrast to thinner TNTs that are mainly actin-based (Dilsizoglu Senol et al. 2019). By combining super-resolution imaging with cryo-electron tomography, researchers have successfully mapped the ultrastructural features of TNTs and validated their open-ended, lumen-bearing nature, which is essential for distinguishing them from other forms of cell–cell contact (Sartori-Rupp et al. 2019; Gousset et al. 2009).

Correlative electron microscopy (EM) and focused ion beam scanning electron microscopy (FIB-SEM) further complement the super-resolution techniques by providing three-dimensional reconstructions of TNT networks. These ultrastructural methods allow for the precise localization of nanotube connections, confirming that the observed structures are continuous with the plasma membranes of the donor and recipient cells. FIB-SEM has been pivotal in identifying key features such as the microtubule content within TMs and distinguishing them from actin-dominated TNTs, yielding quantitative evidence of physical integrity (Weil et al. 2017). Correlative workflows also visualize cargo within the lumen and help exclude the EV-mediated transfer. Taken together, these methods confirm that TNTs are distinct from filopodia and have a defined role in cargo exchange within the TME.

Electrophysiological methods are crucial for assessing the functional coupling of cells connected by TNTs. By using patch-clamp techniques, researchers can measure ion flow and electrical coupling between cells, which is a defining feature of TNT-mediated communication. Connexin proteins, particularly Cx43, at TNT–cell interfaces enable fast ionic and electrical coupling, which can be assessed by blocking the connexin channels with specific inhibitors. Studies using pharmacologic agents or genetic knockdown of Cx43 have demonstrated that these gap junctions are essential for the TNT structural integrity and for propagating stress signals across tumor cells (Osswald et al. 2015; Zoteva et al. 2024). Additionally, the use of electrophysiology combined with fluorescent imaging has allowed for the simultaneous assessment of ion flux and cargo transfer, confirming that TNTs facilitate both electrical and material exchanges (Schneider et al. 2024).

To exclude confounders and ensure that the observed tube-mediated cargo transfer is not due to other forms of cellular interaction, researchers employ several negative controls. A key control is the use of size-restricted tracers that cannot pass through gap junctions but can be efficiently transported through TNTs, thereby confirming that TNTs mediate the movement of cargo without involving classical gap junctions (Pasquier et al. 2013a, b). Additionally, blocking agents that inhibit EV secretion, such as GW4869, can be used to ensure that the observed cargo movement is not due to the EV-mediated transfer (Sáenz-de-Santa-María et al. 2017). Pharmacological or genetic ablation of key TNT regulators, such as M-Sec (TNFAIP2), IRSp53, Myo10, or GAP-43, also serves as a robust control to verify that the observed transport is indeed mediated by TNTs rather than by alternative processes (Osswald et al. 2015, 2016; Hase et al. 2009; Mim and Unger 2012).

To further validate the mechanistic role of TNTs in intercellular communication, evidence grading is used to annotate studies based on their evidence context, such as in vitro, ex vivo, and in vivo, and evidence strength, such as animal models and preclinical. In vitro systems provide initial insights into TNT formation and cargo transfer, but their relevance to human pathology is often limited. Ex vivo models, such as primary human tumor slices or patient-derived xenografts, bridge this gap by preserving more of the TME's complexity. In vivo studies in animal models, especially those using humanized systems, offer the highest level of evidence and can provide more direct translational relevance. Incorporating this evidence grading into future research will allow for a more precise understanding of the TNT roles across tumor types and their potential as therapeutic targets.

Overall, the combination of live-cell imaging, super-resolution microscopy, electrophysiology, genetic perturbation, and ultrastructural methods enables a comprehensive understanding of TNT biology. These methods provide critical insights into the structural, functional, and molecular properties of TNTs, confirming their role in intercellular communication and tumor progression. As research in this field progresses, these advanced methodologies will be instrumental in developing therapeutic strategies that target TNT-mediated communication, offering new avenues for cancer treatment (Li et al. 2025a). Moreover, as evidence accumulates, future research will further clarify which cargoes traffic via TNTs and which signaling pathways regulate their formation and function.

16.7 Therapeutic Implications

Targeting conduit biology begins with a clear therapeutic objective: to reduce the adaptive value of tube-mediated exchange while preserving tissue functions that rely on cytoskeletal dynamics and junctional signaling. In cancer, the desired outcome is to weaken multicellular robustness, limit stress adaptation and drug redistribution, and prevent stromal donations that rescue tumor bioenergetics. In cell therapy, the calculus can be reversed to preserve or even enhance beneficial organelle exchange, for example, to improve the T cell fitness or to rescue ischemic tissues. These opposing objectives underscore the need for context-selective modulators acting on biogenesis, coupling, or cargo flow rather than blunt cytoskeletal poisons.

One rational entry point is biogenesis. The canonical initiator module—M-Sec and RalA with exocyst support—sits near the top of the decision tree for protrusion outgrowth and lumen formation and remains an attractive target. Although no clinical M-Sec inhibitor exists, the pathway is druggable via the Ral arm, where tool compounds such as BQU57 and RBC8 demonstrate on-target activity and antitumor effects in the preclinical models, supporting next-generation Ral inhibitors that could depress conduit initiation without fully ablating actin dynamics (Yin et al. 2023). Mechanistic work further refines the upstream actin logic, showing that Eps8 and IRSp53 cooperate to generate and stabilize slender membrane tubes, which can be tuned by Arp2/3 inhibition with CK-666 in a way that alters protrusion architecture, offering a calibrated rather than catastrophic way to reduce open-ended bridges (Henderson et al. 2022, 2023; Li et al. 2025b). Because myosin motors support luminal patency and tip dynamics, new insights into Myo10 accumulation and function in filopodia suggest an additional axis for future small-molecule discovery, although clinically available Myo10 inhibitors remain to be developed, and specificity will be crucial to avoid global mitotic and adhesion defects (Shangguan and Rock 2024; Mayca Pozo et al. 2023).

A second entry point is electrical and chemical coupling at tube termini. In glioblastoma, the gap junction protein Cx43 is a structural and functional lynchpin for long-lived microtube networks that propagate calcium and distribute injury signals. Plenty of groups now test the pharmacologic modulation of Cx43 in preclinical models. The gap junction modulator tonabersat reduces glioma intercellular communication and enhances the effect of standard treatments in animal models, while Cx43-targeting peptides such as αCT1 reshape channel interactomes in other indications and offer a template for brain-directed designs with better tumor selectivity and delivery (Beichert et al. 2025). Given the physiological roles of Cx43 in the heart, the brain, and wound repair, therapeutic windows will likely depend on local delivery, transient dosing, and tumor-associated uptake, which are active areas for nanoformulation and peptide engineering.

The network perspective motivates a third strategy: disrupt hub nodes and choke points so that a small pharmacologic perturbation yields a large loss of percolation in tumors. In glioma, TMs confer resilience to resection and radiochemotherapy by rapidly rewiring connectivity, and depletion of GAP-43 or reduction of Cx43 undermines this response and sensitizes tumors in vivo, revealing a vulnerability in structural integrity and channel availability (Weil et al. 2017; Wang et al. 2022; Osswald et al. 2015). Extending the logic beyond brain tumors, G protein-coupled receptor (GPCR) cues and small GTPase circuits that bias protrusive states in epithelial cancers may be exploited to shift the system into less connected configurations or to render TMs mechanically fragile under treatment stress (Jing et al. 2024).

A complementary approach targets cargo flow, especially mitochondrial transfer that upgrades respiration and redox control in recipients. Recent in vivo work implicates the Miro family of mitochondrial adapters as rate limiting for horizontal transfer and tumor progression, nominating Miro1 and Miro2 as actionable nodes to restrict organelle trafficking without broadly damaging the cytoskeleton (Novak et al. 2025; Boulton and Caino 2024). Upstream stress signals and nuclear programs also create pharmacologic handles. For example, hypoxia-driven HMGB1 elevates NF-Y activity to upregulate RHOT1 and collaborates with RAC1 to increase mitochondrial transfer in hepatocellular carcinoma, suggesting that HMGB1 antagonism or RAC1 targeting could attenuate pathogenic donations with measurable metabolic consequences in vivo (Jing et al. 2024). In hematologic malignancies, inhibition of oxidative burst generators such as NOX2 is being explored to limit stromal-to-leukemia organelle flow and improve therapy response, and recent translational studies now prioritize lineage-splitting readouts to ensure true conduit blockade rather than general TME toxicity (Marlein et al. 2017; Guan et al. 2024; Borcherding and Brestoff 2023).

In parallel, a growing body of clinical and translational work tests the opposite direction for medical benefit, namely, deliberate augmentation of organelle transfer. These studies demand the same rigor in mechanism assignment because vesicle- and tube-based routes can coexist, yet the clinical momentum is informative for safety and delivery. Autologous mitochondrial transplantation has completed early studies in acute ischemic stroke and is under evaluation for cardiac and extracorporeal

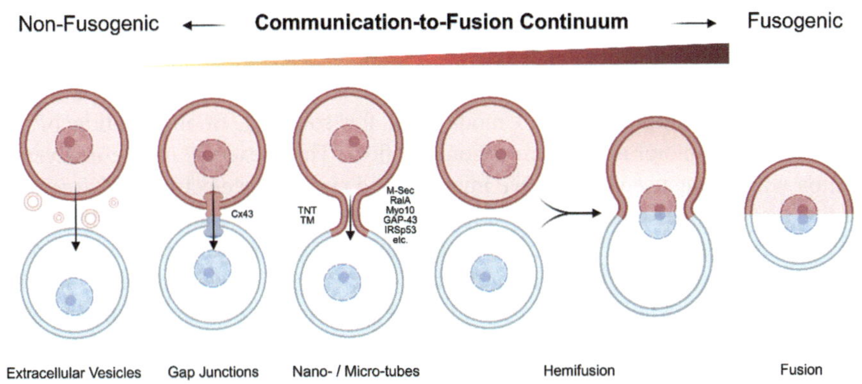

Fig. 16.1 Communication-to-fusion continuum. This schematic depicts intercellular exchange along a spectrum from non-fusogenic to fusogenic, with a gradient bar indicating increasing fusogenic potential. EVs mediate paracrine transfer without membrane continuity. Gap junctions comprise apposed membranes bridged by Cx43 channels, permitting ions and small metabolites (<1 kDa) and enabling electrical coupling while preserving bilayer integrity. Nano-/microtube, TNTs and TMs are open-ended membranous conduits that provide long-range cytoplasmic continuity and bidirectional trafficking of larger cargo, and may couple electrically via Cx43; representative biogenetic drivers include M-Sec, RalA, Myo10, GAP-43, and IRSp53. This mode marks the operational boundary of the continuum because nano-/microtube-mediated exchange lacks nuclear mixing. Progressing rightward, hemifusion merges outer lipid leaflets, and transient pores allow limited content mixing, culminating in fusion, where full bilayer merger yields a single cytoplasm with potential nuclear mixing, known as a syncytium. *Cx43* connexin 43, *EVs* extracellular vesicles, *GAP-43* growth-associated protein 43, *IRSp53* insulin receptor substrate protein of 53 kilodaltons, *Myo*10 myosin-10, *RalA* Ras-like protein A, *TM* tumor microtube, *TNT* tunneling nanotube

membrane oxygenation complications, while several programs explore EV-packaged mitochondria and platelet-derived "mitlets" as delivery vehicles, each with ongoing or recently registered clinical trials reporting feasibility and preliminary safety (Li et al. 2025a; Walker et al. 2024). For oncology, these data argue that systemic exposure to exogenous mitochondria and vesicles is tolerable under controlled conditions, but they also warn that unintended support to tumor cells is plausible unless trafficking is targeted and localized.

Drug delivery through existing conduits remains an attractive concept, especially for macromolecules that rarely cross membranes unaided. The challenge is mechanistic specificity and proof. Any claim of tube-mediated piggybacking must pass negative controls that exclude gap junction-only diffusion and EV transfer—for example, using size-ladder tracers beyond the junctional cutoff, suppressing vesicle secretion with rescue designs, and combining genetic or pharmacologic hits against M-Sec, IRSp53, Myo10, or Cx43 to show convergent loss of delivery (Fig. 16.1). Until such criteria are met in vivo with long-term imaging and correlative ultrastructure, conduit piggybacking should be treated as an experimental possibility rather than a therapeutic platform.

A practical roadmap emerges. First, define a disease-relevant readout known to depend on tube connectivity or cargo exchange, such as network-aided radiochemotherapy resistance in glioblastoma or stromal donations that upgrade oxidative phosphorylation in leukemias. Second, select a proximal intervention justified by mechanism and pharmacology, such as Ral or Eps8–IRSp53–actin interference for initiation, Cx43 modulation for coupling, or interruption of the Miro–RHOT1–motor axis for mitochondrial flow. Third, execute orthogonal verification with lineage separation, live imaging, and correlative electron microscopy to prove that the intended conduit property has been modified rather than that of the general cytoskeleton or the secretory pathways. Finally, build combinatorial regimens that exploit the induced vulnerabilities, for example, lowering the threshold for apoptosis under chemotherapy, preventing network-mediated damage sharing after radiotherapy, or enhancing immunotherapy by depriving tumors of immune-cell mitochondria.

If successful, conduit-targeted therapy will move the field from descriptive connectivity to actionable systems pharmacology. As candidate agents mature, equal care must be applied to safety. Because tubes and junctions support normal tissue repair and homeostasis, dosing should respect the narrow window between network-level tumor fragility and essential physiology. Regionally restricted delivery, peptide therapeutics with tumor-enriched uptake, and intermittent schedules are likely to be decisive for clinical translation.

16.8 Concluding Remarks and Outlook

We situate nanotube-based conduits along a communication-to-fusion continuum (Fig. 16.1), encompassing EVs, synapses or gap junctions, TNTs or TMs, transient pores, and finally true fusion that yields multinucleated syncytia; again, the definitional boundary for this chapter is the absence of nuclear or genomic mixing in tube-mediated exchange (Cordero Cervantes and Zurzolo 2021; Hu et al. 2024; Welsh et al. 2024b; Shultes et al. 2024; Kephart et al. 2024).

Within this spectrum, tubes are defined by open-ended lumens that enable long-range transport of large cargo while often coupling electrically through connexin-enriched interfaces, features that distinguish them from vesicle secretion, narrow gap junction pores, and hemifusion diaphragms that precede membrane merger (Tavakoli et al. 2025). Conceptually, this placement clarifies nomenclature across related protrusions, because cytonemes traffic morphogens along filiform extensions primarily for signaling, whereas nanotubes and microtubes sustain bidirectional material exchange and frequently integrate with junctional coupling at their termini (Chen and Zhao 2024; Hall et al. 2024; Melwani and Pandey 2023).

Rules of partner specificity likely arise from membrane-identity codes and cytoskeletal adapters that stabilize narrow bridges. Recent proteomics nominate the tetraspanins CD9 and CD81 as nonredundant regulators of tube stability and competence, complementing evidence that Cx43 at tube–cell interfaces licenses ionic and small-molecule communication (Zoteva et al. 2024; Manzano et al. 2024; Paunikar and Tamagnone 2024). Cargo prioritization appears to reflect motor availability, membrane potential, and local demand, with mitochondria transfer as the best-characterized case across cancer and immunity, including TNT-mediated donation that upgrades T cell fitness and tumor bioenergetics in vivo (Li et al. 2025a; Baldwin et al. 2024; Chen and Zhao 2024; Zhang et al. 2024; Zurzolo 2025).

Quantitative in vivo flux remains a central gap, but progress is steady via lineage separation, photoconvertible reporters, and correlative ultrastructure that verify open lumens rather than membrane apposition, together with network-level analytics that link connectivity to resilience (Sartori-Rupp et al. 2019; Krishna et al. 2023; Hutchison et al. 2024). Biomarker development should emphasize features marking dependency on connectivity, such as high Cx43 or GAP-43 expression in glioma networks, paired with imaging metrics of percolation and responses to modulators like tonabersat that weaken junctional coupling and lower the threshold for treatment-induced apoptosis.

Finally, a unifying outlook treats tubes as mesoscale actuators that connect molecular modules with tissue-level behavior, implying that the largest clinical gains may come from context-selective modulation that either breaks down malignant networks or, conversely, harnesses controlled organelle transport for repair and regeneration, each guided by standardized reporting and guardrails that separate conduit biology from vesicle traffic or de novo cell fusion.

Acknowledgments This work was supported by research funding from the National Natural Science Foundation of China (82401070, W2523106, 82503348), the Sichuan Province Science and Technology Support Program (2023NSFSC1519), Henan Province Scientific and Technological Project (242102311007), the China Postdoctoral Science Foundation (2022M712240, 2024M752978), and the International Orthodontics Foundation Young Research Grant (IOF2023Y03).

Author Contribution X.Q.: Methodology, Formal Analysis, Investigation, Resources, Data Curation, Writing—Review and Editing, Visualization, Funding Acquisition

B.L.: Conceptualization, Methodology, Formal Analysis, Investigation, Resources, Data Curation, Writing—Original Draft, Writing—Review and Editing, Visualization, Project Administration, Funding Acquisition

Declaration of Competing Interests The authors declare that they have no competing interests.

References

Ady JW et al (2014) Intercellular communication in malignant pleural mesothelioma: properties of tunneling nanotubes. Front Physiol 5:103546

Ahmad T et al (2014) Miro1 regulates intercellular mitochondrial transport & enhances mesenchymal stem cell rescue efficacy. EMBO J 33:994–1010

Austefjord MW, Gerdes HH, Wang X (2014) Tunneling nanotubes: diversity in morphology and structure. Commun Integr Biol 7:e27934

Awanis G et al (2023) HGF/c-met/β1-integrin signalling axis induces tunneling nanotubes in A549 lung adenocarcinoma cells. Life Sci Alliance 6:e202301953

Babenko VA et al (2018) Miro1 enhances mitochondria transfer from multipotent mesenchymal stem cells (MMSC) to neural cells and improves the efficacy of cell recovery. Molecules 23:1–14

Baldwin JG et al (2024) Intercellular nanotube-mediated mitochondrial transfer enhances T cell metabolic fitness and antitumor efficacy. Cell:1–17. https://doi.org/10.1016/j.cell.2024.08.029

Barutta F et al (2022) Protective effect of the tunneling nanotube-TNFAIP2/M-sec system on podocyte autophagy in diabetic nephropathy. Autophagy 19:505

Beichert J, Hoffmann DC, Winkler F, Ratliff M (2025) Innovative therapeutic strategies targeting the network architecture of glioblastoma. Clin Cancer Res 31:2864–2871

Bhat S et al (2020) Rab35 and its effectors promote formation of tunneling nanotubes in neuronal cells. Sci Rep 10:1–14

Bohil AB, Robertson BW, Cheney RE (2006) Myosin-X is a molecular motor that functions in filopodia formation. Proc Natl Acad Sci USA 103:12411–12416

Borcherding N, Brestoff JR (2023) The power and potential of mitochondria transfer. Nature 623: 283–291

Boulton DP, Caino MC (2024) Emerging roles for mitochondrial rho GTPases in tumor biology. J Biol Chem 300:107670

Catalano M, Limatola C, Trettel F (2023) Non-neoplastic astrocytes: key players for brain tumor progression. Front Cell Neurosci 17:1352130

Chakraborty R, Nonaka T, Hasegawa M, Zurzolo C (2023) Tunnelling nanotubes between neuronal and microglial cells allow bi-directional transfer of α-Synuclein and mitochondria. Cell Death Dis 145(14):1–12

Chang M et al (2022) Formation of cellular close-ended tunneling nanotubes through mechanical deformation. Sci Adv 8:3995

Che J et al (2023) αCT1 peptide sensitizes glioma cells to temozolomide in a glioblastoma organoid platform. Biotechnol Bioeng 120:1108–1119

Chen M, Zhao D (2024) Invisible bridges: unveiling the role and prospects of tunneling nanotubes in cancer therapy. Mol Pharm 21:5413–5429

Cole JM, Dahl R, Cowden Dahl KD (2021) MAPK signaling is required for generation of tunneling nanotube-like structures in ovarian cancer cells. Cancer 13:274

Cordero Cervantes D, Zurzolo C (2021) Peering into tunneling nanotubes – the path forward. EMBO J 40:e105789

Dasgupta S et al (2023) Global analysis of contact-dependent human-to-mouse intercellular mRNA and lncRNA transfer in cell culture. elife 12

Davis DM, Sowinski S (2008) Membrane nanotubes: dynamic long-distance connections between animal cells. Nat Rev Mol Cell Biol 9:431–436

Delage E et al (2016) Differential identity of Filopodia and tunneling nanotubes revealed by the opposite functions of actin regulatory complexes. Sci Reports 61(6):1–12

Desir S et al (2018) Chemotherapy-induced tunneling nanotubes mediate intercellular drug efflux in pancreatic cancer. Sci Reports 81(8):1–13

Desir S et al (2019) Intercellular transfer of oncogenic KRAS via tunneling nanotubes introduces intracellular mutational heterogeneity in colon cancer cells. Cancers (Basel) 11

Dilsizoglu Senol A et al (2019) Effect of tolytoxin on tunneling nanotube formation and function. Sci Rep 9:1–15

Dong LF et al (2023) Mitochondria on the move: horizontal mitochondrial transfer in disease and health. J Cell Biol 222:1–27

Driscoll J, Gondaliya P, Patel T (2022) Tunneling nanotube-mediated communication: a mechanism of intercellular nucleic acid transfer. Int J Mol Sci 23:5487

Frisbie L et al (2024) Carcinoma-associated mesenchymal stem cells promote ovarian cancer heterogeneity and metastasis through mitochondrial. Cell Rep 43:114551

Gerdes HH, Carvalho RN (2008) Intercellular transfer mediated by tunneling nanotubes. Curr Opin Cell Biol 20:470–475

Gkerküçük EB, Tramier M, Bertolin G (2020) Imaging mitochondrial functions: from fluorescent dyes to genetically-encoded sensors. Genes (Basel) 11
Goliwas KF et al (2023) Mitochondrial transfer from cancer-associated fibroblasts increases migration in aggressive breast cancer. J Cell Sci 136
Goodenough DA, Paul DL (2009) Gap junctions. Cold Spring Harb Perspect Biol 1:a002576
Gousset K et al (2009) Prions hijack tunnelling nanotubes for intercellular spread. Nat Cell Biol 11: 328–336
Guan F et al (2024) Mitochondrial transfer in tunneling nanotubes – a new target for cancer therapy. J Exp Clin Cancer Res 43:147
Haimovich G et al (2017) Intercellular mRNA trafficking via membrane nanotube-like extensions in mammalian cells. Proc Natl Acad Sci USA 114:E9873–E9882
Hall ET et al (2024) Cytoneme signaling provides essential contributions to mammalian tissue patterning. Cell 187:276–293.e23
Harris AL (2007) Connexin channel permeability to cytoplasmic molecules. Prog Biophys Mol Biol 94:120–143
Hase K et al (2009) M-sec promotes membrane nanotube formation by interacting with Ral and the exocyst complex. Nat Cell Biol 11:1427–1432
Heatlie JK et al (2025) Extracellular vesicles and tunnelling nanotubes as mediators of prostate cancer intercellular communication. Biomolecules 15:23
Henderson JM et al (2022) Network shifts away from branched actin formation promote the uniquely long growth of tunneling nanotubes. Biophys J 121:520a
Henderson JM et al (2023) Tunnelling nanotube formation is driven by Eps8/IRSp53-dependent linear actin polymerization. EMBO J 42:e113761
Hu HT et al (2024) The cellular protrusions for inter-cellular material transfer: similarities between filopodia, cytonemes, tunneling nanotubes, viruses, and extracellular vesicles. Front Cell Dev Biol 12:1422227
Hutchison V, Lynch A, Gutierrez-Gamez AM, Chen J (2024) Inducible tricolor reporter mouse for parallel imaging of lysosomes, mitochondria, and microtubules. J Cell Biol 223
Jansens RJJ, Tishchenko A, Favoreel HW (2020) Bridging the gap: virus long-distance spread via tunneling nanotubes. J Virol 94
Jing M et al (2024) HMGB1 promotes mitochondrial transfer between hepatocellular carcinoma cells through RHOT1 and RAC1 under hypoxia. Cell Death Dis 15:1–15
Kephart SM, Hom N, Lee KK (2024) Visualizing intermediate stages of viral membrane fusion by cryo-electron tomography. Trends Biochem Sci 49:916–931
Kimura S et al (2016) Distinct roles for the N-and C-terminal regions of M-sec in plasma membrane deformation during tunneling nanotube formation. Sci Rep 6:1–12
Kolba MD et al (2019) Tunneling nanotube-mediated intercellular vesicle and protein transfer in the stroma-provided imatinib resistance in chronic myeloid leukemia cells. Cell Death Dis 1011(10):1–16
Korenkova O, Pepe A, Zurzolo C (2020) Fine intercellular connections in development: TNTs, cytonemes, or intercellular bridges? Cell Stress 4:30
Kornberg TB, Roy S (2014) Cytonemes as specialized signaling filopodia. Development 141:729
Kotarba S et al (2024) Potential mechanisms of tunneling nanotube formation and their role in pathology spread in Alzheimer's disease and other Proteinopathies. Int J Mol Sci 25:10797
Kretschmer A et al (2019) Stress-induced tunneling nanotubes support treatment adaptation in prostate cancer. Sci Reports 9(1):1–13
Krishna S et al (2023) Glioblastoma remodelling of human neural circuits decreases survival. Nature 617:599–607
Leduc C et al (2004) Cooperative extraction of membrane nanotubes by molecular motors. Proc Natl Acad Sci USA 101:17096–17101
Li B et al (2025a) Targeting mitochondrial transfer as a promising therapeutic strategy. Trends Mol Med S1471-4914(25):00089–00089

Li L et al (2025b) CK-666 exerts anticancer effects by regulating autophagy, tunneling nanotubes and extracellular vesicles formation. Biomed Pharmacother 183:117825

Liu R et al (2025) Unveiling mitochondrial transfer in tumor immune evasion: mechanisms, challenges, and clinical implications. Front Immunol 16:1625814

Ljubojevic N, Henderson JM, Zurzolo C (2021) The ways of actin: why tunneling nanotubes are unique cell protrusions. Trends Cell Biol 31:130–142

Lou E (2017) Can you hear them now? Tumor microtubes form cellular communication networks that protect gliomas from surgical lesions and chemotherapy treatments. Neuro-Oncology 19: 1289–1291

Lou E et al (2024) Tunneling nanotubes and tumor microtubes – emerging data on their roles in intercellular communication and pathophysiology: summary of an international FASEB catalyst conference October 2023. FASEB J 38:e23514

Lu J et al (2017) Tunneling nanotubes promote intercellular mitochondria transfer followed by increased invasiveness in bladder cancer cells. Oncotarget 8:15539–15552

Lu JJ, Yang WM, Li F, Zhu W, Chen Z (2019) Tunneling nanotubes mediated microRNA-155 intercellular transportation promotes bladder cancer cells' invasive and proliferative capacity. Int J Nanomedicine 14:9731

Manzano RN et al (2024) Proteomic landscape of tunneling nanotubes reveals CD9 and CD81 tetraspanins as key regulators. elife 13

Marabitti V, Vulpis E, Nazio F, Campello S (2024) Mitochondrial transfer as a strategy for enhancing cancer cell fitness:current insights and future directions. Pharmacol Res 208:107382

Marlein CR et al (2017) NADPH oxidase-2 derived superoxide drives mitochondrial transfer from bone marrow stromal cells to leukemic blasts. Blood 130:1649–1660

Marzo L, Gousset K, Zurzolo C (2012) Multifaceted roles of tunneling nanotubes in intercellular communication. Front Physiol 3:1–14

Mayca Pozo F et al (2023) MYO10 regulates genome stability and cancer inflammation through mediating mitosis. Cell Rep 42:112531

Melwani PK, Pandey BN (2023) Tunneling nanotubes: the intercellular conduits contributing to cancer pathogenesis and its therapy. Biochim Biophys Acta Rev Cancer 1878:189028

Mim C, Unger VM (2012) Membrane curvature and its generation by BAR proteins. Trends Biochem Sci 37:526–533

Naphade S et al (2015) Lysosomal cross-correction by hematopoietic stem cell-derived macrophages via tunneling nanotubes. Stem Cells 33:301

Novak J et al (2025) The adaptor protein Miro1 modulates horizontal transfer of mitochondria in mouse melanoma models. Cell Rep 44:115154

Ohno H, Hase K, Kimura S (2010) M-sec: emerging secrets of tunneling nanotube formation. Commun Integr Biol 3:231–233

Önfelt B et al (2006) Structurally distinct membrane nanotubes between human macrophages support long-distance vesicular traffic or surfing of bacteria. J Immunol 177:8476–8483

Osswald M et al (2015) Brain tumour cells interconnect to a functional and resistant network. Nature 528:93–98

Osswald M, Solecki G, Wick W, Winkler F (2016) A malignant cellular network in gliomas: potential clinical implications. Neuro-Oncology 18:479–485

Overholtzer M et al (2007) A nonapoptotic cell death process, Entosis, that occurs by cell-in-cell invasion. Cell 131:966–979

Pasquier J et al (2013a) Preferential transfer of mitochondria from endothelial to cancer cells through tunneling nanotubes modulates chemoresistance. J Transl Med 11:94

Pasquier J et al (2013b) Tunneling nanotubes mediate preferential transfer of mitochondria from endothelial to cancer cells and confer chemoresistance. J Transl Med 11:94

Paunikar S, Tamagnone L (2024) Connexin-43 in cancer: above and beyond gap junctions! Cancers (Basel) 16:4191

Pepe A, Pietropaoli S, Vos M, Barba-Spaeth G, Zurzolo C (2022) Tunneling nanotubes provide a route for SARS-CoV-2 spreading. Sci Adv 8:eabo0171

Pham AH, Mccaffery JM, Chan DC (2012) Mouse lines with photo-activatable mitochondria (PhAM) to study mitochondrial dynamics. Genesis 50:833
Pilling AD, Horiuchi D, Lively CM, Saxton WM (2006) Kinesin-1 and dynein are the primary motors for fast transport of mitochondria in drosophila motor axons. Mol Biol Cell 17:2057–2068
Pinto G, Brou C, Zurzolo C (2020) Tunneling nanotubes: the fuel of tumor progression? Trends in Cancer 6:874–888
Prévost C et al (2015) IRSp53 senses negative membrane curvature and phase separates along membrane tubules. Nat Commun 61(6):1–11
Qiao X et al (2024) Beyond mitochondrial transfer, cell fusion rescues metabolic dysfunction and boosts malignancy in adenoid cystic carcinoma. Cell Rep 43:114652
Roehlecke C, Schmidt MHH (2020) Tunneling nanotubes and tumor microtubes in cancer. Cancers (Basel) 12:857
Rustom A, Saffrich R, Markovic I, Walther P, Gerdes H-HH (2004) Nanotubular highways for intercellular organelle transport. Science 303:1007–1010
Sáenz-de-Santa-María I et al (2017) Control of long-distance cell-to-cell communication and autophagosome transfer in squamous cell carcinoma via tunneling nanotubes. Oncotarget 8: 20939–20960
Sarkari A, Lou E (2024) Do tunneling nanotubes drive chemoresistance in solid tumors and other malignancies? Biochem Soc Trans 52:1757–1764
Sartori-Rupp A et al (2019) Correlative cryo-electron microscopy reveals the structure of TNTs in neuronal cells. Nat Commun 10:1–16
Schiller C et al (2013) LST1 promotes the assembly of a molecular machinery responsible for tunneling nanotube formation. J Cell Sci 126:767–777
Schmidt ENC et al (2025) Tonabersat enhances temozolomide-mediated cytotoxicity in glioblastoma by disrupting intercellular connectivity through connexin 43 inhibition. Mol Oncol 19: 878–898
Schneider M et al (2024) The Alcatraz-strategy: a roadmap to break the connectivity barrier in malignant brain tumours. Mol Oncol 18:2890–2905
Shangguan J, Rock RS (2024) Hundreds of myosin 10s are pushed to the tips of filopodia and could cause traffic jams on actin. elife 12:1–21
Sheng ZH, Cai Q (2012) Mitochondrial transport in neurons: impact on synaptic homeostasis and neurodegeneration. Nat Rev Neurosci 13:77–93
Sherer NM, Mothes W (2008) Cytonemes and tunneling nanotubules in cell-cell communication and viral pathogenesis. Trends Cell Biol 18:414–420
Shultes PV, Weaver DT, Tadele DS, Barker-Clarke RJ, Scott JG (2024) Cell-cell fusion in cancer: the next cancer hallmark? Int J Biochem Cell Biol 175:106649
Simunovic M, Voth GA (2015) Membrane tension controls the assembly of curvature-generating proteins. Nat Commun 6:1–8
Smith IF, Shuai J, Parker I (2011) Active generation and propagation of Ca2+ signals within tunneling membrane nanotubes. Biophys J 100:L37–L39
Smyth JW et al (2025) Cytoplasmic connexin43-microtubule interactions promote glioblastoma stem-like cell maintenance and tumorigenicity. Cell Death Dis 16:388
Solimando AG et al (2025) Tunneling nanotubes between bone marrow stromal cells support transmitophagy and resistance to apoptosis in myeloma. Blood Cancer J 15:1–5
Spees JL, Olson SD, Whitney MJ, Prockop DJ (2006) Mitochondrial transfer between cells can rescue aerobic respiration. Proc Natl Acad Sci USA 103:1283–1288
Tavakoli A, Hu S, Ebrahim S, Kachar B (2025) Hemifusomes and interacting proteolipid nanodroplets mediate multi-vesicular body formation. Nat Commun 16:4609
Thayanithy V, Dickson EL, Steer C, Subramanian S, Lou E (2014) Tumor-stromal cross talk: direct cell-to-cell transfer of oncogenic microRNAs via tunneling nanotubes. Transl Res 164:359–365

Théry C et al (2018) Minimal information for studies of extracellular vesicles 2018 (MISEV2018): a position statement of the International Society for Extracellular Vesicles and update of the MISEV2014 guidelines. J Extracell Vesicles 7:1535750

Tsai F-C, C. et al (2022) Activated I-BAR IRSp53 clustering controls the formation of VASP-actin–based membrane protrusions. Sci Adv 8:8677

Turos-Korgul L, Kolba MD, Chroscicki P, Zieminska A, Piwocka K (2022) Tunneling nanotubes facilitate intercellular protein transfer and cell networks function. Front Cell Dev Biol 10:1–13

Uhl J et al (2018) Myosin-X is essential to the intercellular spread of HIV-1 Nef through tunneling nanotubes. J Cell Commun Signal 13:209

Valdebenito S et al (2021) Tunneling nanotubes, TNT, communicate glioblastoma with surrounding non-tumor astrocytes to adapt them to hypoxic and metabolic tumor conditions. Sci Rep 11: 1–18

Virtuoso A et al (2024) The interplay between glioblastoma cells and tumor microenvironment: new perspectives for early diagnosis and targeted cancer therapy. Brain Sci 14:331

Walker M et al (2024) Autologous mitochondrial transplant for acute cerebral ischemia: phase 1 trial results and review. J Cereb Blood Flow Metab. https://doi.org/10.1177/0271678X241305230

Wang Y, Cui J, Sun X, Zhang Y (2011) Tunneling-nanotube development in astrocytes depends on p53 activation. Cell Death Differ 18:732–742

Wang J et al (2018) Targeting different domains of gap junction protein to control malignant glioma. Neuro-Oncology 20:885–896

Wang X, Liang J, Sun H (2022) The network of tumor microtubes: an improperly reactivated neural cell network with Stemness feature for resistance and recurrence in gliomas. Front Oncol 12: 921975

Weil S et al (2017) Tumor microtubes convey resistance to surgical lesions and chemotherapy in gliomas. Neuro-Oncology 19:1316–1326

Welsh JA et al (2024a) Minimal information for studies of extracellular vesicles (MISEV2023): from basic to advanced approaches. J Extracell Vesicles 13:e12404

Welsh JA, Goberdhan DC, O'Driscoll L, Théry C, Witwer KW (2024b) MISEV2023: an updated guide to EV research and applications. J Extracell Vesicles 13:e12416

Weng Z et al (2021) Therapeutic roles of mesenchymal stem cell-derived extracellular vesicles in cancer. J Hematol Oncol 14:1–22. https://doi.org/10.1186/s13045-021-01141-y

Yin G et al (2023) Targeting small GTPases: emerging grasps on previously untamable targets, pioneered by KRAS. Signal Transduct Target Ther 8:1–39

Zhang M et al (2024) Inter- and intracellular mitochondrial communication: signaling hubs in aging and age-related diseases. Cell Mol Biol Lett 29:153

Zoteva V et al (2024) Integrating and optimizing tonabersat in standard glioblastoma therapy: a preclinical study. PLoS One 19:1–21

Zurzolo C (2025) Transforming the concept of connectivity: unveiling tunneling nanotube biology and their roles in brain development and neurodegeneration. Physiol Rev 105:1823–1865

Chapter 17
Mathematical Modeling of Cell-Cell Fusion Assays to Characterize Virus-Mediated Cell Fusion

Hana M. Dobrovolny

Abstract Syncytia are multinucleated cells created by fusion of normal cells. Some viruses are known to induce syncytia formation as a consequence of infected cells expressing viral surface proteins. The role of syncytia in the spread of viral infections and the resulting disease is still poorly understood. Researchers have developed cell-cell fusion assays to help study the fusion process. These assays use cells transfected to express viral surface proteins to study the fusion process without an active viral infection. Recently, mathematical models have been developed to simulate the dynamics of these assays and have been used to estimate parameters that characterize the fusion process. This review examines the cell-cell fusion assay and the mathematical models that have been developed to reproduce assay dynamics. While modeling of these systems is still in its early stages, further extension of these models will help improve our understanding of the syncytia formation process.

Keywords Syncytia · Cell-cell fusion assay · Mathematical model · Parameter identifiability · Viral infection

17.1 Introduction

Syncytia formation caused by viral infections has been observed in a wide array of viruses. Viruses known to induce cell-cell fusion include varicella-zoster virus (Wang et al. 2017), rotavirus (Diller et al. 2019), human immunodeficiency virus (Symeonides et al. 2015), coronaviruses (Qian et al. 2013; Buchrieser et al. 2020), measles (Ayata et al. 2007), and respiratory syncytial virus (Hu et al. 2020; Gonzalez-Reyes et al. 2001; Porotto et al. 2019). The role of syncytia and how they alter viral infections is still not clear, although there is evidence that they increase the speed of spread of the virus (Richman and Bozzette 1994; Rodrigo

H. M. Dobrovolny (✉)
Department of Physics & Astronomy, Texas Christian University, Fort Worth, TX, USA
e-mail: h.dobrovolny@tcu.edu

© The Author(s), under exclusive license to Springer Nature Switzerland AG 2026

J. Weiler et al. (eds.), *Cell Fusion*, Advances in Experimental Medicine and Biology 1509, https://doi.org/10.1007/978-3-032-22637-2_17

1997) and can help with viral immune evasion (Beach and Dobrovolny 2025; Dobrovolny 2024; Rajah et al. 2022).

The primary tool for understanding virus-mediated cell fusion is the cell-cell fusion assay (Gilbert and Greenberg 1997; Sakamoto et al. 2003). This assay uses two groups of cells, which can be either the same or different types of cells (Hong et al. 1999). The first group, known as donor cells, is transfected to express the viral surface protein. The second group, known as acceptor cells, either naturally expresses the appropriate cell surface receptor or is transfected to express the cell receptor. The two cell groups are also fluorescently labeled with different dyes so that they can be distinguished via imaging (Lin et al. 2020; Saw et al. 2015). The two groups of cells are mixed together allowing for fusion to occur between neighboring donor and acceptor cells. Fusion mixes the dyes originally separated in the two cells, so imaging can distinguish acceptor, donor, and syncytia cells.

These assays have been used to study virus-mediated fusion in a number of viruses, such as HIV (Sakamoto et al. 2003; Vincent and Malvoisin 2012; Whitaker et al. 2019), rotavirus (Gilbert and Greenberg 1997), hepatitis C (Kobayashi et al. 2006), Rift Valley fever (Filone et al. 2006), influenza (Su et al. 2009), parainfluenza virus (Tsurudome et al. 2011), varicella-zoster virus (Yang et al. 2016), Epstein-Barr virus (Mohl et al. 2017), arenavirus (York and Nunberg 2018), and SARS-CoV-2 (Sanders et al. 2021; Rajah et al. 2021; Papa et al. 2021; Buchrieser et al. 2020). The assays are useful for comparison studies, allowing us to determine whether certain strains of viruses are more or less fusogenic (Rajah et al. 2021; Hotard et al. 2015; Wang et al. 2022). They have also been used to characterize potential candidate fusion inhibitors (Filone et al. 2006; York and Nunberg 2018; Xia et al. 2020). Finally they have provided insight into the processes underlying viral cell entry and virus-mediated cell fusion (Hayashi et al. 2014; Ling et al. 2013; Cairns et al. 2005).

More recently, mathematical models have been applied to these assays to extract more quantitative information about the characteristics of syncytia formation. Amidei and Dobrovolny (2022) used a simple ordinary differential equation model of the cell-cell fusion process to fit data from an experiment involving SARS-CoV-2. They were able to estimate the syncytia formation rate from syncytia time courses and compare these estimates for SARS-CoV-2 expressing different proteases. They later applied the same process, although with a more detailed mathematical model, to cell-cell fusion assay data comparing syncytia formation of different SARS-CoV-2 variants of concern (Amidei and Dobrovolny 2024). Finally, Gerg and Dobrovolny (2025) extended the model again to include a density-dependent syncytia formation rate, which they found was necessary to correctly describe syncytia formation in AraC-treated assays where the number of cell receptors is depleted.

In this chapter, we review mathematical models developed for cell-cell fusion assays, highlighting how different model assumptions change the predicted dynamics. We then discuss possible future modeling efforts that can leverage experimental data from cell-cell fusion assays to quantitatively characterize virus-mediated cell fusion.

17.2 Models of the Cell-Cell Fusion Assay

17.2.1 Basic Model

The simplest model of cell-cell fusion assays considers only three different cell types, donor (D), acceptor (A), and syncytia (S) (Amidei and Dobrovolny 2022)

$$\begin{aligned}\frac{\mathrm{d}D}{\mathrm{d}t} &= -\gamma DA - \gamma SD \\ \frac{\mathrm{d}A}{\mathrm{d}t} &= -\gamma DA - \gamma SA \\ \frac{\mathrm{d}S}{\mathrm{d}t} &= 2\gamma DA + \gamma SA + \gamma SD.\end{aligned} \tag{17.1}$$

In this model, donor and acceptor cells can fuse with each other or to existing syncytia. All fusion events are assumed to occur at the same rate. Under the assumption that the total number of cells is fixed ($D + A + S = N$), this system can be solved analytically to yield expressions for the donor and acceptor cells

$$\begin{aligned}D &= \frac{ND_0}{D_0 + (N - D_0)e^{\gamma Nt}} \\ A &= \frac{NA_0}{A_0 + (N - A_0)e^{\gamma Nt}}.\end{aligned}$$

An asymmetric version of this model was also proposed and analyzed (Amidei and Dobrovolny 2022)

$$\begin{aligned}\frac{\mathrm{d}D}{\mathrm{d}t} &= -\gamma DA \\ \frac{\mathrm{d}A}{\mathrm{d}t} &= -\gamma DA - \gamma SA \\ \frac{\mathrm{d}S}{\mathrm{d}t} &= 2\gamma DA + \gamma SA + \gamma SD.\end{aligned} \tag{17.2}$$

In this version of the model, only acceptor cells are allowed to bind to existing syncytia. The biological motivation for this type of asymmetry is unclear, but experiments suggest that acceptor cells are more likely to bind to existing syncytia (Papa et al. 2021). Mathematically, breaking the symmetry between donor and acceptor cells allows for a model that produces syncytia time courses that do not necessarily reach 100% syncytia formation, since acceptor cells get absorbed into syncytia faster than donor cells. This flexibility produces better fits to experimental data (Amidei and Dobrovolny 2022). The asymmetric model can also be solved analytically to yield

$$D = \frac{D_0}{N}\left[N - A_0 + A_0 e^{-\gamma N t}\right]$$
$$A = \frac{N A_0}{A_0 + (N - A_0)e^{\gamma N t}}.$$

Figure 17.1 shows predicted time courses for both the symmetric and asymmetric versions of the basic model under different initial conditions. The symmetric model predicts that all cells will eventually end up in syncytia, while the asymmetric model allows for some donor cells to remain unfused. We can also see that the time course

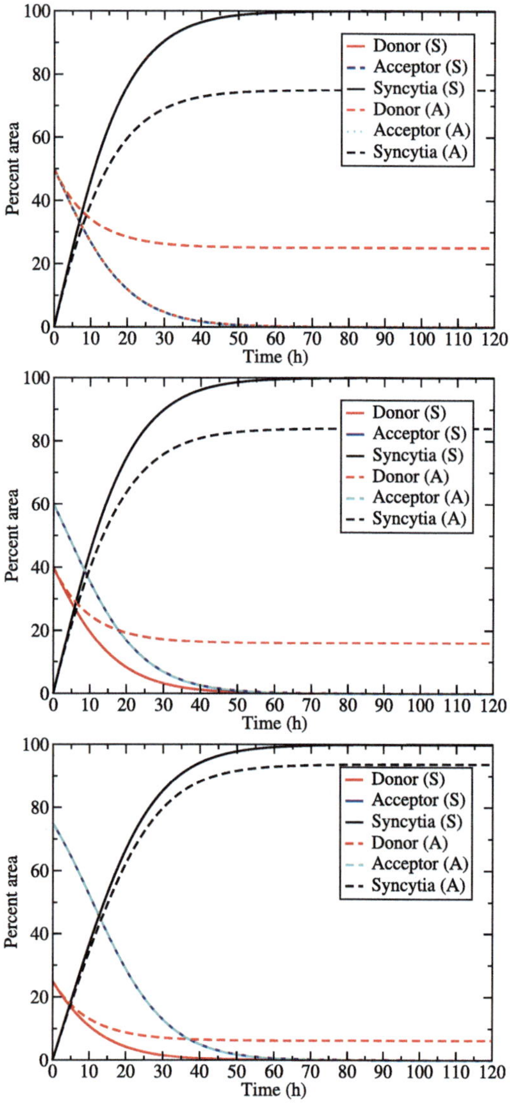

Fig. 17.1 Comparison of the symmetric and asymmetric model predicted time courses for cell-cell fusion assays, assuming a 50/50 split of donor and acceptor cells (top), a 40/60 split of donor and acceptor cells (center), and a 25/75 (bottom) split of donor and acceptor cells

of acceptor cells remains unchanged between the two models, so the breaking of symmetry only affects donor cell and syncytia time courses.

17.2.2 Accounting for Time to Fuse

The first extension of the basic cell-cell fusion assay model incorporated fusing states. The basic model assumes that when acceptor and donor cells fuse, the process is instantaneous. However, fusion is a multistep process involving protein interactions that happen over some span of time (Cohen and Melikyan 2004). A model accounting for these steps is

$$\begin{aligned}
\frac{\mathrm{d}D}{\mathrm{d}t} &= -\gamma DA - \gamma DS \\
\frac{\mathrm{d}A}{\mathrm{d}t} &= -\gamma DA - \gamma SA \\
\frac{\mathrm{d}F_1}{\mathrm{d}t} &= 2\gamma DA + \gamma DS + \gamma SA - kF_1 \\
\frac{\mathrm{d}F_i}{\mathrm{d}t} &= kF_{i-1} - kF_i \quad i = 2 \ldots n \\
\frac{\mathrm{d}S}{\mathrm{d}t} &= kF_n .
\end{aligned} \tag{17.3}$$

In this model, the symmetric version of the basic model is extended so that cells pass through several fusion steps, F_i, before becoming syncytia. The time spent undergoing fusion is given by n/k. Note that an asymmetric version of this model can also be used. Fitting of the model to data from cell-cell fusion assay experiments using SARS-CoV-2 suggested that two fusion compartments were needed to accurately describe SARS-CoV-2 mediated fusion (Amidei and Dobrovolny 2024), but this could differ for other viruses.

Figure 17.2 shows how the number of fusion steps changes the predicted syncytia formation time course for the symmetric model (left) and the asymmetric model (right). Without any fusion compartments, the formation of syncytia occurs immediately, as seen by the steep rise from zero in Fig. 17.1. The addition of even a single fusion compartment ($n = 1$) introduces a delay in the appearance of syncytia. This is consistent with frequently sampled data collected from cell-cell fusion assays (Rajah et al. 2021; Lineberger et al. 2002), which show a clear delay before syncytia start to appear. Changing the number of fusion compartments has other effects on the time course of syncytia formation. There is a change in the slope of the curve, particularly as the number of compartments becomes large. For the asymmetric model, we also see a change in the final number of syncytia, which increases as the number of fusion compartments increases.

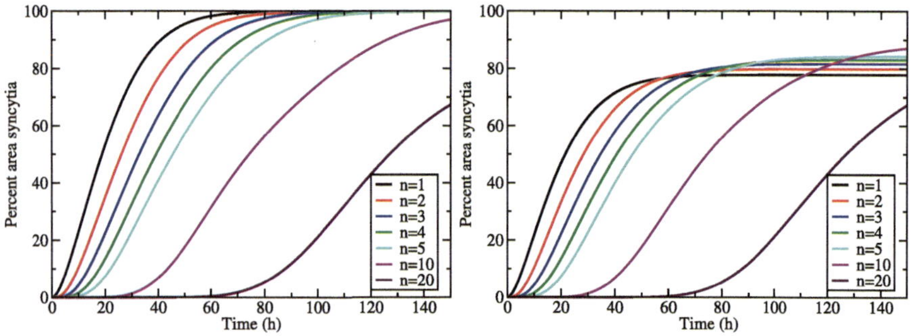

Fig. 17.2 Effect of the number of fusion compartments on time course of syncytia formation in cell-cell fusion assays for the symmetric (left) and asymmetric (right) models. Parameters for these simulations are $\gamma = 1 \times 10^{-3}$ /h, $k = 0.2$ /h, and $D_0 = 50\%$

17.2.3 Density-Dependent Syncytia Formation

The addition of fusion compartments helps to capture the delay in growth of syncytia seen in many cell-cell fusion assay experiments (Amidei and Dobrovolny 2024). Another modification to the basic model is meant to account for the spatial heterogeneity of syncytia formation. Ordinary differential equation (ODE) models assume that components are *well-mixed*, meaning that all objects have access to, and can interact with, all other objects in the system. Syncytia formation occurs via contact with neighboring cells, so the well-mixed assumption of ODE models might not be correctly capturing the dynamics of syncytia formation in the assay. This can be particularly problematic at later times when large syncytia separate any single acceptor and donor cells, so we expect far fewer fusion events at later times. The effect is also noticeable when the density of virus or cell receptors is low, as was observed in an experiment by Lineberger et al. (2002). While fully capturing this effect requires a spatial model, the ODE model can be modified to account for some of this decreased fusion (Gerg and Dobrovolny 2025)

$$\begin{aligned}
\frac{dD}{dt} &= -\frac{\gamma DA}{1+\alpha S} - \frac{\gamma DS}{1+\alpha S} \\
\frac{dA}{dt} &= -\frac{\gamma DA}{1+\alpha S} - \frac{\gamma SA}{1+\alpha S} \\
\frac{dF_1}{dt} &= \frac{2\gamma DA}{1+\alpha S} + \frac{\gamma SA}{1+\alpha S} + \frac{\gamma DA}{1+\alpha S} - kF_1 \\
\frac{dF_i}{dt} &= kF_{i-1} - kF_i \quad i = 2\ldots n \\
\frac{dS}{dt} &= kF_n.
\end{aligned} \tag{17.4}$$

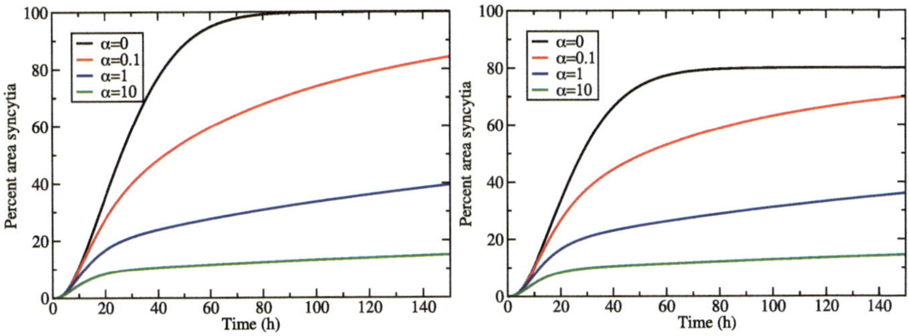

Fig. 17.3 Effect of density-dependent fusion on syncytia formation in cell-cell fusion assays for the symmetric (left) and asymmetric (right) models. Parameters for these simulations are $\gamma = 1 \times 10^{-3}$ /h, $k = 0.2$ /h, $D_0 = 50\%$, and $n = 2$

In this model, the syncytia formation rate is modified to depend on the number of syncytia. As the number of syncytia increases, the syncytia formation rate decreases. The strength of this effect is determined by α.

Figure 17.3 shows how model predictions of syncytia formation vary as the strength of the density dependence varies. Stronger density dependence decreases the number of cells fusing for both the symmetric and asymmetric models. We also see a change in how quickly syncytia form over time, with larger values of α resulting in a more gradual growth in the number of syncytia. Incorporating density dependence does not, however, affect the delay in the appearance of syncytia.

17.2.4 *Syncytia Death*

Since most cell-cell fusion assay experiments are run for a short period of time, typically over a few hours, the models assume that there is no cell death. Death of syncytia is also difficult to detect using standard fluorescence-based measurements since death of the syncytium does not affect the fluorescence emission of the dye. A recent study by Vanhulle et al. (2023) examined a new measurement technique for cell-cell fusion assays based on electrical impedance. Electrical measurements can detect death of syncytia since cell death causes breakdown of the cell membrane changing the impedance. Their measurements suggest that lysis of SARS-CoV-2 spike protein-induced syncytia becomes detectable at ~36 hours after the start of the experiment. We incorporate syncytia death into the model to examine how inclusion of this process affects predicted syncytia time course

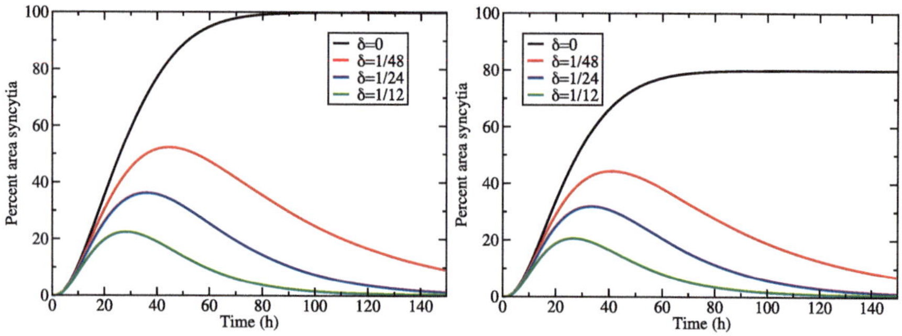

Fig. 17.4 Effect of syncytia death on syncytia formation in cell-cell fusion assays for the symmetric (left) and asymmetric (right) models. Parameters for these simulations are $\gamma = 1 \times 10^{-3}$ /h, $k = 0.2$ /h, $D_0 = 50\%$, and $n = 2$

$$\begin{aligned}
\frac{dD}{dt} &= -\gamma DA - \gamma DS \\
\frac{dA}{dt} &= -\gamma DA - \gamma SA \\
\frac{dF_1}{dt} &= 2\gamma DA + \gamma DS + \gamma SA - kF_1 \\
\frac{dF_i}{dt} &= kF_{i-1} - kF_i \quad i = 2 \ldots n \\
\frac{dS}{dt} &= kF_n - \delta S.
\end{aligned} \tag{17.5}$$

In this model, we assume that only syncytia die, at rate δ, since syncytia appear to have a shortened lifespan as compared to unfused cells (Rajah et al. 2022; Higuchi et al. 2000; Ferri et al. 2000), and the majority of donor and acceptor cells fuse, so few single cells are around long enough to die.

Figure 17.4 shows how model predictions of syncytia formation vary as the death rate changes for both the symmetric and asymmetric versions of the model. In both cases, the syncytia no longer reach a plateau but increase until syncytia death becomes significant. The shape of this curve is consistent with the measurements of Vanhulle et al. (2023) using electrical impedance.

17.3 Using Mathematical Models to Characterize Syncytia Formation

17.3.1 Fitting Mathematical Models to Data

The parameters of the models described in the previous section have biological meaning and can be used to characterize syncytia formation. Table 17.1 lists the

Table 17.1 Parameters in cell-cell fusion assay models

Parameter	Meaning	Unit
γ	Syncytia formation rate	Per unit time
k	Fusion transition rate	Per unit time
α	Strength of density dependence	–
δ	Syncytia death rate	Per unit time

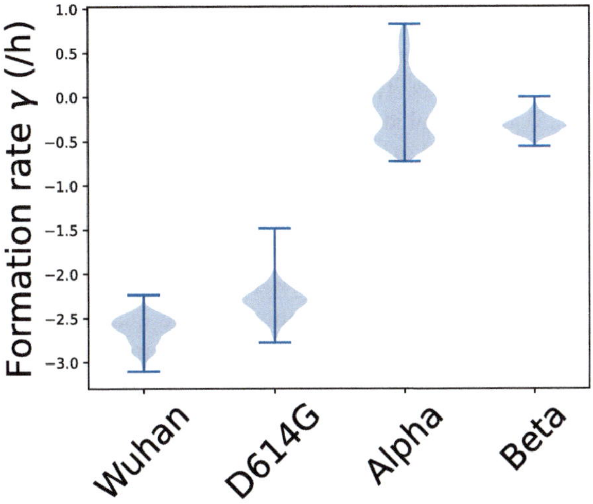

Fig. 17.5 Syncytia formation rates for different strains of SARS-CoV-2, based on estimates from fitting mathematical model Eq. (17.3) to data from Rajah et al. (2021) as described in Amidei and Dobrovolny (2024)

parameters and their meaning. Many of the parameters describe the time scale of different processes that occur in the cell-cell fusion process. The syncytia formation rate is a measure of how easily syncytia form when a donor and an acceptor cell are in contact. As an alternative interpretation, the quantity $1/\gamma N$ gives the average time between fusion events in a population of N cells. The inverse of the fusion transition rate, n/k, also gives important information about the timing of the fusion process. n/k is the average time it takes for the two cells to complete the fusion process. Finally, the inverse of syncytia death rate, $1/\delta$, is the average lifetime of syncytia.

Estimates of the values of these quantities can be obtained by fitting the mathematical models to experimental data collected from the cell-cell fusion assay. The process of fitting involves adjusting parameter values to minimize some measure of distance between the model predicted curve and the experimental data (Cassidy 2023). Given that experimental data is noisy and sparse, parameter estimates derived from mathematical model fits to data have measurement uncertainty, and methods have been developed to quantify this uncertainty (Villaverde et al. 2023). Despite this uncertainty, estimates of the syncytia formation rate determined from fits of the basic model to cell-cell fusion assay data were sufficiently distinct to detect differences in the syncytia formation rate as different proteases were added to the cell-cell fusion assay (Amidei and Dobrovolny 2022). Another study found differences in the syncytia formation rate of different SARS-CoV-2 variants (Fig. 17.5), but that these came with a trade-off in fusion duration (Amidei and Dobrovolny 2024).

SARS-CoV-2 variants with higher syncytia formation rates (those that form syncytia easily) take longer to complete the fusion process.

17.3.2 Parameter Identifiability

While the use of models can help us estimate quantities that characterize syncytia formation, we need to consider whether the data is sufficient to uniquely identify these parameters. This is known as parameter identifiability analysis (Heinrich et al. 2025). Structural identifiability assesses our ability to identify a model's parameters given unlimited, perfect knowledge of a particular variable (Liyanage et al. 2024). For example, the basic model can be solved analytically, giving exact equations for the donor cells, acceptor cells, and the syncytia (since we assume cells are conserved $D + A + S = N_0$). Thus if we measure any one of these quantities, we can identify the syncytia formation rate γ. However, experiments do not produce perfect data. Often data is pooled, and only average quantities are presented; the dynamics of these averaged quantities are not the same as the dynamics of a single system (Hooker and Ganusov 2021; Luo et al. 2022). Additionally, experimental data is noisy, and there can be variability from one experiment to the next (LaBarre and Lowy 2001). Finally, the data is sometimes not sampled at a high enough frequency. For example, the data in Papa et al. (2021) was fit using the basic model to determine syncytia formation rate for SARS-CoV-2 expressing different proteases. Since the data in Papa et al. (2021) was sampled every 6 h, incorporating fusion stages would have yielded a parameter (k the fusion rate) that was unidentifiable since the time to complete fusion is about 4–6 h (Amidei and Dobrovolny 2024). In this case measurements were taken too infrequently to be able to capture the fusion process.

Issues such as these lead to the concept of practical identifiability (Heinrich et al. 2025; Liyanage et al. 2024; Heitzman-Breen et al. 2024). Practical identifiability considers whether we are able to uniquely identify model parameters given the actual experimental data. A number of techniques are used to assess practical identifiability, including looking at parameter correlations and likelihood profiles. Interestingly, even though we have explicit solutions for the basic model, when estimating the syncytia formation rate and the initial number of donor cells, there is correlation between the two parameters (see Fig. 17.6). Figure 17.6 plots bootstrapped values of syncytia formation rate (γ) against the initial number of donor cells (D_0) for all the datasets analyzed in Amidei and Dobrovolny (2022) with each dataset shown in a different color. If parameters are uncorrelated, we expect to see a circle when we plot values of one parameter against the other. In this case, we see a horseshoe-shaped curve with multiple datasets lying along similar curves. This suggests a trade-off in the initial amount of donor cells and the syncytia formation rate particularly when the initial number of donor cells is very high or very low.

Likelihood profiles are used to assess the local identifiability (Kreutz et al. 2013). Fitting of models to data is typically done by minimizing the sum of squared residuals (SSRs). This means that we find the parameter values that yield a model

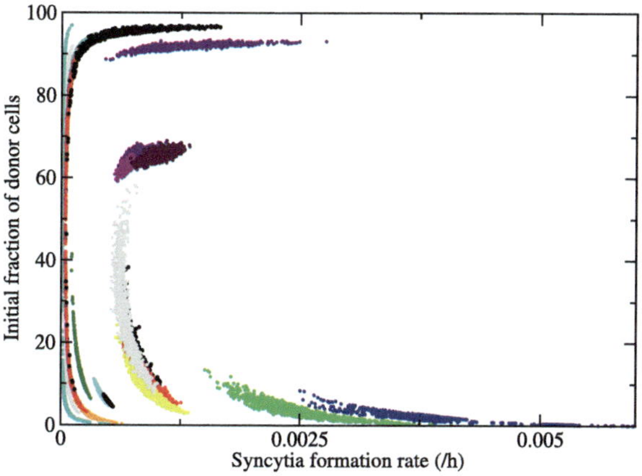

Fig. 17.6 Correlation between bootstrapped parameter estimates for the syncytia formation rate and the initial fraction of donor cells. Each color denotes a different dataset from Papa et al. (2021)

prediction that minimizes the aggregate distance between the predicted curve and the experimental data. Likelihood profiles allow us to verify that we are at a local minimum of the SSR for each of the model parameters. Examples of these types of profiles are shown in Fig. 17.7. This figure shows the parameter likelihood profiles for fits of mathematical model, Eq. (17.3), to cell-cell fusion assay data using the Wuhan strain of SARS-CoV-2. The data is taken from Rajah et al. (2021), and the model fitting is described in Amidei and Dobrovolny (2024). The dashed red line indicates the 90% confidence interval. For both the syncytia formation rate and the fusion rate, there is a fairly narrow parabola with a distinct minimum. The parabola crosses the 90% confidence interval threshold within ±10% of the best fit parameter value. For the initial number of donor cells, the "parabola" is quite flat at its minimum, indicating that there are several possible values of $D(0)$ that would provide equally good fits to the data. We also find that the profile curve does not rise to meet the 90% confidence interval threshold on the right side, indicating that the confidence intervals are quite wide for this parameter.

17.3.3 Implications for Experiments

Now that mathematical models that describe cell-cell fusion assay dynamics have been developed, how can experimental data collection be adjusted to maximize the potential of this new analysis technique? As a start, data from cell-cell fusion assays needs to be measured longitudinally. Measurement of the area fused at one or two time points is insufficient for parameterizing mathematical models. As noted in the previous section, frequent sampling of the system is required. The required frequency is partially dependent on the time scales we are attempting to estimate, but roughly speaking, if a process takes a certain amount of time, we need to sample the system in time steps more frequent than that time. Data also needs to be taken over a sufficiently long period of time. Many of the models predict saturation of the number

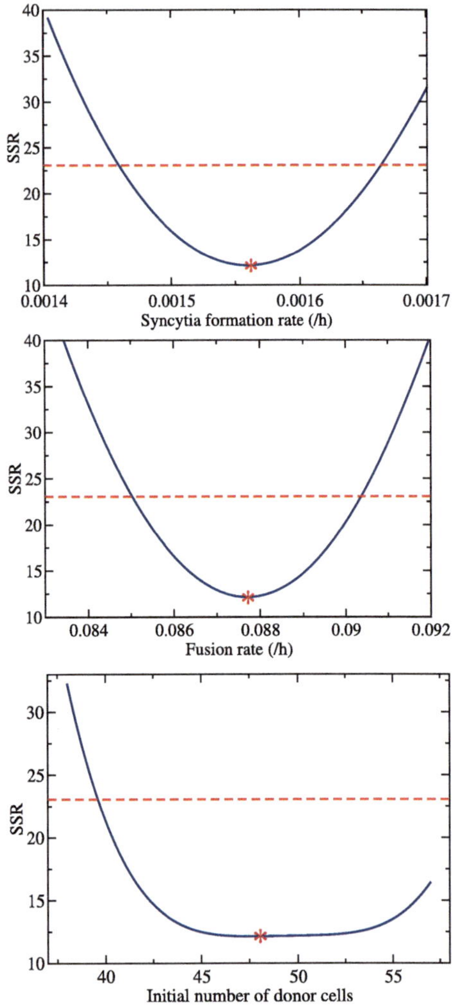

Fig. 17.7 Likelihood profiles derived from fits of mathematical model, Eq. (17.3), to data from the Wuhan strain of SARS-CoV-2 cell-cell fusion assay described in Rajah et al. (2021). Profiles show how the SSR depends on (top) syncytia formation rate, γ, (center) fusion rate, k, and (bottom) the initial number of donor cells D_0. The dashed red line indicates the 90% confidence interval

of syncytia, but longitudinal data is sometimes not taken long enough to capture this plateau (Rajah et al. 2021; Atanasiu et al. 2013; Lineberger et al. 2002). Additionally, if we are to use these experiments to estimate syncytia lifespans, data will need to be taken over an even longer time frame to capture the decay in the number of syncytia. Of course, death of syncytia cannot be detected with the commonly used fluorescence measurements but will require use of electrical impedance measurements as developed in Vanhulle et al. (2023).

To improve model parameter estimates, it would also be helpful to measure the number of (or area covered by) donor and acceptor cells. When donor and acceptor cells are stained with different dyes (Papa et al. 2021), this additional data should be easy to collect. Some cell-cell fusion assays, however, use split protein methods (Rajah et al. 2021; Atanasiu et al. 2013) so that the donor or acceptor cells do not fluoresce, making it difficult to count these two cell populations. Thus if one of the goals is to use the data collected from a cell-cell fusion assay to parameterize a mathematical model, then careful consideration should be given to the measurement technique used to determine the number of cells in syncytia, as well as how often and for how long measurements are made.

17.4 Conclusions

The application of mathematical models to cell-cell fusion assays is fairly new but shows promise in leveraging the data collected in these experiments to quantitatively characterize virus-mediated syncytia formation (Dobrovolny 2024). This review describes early efforts in using mathematical models to analyze cell-cell fusion assays and what we have learned from them with the hope that both modeling and experimental efforts can be combined to help improve our understanding of virus-mediated syncytia formation.

References

Amidei A, Dobrovolny HM (2022) Estimation of virus-mediated cell fusion rate of sars-cov-2. Virol 575:91–100. https://doi.org/10.1016/j.virol.2022.08.016

Amidei A, Dobrovolny HM (2024) Virus-mediated cell fusion of SARS-CoV-2 variants. Math Biosci 369:109144. https://doi.org/10.1016/j.mbs.2024.109144

Atanasiu D, Saw WT, Gallagher JR, Hannah BP, Matsuda Z, Whitbeck JC, Cohen GH, Eisenberg RJ (2013) Dual split protein-based fusion assay reveals that mutations to herpes simplex virus (HSV) glycoprotein gB alter the kinetics of cell-cell fusion induced by HSV entry glycoproteins. J Virol 87(21):11332–11345. https://doi.org/10.1128/JVI.01700-13

Ayata M, Shingai M, Ning X, Matsumoto M, Seya T, Otani S, Seto T, Ohgimoto S, Ogura H (2007) Effect of the alterations in the fusion protein of measles virus isolated from brains of patients with subacute sclerosing panencephalitis on syncytium formation. Virol Res 130(1–2):260–268. https://doi.org/10.1016/j.vir-usres.2007.07.017

Beach I, Dobrovolny HM (2025) The effect of antibodies in the presence of syncytia during viral infections. Bull Math Biol 87(6). https://doi.org/10.1007/s11538-025-01463-9

Buchrieser J, Dufloo J, Hubert M, Monel B, Planas D, Rajah MM, Planchais C, Porrot F, Guivel-Benhassine F, der Werf SV, Casartelli N, Mouquet H, Bruel T, Schwartz O (2020) Syncytia formation by sars-cov-2 infected cells. EMBO J 39(23):e106267. https://doi.org/10.15252/embj.2020106267

Cairns T, Landsburg D, Whitbeek J, Eisenberg R, Cohen G (2005) Contribution of cysteine residues to the structure and function of herpes simplex virus gH/gL. Virology 332(2): 550–562. https://doi.org/10.1016/j.virol.2004.12.006

Cassidy T (2023) A continuation technique for maximum likelihood estimators in biological models. Bull Math Biol 85(10). https://doi.org/10.1007/s11538-023-01200-0

Cohen F, Melikyan G (2004) The energetics of membrane fusion from binding, through hemifusion, pore formation, and pore enlargement. J Membrane Biol 199(1):1–14. https://doi.org/10.1007/s00232-004-0669-8

Diller JR, Parrington HM, Patton JT, Ogden KM (2019) Rotavirus species B encodes a functional fusion-associated small transmembrane protein. J Virol 93(20):e00813–19. https://doi.org/10.1128/JVI.00813-19

Dobrovolny HM (2024) Mathematical modeling of virus-mediated syncytia formation: past successes and future directions. In: Kloc M, Uosef A (eds) Syncytia: origin, structure, and functions, chap 17. Springer, Berlin, pp 345–370

Ferri K, Jacotot E, Blanco J, Esté J, Zamzami N, Susin S, Xie Z, Brothers G, Reed J, Penninger J, Kroemer G (2000) Apoptosis control in syncytia induced by the HIV type 1-envelope glycoprotein complex:: role of mitochondria and caspases. J Exp Med 192(8):1081–1092. https://doi.org/10.1084/jem.192.8.1081

Filone CM, Heise M, Doms RW, Bertolotti-Ciarlet A (2006) Development and characterization of a Rift Valley fever virus cell-cell fusion assay using alphavirus replicon vectors. Virology 356(1–2):155–164. https://doi.org/10.1016/j.virol.2006.07.035

Gerg A, Dobrovolny HM (2025) Quantifying impact of HIV receptor surface density reveals differences in fusion dynamics of HIV strains. Viruses 17:583. https://doi.org/10.3390/v17040583

Gilbert J, Greenberg H (1997) Virus-like particle-induced fusion from without in tissue culture cells: role of outer-layer proteins VP4 and VP7. J Virol 71(6):4555–4563. https://doi.org/10.1128/JVI.71.6.4555-4563.1997

Gonzalez-Reyes L, Ruiz-Arguello M, Garcia-Barreno B, Calder L, Lopez J, Albar J, Skehel J, Wiley D, Melero J (2001) Cleavage of the human respiratory syncytial virus fusion protein at two distinct sites is required for activation of membrane fusion. Proc Natl Acad Sci USA 98(17):9859–9864. https://doi.org/10.1073/pnas.151098198

Hayashi Y, Nemoto-Sasaki Y, Tanikawa T, Oka S, Tsuchiya K, Zama K, Mitsutake S, Sugiura T, Yamashita A (2014) Sphingomyelin synthase 2, but not sphingomyelin synthase 1, is involved in HIV-1 envelope-mediated membrane fusion. J Biol Chem 289(44):30842–30856. https://doi.org/10.1074/jbc.M114.574285

Heinrich M, Rosenblatt M, Wieland FG, Stigter H, Timmer J (2025) On structural and practical identifiability: current status and update of results. Current Opin Syst Biol 41:100546. https://doi.org/10.1016/j.coisb.2025.100546

Heitzman-Breen N, Liyanage YR, Duggal N, Tuncer N, Ciupe SM (2024) The effect of model structure and data availability on Usutu virus dynamics at three biological scales. R Soc Open Sci 11(2). https://doi.org/10.1098/rsos.231146

Higuchi H, Bronk S, Bateman A, Harrington K, Vile R, Gores G (2000) Viral fusogenic membrane glycoprotein expression causes syncytia formation with bioenergetic cell death: implications for gene therapy. Cancer Res 60(22):6396–6402

Hong Y, Wu L, Cui M, McMaster G, Hunt S, Chung F (1999) New reporter cell lines to study macrophage-tropic HIV envelope protein-mediated cell-cell fusion. AIDS Res Human Retrovir 15(18):1667–1672. https://doi.org/10.1089/088922299309702

Hooker KL, Ganusov VV (2021) Impact of oseltamivir treatment on influenza A and B virus dynamics in human volunteers. Front Microbiol 12:631211. https://doi.org/10.3389/fmicb.2021.631211

Hotard AL, Lee S, Currier MG, Crowe JE Jr, Sakamoto K, Newcomb DC, Peebles RS Jr, Plemper RK, Moore ML (2015) Identification of residues in the human respiratory syncytial virus fusion protein that modulate fusion activity and pathogenesis. J Virol 89(1):512–522. https://doi.org/10.1128/JVI.02472-14

Hu M, Bogoyevitch MA, Jans DA (2020) Impact of respiratory syncytial virus on host functions: implications for antiviral strategies. Physiol Rev 100(4):1527–1594. https://doi.org/10.1152/physrev.00030.2019

Kobayashi M, Bennett M, Bercot T, Singh I (2006) Functional analysis of hepatitis C virus envelope proteins, using a cell-cell fusion assay. J Virol 80(4):1817–1825. https://doi.org/10.1128/JVI.80.4.1817-1825.2006

Kreutz C, Raue A, Kaschek D, Timmer J (2013) Profile likelihood in systems biology. FEBS J 280(11):2564–2571. https://doi.org/10.1111/febs.12276

LaBarre D, Lowy R (2001) Improvements in methods for calculating virus titer estimates from $TCID_{50}$ and plaque assays. J Virol Meth 96(2):107–126. https://doi.org/10.1016/S0166-0934(01)00316-0

Lin HH, Huang LM, Wu SC (2020) A quantitative luciferase-based cell-cell fusion assay to measure four-serotype dengue virus E protein-triggered membrane fusion. Human Vaccin Immunother 16(9):2176–2182. https://doi.org/10.1080/21645515.2020.1748989

Lineberger JE, Danzeisen R, Hazuda DJ, Simon AJ, Miller MD (2002) Altering expression levels of human immunodeficiency virus type 1 gp120-gp41 affects efficiency but not kinetics of cell-cell fusion. J Virol 76(7):3522–3533. https://doi.org/10.1128/jvi.76.7.3522-3533.2002. https://journals.asm.org/doi/abs/10.1128/jvi.76.7.3522-3533.2002

Ling Y, Xue H, Jiang X, Cai L, Liu K (2013) Increase of anti-HIV activity of C-peptide fusion inhibitors using a bivalent drug design approach. Bioorg Med Chem Lett 23(17):4770–4773. https://doi.org/10.1016/j.bmcl.2013.07.012

Liyanage YR, Heitzman-Breen N, Tuncer N, Ciupe SM (2024) Identifiability investigation of within-host models of acute virus infection. Math Biosci Eng 21(10):7394–7420. https://doi.org/10.3934/mbe.2024325

Luo MC, Nikolopoulou E, Gevertz JL (2022) From fitting the average to fitting the individual: a cautionary tale for mathematical modelers. Front Oncol 12. https://doi.org/10.3389/fonc.2022.793908

Mohl BS, Chen J, Park SJ, Jardetzky TS, Longnecker R (2017) Epstein-Barr virus fusion with epithelial cells triggered by gB is restricted by a gL glycosylation site. J Virol 91(23). https://doi.org/10.1128/JVI.01255-17

Papa G, Mallery DL, Albecka A, Welch LG, Cattin-Ortola J, Luptak J, Paul D, McMahon HT, Goodfellow IG, Carter A, Munro S, James LC (2021) Furin cleavage of SARS-CoV-2 spike promotes but is not essential for infection and cell-cell fusion. PLoS Pathogens 17(1):e1009246. https://doi.org/10.1371/journal.ppat.1009246

Porotto M, Ferren M, Chen YW, Siu Y, Makhsous N, Rima B, Briese T, Greninger A, Snoeck HW, Moscona A (2019) Authentic modeling of human respiratory virus infection in human pluripotent stem cell-derived lung organoids. MBIO 10(3):e00723–19. https://doi.org/10.1128/mBio.00723-19

Qian Z, Dominguez SR, Holmes KV (2013) Role of the spike glycoprotein of human middle east respiratory syndrome coronavirus (MERS-CoV) in virus entry and syncytia formation. PLoS ONE 8(10):e76469. https://doi.org/10.1371/journal.pone.0076469

Rajah MM, Hubert M, Bishop E, Saunders N, Robinot R, Grzelak L, Planas D, Dufloo J, Gellenoncourt S, Bongers A, Zivaljic M, Planchais C, Guivel-Benhassine F, Porrot F, Mouquet H, Chakrabarti LA, Buchrieser J, Schwartz O (2021) SARS-CoV-2 alpha, beta, and delta variants display enhanced spike-mediated syncytia formation. EMBO J 40(24):e108944. https://doi.org/10.15252/embj.2021108944

Rajah MM, Bernier A, Buchrieser J, Schwartz O (2022) The mechanism and consequences of SARS-CoV-2 spike-mediated fusion and syncytia formation. J Mol Biol 434(6):167280. https://doi.org/10.1016/j.jmb.2021.167280

Richman DD, Bozzette SA (1994) The impact of the syncytium-inducing phenotype of human immunodeficiency virus on disease progression. J Infect Dis 169(5):968–974. https://doi.org/10.1093/infdis/169.5.968

Rodrigo A (1997) Dynamics of syncytium-inducing and non-syncytium-inducing type 1 human immunodeficiency viruses during primary infection. AIDS Res Human Retrovir 13(17):1447–1451. https://doi.org/10.1089/aid.1997.13.1447

Sakamoto T, Ushijima H, Okitsu S, Suzuki E, Sakai K, Morikawa S, Muller W (2003) Establishment of an HIV cell-cell fusion assay by using two genetically modified HeLa cell lines and reporter gene. J Virol Methods 114(2):159–166. https://doi.org/10.1016/j.jviromet.2003.08.016

Sanders DW, Jumper CC, Ackerman PJ, Bracha D, Donlic A, Kim H, Kenney D, Castello-Serrano I, Suzuki S, Tamura T, Tavares AH, Saeed M, Holehouse AS, Ploss A, Levental I, Douam F, Padera RF, Levy BD, Brangwynne CP (2021) SARS-CoV-2 requires cholesterol for viral entry and pathological syncytia formation. eLife 10:e65962. https://doi.org/10.7554/eLife.65962

Saw WT, Matsuda Z, Eisenberg RJ, Cohen GH, Atanasiu D (2015) Using a split luciferase assay (SLA) to measure the kinetics of cell-cell fusion mediated by herpes simplex virus glycoproteins. Methods 90:68–75. https://doi.org/10.1016/j.ymeth.2015.05.021

Su B, Wurtzer S, Rameix-Welti MA, Dwyer D, van der Werf S, Naffakh N, Clavel F, Labrosse B (2009) Enhancement of the influenza A hemagglutinin (HA) mediated cell-cell fusion and virus entry by the viral neuraminidase (NA). Plos One 4(12). https://doi.org/10.1371/journal.pone.0008495

Symeonides M, Murooka TT, Bellfy LN, Roy NH, Mempel TR, Thali M (2015) HIV-1-induced small T cell syncytia can transfer virus particles to target cells through transient contacts. Viruses 7(12):6590–6603. https://doi.org/10.3390/v7122959

Tsurudome M, Ito M, Nishio M, Nakahashi M, Kawano M, Komada H, Nosaka T, Ito Y (2011) Identification of domains on the fusion (F) protein trimer that influence the hemagglutinin-neuraminidase specificity of the F protein in mediating cell-cell fusion. J Virol 85(7): 3153–3161. https://doi.org/10.1128/JVI.01666-10

Vanhulle E, Doijen J, Stroobants J, Provinciael B, Noppen S, Schols D, Stevaert A, Vermeire K (2023) Cellular electrical impedance to profile SARS-CoV-2 fusion inhibitors and to assess the fusogenic potential of spike mutants. Antivir Res 213:105587. https://doi.org/10.1016/j.antiviral.2023.105587

Villaverde AF, Raimundez E, Hasenauer J, Banga JR (2023) Assessment of prediction uncertainty quantification methods in systems biology. IEEE-ACM Trans Comput Biol Bioinf 20(3): 1725–1736. https://doi.org/10.1109/TCBB.2022.3213914

Vincent N, Malvoisin E (2012) Ability of antibodies specific to the hiv-1 envelope glycoprotein to block the fusion inhibitor T20 in a cell-cell fusion assay. Immunobiology 217(10):943–950. https://doi.org/10.1016/j.imbio.2012.01.007

Wang W, Yang L, Huang X, Fu W, Pan D, Cai L, Ye J, Liu J, Xia N, Cheng T, Zhu H (2017) Outer nuclear membrane fusion of adjacent nuclei in varicella-zoster virus-induced syncytia. Virol 512:34–38. https://doi.org/10.1016/j.virol.2017.09.002

Wang XJ, Yao L, Zhang HY, Zhu KL, Zhao J, Zhan BD, Li YK, He XJ, Huang C, Wang ZY, Jiang MD, Yang P, Yang Y, Wang GL, Wang SQ, Dai EH, Gao HX, Ma MJ (2022) Neutralization sensitivity, fusogenicity, and infectivity of omicron subvariants. Genome Med 14(1). https://doi.org/10.1186/s13073-022-01151-6

Whitaker EE, Matheson NJ, Perlee S, Munson PB, Symeonides M, Thali M (2019) EWI-2 inhibits cell-cell fusion at the HIV-1 virological presynapse. Viruses Basel 11(12):1082. https://doi.org/10.3390/v11121082

Xia S, Liu M, Wang C, Xu W, Lan Q, Feng S, Qi F, Bao L, Du L, Liu S, Qin C, Sun F, Shi Z, Zhu Y, Jiang S, Lu L (2020) Inhibition of SARS-CoV-2 (previously 2019-nCoV) infection by a highly potent pan-coronavirus fusion inhibitor targeting its spike protein that harbors a high capacity to mediate membrane fusion. Cell Res 30(4):343–355. https://doi.org/10.1038/s41422-020-0305-x

Yang E, Arvin AM, Oliver SL (2016) Role for the αV integrin subunit in varicella-zoster virus-mediated fusion and infection. J Virol 90(16):7567–7578. https://doi.org/10.1128/JVI.00792-16

York J, Nunberg JH (2018) A cell-cell fusion assay to assess arenavirus envelope glycoprotein membrane-fusion activity. In: Salvato M (ed) Hemorrhagic fever viruses: methods and protocols, methods in molecular biology, vol 1604. Springer, New York, pp 157–167. https://doi.org/10.1007/978-1-4939-6981-4_10